AN INTRODUCTION TO

LUMINESCENCE OF SOLIDS

AN INTRODUCTION TO
LUMINESCENCE
OF
SOLIDS

HUMBOLDT W. LEVERENZ

RCA Laboratories
Princeton, N. J.

DOVER PUBLICATIONS, INC., NEW YORK

This Dover edition, first published in 1968, is an unabridged and corrected republication of the work first published in 1950 by John Wiley & Sons, Inc. The author has written a Preface to the Dover Edition and a new Early History of Luminescence especially for this edition and has replaced the Periodic Chart from the first edition with a new one. Permission given by Encyclopaedia Britannica and Radio Corporation of America for use of certain material is gratefully acknowledged.

Library of Congress Catalog Card Number: 68-55698

Manufactured in the United States of America
Dover Publications, Inc.
180 Varick Street
New York, N. Y. 10014

PREFACE TO THE DOVER EDITION

Luminescence continues to be a rewarding field for scientific study and practical application. It provides fascinating variety even though its basic physical mechanism is conceptually simple: A lone atom, in a vacuum, absorbs energy to become excited, and then disposes of the absorbed energy by emitting light. There is great variety in lone atoms, however, because each kind of atom has many different possible energy states (electron configurations), and there are over a hundred different kinds of atoms. When many atoms combine and interact, as in solid materials, the varieties of luminescence are enormous.

While every kind of lone atom can be made to luminesce, relatively few solid materials luminesce, and very few do so efficiently. The most efficient luminescent materials have singularities, such as a lone atom in a rarefied gas, or a distinct atom or group of atoms in a solid. These singularities, or centers as they are called in solids, temporarily store input energy of the order of a few electron volts while the electrons in the center rearrange in the excited state, after which the excess energy is emitted as light. The temporary storage need be only about 10^{-8} second, yet the probability of luminescence occurring efficiently in solids is low, because there are very frequent opportunities for input energy to be converted into atomic motion (heat) instead of radiated photons (light). A lone excited atom, out of contact with other atoms, does not have the opportunity to dissipate its excess energy as heat. In solids, however, the rapid perturbing vibrations of adjoining atoms can quickly convert stored input energy into additional heat. This dissipative conversion can occur during excitation as well as during temporary energy storage, especially when the input energy is absorbed elsewhere in the solid and must then be transmitted to the center through intervening atoms. In designing efficient luminescent solids, one seeks to contrive centers that emulate lone atoms in that they can be excited readily, and have as little dissipative coupling with their environment as possible.

Scientific inquiry into luminescence began about 1603, after Vincenzo Cascariolo accidentally synthesized a new solid that was seen to glow feebly in the dark. It was subsequently called a phosphor, and was later identified as being chiefly barium sulphide with a trace of bismuth sulphide and some residual "flux" salts. Starting with this mysterious and inordinately complex material, chemists and physicists slowly

vii

evolved better means for purifying, synthesizing, and analyzing other and better phosphors; also, for measuring their luminescence properties and correlating their properties with their compositions and structures.

As our ability to analyze materials increased, it was eventually possible to correlate the luminescences of some materials with significant features of their composition, structure, and defects (including foreign atoms). It is those significant features that determine whether and how well a particular specimen of material will luminesce, and it is the objective of *characterization* to ascertain and describe those features (see Reference 770).

Even though our ability to characterize materials is still inadequate, substantial progress has been made. For example, bismuth has been identified as the activator ingredient that provided luminescence centers in Cascariolo's phosphor, and it has been shown that so-called "flux" salts can provide coactivators, i.e., ions that facilitate incorporation of certain activator ions and compensate their electrical charges to preserve neutrality of the solid. Further, we know that chromium-ion centers emit the intense red light of ruby lasers, and europium-ion centers emit red light in the three-phosphor screens of modern color-television kinescopes. The red emissions of ions of chromium in aluminum oxide crystals, and of europium in yttrium oxysulphide or yttrium vanadate crystals, differ from their emissions as lone ions, because they are altered by the kinds, arrangements, and proximities of the other ions in their solid host. These differences are still largely unpredictable, so those who create new phosphors continue to have their occasional delights and frequent disappointments.

When this book was first published, alternating-current electroluminescence of solids was virtually unknown. Now it is possible to get luminances greater than 1000 millilamberts from special microcrystalline zinc sulphide:copper:iodine phosphors suspended in insulating media between close-spaced condenser plates operated at a few hundred volts and several thousand cycles per second. Maximum efficiency, however, has not risen above about one per cent, and thus the once large research effort on this kind of electroluminescence has greatly decreased. Meanwhile, there is increasing effort on direct-current injection-type electroluminescence. Most of the work is on semiconductor crystals in which P/N junctions can be formed readily, e.g., gallium arsenide, gallium phosphide, and their solid solutions. At room temperature, special gallium phosphide crystals passing two amperes per square centimeter emit about 100 footlamberts of deep red (and infrared) light, with an efficiency of about two per cent. Means are being sought to make P/N junctions in crystals with larger bandgaps to emit light efficiently in other parts of the visible spectrum.

Also, lasers and masers were unknown in 1950, and so the lumines-cence emissions then described were *spontaneous*, occurring randomly and giving incoherent light. Now it is possible to supply very high excitation densities to luminescent materials in precise optical cavities in which luminescence emission is *stimulated* to get coherent, highly monochromatic light. Gaseous, liquid, and solid luminescent materials have been made to lase by raising over half their luminescent centers into the excited state at a given time, and then having cumulative reso-nance in the optical cavity produce stimulated emission that emerges as intense beams of light. At low temperature, special gallium arsenide/phosphide crystals, with P/N junctions, have lased and emitted 10^9 foot-lamberts of red light, at 6440 Å, upon passing 100,000 amperes per square centimeter. About one photon was emitted for every three in-jected electrons, and the overall efficiency was estimated at about ten per cent. Laser beams have been used to drill holes in diamonds, metals, and other materials. By holographic techniques, highly monochromatic and coherent laser beams have been used to make three-dimensional images with adjustable perspective. Other uses include rapid exposure of photosensitive and photochemical materials. Also, efforts are being made to develop high-capacity, high-speed communication, information-handling, and information-storage systems using laser beams.

As mentioned, the chief newcomers in the broad field of luminescence are lasers and alternating-current and direct-current electroluminescence of certain solids. They have had the lion's share of research attention since 1950, with some remarkable results; but they have added to, not superseded, the kinds of luminescence and luminescent materials de-scribed in this book. The republished book has some corrections, a short account of the early history of luminescence, additional references, and a new, revised version of the author's Periodic Chart of the Ele-ments. It provides fundamental and practical information for the preparation, characterization, measurement and correlation of proper-ties, and application of typical and useful luminescent solids.

HUMBOLDT W. LEVERENZ

Princeton, N. J.
February 14, 1968

EARLY HISTORY OF LUMINESCENCE

Many kinds of luminescence emission were observed from lightning, aurorae, fireflies, and certain sea bacteria long before there were written records. Awareness of their kinship, however, did not develop until about three thousand years after the reference to fireflies and glowworms in the Chinese *Shih Ching* (Book of Odes) in the period 1500–1000 BC. In Greece, Aristotle (384–322 BC) observed light being emitted from decaying fish and wrote in his *De Colóribus* "some things though they are not in their nature fire nor any species of fire, yet seem to produce light." Nearly two thousand years later in 1565 a Spaniard, Nicolas Monardes, wrote of the unusually intense blue color of an aqueous extract of a wood called *lignum nephriticum*. About ninety years afterwards, this solution was studied by Athanasius Kircher in Germany, Francesco Grimaldi in Italy, and Robert Boyle and Isaac Newton in England. All reported that when the solution was illuminated with white light it appeared intensely blue by reflection, and yellow by transmission. None identified the intense blue light as luminescence emission. That identification was made by the Englishman George Stokes, in 1852. He used optical filters and prisms to show that incident light of one spectral region was absorbed and transformed by the solution into emitted light of a different spectral region of longer wavelength. This luminescence emission apparently stopped instantaneously when the incident light was shut off, as did the emission from specimens of the mineral fluorspar. Stokes, therefore, elected "to coin a term and call it *fluorescence*, from fluor-spar, as the analogous term opalescence is derived from the name of a mineral."

Luminescence of solids was reportedly first observed in 1603 by the Bolognian Vincenzo Cascariolo, who heated some powdered natural barite (barium sulphate) with coal and found that the cooled porous cake glowed at night. He found that this new "stone" apparently absorbed light from the sun by day, and then glowed for hours in the dark; for this reason it was called *lapis solaris* (sun stone). It came to the attention of Galileo Galilei, who got samples and gave them to Gulio Lagalla of the Collegio Romano. Lagalla wrote about *lapis solaris* and its extraordinary glow in *De Phaenomenis in Orbe Lunae*, 1612. Since the "stone" was porous, it was also called *spongia solis* (sun sponge) under the assumption that it simply absorbed and later released the light of the sun. In 1652, however, Nicolas Zucchi used optical filters to show that the color of the light given off at night was the same whether the "stone" was previously exposed to white light or light of other colors, such as blue or green. Meanwhile, in 1640, Fortunio Liceti wrote the first monograph on the Bolognian "stone," with the Greek title *Litheosphorus,* or "stony phosphorus," where phosphorus means light bearer. Since then, microcrystalline solid luminescent materials have come to be called **phosphors** to distinguish them from the chemical element phosphorus, which was isolated by Hennig Brand in 1669, and labelled phosphorus by Johann Elsholz in 1677. Because Cascariolo's original phosphor glowed long after excitation, long-persistent luminescence emission was called **phosphorescence.**

It was not until 1888 that the German physicist Eilhard Wiedemann introduced the term **luminescence,** of Latin origin, to include both fluorescence and phosphorescence. Wiedemann defined luminescence as denoting "all those phenomena of light which are not caused solely by a rise in temperature." Today, luminescence is considered to be a process by which a material generates nonthermal radiation that is *characteristic of the particular material* (thereby excluding Cerenkov radiation).

In addition to the great variety of luminescence emissions from different materials, there is variety in the kinds of input energies which can excite luminescent materials. Over the years, it has become customary to use prefixes to distinguish between the different causes of excitation.

When the excitation energy comes from particles produced in chemical reactions, as in the oxidation of phosphorus or certain organic compounds, the resulting luminescence is called **chemiluminescence.** The particles may be excited molecules or molecular fragments. Francis Bacon wrote of seeing light emitted from decaying wood before 1627, and Alexander Humboldt showed in 1799 that the emission ceased when decaying wood was placed in carbon dioxide or nitrogen, and resumed when it was placed in air or oxygen. He also showed that a live jellyfish could be made to luminesce under electrical stimulation. The jellyfish luminescence, and that of fireflies, glowworms, and certain bacteria, is called **bioluminescence,** which is chemiluminescence in living things.

When the excitation is by relatively low-energy photons, particularly in visible and near-visible radiation, the luminescence is called **photoluminescence.** In 1792 Johann Goethe and Thomas Seebeck saw light emitted from a phosphor held in the invisible region of rays, adjacent to the violet, from a prism. In 1801 Johann Ritter used phosphors to positively identify the existence of ultraviolet radiation, which is sometimes called "black light." He and Seebeck found, also, that certain phosphorescing materials brightened and others darkened when exposed to infrared radiation.

When high-energy x rays provide the exciting photons, the term **roentgenoluminescence** is used. Wilhelm Röntgen discovered x rays in 1895 by noticing luminescence emission from some barium platinocyanide crystals lying near an operating cathode-ray tube that was wrapped in black paper.

Excitation by electrons provides what is called **electroluminescence** or **cathodoluminescence**. It occurs on a grand scale in lightning, which Benjamin Franklin showed, in 1752, to be caused by an electrical discharge. It can also be made to occur in evacuated glass containers, such as the mercury barometer where Jean Picard observed it in 1675. He noticed that light emission occurred above the mercury when it was moving downward, but not when it was moving upward. In 1700, Johann Bernoulli obtained much greater light emission by shaking clean mercury enclosed in an evacuated glass tube. Christian Ludolff showed, in 1745, that electric effects were involved when he observed deflections of silk threads suspended near the moving mercury/glass interface. At a meeting of the Royal Society of London, in 1860, J. P. Gassiott demonstrated that a high-voltage electric discharge through a glass tube containing carbon dioxide at low pressure gave "a brilliant white light." From further experimentation with electronically excited gases there evolved the familiar neon signs and, in the latter 1930's, very efficient phosphor-coated fluorescent lamps. Today, there are also ac-excited phosphors in electroluminescent cells, and dc-injection P/N-junction semiconductor luminescent devices. In 1858, Julius Plücker noticed that he could use a magnet to deflect invisible rays from the cathode in an electroded vacuum tube; rays that produced luminescence emission from the glass wall of the tube. They were named **cathode rays,** and it was found that they could excite many phosphors to produce what is now called **cathodoluminesence.** Cathode rays were later identified as beams of electrons, and thus cathodoluminescence is a form of electroluminescence. Phosphors were first used in cathode-ray oscilloscope tubes by Karl Braun in 1897. Starting in the 1920's, the oscilloscope, its components, and phosphors were modified and improved to become the kinescope ("picture tube") of present-day television and radar.

There are several other names for luminescence excited by different means. These include **triboluminescence** (Greek *tribo* "to rub") for light emitted when certain solids are scratched, ground, or broken. Francis Bacon reported in *The Advancement of Learning*, 1605, that cakes of sugar emit light when being scratched or broken. This luminescence is really an electroluminescence caused by electric discharge between closely spaced electrically dissimilar faces of the solid as they are separated. Another inappropriate term is **thermoluminescence,** applied to light emission from luminescent solids when they are warmed. Robert Boyle observed such emission from a diamond in 1663. Actually, the additional heat simply intensifies phosphorescence emission from materials that were previously excited by means other than heat. **Anodoluminescence** was used for excitation of phosphors by the rays from anodes (first called canal rays) discovered by E. Goldstein in 1886. These rays are positive ions, and the broader term **ionoluminescence** is applicable for either positive or negative ions used as excitants. Finally, there is **radioluminescence**, where the excitation is by particles emitted from radioactive materials. The discovery of radioactivity was made in 1896 by Antoine Becquerel, who found that a luminescent material, uranyl potassium sulphate, blackened a photographic plate that was enclosed in opaque paper. He was trying to find possible invisible radiations that might be accompanying visible luminescence emission, but he found that the photographic plate was darkened by both the excited and the unexcited phosphor. It was then determined that the new high-energy particles were being emitted by the nuclei of the uranium atoms. The term radioluminescence is not specific about the nature of the excitant, since radioactive materials emit alpha particles (ions), beta-rays (electrons), and gamma-rays (photons).

More of the history of luminescence may be found in the author's chapter on luminescence in *Encyclopaedia Britannica*. There has been dramatic progress in luminescence of solids since Cascariolo saw his first phosphor glow faintly at night, with an estimated intensity of about 10^{-3} millilambert. Now there are over a hundred million television kinescopes operating with luminances of about 10^2 millilamberts, hundreds of millions of fluorescent lamps with luminances exceeding 10^3 millilamberts, and some lasers have produced over 10^9 millilamberts, an intensity greater than that of the sun.

PREFACE TO THE FIRST EDITION

This book is designed to provide an introductory and useful description of luminescent solids, particularly artificial (man-made) phosphors, in language comprehensible to science graduates. Much of the material is drawn from personal experience in synthesizing, studying, and applying luminescent solids since 1931, that is, during the recent era of intensive phosphor research which made possible such modern developments as electronic television, "fluorescent" lighting, radar, electron microscopy, and devices for seeing many otherwise invisible forms of energy. Although the book is intended for nonspecialists in luminescence, it is expected that it will be useful as a text in training future specialists and in aiding scientists who wish (a) to refresh and increase their knowledge of solid matter and its interactions with radiations and charged material particles, and (b) to use phosphors for detecting radiation and material particles.

Those who are unfamiliar with **phosphors** may welcome immediate information about the origin of the word and the distinction between *phosphor* (a luminescent solid) and *phosphorus* (the chemical element). The word *phōsphor* or *phosphoros* (Greek, "light bearer") was used to describe a complex preternatural solid, made by Cascariolo of Bologna, Italy, about 1603: a solid that had the awe-inspiring property of glowing in the dark after exposure to (excitation by) daylight. When the chemical element "phosphorus" was discovered, over 60 years later, it was called a phosphoros or phosphor because it too glowed in the dark *while exposed to moist air*, and the name persists in its present form, even though *phosphorus is not a phosphor*. The luminescence emission from a phosphor, which is usually a specially prepared complex crystalline material, is the result of an internal *physical (electronic) action*, which proceeds best in the absence of chemical change, and so a phosphor may continue to luminesce indefinitely under suitable conditions. The luminescence emission from phosphorus is the result of a surface gas-phase *chemical action* (oxidation), which ceases when the source of oxygen is removed or the phosphorus is consumed.*

With respect to the duration of afterglow after cessation of excitation, the word **fluorescence** was first used to denote the imperceptible *short afterglow* of the mineral fluorspar, and **phosphorescence** was used to denote the *long afterglow* (hours) of the early Bolognese phosphor and later phosphors (also, by misinterpretation from the persistent glow of moist phosphorus). The generic term **luminescence** (Latin, *lumen*, "light" + *-escence*) denotes a process of generating

---

* D. M. Yost and H. Russell, Jr., *Systematic Inorganic Chemistry of the Fifth- and Sixth-Group Nonmetallic Elements*, Prentice-Hall, New York, 1944.

radiation during and after excitation; that is, luminescence includes fluorescence and phosphorescence. The term "luminescence" is here used as an adjective and also as a noun. For example, a luminescent material, when luminescing, emits *luminescence* radiation. The material but not the radiation is luminescent, that is, capable of generating (luminescence) radiation.

Most luminescence phenomena occur when the luminescing material is *not* in equilibrium within itself or with its surroundings. Matter with excess (free) energy, *other than heat*, seeks to attain equilibrium by disposing of the excess energy as luminescence radiation, heat (with concomitant thermal radiation), or electron emission; or by producing structural or chemical changes. Efficient phosphors transform a large proportion of absorbed excitation energy into luminescence radiation rather than into heat, electron emission, or chemical or structural change.

Phosphors are hypersensitive to traces of certain impurities and to changes in local and long-range structure. In most cases, efficient phosphors comprise highly purified and well-crystallized bulk materials, which are called **host crystals,** containing a per cent or less of beneficial impurities, which are called **activators.** The extraordinary impurity and structure sensitivity of phosphors distinguishes them as members of the general class of electronically active nonmetallic solids, including semiconductors, photoconductors, photoemitters, secondary emitters, thermionic emitters, ferroelectrics, and ferromagnetic ferrites. Successful research on the constitutions and properties of any of these unusual solids is certain to promote progress in the others, progress which is measured by an improved understanding of the electronic behaviors of solids as a guide to the devisal of superior electronically active materials for use in the sciences and technology at large.

The luminescence of solids is a hybrid subject whose necessary components include *materials* (chemistry), *structures* (crystallography; geometry), and *energies* (physics). In this respect, luminescence is as challenging as other growing hybrids, such as nucleonics, biophysics, and photochemistry, which are nourished by several major sciences. In the following pages, the subject of luminescence is developed in outline form from a summary of elementary concepts, without detailed derivations and discussions. History is generally omitted, because a discussion of the chronological development of our knowledge of luminescence would be tedious and confusing to the uninitiated. Attempts are made to interweave correlations and interpretations where they exist and are pertinent. It should be kept in mind, however, that this is a *complex and growing subject* whose hypotheses and theories are subject to change. Attempts are made, also, to direct attention to some major experimental and theoretical problems whose solutions would be particularly useful in advancing the present qualitative art of phosphors toward the goal of a quantitative science of phosphors.

During the alchemical infancy and haphazard growth of the phosphor art, the resultant desultory and greatly dispersed literature on luminescence of solids has become cluttered with confused terminologies, incorrect data, and baneful misconceptions. As an antidote, this book attempts to be objective, to provide a rational terminology, and to furnish a critical guide to the general literature. Exemplary information, as complete as possible, is given on the preparations and properties of some of the more interesting and useful types of phosphors, with brief descriptions of their general applications and limitations. Readers who want a capsulized description of phosphors, with particular emphasis on utility, may find Chapters 6 and 7 most useful. Those who intend to become active in the field of luminescent solids should refer frequently and *critically* to the literature. Sufficient *recent* references, many containing extensive bibliographies, are given to locate most of the literature on luminescence of artificial phosphors. Considerable original material from these laboratories is given without references. The majority of materials and properties described herein are familiar to the writer, although the telltale phrase, "it is reported," signifies that the writer has not confirmed the reported results.

The detailed complexity of solids and their luminescence characteristics often obscure some of the salient generalities that should be kept in mind while one is pursuing the subject. For this reason, certain simplifying demarcations are made, and some outstanding general features are repeated frequently throughout the text. This is done to assist the reader in forming broad concepts, which may be applied to correlate different specific phosphors or luminescence characteristics.

Throughout the text, rather complete indexing is provided, especially of terms that readers may find unfamiliar. The page numbers of definitions are denoted by **bold-face type** in the index. Also, the centimeter-gram-second system (or convenient derived units) is used throughout, with uniform symbols, of which the most common are collected in a glossary at the end of the book.

It is a pleasure to acknowledge the friendly advice of many scientists in the RCA Laboratories and Victor Divisions of the Radio Corporation of America. Their helpful suggestions, corrections, and criticisms have been a considerable aid in the preparation of this book. Apart from acknowledgments of source material in the text, the writer wishes to record his appreciation of the contributions of other members of the RCA Laboratories' chemicophysics group, particularly Dr. R. E. Shrader, Mr. S. Lasof, and the late Mr. E. J. Wood; also Dr. S. M. Thomsen, and Messrs. P. R. Celmer, Jr., I. J. Hegyi, S. Larach, and R. H. Bube. For helpful discussions on the theoretical aspects of solids, the writer is particularly indebted to Drs. D. O. North, E. G. Ramberg, and A. Rose, and to Professors L. P. Smith of Cornell University and F. Seitz of the Carnegie Institute of Technology. Further appreciation is expressed to Mr. E. W. Engstrom and Dr. V. K. Zworykin, vice-presidents and directors of research of the Radio Corporation of America, who have long sponsored research on phosphors and other electronically active solids. Finally, the writer gratefully records the patient aid given by his wife, Edith Langmuir Leverenz, especially during the preparation of the indexes.

Some of the data in this book were obtained during work done under contracts NDCrc-150, OEMsr-440, and OEMsr-1031 between the Office of Scientific Research and Development and the Radio Corporation of America and contract N6onr-236 between the Office of Naval Research and the Radio Corporation of America.

HUMBOLDT W. LEVERENZ

Princeton, N. J.
November 14, 1949

CONTENTS

LIST OF TABLES

CHAPTER 1

THE ELEMENTS OF MATTER
AND LUMINESCENCE *

Elementary Concepts

Of all gifts to earth, the first and greatest was darkness. Darkness preceded light, you will remember, in Genesis. Perhaps that is why darkness seems to man natural and universal. It requires no explanation and no cause. We postulate it. Whereas light, being to our minds merely the cleansing vibration that dispels the black, requires some origin, some lamp whence to shine. From the appalling torch of the sun down to the pale belly of the glowworm we deem light a derivative miracle, proceeding from some conceivable source. We can conceive darkness without thought of light; but we cannot conceive light without darkness. [Christopher Morley] [1]

There are two outstanding *processes* by which a material can become a generator or origin of light (radiation) after absorbing suitable extraneous primary energy. In one process the absorbed energy is converted (degraded) into low-quantum-energy heat that *diffuses* through the material which then emits radiation called *thermal* radiation. In the other process an appreciable part of the absorbed energy is temporarily *localized* as relatively high-quantum-energy excitation of atoms or small groups of atoms which then emit radiation called *luminescence* radiation. Strictly speaking, **luminescence** is a *process* whereby matter generates nonthermal radiation which is characteristic of the particular luminescent material. Very often, however, the radiation so generated is also called luminescence. To avoid the misinterpretations that can arise when luminescence and radiation are used synonymously, we shall

---

* Those who find unfamiliar some of the condensed material in the first two chapters may use the following general sources for introductory purposes: F. O. Rice and E. Teller, *The Structure of Matter*, John Wiley & Sons, New York, 1949; W. Hume-Rothery, *Electrons, Atoms, Metals, Alloys*, L. Cassier Co., London, 1948; F. Daniels, *Outlines of Physical Chemistry*, John Wiley & Sons, New York, 1948; K. Lonsdale, *Crystals and X Rays*, G. Bell and Sons, London, 1948; W. Shockley, "The Quantum Physics of Solids," *Bell System Tech. J.*, Vol. 18, 645–723, Oct., 1939; and F. Seitz and R. P. Johnson, "Modern Theory of Solids," *J. Applied Phys.*, Vol. 8, 84–97, 186–199, 246–260, Feb.–Apr., 1937.

use luminescence to denote either the luminescence process as a whole (including excitation, temporary storage of energy, and emission), or sometimes just the final emission part of the luminescence process. Also, we shall use the simple terms *light, ultraviolet,* and *infrared* to denote, respectively, visible, ultraviolet, and infrared radiation.

Luminescence, then, is distinguished by emission of radiation, for example, light and ultraviolet, in *excess* of the thermal radiation produced by heat in a given material. A more quantitative definition is given at the beginning of Chapter 5. Familiar examples of luminescence emission are the narrow spectral lines and bands of radiation emitted by (1) electronically excited gases, such as in lightning and neon lamps, (2) certain oxidizable organic matter in liquids exposed to air, for example in glowworms and fireflies, and (3) coatings of tiny phosphor crystals excited by invisible alpha particles, electrons, and ultraviolet, as in luminescent watch dials, television "picture tubes," and "fluorescent" lamps. In all these cases of luminescence, the temperature of the luminescing material is best maintained near or below room temperature. Also, *the quality and quantity of luminescence radiation are strongly dependent on the nature of the emitting material.* Thermal radiation from solids is generally a broad continuous spectrum of radiation, especially infrared, which is emitted in increasing amount as the temperature of the solid is increased. A familiar example of thermal radiation is the emission of infrared and light from electrically heated incandescent filaments in common lamp bulbs. *The quality and quantity of thermal radiation depend chiefly on the temperature rather than on the nature of the emitting solid material.* Because luminescence is strongly dependent on the nature of the emitting material, the study of luminescence may properly begin with the study of matter.

To evolve the nature of *matter* we might start with a vacuum, introduce and make familiar the presently known elementary entities of matter, for example, *electrons, protons,* and *neutrons,* and then proceed to develop more complex matter, such as atomic nuclei, atoms, ions, molecules, gases, liquids, and solids (crystals). Similarly, in evolving the nature of *radiation* (energy) we might start with darkness, introduce and make familiar the elementary *photons* (quanta) of radiation, and discuss their interactions with matter. Such a building of the eventual science of luminescence from elementary units is outlined in this book. Since the outline must be brief, some of the difficulties and many subsidiary points of interest are not mentioned or adequately discussed. The elementary approach, however, provides a simple introduction to the pertinent sciences of chemistry, crystallography, and physics and serves to demonstrate their fundamental unity.

The chief difficulty encountered in starting with elementary entities is that they are the most remote from our perceptual experience. We have become accustomed to palpable bulk matter, such as a drop of water or a crystal of salt, which is a complex assemblage of innumerable elementary material particles. We imagine we understand liquids or solids when they are explained by analogy with tangibly "comprehensible" water or salt. Our imaginations are aided, in these cases, by being able to visualize mechanical models of distinct groups (atoms, molecules) of elementary material particles randomly and transitorily cohering in formless liquids, or permanently aligned in crystalline solids. With the aid of the electron microscope, it is possible to extend our feeling of personal familiarity with discrete particles of matter down to very large molecules. On going to even smaller structures, however, visual perception fails, and we become unsure about the validity of mental images, such as planetary models, which are sometimes used as aids in studies of subatomic phenomena. Part of our uncertainty comes from the need to ascribe a wave-like behavior to elementary material particles, such as electrons, in order to explain their diffraction by solids. Also, the elementary particles are elusive in that, being submicroscopic, they are strongly influenced by the rather gross experiments we perform in attempts to determine their locations, sizes, and energies so that the experimental information can be appreciated by our relatively coarse human sensibilities. This is a consequence of the epistemological indeterminacy of experimental investigations in the atomic domain where there is always some difference between that which *is* and that which *is determined* (*observed, measured*). Because of indeterminacy we describe the elementary particles and their behaviors in terms of *probabilities*, because perfect certainty is unattainable.

At the outset, then, we confess to an inexact understanding of the fundamental elementary particles, and so it is to be expected that our understanding of the performances of complex aggregates of myriads of these particles in solids is far from satisfying. It *is* satisfying, however, that we are continuing to gain a better understanding of solids as we obtain and correlate more data about (a) the elementary particles and their elementary interactions and (b) the phenomena exhibited by bulk crystals (which generally contain over 10^{24} elementary material particles per cubic centimeter). It *is* satisfying, also, that we are continuing to devise and use electronically active solids despite our imperfect understanding of their natures.

Difficulty and indeterminacy occur also in our understanding of the elements of energy. We think we can appreciate the gross kinetic energy that may be imparted by a moving bulky particle of, say, water or salt, and we believe we can appreciate the gross potential energy of a massive particle that is being held in an elevated position against gravitational attraction. Both these examples, however, illustrate energy as an adjunct of matter, without describing energy alone. Examples of **energy** *per se*, are found in the classical fields of radiation, electricity, magnetism, and gravitation, as well as in the more modern field of quantized radiation.[2] According to *quantum mechanics*, **radiation** is absorbed and emitted as particle-like **photons** having discrete values (quanta) of energy,

$$E = h\nu = hc/\lambda \text{ erg (dyne–cm)} \tag{1}$$

where $h = 6.62 \times 10^{-27}$ erg sec **(Planck's elementary action constant)**
$\nu = $ the oscillation frequency of the radiation, in cycles per second $(= \sec^{-1})$
$c = 2.998 \times 10^{10}$ cm $\sec^{-1}$ (velocity of light in a vacuum)
$\lambda = $ the diffraction wavelength of the radiation, in centimeters

Energy itself is intangible, being apprehended only on interaction with matter which acts as an origin and as a detector of energy. The interactions of light and x rays with solids, to produce both diffraction and photoelectricity, lead to the concept that *radiant energy (photons), like matter, is composed of infinitesimal particles whose probable behaviors are describable by mathematical equations similar to those used for describing waves.*

One mathematical method for describing the *probable* behaviors of elementary particles is known as *wave mechanics*, which was founded when de Broglie associated a diffraction wavelength with every particle. This wavelength λ_d is equal to h divided by the momentum p_m of the particle,

$$\lambda_d = h/p_m \text{ cm} \tag{2}$$

For photons, $p_m = h\nu/c$, whereas, for material particles of mass m and velocity v, $p_m = mv$. According to the **indeterminacy principle** and the *complementarity* (interdependence) of certain properties, accurate determination of the momentum of a particle is possible only at the expense of accuracy in determination of its position. More specifically, as formulated by Heisenberg, the product of the indefiniteness of location Δx and the indefiniteness of momentum Δp_m is equal to or greater than $h/2\pi \equiv \hbar$; that is, $\Delta x \Delta p_m \geq \hbar$. (The corresponding indeterminacy of complementary energy and time is $\Delta E \Delta t \geq \hbar$). In wave mechanics, the motions of particles are described in terms of wave functions, denoted by ψ, where $|\psi|^2$ or $\psi\psi^*$ is the probability distribution function for the location of a particle. The one-dimensional space dependence $\psi(x)$ of a wave function representing a free electron with known momentum (and indeterminate position) has the extended form $\psi(x) = A_1 \epsilon^{2\pi i x/\lambda_d} =$

$A_1(\cos 2\pi x/\lambda_d + i \sin 2\pi x/\lambda_d)$, where A_1 is a constant and x is the distance coordinate. Similarly, the one-dimensional space dependence of the wave function, at a given time, for a free electron known to be localized within a wave packet of width α^{-1}, may be represented by the Gaussian error function $\psi(x) = A_2\epsilon^{-\alpha^2 x^2}$, where α ($\propto \lambda_d^{-1}$) is a decreasing function of time. For narrow wave packets, λ_d must be made correspondingly small, and the wave packet diffuses away more rapidly from the initial locale as its width and the wavelength of the electron are decreased. An important consequence of the indeterminacy principle and the foregoing formulations is that a particle cannot be completely localized for long times within an infinitesimal volume. If an electron, for example, were to be initially located within a spherical domain of diameter x, then the wavelength λ_d of the electron would have to be made sufficiently small to make the probability distribution function $|\psi|^2$ negligible outside the domain. Therefore, λ_d decreases as x decreases and, by eq. 2, the momentum p_m of the electron increases as x decreases. This results in the electron remaining localized for shorter times as the diameter of the domain is decreased, unless the electron is bound to a stationary positively charged particle whose charge is increased as the domain is decreased.

There are five fundamental concepts, regarding the constancy of certain properties of matter and energy, which are pertinent in the study of luminescence. These are: (1) The number of elementary material particles (particles with rest masses) remains constant; that is, elementary material particles are conserved; (2) electric charge is conserved; (3) momentum is conserved; (4) the maximum rate of propagation of energy (or mass) is c; and (5) energy is conserved, it being borne in mind that energy E and mass m are related to each other according to the Einstein relation,

$$E = mc^2 \tag{3}$$

that is, mass is a measure of energy (and energy is a measure of mass).

The number of elementary material particles presumably remains constant even when a gamma-ray photon produces an electron-positron pair in the near vicinity of a material particle, or when an electron-positron pair apparently disappears with concomitant emission of one or two photons.[2] According to Dirac, "empty" space is permeated with normally unobservable **negative-energy electrons** having *allowed negative energies* ranging from $-m_0^-c^2$ to $-\infty$, where m_0^- is the rest mass of the electron (Fig. 1) [The superscript $(-)$ denotes the negatively charged electron]. A photon having energy greater than $2m_0^-c^2$ can transform an unobservable negative-energy electron into an observable **positive-energy electron** (conventional electron) by raising the energy of the former into the range of *allowed positive energies* from $+m_0^-c^2$ to $+\infty$. The

positron is then a residual vacancy (**positive hole**) in the negative-energy band. Recombination of an electron and a positron in the vicinity of another particle results in the emission of a photon, whereas the recombination in the absence of a third particle results in the emission of *two* smaller photons (to conserve momentum). This process, while similar to luminescence, produces radiations which are *not* characteristic of the matter involved.

Figure 1 depicts several basic concepts, such as allowed and forbidden energy levels or bands, filled and unfilled bands, and holes in normally filled bands, which are used frequently in discussions of luminescence. The energy-level diagrams

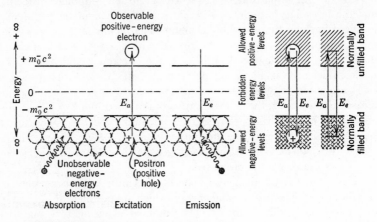

◉ = absorbed photon, $h\nu_a = E_a$
● = emitted photon(s), $E_e = h\nu_e$ (or $h\nu_{e_1} + h\nu_{e_2}$)

Fig. 1. Pair formation and annihilation. Ordinates = energy, abscissas = distance in generalized space coordinates. Either of the two simplified diagrams at the right may be used in place of the sequence of three detailed diagrams at the left.

to be used hereafter, however, lie entirely in the familiar positive-energy region above $+m_0c^2$ in Fig. 1. The magnitude of $m_0^-c^2$ is 8.19×10^{-7} erg, or 512,000 electron volts (ev). (An **electron volt** is a unit of energy which is equal to the energy acquired by an electron falling through a potential difference of one volt. 1 ev = 1.60×10^{-12} erg = 3.82×10^{-20} cal). Also, the self energies $E_s = m_0c^2$ of particles are neglected, because conventional luminescence (Appendix 4) is excited by excess kinetic and potential energy (free energy).

Elementary Particles

Although photons and electrons are the quintessential performers in conventional luminescence, at least protons and neutrons must be considered also to carry out our projected outline development of the nature of matter. Table 1 shows some of the similarities and differences among these four particles. Since neutrons presumably dissociate radioactively into electrons and protons (and neutrinos?), they are not included in the following comparative summary of the properties of *photons, electrons, and protons.*

There is no property common to all three particles at rest, since photons do not exist with zero velocity. Outstanding *common properties* of all three particles in motion are: (1) Diffraction wavelength, (2) frequency, (3) polarization (anisotropy), (4) momentum, (5) kinetic energy, and (6) upper limit of velocity c. Outstanding *differences* between photons and the listed material particles are: (1) Photons may be created and annihilated, whereas the material particles are conserved; (2) photons do not interact with each other, and they obey Bose–Einstein statistics, whereas the listed material particles do interact with each other (even apart from electrostatic forces) and obey Fermi–Dirac statistics; (3) photons do not have the constant electrostatic charges, spin moments, and rest masses possessed by electrons and protons; (4) all photons have a constant common velocity c, whereas the velocities of electrons and protons are variable, with c as an upper limit; (5) photons have diffraction wavelengths λ_d equal to their radiation (conversion) wavelengths λ_E, whereas electrons and protons have $\lambda_d \propto V^{-\frac{1}{2}}$ but $\lambda_E \propto V^{-1}$ (V is the accelerating voltage); and (6) photons have momentum and kinetic energy E_k dependent on variable frequency, whereas electrons and protons have momentum and kinetic energy dependent on variable velocity. The chief differences between electrons and protons are the opposite signs of their electrostatic charges and the different magnitudes of their rest masses and magnetic spin moments.

It is not possible to assign definite sizes to the infinitesimal elementary particles, because indeterminacy makes each as much nebulous wave packet as discrete particle.[4] An *approximate* notion of the "radius," r_0^-, of an accelerated slow isolated electron is obtained by correlating the apparent increase in electron mass with the increased electromagnetic momentum in the field of the electron, or by relating the total coulomb energy of an electronic charge distributed on the surface of a sphere, e^2/r_0^-, to the energy equivalent of its mass $m_0^- c^2$. Either relation gives $r_0^- = 0.667 e^2/m_0^- c^2 = 1.9 \times 10^{-13}$ cm.[2,5,6]

A summary of the sizes, masses, binding energies, shapes and structures (where known), and other characteristics of the elementary particles and of matter built with these particles is given in Appendix 5. It is recommended that Table 23 in Appendix 5 be scrutinized frequently during the reading of the first three chapters. The table serves as a reminder of important structural details which are often omitted in discussions of solids.

TABLE 1
CORRELATIONS BETWEEN SOME OF THE OUTSTANDING PROPERTIES OF PHOTONS AND THREE ELEMENTARY MATERIAL PARTICLES

	Property	Photons	Material Particles		
			Electrons	Protons	Neutrons
1	Charge (e)	....	$-e$ ($= 4.80 \times 10^{-10}$ esu)	$+e$ ($= 4.80 \times 10^{-10}$ esu)	....
2	Mechanical spin moment (μ_s)	....	$\frac{1}{2}\frac{h}{2\pi}$	$\frac{1}{2}\frac{h}{2\pi}$	$\frac{1}{2}\frac{h}{2\pi}$
3	Magnetic spin moment (μ_m)	....	$eh/4\pi m^- c$ ($= 9.3 \times 10^{-21}$ erg oersted^{-1})	$2.8eh/4\pi m^+ c$ ($= 2.8 \times 5 \times 10^{-24}$ erg oersted^{-1})	$-1.9eh/4\pi m^+ c$
4	Rest mass (m_0)	....	m_0^- ($= 9.1 \times 10^{-28}$ g)	m_0^+ ($= 1.672 \times 10^{-24}$ g)	$m_0^0 = 1.00135 m_0^+$
5	Self-energy (E_s)	....	$m_0^- c^2$ ($= 5.12 \times 10^5$ ev)	$m_0^+ c^2$ ($= 9.4 \times 10^8$ ev)	$m_0^0 c^2 = 1.00135 m_0^+ c^2$
6	Moving mass (m)	$h\nu/c^2$	$m^- = m_0^- (1 - \beta^2)^{-1/2}$ *	$m^+ = m_0^+ (1 - \beta^2)^{-1/2}$ *	$m^0 = m_0^0 (1 - \beta^2)^{-1/2}$ *
7	Velocity (v)	c	$(eV/150m_0^-)^{1/2}(1 + eV/600m_0^- c^2)^{1/2}(1 + eV/300m_0^- c^2)^{-1}$	$(eV/150m_0^+)^{1/2}(1 + eV/600m_0^+ c^2)^{1/2}(1 + eV/300m_0^+ c^2)^{-1}$	Independent of V (no charge)
8	Diffraction wavelength (λ_d)	c/ν	$h/m^- v$ ($= 1.22 \times 10^{-7} V^{-1/2}$; $v \ll c$)	$h/m^+ v$	$h/m^0 v$
9	Radiation wavelength (λ_R) †	c/ν	$(2hc/m^- v^2)$ † ($= 1.24 \times 10^{-4} V^{-1}$; $v \ll c$) †	$(2hc/m^+ v^2)$ †	$(2hc/m^0 v^2)$ †

		c/λ	$e^2/\hbar\lambda_d = m^-c^2/\hbar$ ($\nu_E = c/\lambda_E$) †	$e^2/\hbar\lambda_d = m^+c^2/\hbar$	$e^2/\hbar\lambda_d = m^0c^2/\hbar$
10	Frequency (ν_d)	c/λ			
11	Momentum (p_m)	$h\nu/c$	m^-v	m^+v	m^0v
12	Kinetic energy (E_k)	$h\nu$	$m_0^-c^2[(1-\beta^2)^{-1/2}-1]$ $= m_0^-c^2(1 + \tfrac{1}{2}\beta^2 + \tfrac{3}{8}\beta^4 + \cdots - 1)$	$m_0^+c^2[(1-\beta^2)^{-1/2}-1]$	$m_0^0c^2[(1-\beta^2)^{-1/2}-1]$
13	Total energy (E_T)	$h\nu$	$m_0^-c^2(1 + [(1-\beta^2)^{-1/2}-1]) + E_p$ ‡	$m_0^+c^2(1 + [(1-\beta^2)^{-1/2}-1]) + E_p$ ‡	$m_0^0c^2(1 + [(1-\beta^2)^{-1/2}-1]) + E_p$ ‡
14	Mutual interaction	....	Coulomb and exchange	Coulomb and exchange	Exchange
15	Statistics: § $\psi(P_1, P_2) =$	Bose-Einstein $\psi(P_1, P_2)$	Fermi-Dirac $-\psi(P_2, P_1)$	Fermi-Dirac $-\psi(P_2, P_1)$	Fermi-Dirac $-\psi(P_2, P_1)$
16	Dissociability	....	....	....	Neutron → electron + proton (Half-life ≈ 2×10^3 sec)

$h = 6.624 \times 10^{-27}$ erg sec
$c = 2.998 \times 10^{10}$ cm sec⁻¹

1 erg = 6.25×10^{11} ev = 2.39×10^{-8} cal
1 erg/molecule = 1.44×10^{13} kcal mole⁻¹

1 amp = 1 coulomb sec⁻¹ = 6.24×10^{18} electrons sec⁻¹
1 watt = 10^7 ergs sec⁻¹ = **685** lumens (2.22-ev photons)

1 esu = 1 statcoulomb = 2.08×10^9 electronic charges

1 erg of 2.22-ev photons sec⁻¹ = 6.85×10^{-5} lumen = 2.17×10^{-5} candle (cosine-law luminator)
1 millilambert (mL) = 3.2×10^{-4} candle cm⁻² = 10^{-3} lumen cm⁻² = 4.1×10^{12} 2.22-ev photons cm⁻² sec⁻¹

* $\beta = v/c$; V = voltage.
† λ_E is the wavelength of a photon whose energy $h\nu$ is equal to the kinetic energy of the primary particle.
‡ E_p is the potential energy of a particle (Note: $E_T = E_s + E_k + E_p$).
§ P_1 and P_2 represent particle number 1 and particle number 2.

Atomic Nuclei

With the notable exception of gamma-ray emission, most luminescence is extranuclear in origin (see Appendix 4). To complete the outline of building matter from elementary particles, however, a brief
description of atomic nuclei is included.

Atomic nuclei are believed to be composed of protons and neutrons,
both of which are called **nucleons**. Table 2 shows that the nucleus of
each element has a fixed number of protons Z, but a variable number of
neutrons. The number Z is called the **atomic number** of the element.
Atoms with the same Z but different numbers of neutrons are called
isotopes. Isotopes have different masses but practically identical chemical properties, since chemical properties are determined by Z. The
largest number of known isotopes per element is over 20 for tin; whereas
about 20 elements have no known isotopes.[103] The ratio of neutrons to
protons in the nucleus increases from 1 to about 1.5 on going from
hydrogen ($Z = 1$) to curium ($Z = 96$), with the exception of $_1H^1$, $_1H^3$,
and $_2He^3$ (Table 2). Correspondingly, the **mass numbers,** N_n, that is,
the number of nucleons per nucleus, increase from 1 ($_1H^1$) to
242 ($_{96}Cm^{242}$). The simplest nucleus $_1H^1$ has a mass of 1.627×10^{-24}
g (m_0^+), whereas the mass of the uranium nucleus $_{92}U^{238}$ is approximately $238m_0^+$ g.

TABLE 2

Examples of Nuclear Constitutions, Symbols, and Chief Isotopes of
Several of the Elements of Low Atomic Number

| Element | Symbol | Number of | | | Nuclear Symbol $_Z[\ ]^{N_n}$ | Number of Known Isotopes |
		Protons (Atomic Number) Z	Neutrons N_n-Z	Nucleons (Mass Number) N_n		
Hydrogen, common	H	1	0	1	$_1H^1$	
Hydrogen, heavy * (D)	H	1	1	2	$_1H^2$	3
Hydrogen, superheavy † (T)	H	1	2	3	$_1H^3$ ‡	
Helium, light	He	2	1	3	$_2He^3$	
Helium, common	He	2	2	4	$_2He^4$	3
Helium, heavy	He	2	4	6	$_2He^6$ ‡	
Lithium, light	Li	3	3	6	$_3Li^6$	
Lithium, common	Li	3	4	7	$_3Li^7$	3
Lithium, heavy	Li	3	5	8	$_3Li^8$ ‡	
Beryllium, light	Be	4	3	7	$_4Be^7$ ‡	
Beryllium, common	Be	4	5	9	$_4Be^9$	3
Beryllium, heavy	Be	4	6	10	$_4Be^{10}$ ‡	

* Deuterium.   † Tritium.   ‡ Radioactive.[103]

The mass of a complex nucleus is not exactly equal to the sums of the masses of its nucleons, but is about 1 per cent less. The loss in mass is known as the **mass defect** and implies a binding energy (of the nucleons in the nucleus) according to eq. 3. In the case of $_2He^4$, the binding energy is 28×10^6 ev per nucleus, whereas for $_{92}U^{238}$ it amounts to about 1780×10^6 ev per nucleus. There are 4 nucleons in $_2He^4$ and 238 nucleons in $_{92}U^{238}$; hence, the mass defect represents about 7×10^6 ev average binding energy per nucleon.

The volumes of atomic nuclei have been determined to be directly proportional to the number of nucleons, being about 10^{-36} cm^3 for $_{92}U^{238}$. The corresponding radii of atomic nuclei are, therefore, less than about 10^{-12} cm.

The total long-range coulomb *repulsion energy* between all the protons in a nucleus is proportional to approximately $N_n^{5/3}$. The short-range *attraction energies* among the swarming nucleons in a nucleus, considered as a droplet, are believed to be strongest between adjacent neutrons and protons, and comprise: [7-10]

1. **Exchange energies,** that is, attraction energies arising from interchanging and mutual sharing of "bits" of charge, mass, and energy. These bits are at present associated with mesons and may be visualized as items of shared community property which shuttle rapidly back and forth between adjacent nucleons to give them a common bond. This bonding by exchange energy, which involves sharing of particles, is a *quantum* phenomenon.

2. **Van der Waals energies,** for example, attraction energies arising between electric dipoles, or between dipoles and monopoles or multipoles. Although these energies have not been proved to exist in the nucleus, they may arise from polarization (electric deformation) of a neutron in the field of a proton, or by interaction with another neutron.

Other manifestations of these fundamental attraction energies include:

3. **Saturation energies,** that is, attempts to minimize the potential energy of the nucleus by the formation of particularly stable small groups or configurations of nucleons. Stable groups are usually characterized by arrangements of even small numbers of nucleons, as in the helium nucleus, $_2He^4$ **(alpha particle).**

4. **Surface-tension energies** which arise because nucleons at the surface of a stable nuclear droplet experience a net inward attraction which operates to minimize the surface of the drop. Gamow [11] has calculated the density of the nuclear "fluid" to be about 2.3×10^{14} g cm^{-3}, and the surface tension to be about 9.3×10^{19} dynes cm^{-1}. These figures may be contrasted with those for water which has a density of 1 g cm^{-3}, and a surface tension of 75 dynes cm^{-1}.

The short-range exchange, van der Waals, and saturation energies increase in proportion to N_n, whereas the surface-tension energy increases as $N_n^{2/3}$. These attraction energies outweigh the long-range coulomb-repulsion energy for the naturally occurring elements from hydrogen ($N_n = 1$) to uranium ($N_n = 238$), but the more rapid increase of the repulsion energy ($\propto N_n^{5/3}$) eventually dominates the situation and makes the very heavy atomic nuclei radioactively unstable.

Atoms and Ions

A *neutral* **atom** is an atomic nucleus with Z net positive charges electrically neutralized by Z attached (bound) electrons. The Z electrons occupy a "cloud" around the nucleus, where the electron-cloud density expresses the probability of finding an electron. The left side of Table 3

TABLE 3

EXAMPLES OF THE BUILDING OF ATOMS FROM ATOMIC NUCLEI AND ELECTRONS

The Most Abundant Naturally Occurring Isotopes Are Indicated by **Bold-faced** Type

Atomic			Element			Extranuclear Electrons
Number (Nuclear Charge) Z	Mass Numbers (Common Isotopes) N_n	Weight (O = 16) W	Name	Symbol	Formal Valence	$K\ L\ \ \ M$ $1\ 2\ \ \ 3\quad \cdots = n$ $0, 0\ 1, 0\ 1\ 2,\ \cdots = l$
1	**1**, 2	1.008	Hydrogen	H	+1	s
2	3, **4**,	4.003	Helium	He	0	s^2
3	6, **7**	6.94	Lithium	Li	+1	s^2, s
4	**9**, 10	9.02	Beryllium	Be	+2	s^2, s^2
5	10, **11**	10.82	Boron	B	+3	s^2, s^2p
6	**12**, 13	12.01	Carbon	C	±4 (+2)	s^2, s^2p^2
7	**14**, 15	14.008	Nitrogen	N	−3 (+1, 2, 3, 4, 5)	s^2, s^2p^3
8	**16**, 17, 18	16.000 *	Oxygen	O	−2 (+6)	s^2, s^2p^4
9	**19**	19.00	Fluorine	F	−1 (+7)	s^2, s^2p^5
10	**20**, 21, 22	20.183	Neon	Ne	0	s^2, s^2p^6
11	**23**	22.997	Sodium	Na	+1	s^2, s^2p^6, s

* Reference standard.

shows a continuation of the process of building atomic nuclei as outlined in connection with Table 2. Each element has a fixed nuclear charge Z but a variable number and proportion of isotopes. The isotopic proportions were determined by the violent circumstances during the synthesis of this earth and by the subsequent time during which the more radioactive isotopes disintegrated. Long before isotopes were discovered, chemists assigned the arbitrary **atomic weight** W of 16.0000 to *naturally occurring* oxygen as a reference standard for all the other elements. It is known now that the oxygen isotope of mass number 16 comprises only 99.76 per cent of naturally occurring oxygen, the remainder being 0.04 per cent O^{17} and 0.2 per cent O^{18}. (In addition to these naturally occurring isotopes, the short-lived radioactive isotopes O^{14}, O^{15}, and

O^{19} have been produced artificially.) The electron mass is only 1/1836 the mass of a nucleon, hence, W is an approximate average of weighted proportions of the different N_n's for each element. In practice, an atomic weight W in grams corresponds to the weight of 6.02×10^{23} atoms (**Avogadro's number** N_A).

A *charged* ion is simply an atom, or molecular aggregate of atoms, with more or less electrons than Z, or the sum of the Z's in the case of a molecular ion (or a so-called radical). Ions are denoted by atomic symbols with superscripts giving their net charges, for example, the free *positive* **cation** H^+ has lost one electron ($Z - 1$), whereas the free *negative* **anion** $O^=$ (or O^{2-}) has gained two electrons ($Z + 2$). (The charges of ions in *solids* are usually *not* exact integers because neighboring ions tend to share electrons.)

The radii of atoms (in solids), considered as idealized spheres, range from about 0.5 Å (H) to about 2.7 Å (Cs), and the radii of ions range from nuclear dimensions for ions devoid of all their electrons (H^+, F^{7+}) to about 2.45 Å (Sb^{3-}), where 1 Å $= 10^{-8}$ cm.[12,13] These radii may be used to describe idealized spheres which contain most, but not all, of the electron "clouds" of isolated ions, the radii being based on $r_{O^{2-}} = 1.32$ Å and $r_{F^-} = 1.33$ Å. (Pauling advocates $r_{O^{2-}} = 1.40$ Å; see later Table 6.) When atoms and ions are brought in intimate contact, as in molecules and solids, these radii are somewhat indefinite because the atoms and ions frequently share electrons and become distorted by electrostatic interaction. It is to be noted that the effective volumes of atoms increase at a rate which is less than proportional to the increase in the number of electrons added on going from $Z = 1$ to $Z = 96$; for example, the volume of He ($Z = 2$) is about 4×10^{-24} cm^3, whereas the volume of Xe ($Z = 54$) is about 20×10^{-24} cm^3. In other words, the electron densities of the elements increase rapidly with increasing atomic number. As is discussed later, the electron density of matter is important in determining the absorption of primary particles, and elements with high Z are useful components of materials intended as absorbers of high-energy primary photons or charged material particles.

The outlined progressive building of matter has been, thus far, fairly simple arithmetic with elementary particles. If we were to continue with numbers, this book might be simpler for writer and reader alike. From this point on, however, there is increasing use of the traditional cabalistic symbols of chemistry, spectroscopy, and (later) crystallography. *It should be kept in mind that each symbol is a shorthand notation for a number, or group of numbers, of elementary significance.* Each chemical symbol, for example, denotes an atom having a particular value of Z, for example, S (sulphur) has $Z = 16$, Zn (zinc) has $Z = 30$, and

Ag (*argentum* = silver) has $Z = 47$, as may be seen from the periodic chart in the back of this volume. Were it not for long-established usage, the chemical formula for zinc sulphide (ZnS) might be written **30·16**, silver sulphide (Ag_2S) might be written **47₂·16**, and zinc orthosilicate (Zn_2SiO_4) might be written **30₂·14·84.** It is a matter of speculation whether such numerical designations would be more or less convenient than the present symbolic designations, particularly because the chemical and optical properties of the elements are nonlinear functions of atomic number. This is shown by the periodic variations of several properties plotted as a function of atomic number in the periodic chart in the back of this book. After the following explanation, it is expected that the reader should be able to use the periodic chart to determine the numerical import of chemical and spectroscopic symbols to be encountered later.

THE ENERGY STATES OF ATOMS, AND SPECTROSCOPIC SYMBOLS. One of the most useful contributions of wave mechanics has been a rationalization of the allowed stationary states of electrons in atoms. By a **stationary state** is meant a set of circumstances in which a bound electron may remain at least temporarily without radiating, that is, without instantly disposing of some of its energy as a photon. This set of circumstances is described by a wave function, symbolized by ψ, which is everywhere finite and single-valued, and with which is associated an energy value for the system. According to an earlier pictorial view, occasionally still useful, a nonradiating electron bound in a circular orbit in the field of a nucleus must move such that the circumference of the orbit is an integral multiple of the wavelength λ_d of the electron. If this inphase or standing-wave condition is not fulfilled, the waves interfere and the system is unstable. For circular orbits of radius r, the stationary-state condition is given by $2\pi r = n\lambda_d$, where the **principal quantum number** n may be $1, 2, 3, \cdots \infty$. Schrödinger's wave-mechanical formulation of the possible stationary states of the single electron in a hydrogen atom is, for example,[3, 5, 14–16]

$$\frac{\partial^2\psi}{\partial x^2} + \frac{\partial^2\psi}{\partial y^2} + \frac{\partial^2\psi}{\partial z^2} + \frac{8\pi^2 m_0^-}{h^2}\left(E + \frac{Ze^2}{r}\right)\psi = 0 \tag{4}$$

where the complex wave function ψ "describes" the *probable* (not necessarily circularly orbital) path of the nonradiating bound electron, and $|\psi|^2\,dxdydz$ represents the probability of the electron being in a particular volume element $dxdydz$. Proper solution of eq. 4, that is, using only solutions which correspond to experimental observations, gives a series of discrete allowed (observed) stationary-state energy levels according to

$$E = -2\pi^2 e^4 m_0^- Z^2/h^2 n^2 = -2.15 n^{-2} \times 10^{-15} \text{ erg} = -13.5 n^{-2} \text{ ev} \quad (5)$$

In the hydrogen atom, the total energy of the lone electron (neglecting self energy) is the sum of its potential energy of electrostatic attraction to the nucleus $(Z = 1)$ and its kinetic energy of motion about the nucleus,

$$E = E_p + E_k = -\frac{Ze^2}{r} + \frac{m_0^- v^2}{2} = -\frac{Ze^2}{r} + \frac{p_m^2}{2m_0^-}$$

Accordingly, the energy of the system is lowest when r and p_m are as small as possible. The limit for simultaneous reduction of r and p_m is set by the indeterminacy principle, whereby the smallest possible average values of r and p_m are related by $\bar{r}\bar{p}_m \approx h/2\pi$, or $\bar{p}_m \approx h/2\pi\bar{r}$. By substituting the latter relation in the equation for the total energy, one obtains for E as a function of $\bar{r}$ alone

$$E = -\frac{Ze^2}{\bar{r}} + \frac{h^2}{8\pi^2 m_0^- \bar{r}^2}$$

The rate of change of E with change of $\bar{r}$ is then given by

$$\frac{dE}{d\bar{r}} = Ze^2 - \frac{h^2}{4\pi^2 m_0^- \bar{r}}$$

For minimum E, $dE/d\bar{r} = 0$, and so the average radius of the electron in an unexcited hydrogen atom is (see Table 23)

$$\bar{r}_0 = h^2/4\pi^2 m_0^- Ze^2 = 0.53 \times 10^{-8} \text{ cm} = 0.53 \text{ Å}$$

By substituting $\bar{r}_0$ in the original equation for E, one obtains, for the total binding energy of the electron in an unexcited hydrogen atom,

$$E_0 = -2\pi^2 m_0^- Z^2 e^4 h^{-2} = -2.15 \times 10^{-15} \text{ erg} = -13.5 \text{ ev}$$

This value agrees well with the experimental value for the **ionization potential,** that is, the minimum energy required to separate completely the electron and proton of an isolated hydrogen atom.

According to the last equation, the Zth ionization potential, that is, the minimum energy to remove the last (K) electron nearest the nucleus, increases as Z^2. For example, the minimum energy to remove the most strongly bound K electron, even when all the other electrons are present to reduce the effective Z, is 1070 ev for Na $(Z = 11)$, 8860 ev for Cu $(Z = 29)$, 69,300 ev for W $(Z = 74)$, and 115,000 ev for U $(Z = 92)$.

Additional quantum numbers, evolved in the solution of eq. 4 and used to specify completely the allowed (observed) energy levels, are (1) the **angular-momentum quantum number,** $l = 0, 1, 2, \cdots n - 1$, (2) the correlated **magnetic-moment quantum number,** $m_\mu = -l$,

$-l + 1, \cdots 0 \cdots l - 1, l$, and (3) the independent* mechanical **spin-moment quantum number,** $s_\mu = \pm\frac{1}{2}$. These momenta are expressed in units of $h/2\pi \equiv \hbar$; for example, the angular momentum corresponding to $l = 3$ is $\hbar[l(l + 1)]^{\frac{1}{2}} = 3.46\hbar = 0.55h$. The principal quantum number n denotes the energy of a bound electron in a stationary state; l denotes the possible angular momenta of the electron in motion about the nucleus in state n; m_μ denotes the projection of these angular momenta, given by l, on an axis corresponding to an external magnetic field, and s_μ denotes the independent spin of the electron along this axis. In brief, then, a stationary state is defined by four quantum numbers which may be only integers or half-integers according to

$$s_\mu = \pm\frac{1}{2} \quad \text{and} \quad |m_\mu| \leq l < n = 1, 2, 3, \cdots\infty \qquad (6)$$

In spectroscopy, $n = 1, 2, 3, \cdots$ is denoted by $K, L, M, N, \cdots$, and $l = 0, 1, 2, 3, 4, 5$ is denoted by s, p, d, f, g, h for individual electrons. The resultant total angular momentum of all the electrons in an atom is assigned a quantum number $L = 0, 1, 2, 3, 4, 5$ which is denoted by S, P, D, F, G, H.

THE PERIODIC CHART OF THE ELEMENTS. If, now, one *postulates* the **Pauli exclusion principle** which states that no two electrons in a system, for example in an atom, can have identical n, l, m_μ, and s_μ, it is possible to use eq. 6 to derive the lowest stationary states of isolated atoms of the chemical elements. From eq. 6 and the exclusion principle, the maximum number of bound electrons with a given l is $2(2l + 1)$, that is, 2, 6, 10, 14, $\cdots$, according to the value of n. The quantity $2(2l + 1)$ is called the total **degeneracy** of an allowed state, being the maximum number of electrons which can have identical n and l in a given system.

The lowest stationary states **(ground states)** of the atoms are derived as follows (see Table 3 and the periodic chart):

1. The lowest possible state for the single electron of H is $n = 1$ (K), $l = 0$ (s), $m_\mu = 0$, $s_\mu = \pm\frac{1}{2}$, that is, an s state in the first or K "shell." This s electron may be ejected from a hydrogen atom by a minimum energy of 13.5 ev to leave a positive hydrogen ion, H^+.

2. The lowest possible states for the two electrons of He are $n = 1$, $l = 0, m_\mu = 0$, with one electron having $s_\mu = +\frac{1}{2}$ and the other electron

---

* The independent empirically introduced spin-moment quantum number s_μ is not obtainable from the nonrelativistic eq. 4, although s_μ has been derived from Dirac's relativistic wave equation which is unsymmetrical in its space and time derivatives. According to Dirac, the presence of electron spin may be thought of as (1) giving the 3-variable (n,l,m_μ) representation of a stationary state two components, $\pm(n,l,m_\mu)$, or (2) adding a fourth variable, $s_\mu = \pm\frac{1}{2}$, as in eq. 6.[2, 3, 5]

having $s_\mu = -\frac{1}{2}$ **(paired spins).** The number of electrons having a given l is denoted by a superscript, and so the two electrons in He are represented by s^2. According to the exclusion principle and eq. 6, this exhausts the possibilities of state $n = 1$ and, incidentally, provides a very stable atom, as evidenced by the high first ionization potential (24.5 ev) of helium. The second ionization potential, to produce He^{2+}, is 54.2 ev.

3. On proceeding to Li, the third electron must be placed in the next lowest state $n = 2$, $l = 0$, $m_\mu = 0$, $s_\mu = \pm\frac{1}{2}$. The 2s electron of lithium can be removed with only 5.4 ev to leave Li^+ which has the stable He configuration.

4. The fourth electron, added at Be, also has $n = 2$, $l = 0$, $m_\mu = 0$, but its spin is of opposite sign to that of the previously added 2s electron. This completes the $2s^2$ shell, and the first ionization potential of Be is 9.3 ev.

5. On proceeding to build B, C, N, O, F, and Ne, the first three added electrons go into the $2p$ ($l = 1$) shell with *parallel* spins because of an exchange effect whereby the p electrons remain as far from each other as possible to minimize electrostatic repulsion.[37] The resultant ground-state spins of this series of elements, therefore, increase to a maximum of $\frac{3}{2}$ at N, whereafter further added electrons must pair with previously added electrons to satisfy eq. 6, so the resultant spin decreases to zero at Ne when the $2p^6$ shell is completely filled. In the nitrogen atom, the 3 electrons in the $2p$ shell differ in that one electron has $m_\mu = 0$, another has $m_\mu = +1$, and the third has $m_\mu = -1$. This atom-building process may be continued through the entire periodic chart of the elements, with some irregularities incidental to the d, and f electrons, it being kept in mind that the sum of the $s, p, d, f, \cdots$ superscripts is equal to Z.

As an example of the use of the periodic chart in the back of this book, the electronic configuration of zinc (Zn), $Z = 30$, is determined as follows:[13,17] The vertical line for Zn intersects the antepenultimate-group list (at the top of the lower chart) in the space containing s^2p^6 which is preceded (at the left) by the completed group s^2, giving an inner configuration of s^2, s^2p^6. The next higher-energy state is the penultimate group (only the group intersected!) $s^2p^6d^{10}$. Finally, the outermost (ultimate) group for Zn is found at the left of the top chart to be s^2, making the total configuration s^2, s^2p^6, $s^2p^6d^{10}$, s^2 where the superscripts total 30, which is the atomic number of zinc. Each comma marks the completion of a particularly stable principal quantum state, and so the last s^2 electrons of Zn have $n = 4$ (sometimes written $4s^2$). The preceding completed and the following to-be-completed quantum states of an atom largely determine its **ionic valence (formal valence),** that is, the number of electrons it may lose or gain to become an ion with a particularly stable electron configuration. In Table 3 it may be seen, for example, that Be tends to lose two electrons to become Be^{2+} (s^2, $s^2 \to s^2$), B loses three electrons to become B^{3+} (s^2, $s^2p \to s^2$), and C may lose four electrons to become C^{4+} (s^2, $s^2p^2 \to s^2$), or it may gain four

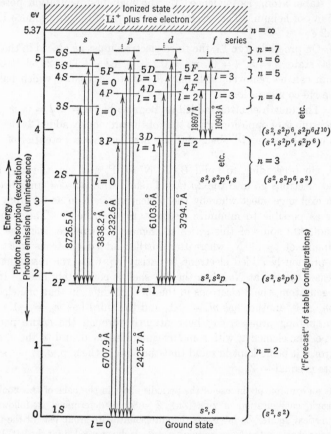

FIG. 2a. Some allowed energy levels and electronic transitions of a free lithium atom. [17, 320]   1 Å = 10^{-8} cm.

electrons to become C^{4-} (s^2, $s^2p^2 \to s^2$, s^2p^6). Some elements, such as N, exhibit several valencies with varying probabilities of existence according to the conditions under which the element gains, loses, or shares electrons upon reaction and combination with other elements.

In addition to the Z normally occupied ground-state levels of the atom, there are numerous normally unoccupied higher-energy levels

(excited-state energy levels) in which an atom has optical lifetimes of the order of 10^{-8} sec. That is, the excited atom generally remains in an excited-state energy level for about 10^{-8} sec before dropping to a state of lower energy and radiating the energy difference between the two levels as a photon. This process is called a **radiative transition.** Figure 2a shows some of the excited-state levels and possible energy transitions of the outermost 2s electron of a neutral lithium atom.[17, 320] The ground-state level of the 2s electron is assigned the energy value zero, because transitions from the smaller and lower-lying 1s "shell" require much

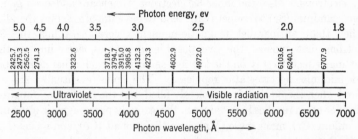

FIG. 2b. Some prominent lines in the arc-emission spectrum of lithium. Exemplary transitions (Fig. 2a) are: 2425.7 ($7^2P \rightarrow 1^2S$), 2741.3 ($4^2D \rightarrow 1^2S$), 3232.6 ($3^2P \rightarrow 1^2S$), 3794.7 ($7^2D \rightarrow 2^2P$), 4602.9 ($4^2D \rightarrow 2^2P$), 4972 ($3^2S \rightarrow 2^2P$), 6103.6 ($3^2D \rightarrow 2^2P$), 6707.9 ($2^2P \rightarrow 1^2S$), where the superscript [2] indicates that the level is split into two closely spaced levels (doublets) attributed to different spin values.

more excitation energy and have correspondingly less probability of occurring.* In addition to the excited-state energy levels shown in Fig. 2a, there are many more higher discrete levels lying between $n = 7$ and the **ionization continuum,** $n = \infty$. An electron which has been raised, in energy, into the ionization continuum is free to move away from the residual positive ion. Any excess energy above $n = \infty$ appears as kinetic energy of the electron and ion moving away from each other. Figure 2b shows some of the observed emission lines of Li, plotted on a linear wavelength scale, and gives several correlations with the allowed transitions shown in Fig. 2a.

It may be seen from Fig. 2a that the pattern of allowed excited-state energy levels constitutes a qualitative forecast of the stable electron configurations of atoms with more electrons than lithium. A quantitative forecast would have to take into account the complex interactions

* The word "shell" is used loosely to denote an electron "cloud" corresponding to a given value of n and l. The term is descriptive of the relative impenetrability (exclusiveness when filled) of the completed electron-cloud strata rather than their obscure structures (ψ^2 distributions) which are *not* sharply limited (see Table 23).

of the electrons. Because of these interactions, the electrons added during the process of "building" atoms of successively higher atomic number do not always go into the next higher unoccupied allowed level but sometimes skip a level ("shell") with $l > 1$ which is filled later. The actual order of filling is $1s$, $2s$, $2p$, $3s$, $3p$, $4s$, $3d$, $4p$, $5s$, $4d$, $5p$, $6s$, $4f$, $5d$, $6p$, $7s$, $5f$, $6d$. Examples of unfilled inner shells are found in the first transition-group elements starting at potassium (K) and ending at zinc (Zn). The electron added at K goes into the $4s$ shell instead of the $3d$ shell, and the $4s$ shell fills at calcium (Ca) before the $3d$ shell acquires any electrons. Also, the inner $3d$ electrons, which are added on going from scandium (Sc) to copper (Cu), tend to have unpaired spins (parallel rather than antiparallel magnetic-moment vectors). Thus the $3d$ electrons first exhaust the five possible m_μ values, according to the exchange effect, to stay on the average as far from each other as possible and minimize electrostatic repulsion. The spins eventually are all paired when the $3d$ shell is completely filled, but the number of unpaired spins per atom reaches high values in the strongly paramagnetic elements chromium (Cr), manganese (Mn), iron (Fe), cobalt (Co), and nickel (Ni) (note the high resultant ground-state spins of these elements in the periodic chart). Another example of belated filling of an inner shell is found in the incompleted $4f$ shell of the rare-earth elements (atomic numbers 57 to 71).

It may be recalled that atomic **diamagnetism** is attributed to precession of presumed electron orbits about the direction of the applied magnetic field, without a change being required in energy or shape of any orbit. The direction of the precession is such as to oppose the applied field, and, hence, the atom experiences a force which moves it in the direction of the lowest density (intensity) of the field. Diamagnetism is particularly strong in elements with just completed d shells. Atomic **paramagnetism** is attributed to a predominant orientation of the permanent or induced resultant magnetic vector, arising in part from electrons with $l > 0$ but chiefly due to unpaired electron spins, in the direction of an applied magnetic field. This lowers the potential energy of the atom in the field and moves the atom in the direction of the highest density of the applied magnetic field. All atoms exhibit diamagnetism, but in some cases there is sufficient paramagnetism produced by one or more electrons with unpaired spins (especially unpaired electrons in incompleted shells) to overwhelm the resultant diamagnetism occasioned by the majority of electrons with paired spins. The property of **ferromagnetism,** which is exhibited by certain crystalline materials, is attributed to group action of permanently unpaired spins of electrons in inner incompleted shells. Within $\approx 10^{-6}$ to >2-cm volume elements, called *domains*, the unpaired spins align in parallel along major crystallographic axes in the absence of an external magnetic field. (This alignment is perturbed by thermal agitation of the atoms).[13–22] An applied magnetic field acts to orient the magnetic moments of the domains in the direction of the field. Then, domains initially oriented in the field direction may grow at the expense of others, and the stepwise irregular movement of the thin spin-reversal region (Bloch wall) between domains gives rise to the Barkhausen effect. When the material retains a preponderant volume of domains oriented in a given direction after the external field is removed, it becomes a permanent magnet.

The absorption and emission of (photon) energy by a bound electron occurs between the allowed energy levels according to certain **selection rules** which describe the allowed (observed) changes in the various quantum numbers. An important first-order selection rule is that, when l changes, the change must not exceed ± 1 during a transition. In the absence of interaction between spin and orbital motions, the selection rule for the quantum number S_s denoting the resultant angular momentum of electron spin due to all the electrons in an atom is $\Delta S_s = 0$, and the selection rule for the quantum number L denoting the resultant angular momentum of orbital motion of all the electrons in an atom is $\Delta L = 0$ or ± 1, but not $L = 0 \rightarrow L = 0$. The values of S_s are called **multiplicities**; for example, $S_s = 0$ (singlet), $S_s = \frac{1}{2}$ (doublet), $S_s = 1$ (triplet). The total angular momentum J is the resultant of S_s and L, such that $\left| L - S_s \right| \leq J \leq L + S_s$. The selection rule for the quantum number J is that J may change by $1, 0$, or -1, except for $J = 0 \rightarrow J = 0$ which is forbidden. In spectroscopic notation, the symbol representing L is given a preceding superscript denoting the **multiplicity** $(2S_s + 1)$ and a following subscript denoting J, for example, $^3D_1$ signifies $S_s = 1$ (triplet state), $L = 2$, and $J = 1$.

The spectra of ions having only one electron are similar to those of atoms having the same number of electrons, except that the discrete energy differences (**characteristic frequencies** according to eq. 1) for corresponding transitions between allowed energy levels are increased according to Z^2. For corresponding transitions,

$$\nu_{\text{H}} \approx \frac{\nu_{\text{He}^+}}{2^2} \approx \frac{\nu_{\text{Li}^{2+}}}{3^2} \cdots \tag{7}$$

Under certain conditions, an atom or ion can produce **resonance radiation**; that is, the energy of the emitted photon equals the energy of the absorbed photon which produces the electronic excitation. When each absorbed primary photon produces a secondary emitted photon, the process has 100 per cent **quantum efficiency**. If, in addition, there is only resonance radiation ($h\nu_{\text{absorbed}} = h\nu_{\text{emitted}}$), then there is zero **energy deficit**; that is, no surplus energy remains in the system after the luminescence-radiation process has ceased.

Molecules

A **molecule** is generally a stable electrically neutral combination of two or more atoms or ions. The formation of **polar molecules (ionic molecules)** from ions is easily grasped; it may be accomplished by the mutual neutralization of unlike charges during the combination of ions to form isolated simple molecules, as in $K^+ + Cl^- \rightarrow KCl$, and $Ca^{2+} + 2Cl^- \rightarrow CaCl_2$. In these chemical compounds, two or more simple ions with stable electron configurations combine to retain most of their individual stable configurations; the resultant molecule being polar without having a net electrostatic charge. For example, a free neutral K^+Cl^- molecule has a dipole moment of about 10^{-17} esu-cm (*ex* in statcoulomb-cm) which corresponds to two unit electronic charges e of opposite sign separated by about 2×10^{-8} cm (2 Å). Here, then, the two ions are held together largely by a **polar bond (ionic bond).** Polar molecules may be formed also when two neutral atoms with strong proclivities to ionize, such as K and Cl, are brought together under suitable circumstances,[23] so that electron transfer may take place with the formation of an ionic molecule; for example, $K + Cl \rightarrow K^+Cl^-$ (usually written simply KCl). This process is particularly probable when one atom readily gives up electrons (has a low ionization energy) and the other atom readily acquires electrons (has a high electron affinity; see Table 3).[24,25] As an example, the strongly electropositive K atom has an ionization energy of 4.3 ev, whereas the strongly electronegative Cl atom has an electron affinity of 3.8 ev, so that only about 0.5 ev of additional energy has to be furnished to effect the transfer of an electron from K to Cl during the formation of ionic KCl from the elements. The most electropositive metal atom Cs has an ionization potential of 3.9 ev, and the most electronegative nonmetal atom F has an electron affinity of 4.1 ev. The formation of the ionic molecule CsF from its atomic constituents, therefore, proceeds exothermically (gives up energy as heat).

When a molecule has a lower potential energy than its separate constituent atoms, the reaction may proceed exothermically and sometimes produce luminescence radiation **(chemiluminescence)** in order to dispose of part of the excess energy. In general, the energy balance of a spontaneous chemical reaction may be written

$$E_{p_1} + E_{k_1} = E_{p_2} + E_{k_2} + \qquad \Sigma h\nu \qquad (8)$$
$$\text{(reactants)} \quad = \quad \text{(products)} \quad + \text{ (luminescence radiation)}$$

where E_p and E_k are the total potential and kinetic energies, respectively, of the atoms, ions, or molecules on each side of the equation, and $\Sigma h\nu$ is the total energy emitted as luminescence radiation. The reaction may proceed spontaneously only when $E_{p_1} > E_{p_2}$, that is, when the free energy of the reactants is greater than that of the products. If a chemical reaction requires additional energy (endothermic reaction), which may be furnished by heat, photons, or bombardment by material particles, then eq. 8 may be written

$$E_{p_1} + E_{k_1} + E_a = E_{p_2} + E_{k_2} + \Sigma h\nu \tag{9}$$

where E_a is the additional energy absorbed from external sources of heat, light, or material particles.

On a quantum basis, the minimum extraneous energy required to initiate an exothermic reaction or to effect an endothermic reaction is called the **activation energy**.[25] The activation energy may be thought of as the energy required to lift the system over a potential hill between the reactants and products, or as the energy required to form from the reactants an unavoidable intermediate activated molecular complex (or excited transition state of the system), which dissociates exothermically to form the products of the reaction. An activation energy, just as any energy, is measured as a *difference* between two (usually unknown) energy levels, and so an activation energy may be symbolized by either E^* or ΔE^*. The symbol ΔE^* is convenient when E^* is used to represent an excited-state or activated-state energy level, for example, letting $\Delta E^* = E^* - E_0$, but E^* is often used to represent an activation energy when there is no such danger of confusion.

When the kinetic energies of the reactants or products are very high, there may be an appreciable intensity of thermal radiation in addition to the luminescence radiation during the reaction. **Thermal radiation** is produced by inelastic collisions of the atoms, ions, or molecules, whereby part of the kinetic energy before collision is expended in effecting electronic excitation of one or more of the colliding particles; the radiative transition from the excited-state level then proceeds with the emission of a photon as in the case of luminescence. Excess thermal radiation may continue long after the reaction and the chemiluminescence have ceased, that is, as long as the reactants or products have E_k in excess of the E_k corresponding to the temperature of the surroundings. Chemiluminescence emission and thermal radiation have quite different **spectral-distribution characteristics** [rate of emission of radiated energy **(radiance)** as a function of wavelength or frequency], and, when their spectral-distribution characteristics overlap, the luminescence contribution may be recognized as emission in excess of thermal radiation. The thermal radiation from an object at a given temperature cannot exceed the radiation from an ideal black body at the given temperature, *at any frequency* (see the beginning of Chapter 5).

The formation of **nonpolar molecules** (homopolar, or nonionic, molecules) from neutral atoms is a more subtle process than that based on coulomb attraction energies as outlined for polar molecules. In chiefly nonpolar molecules, the predominant binding energies apparently comprise exchange energies (resonance energies), including electron-spin-coupling, and dipole-type attractions between the atoms. In general, the association of atoms to form molecules occurs when such combinations produce over-all electron configurations of greater stability (lower E_p). The stablest configurations are the completed n states ("shells"), s^2, s^2p^6, etc., possessed by the inert gases He, Ne, A, Kr, Xe, and Rn. On this basis, it is possible to visualize the coherence of two H atoms, each having a single s electron, to form the molecule H_2 which has the stable s^2 configuration of He. In the H_2 molecule, the two electrons move about *both* H^+ nuclei, and with this electron sharing there is associated an exchange energy, the spin-moment vectors of the two electrons being oriented in opposite directions (**antiparallel spins**, or **paired spins**) to allow close approach of the electrons and promote coupling between the atoms. If two colliding H atoms happen to have electrons with parallel spins, the electrons (and their atoms) will repel rather than attract each other, as a consequence of the exclusion principle, and no molecule will form. After s^2, the next stable configuration is s^2p^6. This stable completed shell has 8 electrons, and many nonpolar molecules appear to have their atoms arranged so that the valence electrons participate in communal groups of 8, as in

$$Cl_2 = \; :\overset{\cdot\cdot}{Cl}:\overset{\cdot\cdot}{Cl}: \quad \text{and} \quad CCl_4 = \; \begin{matrix} :\overset{\cdot\cdot}{Cl}: \\ :\overset{\cdot\cdot}{Cl}:\overset{\cdot\cdot}{C}:\overset{\cdot\cdot}{Cl}: \\ :\overset{\cdot\cdot}{Cl}: \end{matrix} \quad (10)$$

where the dots represent the outermost valence electrons. In these cases, also, the valence electrons move about all the atoms in the simple molecules, to provide exchange-type attractions, and the bonding process is facilitated by the formation of paired spins, that is, **electron-pair bonds** (or **covalent bonds**) where bound electrons with antiparallel spins couple to each other.[24]

In its elementary form the general concept of **valence** pertained simply to the proportions in which atoms combined with each other. The reference standards are usually univalent H^+ (unipositive) and Cl^- (uninegative). Many chemical substances are distinguished by simple integral combining proportions of their constituent atoms, for example, N hydrogen atoms or ions will combine with exactly N chlorine atoms or ions, no more or less, to form hydrogen chloride (HCl) in a gaseous mixture containing an excess of chlorine. Similarly, N simple molecules of KOH in solution will combine with up to but not more than N simple molecules of HCl to form N molecules of KCl and N molecules of H_2O. In these examples, potassium (K) is univalent, and oxygen (O) is bivalent with reference to the standards of hydrogen and chlorine. These examples of the simplicity of the valence concept, which is unequivocally useful for many reactions in gases and liquids, stand in contrast to the complexities encountered in reactions in many solids. In the formation of solids, as described later, there are many occasions where the combining proportions of the constituent atoms are *not* simple integers. This is particularly true when there is a variable covalent contribution to the bonding energy. For example, solid titanium oxides may be formed with the ratio Ti/O ranging from about $\frac{1}{2}$ to over 2, with any intermediate value obtainable by varying the conditions of oxidation or reduction during high-temperature crystallization. A similar variability exists in the case of the palladium hydrides, Pd_xH. The concept of valence, especially in solids, is complicated by the joint action of variable proportions of both ionic (electrostatic) and covalent bonds in effecting the combination of the atoms, and these proportions are sensitive to the crystal structure of the solid. The general simple integral valencies (combining proportions) of isolated atoms, therefore, may become variable nonintegers in solids where the atomic proportions and configurations are influenced by other factors such as the relative sizes, shapes (electron-cloud distributions), deformabilities, and number of species of the combining atoms. In silica (SiO_2), for example, Pauling estimates the Si–O bonds to be approximately 50 per cent ionic and 50 per cent covalent, whereas the Be–O bond in beryllia (BeO) is estimated to be 63 per cent ionic.[24] An improvement in our quantitative understanding of the valence concept as applied in solids would be a material aid in increasing our understanding of the mechanism of luminescence in phosphor crystals.

In a simple molecule, a particular electron cannot be said to belong solely to one nucleus, but must be considered as being in the resultant field of all the nuclei and "belonging" to each nucleus part of the time. The wave function ψ of each electron extends over the entire molecule, but the *degree* of an electron's belonging to a particular nucleus increases on going to the inner electrons (denser shells nearer the nucleus) which have lower potential energy. Statistically speaking, there are instants of time when one or another nucleus in a molecule has more than its share of electrons, whereas at other instants of time the electrons are distributed quite uniformly among the nuclei. The molecular assemblage then fluctuates between instantaneous states of being polar and nonpolar.

Molecules which have "permanent" dipole moments spend a predominant amount of time in a particular state of nonuniform charge distribution within the volume occupied by the molecule, whereas nonpolar molecules spend, on the average, equal times in oppositely disposed nonuniform charge distributions and, hence, do not have a resultant polarity. The following schematic examples show how electronic fluctuations may occur in simple molecules with and without resultant dipole moments:

$$H^+\overset{\cdot}{\underset{\cdot}{H}}{}^- \rightleftarrows H:H \rightleftarrows \overset{\cdot}{\underset{\cdot}{H}}{}^-H^+ \qquad \text{No resultant dipole moment} \qquad (11)$$

$$K^+\overset{\cdot}{\underset{\cdot}{Cl}}{}^- \rightleftharpoons \overset{\cdot}{\underset{\cdot}{K}}Cl \rightleftharpoons \overset{\cdot}{\underset{\cdot}{K}}{}^-Cl^+ \qquad \text{``Permanent'' dipole moment} \qquad (12)$$

In any assemblage of combined atoms or ions, the individual mononuclear units constantly fluctuate to some extent between the extremes of being neutral atoms and ionized atoms (ions).

As a convenience, the word **atom** *is used to include both atoms and ions*, except in cases where it is desirable to emphasize a preponderance of the extreme ionic condition. Also, the symbols for atoms and atomic groups are given wherever possible without indicating their ionic charge (formal valence), because the magnitude of this charge is often unknown and variable in solids.

There are several experimental methods for determining whether the bond between two atoms is predominantly polar or nonpolar.[24-28,284,309] In crystals, the information may be obtained from plots of average electron density as a function of internuclear spacing. These plots can be made from structure-factor data determined from measurements of the diffraction of x rays, electrons, or neutrons by the crystal.[91,685] When the electron density is practically zero for much of the region between the atoms (ions), and integration of the curves reveals that electron transfer has occurred, the bonding is chiefly polar. When the electron-density curve has a large hump between the atoms, signifying shared electrons, the bonding is chiefly nonpolar.

Electronic fluctuations of the types discussed in connection with eqs. 10, 11, and 12 are sometimes called **resonance,** a term descriptive of the oscillatory exchanging of molecular systems between their various possible electronic configurations.[24,25] According to the wave-mechanical theory of resonance, any system (for example, a molecule) which resonates between two or more electronic configurations (states), with practically identical atomic positions and energies, exists as a *hybrid* of the possible configurations and is more stable (has lower potential energy) than the same system "frozen" in any one of the particular states. A resonance or exchange effect occurs during the filling of atomic shells with $l \geq 1$. For example, the three $2p$ electrons in the nitrogen atom make full use of the three possible values of m_μ to stay as far apart on the average as the configurational possibilities allow, thereby reducing electrostatic repulsion and attaining a low potential energy for the system. The energy lowering follows also from the indeterminacy principle whereby the momentum (and kinetic energy) of an electron may decrease as its wave function is extended in space. This extension permits an increase in the wavelength of the electron and, because momentum and wavelength are inversely related, lowers the energy of the electron. Atoms in a multiconfigurational resonating system are bound together more strongly, that is, have shorter interatomic spacings, than in the same system with fewer degrees of configurational equivalence.

A macroscopic example of resonance is found in the coupling of two simple harmonic oscillators operating in the uncoupled state with the same characteristic frequency ν_{c_0}. A common general relationship between ν_{c_0} and the characteristic frequencies ν_c arising in the resultant system when the oscillators are nondissipatively coupled, with a coupling coefficient γ, is

$$(1 - \nu_{c_0}{}^2/\nu_c{}^2)^2 = \gamma^2 \qquad (13)$$

whence

$$\nu_c = \nu_{c_0}(1 \pm \gamma)^{-\frac{1}{2}} \qquad (14)$$

The lower frequency represents a new minimum energy in terms of the characteristic oscillations of the system.

Another example of resonance is found in the wave-mechanical explanation of the occurrence of ortho- and parahelium spectra, where orthohelium has unpaired electron spins ($S_s = 1$) and parahelium has paired spins ($S_s = 0$). The ortho form corresponds to an antisymmetric linear combination (ψ_A), and the para form to a symmetric linear combination (ψ_S), of the possible ψ's of the two electrons in normal and excited helium.[25] The two new wave functions, ψ_A and ψ_S, occur as the result of wave-mechanical resonance and are analogous to the two ν_c's produced according to eq. 14. The energies E_A and E_S, corresponding to ψ_A and ψ_S, differ according to

$$E_A = E_N - E_R \quad \text{and} \quad E_S = E_N + E_R \qquad (15)$$

where E_N is the energy of the nonresonating (uncoupled) system, and E_R is called the **resonance energy,** or **exchange energy.**

In chemical parlance, it is common to speak of bonding electrons, or simply bonds, as *resonating* among neighboring bonded atoms (**ligands**). This concept has been developed, particularly by Pauling, as a means for correlating many properties of molecules and bulk matter.[12,24]

DIRECTED BONDS. The spatial distributions of the wave functions (ψ's) or electron-cloud densities of simple s states ("orbitals" or "shells") are spherical about an atomic nucleus and have $n - 1$ spherical nodes (regions of vanishingly small $|\psi|^2$). The ψ's of the p states, however, make use of a previously mentioned exchange effect to form three mutually perpendicular double ellipsoids ("dumbbells") which intersect with their central nodes at the nucleus, and the ψ's of the d and f states form even more complex angular distributions. Thus, atoms and ions whose outermost bonding (valence) electrons are *not* in simple s states have knobby outer electron "clouds." When atoms and ions have such nonspherical electron "clouds," they may exhibit **directional bonding** (directed valence) which often has a profound influence on the spatial arrangements of atoms in molecules and solids. An isolated simple molecule of water, for example, has the approximate arrangement H

:Ö:H, although mutual repulsion of the two H atoms (in part, H^+ ions) expands their relative bond angle to about 104° instead of the 90° that would obtain without the H–H interaction (the dipole moment of an H_2O molecule is 1.8×10^{-18} esu-cm). Also, the angles between the C—Cl bonds in nonpolar CCl_4 (eq. 10) are almost exactly 109° 28′; that is, the central carbon atom is surrounded by chlorine atoms situated at the corners of a regular tetrahedron. Here, the four equal C:Cl shared-electron bonds are apparently produced by hybridization of three p states (orbitals) and one s state (orbital); or the bond between a central C atom and any Cl atom may be thought of as rapidly exchanging (resonating) between s and (mostly) p states.[24, 25]

Silicon (Si), which is chemically similar to carbon, readily forms similar hybridized and tetrahedrally oriented s—p bonds, such as those in the SiO_4 groups in silicate crystals where each Si atom is surrounded by four O atoms arranged at the corners of a tetrahedron. In a crystal of beryllium orthosilicate (Fig. 11b), for example, there are corner-joined BeO_4 and SiO_4 tetrahedra wherein each Be atom and each Si atom is bonded to four O atoms and each O atom is bonded to two Be atoms and one Si atom. The Si—O bonds are formed by exchanging of the *four* $3s^2 + 3p^2$ electrons of Si and the four $2p^4$ electrons of O between the hybridized s and p states, whereas the Be—O bonds are formed by exchanging of the *two* $2s^2 \rightarrow 2s^1 + 2p^1$ electrons of Be and the four $2p^4$ electrons of O between the hybridized s and p states.

As is described later, the normal directed covalent bonds of some isolated simple atoms (ions, molecules) can be warped and deformed when the atoms are incorporated as impurities in solids where the configurations (geometric arrangements and spacings) of the surrounding host-crystal atoms do not coincide with the directional characteristics of the impurity atom. This distortion effect can profoundly influence the energy levels and degrees of ionic and covalent bonding of impurity atoms.

GENERAL ENERGY-LEVEL DIAGRAMS OF ISOLATED MOLECULES. When two like atoms combine to form an isolated simple molecule, as in H + H $\rightarrow H_2$ and Cl + Cl $\rightarrow Cl_2$, the energy of the system changes as sketched in Fig. 3. As the atoms come together, their various energy levels are altered independently, the highest levels being the first and the lowest levels the last to alter with decreasing nuclear separation. Each level may have a minimum potential energy at a different internuclear distance. The difference in potential energy between the ground states of the component isolated atoms and the equilibrium state of the resultant molecule (at constant ambient temperature T) is transformed into kinetic energy of the combined atoms, that is, **thermal agitation,** and appears as the heat of reaction. On very close approach of the atoms, the nuclei repel each other, and there is a strong repulsion energy due to the mutual deformation of inner completed electron shells (s^2, s^2p^6, etc.) as a consequence of the exclusion principle. According to the indeterminacy principle, a molecule has a finite energy **(zero-point energy)** even in its lowest (ground) state at $T = 0$. In other words, even at the absolute zero of temperature, there are fluctuating attractions and repulsions of the atoms whose nuclei, therefore, vibrate relative to each other. For moderate amplitudes of vibration there is simple harmonic motion with vibrational frequencies (ν_a) of the order of 3×10^{13} sec^{-1}. The amplitudes of vibration of the nuclei are generally of the order of 0.03 Å and increase with increasing temperature. In Fig. 3 the **vibrational states** are indicated by horizontal lines (energy levels) in the **potential wells.** The lowest vibrational level, for zero-point energy, is $h\nu_a/2$ above the bottom of the potential well. Vibrational motion from a_2' to c_2' corresponds to a fluctuation of potential energy in the ground state of the atom from a_2 to b to c_2, where the kinetic energy is a maximum at b. The energy differences between adjacent vibrational levels are generally 1 to 10 per cent of those between neighboring electronic-transition states ($E^*_2 - E_1$). Still finer sublevels (not shown in Fig. 3), having energy differences about 1 per cent of those between vibrational levels, may be produced by rotational motions of the nuclei.[25] The additional sublevels of vibration and rotation change the spectral lines of the atoms in the molecule into groups of many closely spaced lines which are known as **bands.**[320c]

According to the **Franck–Condon principle,** electronic transitions in molecules occur in times which are very short compared with the period of vibration of the nuclei, that is, much less than 3×10^{-14} sec.[26] Hence, the most probable electronic transitions proceed vertically in the energy-level diagram, and the transitions are most probable when the kinetic energies of the nuclei are a minimum, that is, at the end points of their vibrations (for example, a_2 and c_2). The lowest vibrational levels are possible exceptions to this principle. Some possible transitions are shown in Fig. 3. If the molecule is in its ground state E_1 with vibrational energy corresponding to sublevel a_2, the system may be raised (excited) to state E^*_2, vibrational level a^*_2, by absorption of a photon with energy $h\nu_a = E_a$. The excited molecule commences vibrating between $a^*_2 - d - c^*_2$ and may emit a low-energy photon to reduce the energy of the system to the lower vibrational level a^*_1, whence it

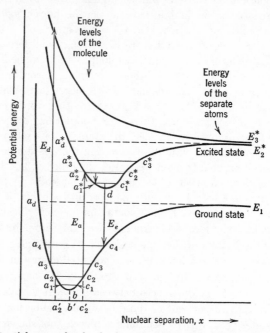

FIG. 3. Potential-energy levels of a hypothetical free diatomic molecule composed of atoms of the same species.

may proceed back to state E_1, sublevel a_4, by emitting a photon of energy $h\nu_e = E_e$. Further energy losses by emission of low-energy photons, or by inelastic collisions with other molecules, are necessary before the molecule can return to its original energy level. As mentioned before, photon emissions which are occasioned by thermal agitation are known as **thermal radiation.** The excitation process of thermal radiation involves the transformation of thermally induced kinetic energy of an atom or a molecule into increased internal energy of another molecule upon inelastic collisions (**collisions of the first kind**). Subatomic photon emissions occasioned by other means of excitation are classified *broadly* as **luminescence,** especially when the primary excitant particle gives up quanta of energy larger than about 1 ev to the system (see Appendix 4).

An *isolated* excited atom, ion, or simple molecule may return to the ground state only by radiation of energy as one or more photons (Fig. 2). A molecule, with its additional closely spaced vibrational and rotational energy levels, however, may emit a much larger number of different photons (that is, photons of different energy) than the isolated atoms of which the molecule is composed. In addition, if a molecule acquires sufficient vibrational energy to raise it to a level above the corresponding energy level of the isolated atoms, for example, to a_d or a^*_d in Fig. 3, the molecule has a high probability of dissociating. The nuclei can fly apart in about 10^{-10} sec, whereas the lifetime of an excited state before optical photon emission is about 10^{-8} sec. High-energy photons or material particles can also dissociate the molecule by raising it into a state, for example, E^*_3 in Fig. 3, where the molecule is unstable (no potential well). The number of possible means for an atom to lose excitation energy generally increases as the atom is combined with other atoms to form simple isolated molecules and increases further when large numbers of isolated atoms or molecules are condensed to form liquids and solids.

Gases, Liquids, and Solids

The sizes of molecules were not mentioned in the preceding section, because there is no upper limit to the number of atoms which may form stable associations. In general, however, the sizes of molecules are roughly proportional to the numbers and sizes of their constituent atoms (ions).

GASES. The next step in our outline development of matter, is a brief discussion of the aggregation of molecules (atoms) to form more complex matter. Consider separate simple molecules of vaporized water (H_2O) or potassium chloride (KCl) moving about as in a rarefied gas. Our best man-made vacua contain about 10^8 molecules cm^{-3}, and so the average distance between molecules in this case would be less than 10^{-3} cm. This spacing is over 10^4 times larger than the sizes of the simple molecules. Under these conditions, it can be assumed that intermolecular forces are *practically* negligible except when the molecules collide. If this is assumed to be an **ideal gas,** the total kinetic energy of N_A molecules of the gas at temperature T is given by

$$E_k = \tfrac{3}{2}N_A kT = \tfrac{1}{2}m_m N_A \bar{v}_m^2 \tag{16}$$

where k is **Boltzmann's constant,** 1.38×10^{-16} erg deg^{-1} (8.62×10^{-5} ev deg^{-1}), m_m is the mass of the molecule, and $\bar{v}_m^2$ is the mean square velocity of the molecules. As the temperature is decreased, the gas will liquefy at its condensation (boiling) point and solidify at its freezing (melting) point to form *voluntarily* a coherent mass with about 10^{22} simple molecules cm^{-3}. The words *practically* and *voluntarily* are emphasized to indicate that there are intermolecular attraction forces which become dominant when the kinetic energies of the molecules are decreased to the point where the molecules can associate for long times without being sundered by collisions. The chief attraction forces between uncombined simple molecules apparently comprise dipole interactions, that is, electrostatic attractions of bodies whose opposite ends have different electric charges.[25, 27-29] Molecules without "permanent" dipole moments may have moments induced by adjacent "permanently" dipolar molecules. Even in gases without any "permanently" dipolar molecules, the fluctuations of the molecules between polar and nonpolar states (eq. 11) give rise to the London **dispersion effect** whereby molecules synchronize to some extent their polar fluctuations and spatial orientations for a net attraction. The potential energy E_p of attraction caused by the dispersion effect between molecules of polarizability χ spaced a distance x apart, is given by

$$E_p \approx -\tfrac{3}{4}h\nu_i\chi^2 x^{-6} \tag{17}$$

where ν_i is the oscillation frequency corresponding to ionization in the case of simple molecules. The dispersion effect is so named because the fluctuations which enable the intermolecular attraction are responsible for the dispersion of light by the molecules. An important feature of the dispersion effect, as distinguished from attractions involving "permanent" dipoles, is that the dispersion-effect attraction is *additive* for all the molecules in a gas; that is, any number of molecules may interact without saturating the effect.

CONDENSATION (LIQUIDS). Imagine space in a closed vessel to be partitioned into elemental volumes (which we shall call "holes") of molecular size, with a few of the "holes" transitorily occupied by simple molecules. As the temperature ($T \approx E_k$ of the molecules) of the gas is decreased, the attractions between molecules (and gravitational attraction) draw the molecules closer together at the bottom of the vessel. At the **critical temperature** (above which the liquid cannot exist), about half the "holes" at the bottom of the vessel are occupied by molecules. When the temperature is lowered further, condensation to the liquid state occurs, that is, the molecules have high enough potential energy of attraction to near neighbors and low enough kinetic energy to remain attracted together. In the liquid, the attraction energies comprise dipole interactions, exchange (resonance) interactions, and ionic interactions (if electron transfer takes place on close approach of *different species* of molecules). Very close approach of the molecules is necessary to make effective use of the **van der Waals intermolecular attraction forces,** which vary as $dE_p/dx \propto -x^{-7}$ for dipole interactions, as $-x^{-9}$ for dipole–quadripole interactions, and as $-x^{-11}$ for quadripole–quadripole interactions. A manifestation of the cohesion of the liquid phase is surface tension, which helps to increase the density as the temperature is lowered and the molecules huddle closer together to effectively "squeeze out the 'holes'." The cohesion of molecules in liquids produces internal pressures amounting to thousands of times normal atmospheric pressure (10^6 dynes cm^{-2}). Eventually, in the condensation process, the repulsion potentials between the nuclei and between the completed electron shells of the different atoms become important in limiting the density of the molecular aggregate. According to wave-mechanical calculations, the repulsion energy has the form $\phi(x)\epsilon^{-x/a}$, where $\phi(x)$ is a polynomial function of the intermolecular distance x, and a is a constant for the particular molecule. A more tractable *approximation* is obtained by assuming the repulsion energy to vary as x^{-6} to x^{-12}, and so x^{-9} is a fair approximation for a number of molecules. A general simplified expression for the total energy of interaction between molecules is then

$$E_p \approx -Ax^{-6} + Bx^{-9} \tag{18}$$

where A and B are characteristic constants of the molecules. In a liquid at equilibrium, the average spacing $\bar{x}$ between the simple molecules corresponds to the intermolecular distance where E_p is a minimum, except that $\bar{x}$ increases as T increases (thermal expansion, corresponding to the asymmetry of potential wells of the type shown in Fig. 3).

An important distinction between gases and liquids is their different dependence of viscosity (internal friction) on temperature. The **coefficient of viscosity** η is the force in dynes per square centimeter required to sustain a velocity gradient of one centimeter per second per centimeter. That is, η is the force which must be applied to propel a layer of fluid at the constant rate of one centimeter per second relative to a parallel layer one centimeter distant. For *ideal* gases, η varies as $T^{\frac{1}{2}}$, because the only interaction between the layers is due to interchange of molecules through practically free space ($v_m \propto T^{\frac{1}{2}}$, according to eq. 16). For liquids, however, η is found to vary as $\epsilon^{E^*/kT}$, where the activation energy E^* is the work which must be done to create a "hole" in the liquid in order to allow the molecules to move to new sites relative to their neighbors. The number of molecules having sufficient energy E^* to jostle a "hole" large enough for interchange of molecules is proportional to the **Boltzmann factor,** $\epsilon^{-E^*/kT}$. The **fluidity** of the liquid, then, is proportional to $\epsilon^{-E^*/kT}$, and the viscosity is proportional to the reciprocal of fluidity, or $\epsilon^{E^*/kT}$. There is a further relation between E^* and the **latent heat of vaporization** H_L, which is the energy required to eject a molecule from the liquid without leaving a "hole." Experimentally, the ratio H_L/E^* varies from about 3 for liquids composed of symmetrical molecules (A, N_2, CCl_4) to as much as 25 for liquid metals. Consequently, the size of the "hole" required to allow molecular interchange need be only $\frac{1}{3}$ to $\frac{1}{25}$ the size of the molecule, atom (monatomic molecule), or ion (as in metals).[25]

CRYSTALLIZATION (SOLIDS). Returning to consideration of the transitions among gases, liquids, and solids, we may suspect that gases should exhibit some degree of crystallinity as a result of molecular alignments induced by the dispersion effect. There are, undoubtedly, some temporary local alignments, just as there are temporary associations of molecules, but these are of fleeting duration owing to the random molecular motions and the disruptive effects of collisions. **Crystallinity** ideally requires perfect rectilinear alignment of *all* the atoms in a solid body. That there is a large degree of crystallinity in liquids, however, is evidenced by their x-ray diffraction patterns and high specific heats.[25, 29–31] The molecules in liquids may be pictured as being in or fluctuating between the following states: (1) the **quasicrystalline state,** wherein the distribution of molecules at any *instant* is apparently random owing to thermal agitation, but the *average* arrangement is crystalline to some degree, and (2) the **microcrystalline state,** wherein molecules in local groups are found to be arranged in crystalline order at any instant, but there are randomly distributed molecules between the microcrystalline groups, and the groups are randomly growing and dissolving.

The attraction forces between molecules often promote rectilinear alignment, but rectilinear crystalline order is determined chiefly by geometric factors to be discussed in the next chapter (magnetic interactions, that is, alignment of resultant spin and orbital magnetic-moment vectors, may play a part in some crystallizations, especially when the atoms have unpaired electron spins [32]). As the temperature of a liquid is decreased, molecular motion decreases, and the number of "holes" likewise decreases. At the freezing point, molecular motion has been reduced to the point where rectilinear alignment of atoms may be maintained throughout the entire solid body **(crystal);** that is, long-range order is attained and preserved.[33] As the temperature is decreased below the freezing point, the atoms in the crystal oscillate with decreasing amplitudes and in fewer modes about their aligned equilibrium positions. Correspondingly, there is a decreasing probability of an atom chancing to acquire sufficient thermal activation energy E^* to leave its equilibrium position. Meanwhile, surface tension remains as a dissenting effect which strives to make the solid spherical (curvilinear) rather than multihedral (rectilinear). In the interior of large crystals, however, the effect of surface tension is small relative to the forces promoting rectilinear alignment.

Glasses (vitreous materials) are exceptionally viscous liquids whose large degrees of liquid-like amorphousness, as evidenced by their curvilinear boundaries, optical isotropy (when unstrained), diffuse x-ray diffraction patterns, and lack of definite melting points, belie their

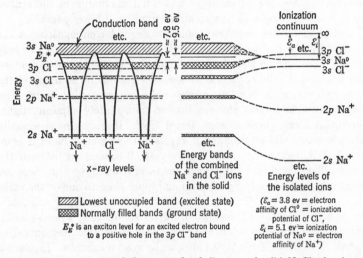

FIG. 4. The upper part of the energy-level diagram of solid NaCl, showing the splitting of energy levels of the ions into bands which extend throughout the ideal crystal (not to scale). The exciton level E^*_E in cub.-NaCl is actually a band of levels.

apparent solidity. **Crystals** (solids) usually have rectilinear boundaries as well as optical anisotropy (except for cubic crystals), sharp x-ray diffraction patterns, and definite melting points, which bespeak the long-range rectilinear symmetrical arrangements of their atoms. It should be noted that the words *solid* and *crystal* are used synonymously.

Energy Levels in Solids. Figure 4 is a simplified sketch of some of the energy-level changes which occur when a number $2N$ of each of two species of atoms (ions) combine and solidify to form a simple ionic **electrical-insulator crystal,** such as that of cubic sodium chloride (NaCl).[34] Reading from *right to left*, each of the practically identical (degenerate) energy levels of the separate atoms is proliferated into N closely spaced levels to form **energy bands** in the solid. Each *filled* level in a band is occupied by two electrons with paired spins; hence, there are $2N$ electrons in a completely filled band with N levels. The N

levels extend over the entire (ideal) crystal which may be considered as a single vast molecule. The proliferation of the discrete atomic-energy levels into bands in the solid is in accord with the exclusion principle, it being taken into account that the wave function ψ of an outer electron, in particular, extends throughout the entire (ideal) crystal.[35-37] In non-overlapping bands, the density of levels is a maximum at the center of each band, falling off to zero at the upper and lower edges of the band. Likewise, the velocity of electrons has a flat maximum in the central part of the band and falls to zero at the upper and lower edges.

According to the electron configurations and combinations of the atoms in a solid, the occupied energy bands may be completely or only partially filled. In an **electrical-conductor crystal,** the topmost occupied band is only partially occupied. Then the electrons fill as many of the low-lying levels in the band as they are able, and an applied electric field easily raises a high-lying electron into an immediately adjacent higher unoccupied level where it may travel freely through the assumedly perfect crystal. In **semiconductor crystals,** the topmost band is completely filled, but either there is a very small separation between the topmost filled band and the lowest unoccupied band, or else there are special impurity levels lying between and quite close to one or the other of the two bands.[18, 19, 35-37] Solids composed entirely of metallic elements (for example, K, Zn, Al) are usually good conductors, whereas solids composed of compounds of metallic and nonmetallic elements (for example, KCl, ZnO, ZnS, Al_2O_3) are usually good insulators. Insulating crystals of ZnO which have been heated to sufficiently high temperatures to effect partial decomposition (leaving some free Zn in the crystals) become semiconductors; likewise, crystals of ZnO or TiO_2 heated in reducing atmospheres, or heated mixtures of conductors and non-conductors (for example, Ni + TiO_2, or Cu + ZrO_2 + Fe_2O_3), may be made to produce semiconductors.[35-38]

Very narrow energy levels are designated as **discrete levels,** whereas levels and bands, in general, are denoted by the generic term **level,** except when the band property is to be emphasized. An **energy state of an atom** comprises many levels; for example, the *ground state* of Li in Fig. 2 comprises the energy level s^2, s ($2s$ Li) and the lower-lying s^2 ($1s$ Li) level (not shown in the figure), whereas the general *excited state* of Li comprises all the ground-state levels plus any or all the excitation levels above s^2, s. A *particular* excited state would be specified by denoting the particular occupied excitation levels and unoccupied ground-state levels. The ground state of Na^+ in NaCl (Fig. 4) comprises the occupied energy levels $2p$ Na^+, $2s$ Na^+, and lower occupied levels which are not shown, whereas the general excited state obtained by raising an electron from any occupied level to any unoccupied level, comprises the ground-state levels plus the 3s Na (removing an electron from Cl^- to Na^+ makes the latter Na) and any or all of the higher excitation levels (the latter are not shown in the figure). An **energy state of a bound electron** (that is, of the *system* containing the electron), at any instant, is a particular energy level, although it is common to refer to an electron as being in a ground state or excited state without specifying which ground-state level or excited-state level is meant. This is particularly true in speaking of electrons in solids where there is often insufficient quantitative information about the various energy levels.

The left side of Fig. 4 shows an energy-level diagram for an electron moved along the center (through the atoms) of part of a rectilinear row of $\cdots Na^+$-Cl^--Na^+-$Cl^- \cdots$ ions in a crystal of cubic NaCl. The outer electrons of the Na^+ and Cl^- ions, though "fenced-in" by intervening potential barriers (regions of high repulsion potential for electrons) belong to the whole ideal crystal rather than to their parent ions alone. The degree of belonging to the entire crystal is proportional to the widths of the allowed energy bands. The inner x-ray electrons may be considered as practically belonging to the individual atoms or ions, but the outer electrons have a high probability of penetrating the potential barriers by a wave-mechanical diffraction process which is commonly known as the **tunnel effect.** The tunnel effect is a consequence of the wavelike behavior of the electron and its indeterminacy of location and momentum. The probability of tunneling is inversely proportional to the area of the potential barrier above the energy level of the electron. Hence, the higher the energy of the electron and the lower and narrower the potential barrier, the greater the probability of the electron tunneling ("leaking") through the barrier. As shown in Fig. 4, the widths of the allowed energy bands increase and the separations between the allowed bands decrease on going upward to higher energies in the diagram. This trend results in considerable overlapping of the higher energy bands of many crystals. Each of the energy levels in the solid is split further into fine sublevels (not shown in Fig. 4) corresponding to the nuclear vibrations shown in Fig. 3.

An electron moved along *between* the rectilinear rows of alternating Na^+ and Cl^- ions in cubic NaCl experiences an undulating potential of much lower amplitude and does not encounter the discrete energy levels shown below 3s Na^0 in Fig. 4. The amplitude of potential variation is particularly small when an electron is moved through large-diameter "tubes" of low electron density between widely separated rows of atoms in loose-packed crystals [for example, the hexagonal "tubes" running through crystals of rhombohedral zinc silicate (Fig. 11b)]. For our purposes, however, conventional energy-level diagrams are drawn with reference to an electron or unit negative charge moved along a line passing through the *centers* of aligned atoms. In many solids, especially those that do not have simple cubic structures, this line of motion along a row of atoms is *not* a straight line.

An important difference between the vibrational energy levels of *isolated* molecules, as represented in Fig. 3, and those of closely coupled atoms and simple or complex ions in solids, is that transfer of vibrational energy takes place readily from atom to atom in the solid, whereas an isolated molecule cannot gain or lose energy by this process of heat conduction (except on collision with another molecule). In solids, therefore, there is considerable opportunity to dissipate localized excitation energy as heat rather than radiation, and it is shown later that the degradation of excitation energy into heat in crystals increases with (1) increasing number of different species of atoms or ions in the crystals, and (2) increasing degree of imperfection of the crystals, that is, increasing irregularity in the energy-level diagrams of the solids. Another point is that the amplitudes of vibration of different species of atoms (or ions), or of the same species of atom with different neighbors, in a given crystal are different; for example, at room temperature, the relatively light sodium ion in a sodium chloride crystal has a larger root-mean square amplitude of vibration (0.245 Å) than the chlorine ion (0.235 Å); furthermore, these vibrations may be quite anisotropic in some crystals, depending on the symmetry or lack of symmetry of the immediate region within which a particular atom or ion is located.[47, 57] (In general, the amplitude of vibration decreases and the frequency of vibration increases as the bonding energy of an atom increases.) The energy levels in solids are complicated also by statistical fluctuations in thermal agitation from point to point in the crystal (even at equilibrium) as well as by various crystal imperfections which are discussed in the next chapter.

It should be emphasized that, although the *band* theory of energy levels in solids has broad applicability, it becomes increasingly difficult to apply as the solids depart from being simple cubic crystals composed of binary combinations of univalent ions. In more complex crystals, where the type of interatomic bonding differs with the different species and crystallographic locations of the atoms, it is sometimes more convenient to speak in terms of *bond* theory, especially when the bonding is largely non-ionic. The bond picture (eqs. 10–12) emphasizes the local conditions in the immediate vicinity of an atom or small distinctive group of atoms in a solid, whereas the band picture emphasizes the fact that the individual atoms and distinctive atomic groups are cooperative members of the entire crystal. As an example, considerable attention will be given in following sections to crystals of rhombohedral zinc orthosilicate (rbhdl.-Zn_2SiO_4, Fig. 11b) which are made of corner-joined tetrahedral groups of ZnO_4 and SiO_4. In these crystals, the ZnO_4 tetrahedra are all interconnected and have weak semi-ionic Zn—O bonds, whereas the SiO_4 tetrahedra are isolated from each other and have stronger and apparently more covalent Si—O bonds. Here, then, the band picture may be used to describe the energy levels pertaining to the continuous ZnO_4 part of the structure, while the bond picture may be used to focus on local conditions in the discontinuous SiO_4 part of the structure. In the band picture, the width of a band is a measure of the degree to which electrons in that band exchange among all the identical atoms in a crystal. This exchanging increases the interatomic bonding energy of the crystal by adding exchange energy to the system.

CHAPTER 2

SOLIDS (CRYSTALS)

Ideal Crystals

When rigid spheres of *one* size are packed together as tightly as possible, they automatically assume one of two rectilinear arrays.[24, 25, 39, 40] Figure 5 shows that these crystalline arrays are (1) the face-centered cubic structure and (2) the hexagonal close-packed structure. In both structures, each sphere is in contact with 12 like neighbors; that is, both

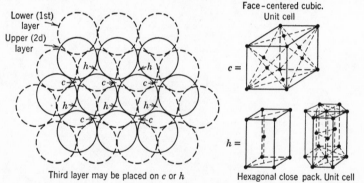

Fig. 5. Close packing of identical rigid spheres affords two possible crystalline arrays.

structures have a **coordination number** of 12, and each sphere is said to be 12-coordinated. As examples of coordination numbers, (1) a 3-coordinated sphere would be located in the center of a group of three other spheres placed at the corners of an equilateral triangle, (2) a 4-coordinated sphere would occupy the center of a group of four other spheres located at the corners of a tetrahedron, (3) a 6-coordinated sphere would be at the center of a group of six spheres occupying the corners of an octahedron, and (4) an 8-coordinated sphere would be at the center of a group of eight spheres occupying the corners of a hexahedron (cube). If certain simple integral proportions of *two* sizes of rigid spheres with radii r_1 and r_2 are distributed homogeneously and packed together, the maximum coordination number (in this case, the number of like spheres immediately surrounding a sphere of *different* size) is generally 8 for the **radius ratio,** $1 > r_1/r_2 > 0.73$, 6 for $0.73 > r_1/r_2 > 0.41$, 4 for $0.41 > r_1/r_2 > 0.22$, and 3 for $0.22 > r_1/r_2 > 0.15$. In all these cases, the close-packed spheres still naturally assume crystalline arrays, except at the surface where they may be influenced by the shape of the container.

In principle, at least, it is easy to substitute the word *atom,* or *ion,* for the word *sphere* and visualize idealized spherical atoms aggregating voluntarily into crystalline arrays at temperatures below the freezing point. The fact that atoms and ions are neither perfectly rigid nor always spheres, and do not have sharp boundaries, often leads to considerable departures from the crystalline arrays which might be expected on the basis of close-packing of rigid spheres. Some compounds crystallize as relatively open structures; that is, the system has a lower potential energy in the loose-packed state than it has in the close-packed state. Such structures sometimes have permanent voids (regions of very low electron-cloud density) which are larger than the apparent volumes occupied by their component atoms or ions. The major factors influencing the type of crystal structure assumed by actual combinations of atoms (ions) are (1) their radius ratios, (2) their effective ionic charges, (3) their polarizabilities (deformabilities), (4) the number and proportions of different species of atoms in the structure, (5) the type of bonding, for example, the proportions and directivities of ionic and covalent bonding, and (6) the ambient temperature and pressure.[41, 47, 51]

Despite the irregularities in real crystals, it is generally possible to describe the arrangements of their atoms, as a first approximation, in terms of *ideal* crystal lattices of infinite extent. **Crystal lattices** are geometric fictions which are *assumed* to describe idealized atomic arrangements having no imperfections and no thermal agitation. A basic concept of the crystal lattice is the **minimum identity translation** x_{τ_0}, which is the shortest distance a **lattice point** (intersection of imaginary rectilinear lattice lines) can be moved to arrive in surroundings identical with those in the place where it started. In a simple cubic lattice, x_{τ_0} is the edge of the unit cell; whereas, in a face-centered cubic lattice (Fig. 5), x_{τ_0} is half the distance along a face diagonal of the unit cell. The **unit cell** is generally the smallest portion of the lattice which may be used to reproduce the entire crystal lattice by simple duplication and ordered aggregation.

Ideal crystal lattices are derived by the operation of **symmetry elements** of (1) *rotation* of a point about (*a*) a center of symmetry,

TABLE 4

Some of the Elements of the Formal Geometry of Ideal Crystal Lattices [44]

System, Optical Anisotropy	Crystallographic: Elements, Axial Ratios and Angles $\angle bc = \alpha, ca = \beta, ab = \gamma$		Essential Symmetry, Hermann-Mauguin Symbol	Space Lattice, Hermann-Mauguin Symbol	Lattice Points per Unit Cell	Point Group — Schoenflies Symbol	Point Group — Hermann-Mauguin Symbol	Activity, Electro-optical and Piezoelectric	No. of Space Groups	Symmetry Group
1. Triclinic or anorthic (biaxial)	$a:b:c = x:1:y$ $\alpha, \beta, \gamma \neq 90°$		Point (1) No axes or planes of symmetry	Simple (P)	1	C_1	1	+	1	1
						$C_i, (S_2)$	$\bar{1}$	0	1	
2. Monoclinic (biaxial)	$a:b:c = x:1:y$ $\alpha = \gamma = 90°, \beta \neq 90°$		Plane (m) or diad axis (2)	Simple (P)	1	$C_s, (C_{1h})$	m	+	4	2
				Side-centered (A, C)	2	C_2	2	+	3	
						C_{2h}	2/m	0	6	
3. Orthorhombic or rhombic (biaxial)	$a:b:c = x:1:y$ $\alpha = \beta = \gamma = 90°$		3 diad axes (222) or 1 diad axis and 2 ⊥ planes intersecting in a diad axis	Simple (P)	1	C_{2v}	mm	+	22	3
				Base- or side-centered (A, B, C)	2	$D_2, (V)$	222	+	9	
				Face-centered (F)	4	$D_{2h}, (V_h)$	mmm	0	28	
				Body-centered (I)	2					

System	Axial elements	Axis	Lattice		S (Schoenflies)	HM	sign	n	no.
4. Tetragonal (uniaxial)	$a:b:c = 1:1:y$ $\alpha = \beta = \gamma = 90°$	Tetrad axis (4)	Simple (P) or base-centered (A,B,C) Face-centered (F) or body-centered (I)	1 2	S_4	$\bar{4}$	+	2	4
					C_4	4	+	6	
					C_{4h}	$4/m$	0	6	5
					D_{2d}, V_d	$\bar{4}2m$	+	12	
					C_{4v}	$4mm$	+	12	
					D_4	42	+	10	
					D_{4h}	$4/mmm$	0	20	
5. Cubic ("isotropic")	$a:b:c = 1:1:1$ $\alpha = \beta = \gamma = 90°$	3 tetrad axes (444) (or 3 diad axes) and 4 triad axes	Simple (P), Face-centered (F), Body-centered (I)	1 4 2	T	23	+	5	
					T_h	$m3$	0	7	6
					T_d	$\bar{4}3m$	+	6	7
					O	43	+	8	
					O_h	$m3m$	0	10	8
6. Rhombohedral or trigonal (uniaxial)	$a:b:c = 1:1:1$ $\alpha = \beta = \gamma \neq 90°$	Triad axis (3)	Simple (R)	1	C_3	3	+	4	
					$C_{3i}, (S_6)$	$\bar{3}$	0	2	
					C_{3v}	$3m$	+	6	9
					D_3	32	+	7	
					D_{3d}	$\bar{3}m$	0	6	
7. Hexagonal (uniaxial)	$a_1:a_2:a_3:c = 1:1:1:x$ $\angle a_1a_2 = a_2a_3 = a_3a_1 = 120°$ $c \perp a_1a_2a_3$	Hexad axis (6)	Simple (C or H)	1	C_{3h}	$\bar{6}$	+	1	10
					C_6	6	+	6	
					C_{6h}	$6/m$	0	2	
					D_{3h}	$\bar{6}2m$	+	4	11
					C_{6v}	$6mm$	+	4	
					D_6	62	+	6	
					D_{6h}	$6/mmm$	0	4	

and/or (*b*) an axis of symmetry; (2) *reflection* of a point through a plane of symmetry; and (3) *translation* (displacement) of a point along a straight line until an equivalent point is encountered. The maximum number of *different* operations of rotational and reflectional symmetry which can be performed without violating the concept of the minimum identity translation, affords 32 **point groups** which are classifiable in terms of 11 **symmetry groups.** The maximum number of *different* ways of repetitively translating like points in rectilinear three-dimensional order affords 14 **space lattices,** which are catalogued under 7 **crystallographic systems.**[39-50] Table 4 lists the elemental crystallographic systems, unit cells, space lattices, and point groups, with some information concerning symmetry, electro-optical (piezoelectric) activity, and notation.[44] The different combinations of the 14 space lattices with their corresponding 32 point groups afford a maximum of 230 **space groups;** that is, there are at most 230 different ways of arranging like points in *ideal* crystal lattices. Members of the 11 different Laue symmetry (point) groups are convertible into a common point group by the operation of a center of symmetry; for example, when point group *m* or 2 in the monoclinic system is operated on by a center of symmetry it is transformed into the centrosymmetrical point group 2/*m*. Note that point groups having centers of symmetry are electrooptically and piezoelectrically inactive, whereas the other point groups are active (except subgroup 432 of cubic point group 43). This recital of the elements of formal crystallography is intended as only a bare outline of some of the chief features of the geometry of ideal crystal lattices. Serious students of luminescence of solids who are not already acquainted with crystallographic notations and their significance should study carefully the references cited in this chapter.

Although there are only 230 possible space groups, there are innumerable different real crystals which can be made by variations in the number, proportions, and conditions of chemical combination and crystallization of the 96 known elements. Crystals, like molecules, are in general electrically neutral, so that even in strongly ionic crystals the total anionic charge equals the total cationic charge. It is possible, however, to have local regions or surfaces with excess positive or negative charge as long as the electrical neutrality of the crystal as a whole is maintained. In some cases, it is possible to interchange ions of different valence in a crystal without altering the structure, as long as the overall neutrality is maintained. As examples, the minerals diopside, $CaMg(SiO_3)_2$; jadeite, $NaAl(SiO_3)_2$; and spodumene, $LiAl(SiO_3)_2$; all have the same monoclinic structure (*C*2/*c*). In these crystals, the two bivalent Ca^{2+} and Mg^{2+} ions are replaced by one univalent Na^+

(or Li^+) ion and one tervalent Al^{3+} ion to provide the same total cationic charge.[39] It may be noted that the radius ratios of Ca^{2+}/Mg^{2+} (1.06 Å/0.78 Å = 1.36) and Li^+/Al^{3+} (0.78 Å/0.57 Å = 1.37) are practically identical, and the radius ratio of Na^+/Al^{3+} (0.98 Å/0.57 Å = 1.72) is only 25 per cent larger than that of the other two pairs of ions.

Some crystals have insufficient atoms to occupy all the geometric sites allowed by their space groups, giving rise to **defect structures** or **deficiency structures.** In some crystals which have two or more species of cations or anions (1) the different species may occupy only structurally different sites **(ordered structure),** or (2) each may occupy several different sites in various proportions **(disordered structure).** Examples of these structures are found in $Pm3m$-Na_xWO_3 (perovskite-type sodium–tungsten bronze, with $x < 1$), and various normal and inverse spinels ($Fd3m$) which are described in detail in the literature.[39, 43, 47]

In certain crystals, also, some of the atoms may be arranged, displaced, or deformed (polarized) in such a manner that relatively simple molecular units may be partially distinct. For example, the reversal of intensities of some electron and x-ray diffraction lines of hex.-ZnO has been interpreted as indicating that there is an asymmetric charge distribution around the Zn and O atoms, such that these atoms are linked together along the c axis of the crystal as pseudodiatomic molecules.[50a] There is, apparently, a different bonding between Zn and O atoms along the c axis from that in other directions in uniaxial *real* crystals of hex.-ZnO, even though the *ideal* lattice structure of $C6mc$-ZnO would have both Zn and O atoms symmetrically 4-coordinated by their opposites in tetrahedral arrangement. The sublimation of zinc oxide at temperatures well below its melting point may be pictured as a thermal breaking of the weaker bonds which do not lie along the c axis, and volatilization of simple ZnO molecules which retain the stronger Zn–O bondings along the c axis. For complete melting, the bonds along the c axis would also have to be broken by thermal agitation. It is possible that the asymmetry of charge distribution and bonding reported for zinc oxide exists also in hexagonal zinc sulphide and cadmium sulphide.

There are two major systems of crystallographic notation, the older Schoenflies system, and the newer Hermann–Mauguin system.[46-48] Although considerable memory work is necessary for rapid use of either system, the Hermann–Mauguin-system symbols are more easily translated into their symmetry elements. In the Hermann–Mauguin system, the elements of rotational and reflectional symmetry are designated as follows: 1 denotes no symmetry; $\bar{1}$ is a center of symmetry; 2, 3, 4, and 6 are diad, triad, tetrad, and hexad axes of rotational symmetry; $\bar{2}$, $\bar{3}$, $\bar{4}$, and $\bar{6}$ are the indicated multiple axes of rotary inversion (rotation followed by reflection through a plane perpendicular to the axis of rotation); and m (mirror) is a plane of reflection symmetry. Translations are represented by the capital letters P, A, B, C, H, F, R, and I whose connotations should be apparent from Table 4. Combinations of rotation (reflection) and translation are: (1) **screw axes**, which are denoted by ζ_ξ (for example, 2_1, 3_1, 3_2, *etc.*) indicating a ζ-fold axis with its $2\pi/\zeta$ rotation plus a displacement by a fraction ξ/ζ of the identity translation x_{τ_0}, parallel to the axis, and (2) **glide planes,** which are denoted by a, b, c, d, and n, according to the amount and direction of glide.[46] The degree of point-group symmetry commences with the triclinic space group $P1$ (no symmetry), at the top of Table 4, and generally increases on going down through the table to the highly symmetrical cubic, hexagonal, and rhombohedral structures. As an example of the significance of crystallographic notation, the atoms in zinc selenide (ZnSe) crystals are arranged as an approximation of the point group $\bar{4}3m$ in the space group $F\bar{4}3m$ (compare Figs. 11*b* and 12), which is a face-centered (F) cubic lattice wherein the lattice points are generated by the combined operation of a tetrad axis of rotary inversion $(\bar{4})$, a triad axis of rotation (3), and a plane of symmetry(m). This space group corresponds to $T_d{}^2$ in the Schoenflies system.[42,46-48]

It should be emphasized that (1) crystallographic symbols are primarily concerned with the *symmetry* aspects of crystal *lattices,* and (2) lattice points are centers of reference for the purpose of describing point-group (local) symmetries. In an actual crystal, therefore, *lattice points may or may not be occupied, and a space group alone does not necessarily describe the positions of all the atoms.* In a few simple ideal crystals, such as face-centered-cubic *Fm3m*–KCl, in which the K atoms and the Cl atoms are each arranged in a face-centered-cubic array interpenetrating the other array, all the lattice points and only the lattice points described by the space group are occupied. In other ideal crystals, however, such as those containing complex molecular or radical groups without central atoms in such groups (for example, the chlorate, $ClO_3{}^{2-}$, group whose point group is $3m$), groups of atoms may be located around a lattice point without an atom actually occupying the lattice point. Hence, the *combination* of crystallographic and chemical symbols is necessary (but not always sufficient) to define a particular structure. For example, cub.-KCl and cub.-CaF$_2$ both crystallize in space group *Fm3m*, but the fluorite (cub.-CaF$_2$) structure has a face-centered unit cell of Ca atoms enclosing an inner simple cube of F atoms, as distinguished from the two interpenetrating face-centered-cubic arrays of K and Cl atoms (ions) in cub.-KCl.

Incidentally, the $F\bar{4}3m$-ZnS type of structure may be derived (1) from the *Fm3m*-CaF$_2$ type of structure by omitting every other F atom, or (2) from the *Fm3m*-KCl type of structure by displacing the normally centered origin of one of the interpenetrating face-centered-cubic structures so this origin is located one quarter of the distance along the body diagonal of a unit cube of the other face-centered-cubic structure (see Fig. 11b). When all the atoms in the $F\bar{4}3m$-ZnS type of structure are made alike, one obtains crystals of the diamond type (that is, $Fd3m$-C). A further point is that the *unit cell* of the low-temperature form of cesium chloride is a *body-centered cube* ($Pm3m$-CsCl), that is, each of the eight corners of the unit cube is occupied by a member of one species of atom (for example, Cs), and the center of the cube is occupied by one member of the other species of atom (for example, Cl), but the *crystal* is called *simple cubic* because it is necessary to translate along a simple cubic structure to arrive at identical surroundings on proceeding from any of the atoms (for example, a translation from a Cs atom at a corner of the unit cell to a Cl atom at the center would amount to changing from a position 8-coordinated by Cl to a position 8-coordinated by Cs). Similarly, the basic lattice of $Fm3m$-KCl is simple cubic, but the occupation of lattice points by alternating K and Cl atoms in an actual crystal necessitates calling cub.-KCl a face-centered-cubic structure to describe the translation pattern which must be followed to arrive at positions with equivalent surroundings in such a crystal.

Growth of Real Crystals

From the brief discussion in Chapter 1 it should be apparent that a cooled liquid may *approach* the ideal crystalline state as the temperature is lowered. *No real crystal, however, ever attains the static perfection of an ideal crystal lattice!* Apart from atomic motions caused by thermal agitation, some of the unavoidable imperfections which are occasioned in any real growth process are (1) residual holes **(omission defects),** (2) misaligned microcrystals **(mosaic structure),** (3) impurities, and (4) local variations in the compositions of polyatomic crystals, that is, crystals containing more than one species of atom. The growth of a crystal by accretion of molecules from a surrounding fluid (for example, a vapor, or melt) is a discontinuous process, proceeding tangentially, plane by plane, the rate of growth perpendicular to a plane being inversely proportional to the population density of atoms in the plane.[29, 47, 51] The exposed faces of real crystals, therefore, are usually those which grow most slowly. Among the various methods of growing crystals, including (1) cooling a melt, (2) direct condensation from a gas, (3) growth from solution, (4) decomposition from more complex chemical compounds, and (5) reaction of simpler chemical compounds, none has been found to produce crystals without imperfections. Even diamonds, which are sometimes considered to be "perfect," are known to have many detectable imperfections.[52-57] The nearest approach to a perfect crystal would probably be obtained by mononucleating and cooling an *impurity-free* melt of a substance so slowly that the process could be considered reversible at all times. This process would require cooling times of the order of geologic eras, however, and is generally out of the question for the synthesis of artificial (man-made) crystals. Practically none of the natural minerals is satisfactory as a luminescent solid, even though many are well crystallized, because (1) the few moderately efficient luminescent minerals, such as manganese-containing willemite (rbhdl.-$(Zn:Mn)_2SiO_4$) and scheelite (tetr.-$CaWO_4$), are generally contaminated with detrimental impurities, and (2) nature has not produced most of the superior (and exactingly complex) luminescent solids, such as rbhdl.-$(Zn:Be:Mn)_2SiO_4$, hex.-$(Zn:Cd)S:Ag$, and cub.-$Sr(S:Se):$ $Sm:Eu$, which are described in later sections.

There is an inherent conflict in the role of temperature in growing crystals, especially large crystals. High temperature is necessary to provide enough mobility to allow disoriented mosaic (or micro-) crystals to align, but high temperature increases (1) the number of holes, (2) the dissolving of impurities from the container, (3) the selective decomposition of polyatomic crystals, and (4) the probability of dislocating parts of the crystal by statistical fluctuations in thermal agitation. Almost all the artificial luminescent solids (phosphors) are synthesized by reactions and/or decompositions below the melting points of the substances involved. This is done because most of the useful phosphors melt at temperatures higher than can be attained in furnaces with suitable atmospheres, and many phosphors react unduly with their containers, or sublime or dissociate unduly, at temperatures near or above their melting points.[58,678] Most phosphors are crystallized in the course of solid-state reactions which afford tiny crystals, averaging less than about 20 microns (1 μ = 10^{-4} cm) in diameter.[678] The development of high-pressure furnaces and nonreacting containers which can be used with certain useful atmospheres, such as oxygen, hydrogen sulphide, and carbon monoxide, would be quite beneficial in facilitating the growth of larger crystals of many refractory materials, such as phosphors, which are sensitive to changes in structure and impurity content.[690] Also, the use of applied electric and magnetic fields during crystallization and cooling of phosphors deserves investigation.[21,32,59]

Figure 6 illustrates the growth of an idealized crystal of rhombohedral zinc orthosilicate ($R\bar{3}$- or rbhdl.-Zn_2SiO_4) by the solid-state reaction of zinc oxide (ZnO) and silica (SiO_2). The initially amorphous ZnO assumes the hexagonal ($C6mc$) structure on being heated, whereas amorphous SiO_2 undergoes a number of crystalline transitions which promote reactivity.[60,61,71b] There is practically no reaction between intimately mixed micron-size particles of ZnO and SiO_2 at temperatures up to about 650°C and heating times of several hours. At about 850°C, there is apparently a crystalline transition of silica, and the rate of reaction with zinc oxide increases rapidly because the rearrangement of atoms in the silica crystals allows zinc oxide molecules to diffuse into the changing structure. The rate of reaction increases very rapidly with increasing temperature above 850°C (see Figs. 90–91). Near 1000°C, the reaction between ZnO and SiO_2 becomes exothermic,[62] indicating that the simple molecules have acquired sufficient activation energy E^* to break their bonds with like neighbors and react rapidly with unlike molecules to form the new crystal of rbhdl.-Zn_2SiO_4. The initial reaction occurs through the vicinal faces of the ZnO and SiO_2

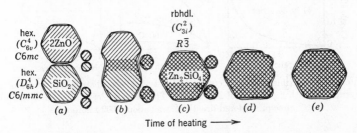

FIG. 6. Schema of the solid-state reaction $2ZnO + SiO_2 \rightarrow Zn_2SiO_4$, showing the growth of large crystals at the expense of smaller crystals. Real rbhdl.-Zn_2SiO_4 crystals grown in this manner have very irregular shapes.[678]

crystals which are distorted by surface tension.[30] Thereafter, the ZnO molecules diffuse into the crystals of SiO_2 by passing through the intervening layer of rbhdl.-Zn_2SiO_4. Ordinary pure silica, with particles averaging about 1 micron, reacts with ZnO (from precipitated $Zn(OH)_2$) to form rbhdl.-Zn_2SiO_4 crystals averaging about 2 microns in size; whereas, with the same ZnO and the same conditions of crystallization, very fine particles (0.03 micron) of bulky silica from dried silicic-acid gels react to form rbhdl.-Zn_2SiO_4 crystals averaging less than 0.1 micron in size.[58, 63] Because the energies of crystallization are additive, atoms on the surface of a large crystal are bound more securely (have lower potential energy) than those on a small crystal. This explains the growth of large crystals at the expense of smaller crystals (Fig. 6). Figure 7 shows the influence of temperature and time of crystallization on the growth of crystals of zinc sulphide (ZnS) from a mixture of amorphous ZnS and 2 per cent of sodium chloride (NaCl). It may be seen that increasing temperature exerts the most pronounced influence, in that the crystals grow in approximate proportion to $\epsilon^{-E^*/kT}$. The rate of growth decreases with increasing time, because the crystals approach a size distribution which is determined largely by the temperature and by the natures and sizes of the initial particles. A typical crystal-size distribution curve for a completed phosphor batch is shown at the right of Fig. 7. With low crystallization temperatures ($< 800°C$) and low flux proportions, the average crystal size $\bar{x}_c$ of the product may be made smaller than 1 micron; whereas, with high temperatures ($> 1200°C$) and high flux proportions, $\bar{x}_c$ may be made larger than 100 microns.

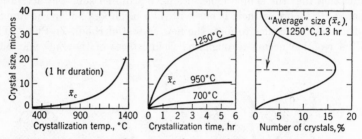

Fig. 7. Approximate growth curves and a crystal-size distribution curve of ZnS crystallized with 2 per cent of NaCl flux. The average "crystal" size of the initial precipitated-and-dried ZnS was about 0.02 micron.[678]

Real Crystals

Real crystals attempt to attain the perfection of ideal lattices but are hindered by imperfections introduced during crystallization and by inherent imperfections arising from surface discontinuities and thermal agitation. Concerning the influence of thermal agitation, if N_S be the total number of occupied lattice sites in an initially ideal crystal *in equilibrium* at temperature T, and N_D be the number of atoms displaced by thermal agitation, then the **entropy** (randomness) S_E of the crystal is [29, 35-37, 58]

$$S_E = k \log_\epsilon N_S! / N_D! (N_S - N_D)! \approx -k N_D \log_\epsilon (N_D/N_S) \quad (19)$$

In order for the crystal to be stable, the **free energy,**

$$E_F = E - T S_E \approx N_D E^* + k T N_D \log_\epsilon (N_D/N_S) + \text{constant} \quad (20)$$

must be a minimum; where E is the total kinetic and potential energy in the crystal, and E^* is the activation energy necessary to displace *thermally* an atom from its lattice site in the crystal and place it in a lattice site at the surface of the crystal. At $T = 0$, E_F is a minimum when N_D is zero (E is a minimum); that is, the crystal most nearly approximates a perfect crystal. At high temperatures, when $E \ll T S_E$, E_F is a minimum when S_E is a maximum; that is, the crystal is more stable with a number of different distributions of N_D vacant sites among the N_S lattice sites. At intermediate temperatures, there is a compromise between the tendency for the energy of the crystal to decrease (perfection increase) and the entropy of the crystal to increase (randomness increase). For a crystal in equilibrium, it may be seen from eq. 20 that E_F is a minimum when

$$N_D/N_S = \epsilon^{-E*/kT} \tag{21}$$

For many high-melting phosphor crystals, the thermal activation energy $E*$ appears to be about 2 ev, so the equilibrium value of $N_D/N_S \approx \epsilon^{2.3 \times 10^4/T}$ is about 3×10^{-7} at a crystallization temperature of 1200°C ($T = 1473°K$). In a crystal with 3×10^{22} regular lattice sites cm^{-3}, about 10^{16} sites cm^{-3} would be vacant at 1200°C, and many of these omission defects may be frozen in by rapid cooling. Some of the displaced atoms may reach and remain on the surface, to expand the crystal and decrease its density, but a number may lodge in interstitial positions between the regular lattice sites, especially in loose-packed structures. When a displaced (or misplaced) host-crystal atom lodges interstitially, it functions as an *impurity* in luminescent solids.

Predominantly ionic crystals which are composed of *univalent* or *bivalent* ions, as in NaCl, KCl, or SrS, generally crystallize in the cubic system and are closely packed. The potential barriers (Fig. 4) between univalent ions are usually relatively broad compared with the more peaked potential barriers between polyvalent ions.[19] Certain predominantly ionic crystals with *polyvalent* ions may assume relatively loose-packed structures, such as the perovskite (cubic, $Pm3m = O_h^1$) structure of certain alkaline-earth titanates (for example, (Sr:Ba)TiO$_3$) which have extraordinarily high dielectric constants (high polarizabilities).[64] Crystals which are not predominantly ionic, such as hex.-ZnO, rbhdl.-Zn$_2$SiO$_4$, cub.-ZnSe, and hex.-CdS (these usually comprise polyvalent elements), also may have loose-packed structures with large interstitial spaces in which displaced atoms or impurities may lodge (see Figs. 11 and 12).

It is noteworthy that the energy required to remove an electron from an anion in an ionic alkali-halide crystal is usually considerably larger than that required to remove an electron from an anion in a relatively nonionic crystal. For example, this excitation energy is over 6 ev for almost all the alkali halides (strong absorption usually less than 2000 Å), whereas it is less than 3.5 ev for hex.-ZnO, cub.-ZnSe, cub.-ZnS, hex.-ZnS, and hex.-CdS (long-wave limit of strong absorption ranges from 3500 Å to 5200 Å). On the other hand, the energy required to volatilize or decompose the relatively nonionic crystals is usually much higher than that required in the case of the predominantly ionic crystals. Hence, the predominantly nonionic crystals usually make the most useful and durable phosphors because (1) there is relatively little energy deficit between the quantum energies of excitation and emission (visible light $h\nu \approx 2$ ev), and

(2) the larger exchange-type bonding energies of predominantly nonionic crystals render them more stable than the ionic crystals whose predominantly electrostatic bonds are broken, except for polarization energies, by the removal of excess electrons from an anion.

Any real crystal may have various concentrations and distributions of the following types of imperfections:

1. Distorted surfaces and cracks due to discontinuities and surface tension.

2. Distorted internal structure between mosaic crystals, or between crystals which have partially interpenetrated during growth.

3. Displaced atoms (ions) in (a) interstitial sites, and (b) surface sites.

4. Omission defects, which may be simple or compound according to whether single atoms or groups of atoms or simple molecules are omitted from regular lattice sites.

5. Inhomogeneities of composition between different parts of the crystal.

6. Impurities in (a) surface sites, (b) regular lattice sites **(substitutional sites)**, and (c) interstitial sites.

7. Charge displacements due to separated anion–cation pairs, or abnormal ionized atoms and trapped electrons.

An exaggerated schema of some crystal imperfections is shown in Fig. 8.[678] There is a local departure from crystallinity at the boundary of a crystal (a), or in the vicinity of an impurity (b, c, d) or omission defect (not shown), which may be visualized as a quasiamorphous or quasicrystalline region of limited extent. The degree of departure from the regular crystal structure decreases exponentially with distance away from the boundary, or other imperfection, in the direction of the more perfect regions of the crystal. Impurity atoms may sometimes change greatly the configurations as well as the spacings of nearby host-crystal atoms. When, in the case of a substitutional impurity, the foreign and the normal atoms have identical effective valencies and have radii which do not differ by more than about ±15 per cent, the small uniformly distributed crystal perturbation (b, c) may decrease to a negligibly low value in perhaps 3 to 10 atomic spacings. In this case, there is a close coupling (bonding) of the substituted atom to the other atoms in the crystal. In the case of interstitial impurities, the foreign atoms may differ considerably in effective valency and size from the normal crystal atoms and may produce large nonuniformly distributed distortions (d) when not neutralized by an adjoining interstitial ion of opposite sign and equal charge. The crystal perturbation produced by a highly charged unneutralized interstitial ion may not extend much farther than that produced by a substitutional ion, but the degree of local distortion, caused by the unbalancing effect of the interstitial ionic charge may be much greater than in the case of the substitutional impurity. The coupling between the regular crystal atoms (ions) and an interstitial atom (ion) is generally relatively low, and the neighboring region of strong local distortion serves to attenuate energy being transmitted to

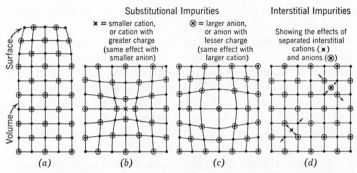

Fig. 8. Schematic examples of planes through mosaic crystals distorted by (*a*) surface tension (in some cases the atoms may be spaced *farther* apart vertically in the surface layers), (*b*) a smaller or more highly charged cation (or anion), (*c*) a larger or less highly charged anion (or cation), and (*d*) an interstitial anion or cation. [678] Omission defects may cause an expansion or contraction of the crystal in the region of the defect, according to the type of binding and the structure of the crystal. The interstitial impurities are located one-half lattice spacing above or below the plane of the paper (assuming a cubic array). In real crystals, the separate mosaic crystals vary in size and are slightly disoriented relative to each other. In real crystals, also, the atoms (ions) are generally *not spherical* and may have both intrinsic knobby structure and mutually induced deformations.

the impurity from the atoms of the regular crystal. In some cases, however, the distortion and attenuation may be quite small, as in the ZnS-type crystals where the interstitial and regular sites are geometrically identical (see Fig. 12). In such cases, it is essential that the interstitial atom have an effective size which is not greatly different from the corresponding regular-crystal atom, and the interstitial impurity should be preferably as nonionic as possible, or of the same degree of ionization as the corresponding regular-crystal atom.

At any temperature, a real crystal is in a state of dynamic unrest, even at equilibrium, and so the various atoms and electrons in the crystal have a certain probability of changing their positions and energies without being additionally stimulated by an external source of energy. Changes in position of atoms and omission defects give rise to a **self-diffusion coefficient** D, which increases exponentially with temperature according to

$$D = A\epsilon^{-B/kT} \propto \nu_a \epsilon^{-\bar{E}*/kT} \tag{22}$$

where A and B depend on the size and nature of the atom or omission defect, and on the structure and composition of the crystal. Also, ν_a is the frequency of atomic vibration, which is about 10^{12} sec^{-1}, and $\bar{E}*$ is a composite activation energy for atomic displacement by thermal energy. If one could start with a perfect crystal, the thermal activation energy for interchange of two normal atoms would be very high (≈ 10 ev), so $\bar{E}*$ would generally represent the *sum* of the activation energies to produce (*a*) omission defects or interstitial atoms and (*b*) displacement of the vacancies and interstitial atoms from site to site through the crystal. In real crystals, vacancies and interstitial atoms are always present to some degree which is dependent on the temperature, the impurity content, the sizes, charges, and bonding energies of the atoms, the crystal structure and lattice spacings, and the amount of departure from exact integral combining proportions of the constituent atoms. A large D represents a large departure from the ideal crystalline state toward the liquid state. The most useful and durable phosphors have high melting points and sufficiently high values of $\bar{E}*$ to have very low values of D at their operating temperatures. In other words, any long-range mobile displacements in crystals of the *better* phosphors are predominantly electronic, rather than ionic or atomic.

Energy Levels in Real Crystals

The conventional energy-level diagrams of Fig. 9 indicate the general effects which two types of imperfections may have on the energy levels of an insulator solid of the kind which is found most suitable for obtaining efficient luminescence. According to the band picture, conduction is not possible upon application of a weak electric field to an insulator crystal in the ground state. This is so, because there is a fixed width and density of the filled band throughout the crystal; hence, an electron moving in one direction in a *completely filled* band must be balanced by an electron moving in the opposite direction, or the exclusion principle would be violated. It is only when an electron is raised into a normally unoccupied level, as in the conduction band, for example, by absorption of excitation energy $E*_C - E_H$ (*a*), that the system may become conducting. An electron in a partially occupied band could move freely through an *ideal* crystal and be attracted to an anode used to apply an electric field to the crystal. In *real* insulator crystals, however, the insulator quality may be increased greatly by electrons trapped near some of the unavoidable imperfections. These trapped electrons produce an electrostatic space charge in the volume of the crystal such that electrons being brought in from an external electrode are repelled.[676] [Conductor crystals, having partially filled bands, are unsuitable for the production of efficient visible or near-visible luminescence emission, because excitation energy is consumed chiefly in raising electrons within and from the partially filled bands, where the excited electrons make very small stepwise nonradiative return transitions (between the closely spaced vibrational energy levels) to the ground state, giving up their excitation energy to the solid as heat, instead of being emitted as light.] The raising of an electron from a normally filled band, for example, E_H, corresponds to removing an electron from an atom (ion) which becomes more positively charged. This act, in the band picture, leaves a **positive hole** (electron deficiency) in the normally filled band, and positive holes are attracted to a cathode used to apply an electric field to the crystal. The motion of a positive hole proceeds as a stepwise process, such that the positive hole becomes filled by taking an electron from a neighboring atom, whereupon the positive hole moves to the neighboring atom, and so on. If a potential be applied across a crystal having an excited electron in the conduction band and a residual positive hole in the uppermost normally filled band, the electron will travel toward the anode, and the positive hole will travel toward the cathode, thereby separating the two and preventing their recombination.

Just below the conduction band in Fig. 9, is shown an exciton level, E^*_E. An **exciton** is visualized as a positive hole with an excited electron moving around it, much as the hydrogen electron moves about its nucleus, and the exciton may move through the crystal. The wave function ψ_e of the exciton's electron may extend over several atomic spacings around the hole, and the exciton as a whole has a ψ_E extending throughout the entire (ideal) crystal. An exciton differs from an ordinary excited atom in that the exciton is mobile, whereas the excited atom is fixed in its position in a crystal. Since the exciton as a whole is electrically neutral, it does not contribute to the conductivity of the crystal, unless the exciton is polarized and dissociated by the electric field.[35, 36]

In Fig. 9, it is seen that an omission defect and a substitutional impurity may introduce additional discrete normally unoccupied energy levels (or narrow bands) in the immediate vicinity of the imperfection. Because the imperfections are foreign to the general pattern of the crystal, there is often relatively weak coupling between the imperfections and the crystal, and so the ψ's of electrons in the imperfection levels

generally decrease rapidly with distance away from the imperfection.
This minimizes interaction between remote imperfections and helps to

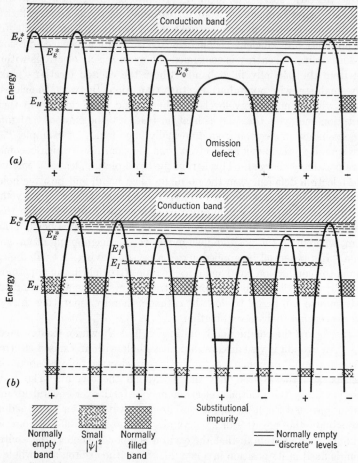

FIG. 9. Simplified diagrams illustrating discrete normally empty energy levels localized near crystal imperfections and additional filled energy levels introduced by impurities. Solid lines indicate large (strong) wave functions; dotted lines indicate small
(weak) wave functions (compare with Fig. 4). The energy levels and interatomic
spacings change somewhat after every electronic transition during luminescence.

maintain the discreteness of the local atomic energy levels. The local
levels can broaden with height above E_H or E_I, however, and so the upper
levels may overlap each other and E^*_C. There may be appreciable interaction (ψ–overlap) between the upper excited-state levels of closely spaced

impurity atoms, and this interaction limits in part the concentration of impurities which can act as virtually independent radiative oscillators in phosphor crystals. Magnetic interaction between impurity atoms may also be a factor, particularly when the atoms have unfilled inner shells. As shown in Fig. 9, there is a considerable difference between the excitation energies associated with the exciton levels and the excitation energies associated with the discrete imperfection energy levels. Excited but not ionized atoms in the immediate neighborhood of an imperfection often have a high probability of retaining the excitation energy in the vicinity of the imperfection, whereas excitons produced by excitation of the regular crystal atoms may wander throughout the nearly perfect portions of the crystal.

Both the imperfections shown in Fig. 9 lower the potential barriers in their vicinities. This perturbation creates a small localized domain, which we shall call a **center,** containing most of the ψ's of excited electrons in the imperfection. A center which can capture an excited electron, produced in the center or elsewhere in the crystal, *without allowing the electron to fall directly into a normally filled level,* is called an **electron trap,** or just a **trap.** A trap may be visualized as an excited-state potential well (Fig. 3), which is generally remote from the original ionized atom, and which requires additional energy input so that a trapped excited electron may be raised out of the well in order to return to its point of origin. In addition to discrete excitation levels, impurity atoms (ions) introduce one or more local energy levels or narrow bands which may be completely or only partially filled. It has been found that multivalent impurities which have a normally occupied ground state, for example, Cu^{2+}, Sm^{3+}, and Sn^{4+}, and which can assume another occupied state, for example, Cu^{+}, Sm^{2+}, and Sn^{2+}, are particularly advantageous for forming trapping centers in phosphor crystals. In this respect, it may be noted that different atoms of a multivalent element may exhibit several valencies in the same crystal; for example, $Fe_3O_4 = Fe^{2+}O^{2-} \cdot Fe_2^{3+}O_3^{2-}$, and Mn, Cr, and Sn each exhibit at least two valencies even in the metallic state.[12, 22, 24, 39]

When an impurity (or, perhaps, an isolated missing positive ion) introduces localized filled levels above the topmost normally filled band E_H of the host crystal, these local levels may act as **positive-hole traps.** The trapping of positive holes proceeds by having the positive holes jump *up* from E_H to the normally filled local levels, which means that electrons drop down from the local filled levels to fill positive holes which approach through E_H.

In an ideally perfect insulator crystal, the minimum energy for an electronic excitation would be $E^*_E - E_H$ without electric conduction, and $E^*_C - E_H$ with conduction; where E_H is the topmost filled band of the unexcited solid. These minimum excitation energies do not seem to be affected by the presence of omission defects,[65] although anion omission defects may trap electrons which have been raised from E_H into the conduction band to provide low *stimulation* energies. (Electrons trapped in anion vacancies in alkali-halide crystals are known as **F centers.**[35-37, 58, 66, 67]) In some cases it is possible to lower the minimum excitation energy by introducing certain impurities into the crystal, particularly multivalent impurities such as Cu^+, Cu^{2+}; Eu^{2+}, Eu^{3+}; Mn^{2+}, Mn^{4+}.[34, 61] Some of the useful impurity elements, such as Cr and the rare-earth elements, have occupied but unfilled inner shells, for example the $3d$ shell of Cr, and the $4f$ shells of the rare earths.[68-70] These incomplete inner shells may introduce new discrete occupied and excitation levels lying below or astride the topmost normally filled band of the host crystal. In such cases, electronic transitions between the shielded discrete inner levels, corresponding to different resultant electron-spin configurations of the impurities, may give rise to sharp lines of excitation and emission, similar to those obtained from isolated atoms.[65, 68]

CHAPTER 3

SYNTHESES OF LUMINESCENT SOLIDS
(PHOSPHORS)

Elementary Considerations

The preceding outline of the genesis of solids began with the *physics* of three elementary material particles and their combinations as 96 presently known species of atoms (chemical elements), proceeded briefly through the *chemistry* of combining atoms to form an unstated number of molecules, and concluded with the *crystallography* of atoms (monatomic molecules) and molecules arranged in solids according to the 230 ideal space groups. The magnitude of the "unstated number" of known simple molecules is about 10^6, of which a minority of about 10^5 are inorganic substances and the rest are organic substances, that is, most of the compounds comprising carbon. A formal distinction is made here between *substances* and *materials*. A pure nonelemental **substance,** that is, a chemical compound, consistently contains the same simple integral proportions **(stoichiometric proportions)** of its ingredient atoms.[71a] For example, N atoms of zinc (Zn) consistently combine with almost exactly N atoms of oxygen (O) to form N simple molecules of the substance zinc oxide (ZnO) when zinc is vaporized and burned in oxygen (which exists as O_2 molecules). The substance zinc peroxide ZnO_2 may be formed under special circumstances, but it is not common. Similarly, N ions of Zn^{2+} consistently combine with almost exactly N ions of S^{2-}, or $2N$ ions of F^-, in aqueous solution to form N simple molecules of ZnS, or ZnF_2, as in the reactions,

$$ZnSO_4 + H_2S \; \overset{\leftarrow}{\longrightarrow} \; ZnS \downarrow + H_2SO_4 \tag{23}$$

and

$$ZnO + 2HF \; \overset{\leftarrow}{\longrightarrow} \; ZnF_2 \downarrow + H_2O \tag{24}$$

where the $\downarrow$ sign indicates that the particular reaction product is so insoluble that it precipitates out of the solution. Deviations from equality in the number of Zn and O atoms in common ZnO, or in the number of Zn and S atoms in common ZnS, are probably less than about 0.0001 per cent.

In **chemical equations,** the **chemical symbol** of each element represents a weight *proportional* to the atomic weight of the element. Thus,
the equation

$$Zn + O = ZnO \tag{25}$$

or, more correctly,

$$Zn + \tfrac{1}{2}O_2 \overset{\longleftarrow}{\longrightarrow} ZnO \tag{26}$$

means N atoms of Zn plus N atoms of O (or $N/2$ molecules of O_2) equals
N simple molecules of ZnO, where $N = 1, 2, 3, \cdots\infty$. When $N = N_A$,
as is conventional in the chemistry of fluids, the equation means 1
gram-atomic weight of Zn (65.38 g) plus 1 gram-atomic weight of
O (16.00 g) [or $\tfrac{1}{2}$ **gram-molecular weight** of O_2 (32.00/2 = 16.00 g)]
equals 1 gram-molecular weight of ZnO (81.38 g), or

$$1 \text{ mole Zn} + \tfrac{1}{2} \text{ mole } O_2 = 1 \text{ mole ZnO} \tag{27}$$

where a **mole** is equivalent to a gram-molecular weight. The = sign
means that the reaction may go either way, although most chemical
reactions are so irreversible that oftentimes a single arrow is used to
designate the predominant direction of the reaction. In eq. 26, the
reverse process is practically negligible, except at temperatures above
about 800°C when appreciable dissociation of the ZnO occurs. In
general, the relative proportions of the substances on the two sides of a
reaction are determined by the temperature, pressure, time of reaction,
the natures of the reactants, and the presence of extraneous substances
(catalysts or inhibitors) which may accelerate or retard the rate of
reaction in a given direction. (At this point, it may be noted that the
energy equivalent in calories per mole for one electron volt per simple
molecule is 1 ev/molecule = $N_A 3.824 \times 10^{-20}$ = 23,050 cal mole^{-1}
= 23.05 kcal mole^{-1}).

A **material** does not consistently contain the same integral proportions of its ingredient atoms. For example, *pure* common salt (NaCl)
is a substance, but when it is crystallized with an arbitrary amount of
an alkali iodide (for example, KI) it becomes a material. Also, when a
crystal of pure ZnO is heated at about 1200°C, some of the substance
dissociates, and, because the liberated oxygen and zinc diffuse and
volatilize at different rates, the crystal becomes a material with an excess
of one or the other of the constituents. As a further example, *if* exactly
$2N$ simple molecules of ZnO react and combine completely with exactly
N simple molecules of SiO_2, the substance Zn_2SiO_4 is formed. In this
case, however, a large or small excess of either the ZnO or the SiO_2 will
not be rejected, but will be included in the resultant solid material.[71b, 72]
As a generic term, the word material includes both substances and

materials, but it should be kept in mind that materials permit non-stoichiometric proportions and are usually heterogeneous. The number of possible materials is infinitely greater than the number of possible substances. Strictly speaking, any substance in bulk (very large N) is just a "practically pure" material, because complete purity is as un-obtainable as a perfect crystal. The constant general increase of entropy (randomness) with time militates against complete segregation and perfect order of large numbers of any thing.[73] Phosphors, as will shortly be shown, are generally materials.

Some organic substances and materials, such as acetone, benzene, trypaflavin, phenanthrene, stilbene, naphthalene, and anthracene (especially when containing naphthacene), exhibit appreciable luminescence in the solid state.[74-84, 145] Most luminescent organic dyes and other organic materials, however, require traces of moisture to promote appreciable luminescence, for example, dyes in moisture-containing textiles,[82] plastics, and boric-acid glasses (similarly, hydrated uranium trioxide luminesces, but not the anhydrous oxide [670]). The uses of these luminescent organic materials are limited by instability at elevated temperatures, by low absorbing power for high-energy charged material particles, and by their usually short afterglow (generally less than 10^{-7} sec). We are concerned chiefly with the more versatile and stable luminescent solids (phosphors) which are chiefly inorganic materials. The word *phosphor* is used synonymously with *luminescent solid,* in this book, and emphasis is placed on artificial (man-made) inorganic phosphors. Artificial phosphors are not only much more efficient and versatile than the naturally occurring luminescent minerals, but they are also better defined and, hence, more suitable for study.

Syntheses and Symbolism of Phosphors

Efficient general research on phosphors requires (1) an exceptionally clean chemical laboratory equipped to synthesize very pure inorganic substances, (2) one or more furnaces capable of attaining at least 1600°C with either oxidizing or reducing atmospheres, and (3) a physics laboratory having suitable sources of photons and charged material particles and means for controlling and determining the energies and numbers of these particles.[58,72, 85-90]

The *chemical facilities* are necessary to prepare luminescence-pure (LP) substances, whose approximate degree of purity is indicated in the following series:

$\approx 1\%$	$\approx 90\%$	99.9%	99.99%
··· ore,	technically pure,	chemically pure (CP),	reagent-grade pure(RP)

99.999%	99.9999%	100%
spectroscopically pure(SP),	luminescence pure(LP),	··· completely pure

It has been proved that as little as 10^{-4} *per cent* of certain impurities can affect the luminescences of some phosphors; hence, the substances used in synthesizing phosphors should be at least LP if all the results are to be above suspicion.[58]

The *furnace* is necessary for the reaction and crystallization of phosphor ingredients, most of which melt well above 1600°C. The temperature of the furnace chamber should be as uniform as possible and should be controllable within at least $\mp 10°C$, because many phosphors are very sensitive to small changes in temperature during reaction and crystallization.[61] A means for *gradually* varying the power input to the furnace, such as by (1) thermocouple-controlled electronically operated saturable reactors, or (2) motor-driven variable-voltage transformers, is vastly superior to the usual "on-off" switch type of control. Atmosphere control is essential to assist in controlling the anion–cation ratio and effective valence states in phosphors. Complete control of this ratio, however, requires controllable high pressure during heating and cooling.

The *physical facilities* should include (1) the following sources for exciting or stimulating phosphors to luminescence: (*a*) an **infrared (IR) source,** for example, a 250-watt heat lamp with a Corning 2540 filter (passes photons with wavelengths longer than 7000 Å), (*b*) **ultraviolet (UV) sources,** for example, a 360 BL "fluorescent" lamp (emits a band near 3600 Å), or an H4 mercury-discharge lamp (emits strongly 3650 Å) with Corning 5840 and 5970 filters, and a "Germicidal lamp" or "Sterilamp" (emits mostly 2537 Å) with Corning 9863 filter,* (*c*) an **x-ray source,** [91, 92] (*d*) a **cathode-ray (CR) source,** which in its

---

* The intensities of sources of UV are generally expressed in microwatts per square centimeter (μw cm^{-2}) measured one meter from the lamp. On this basis, the approximate outputs of some representative UV lamps are (from bulletins of General Electric Company, Westinghouse Electric Corporation, Sylvania Electric Company, Hanovia Chemical and Manufacturing Company, and George W. Gates and Company):

1. Germicidal lamps or sterilamps, approximately 10 microns pressure of mercury,

4 watts, $\frac{1}{2}$ in. diameter $\times$ 5$\frac{3}{8}$ in. long, 6 μw cm^{-2} of 2537-Å UV at 1 meter
8 watts, $\frac{5}{8}$ in. diameter $\times$ 12 in. long, 25 μw cm^{-2} of 2537-Å UV at 1 meter
30 watts, 1 in. diameter $\times$ 36 in. long, 75 μw cm^{-2} of 2537-Å UV at 1 meter

2. Uviarc lamp, approximately 2 atmospheres pressure of mercury,

360 watts, $\frac{3}{4}$ in. diameter $\times$ 6 in. long,　39 μw cm^{-2} of 2537-Å UV at 1 meter
26 μw cm^{-2} of 2652-Å UV at 1 meter
12 μw cm^{-2} of 2804-Å UV at 1 meter
7 μw cm^{-2} of 2894-Å UV at 1 meter
17 μw cm^{-2} of 2967-Å UV at 1 meter
32 μw cm^{-2} of 3022-Å UV at 1 meter
71 μw cm^{-2} of 3129-Å UV at 1 meter
100 μw cm^{-2} of 3650-Å UV at 1 meter
300 μw cm^{-2} of 4000–7000-Å UV at 1 meter
(25 lumens/watt)

3. A 15-watt 360 BL lamp emits 6.5 μw cm^{-2} at 1 meter, measured at the peak of the phosphor emission band plus some 3650-Å UV.

simplest form may be a glass vessel evacuated to about 10 microns pressure in which an electronic discharge may be produced by applying a high voltage spark coil; [93-101] and (e) **proton, alpha-particle, and gamma-ray sources,** for example, radioactive materials; [102-104] and (2) the following apparatus for measuring luminescence photons: (a) visual and photoelectric **photometers,**[105, 106] (b) **spectroscopic and densitometric equipment** for wavelength and intensity measurements from at least 1800 to over 15,000 Å,[93-98, 107-109] and (c) **phosphoroscopes** capable of measuring rates of photon emission from times less than 10^{-7} to more than 10^5 sec after cessation of excitation of a phosphor.[95, 110-120] Still other equipment is necessary to measure other phosphor properties such as crystal structure and particle size,[58, 91, 92, 101] photoconduction,[121-123] and secondary electron emission.[124-134]

HEALTH PRECAUTIONS. Before proceeding to a brief description of the syntheses of several representative phosphors, a few words of caution may be interjected with respect to *health precautions* in handling some of the chemicals used in making phosphors. Most of the dry chemicals used in making phosphors are harmless, and, among the wet chemicals, the strong acids and alkalis should be handled with the same respect as in elementary chemistry. If good hoods are provided (and used!) and dangerous chemicals are handled with rubber gloves or tongs, the chemical hazards of phosphor research should be no greater than those encountered in a normal laboratory course in quantitative analysis. The chief dangers lie in respiratory infections or dermatitis which may be caused by excessive inhalation of, or bodily contact with, *toxic vapors and gases*, such as H_2S, H_2Se, HF, and SO_2, or *toxic dry dusts*, such as BeO, $BeSO_4$, SiO_2, and oxyfluorides.[135-142] Explosion hazards are practically absent as long as reasonable care is exercised in handling the few explosives encountered, for example, during heating of thermodynamically unstable NH_4NO_3 (large free energy![143]), and during the combining of elements, such as mixtures of powdered Zn + S, or 2Al + 3Se, by the familiar magnesium-ribbon technique. A strong-walled hood, with external controls and thick safety-glass door, should generally be used in handling any explosives, although it is possible to decompose and volatilize 100-g batches of NH_4NO_3 safely by heating the substance in tall fused-silica beakers *from the top down* with blast burners.

That phosphors are *materials*, not pure substances, is illustrated by the following examples of their preparation, assuming the purest possible fine-particle ingredients which have not been heated at temperatures above about 200°C (see Appendix 1). The examples are chosen to

---

4. The 400-watt DH1 lamp, operating at about 2.5 atmospheres, emits about 25 watts as 3650-Å UV from a concentrated arc.[90]

5. A 100-watt S-4 mercury lamp operates at 8 atmospheres pressure and emits (at 1 meter) 12 μw cm^{-2} from 2638 to 3050 Å (practically no 2537 Å), 80 μw cm^{-2} from 3050 to 4110 Å, 105 μw cm^{-2} from 4110 to 6830 Å, and 94 μw cm^{-2} from 6830 to 17,390 Å.

typify the considerable variety of possible preparative procedures and resultant phosphors.

1*a*. Scatter some crystals of KCl near one edge of a glass plate, and place a crystal of $SbCl_3$ near the edge of another glass plate. Bring the two edges within about 1 cm of each other under 3650-Å UV (in the dark), and the nonluminescent crystals of KCl nearest the nonluminescent $SbCl_3$ may be observed to commence luminescing yellow with an imperceptibly short afterglow. Molecules evidently volatilize from the $SbCl_3$ crystal at room temperature and, on striking the KCl crystals, implant the property of visible luminescence. This preparation may be expedited by simply grinding together the ingredients.[65] The resultant (deliquescent) phosphor may be symbolized by $Fm3m$-KCl:Sb, 30°C, or simply by cub.-KCl:Sb, where the KCl is called the **host crystal** (or *base material*, or *matrix*) and the relatively small proportion of luminescence-promoting Sb is called the **activator,** or *phosphorogen* (Greek: phosphor-producing), or *luminogen*. The **colon** (:) indicates variable generally nonstoichiometric proportions.

1*b*. Vigorously shake or grind together 0.1 mole (7.455 g) of non-luminescent KCl and 0.003 mole (0.720 g) of nonluminescent TlCl.[165] This affords a product with a cream-yellow luminescence emission under 2537-Å UV, with no perceptible afterglow, and with no perceptible emission under 3650-Å UV. Under x-ray and CR excitation there is the same cream-yellow luminescence emission during excitation and a long afterglow of the same yellow color after cessation of excitation. The resultant phosphor may be symbolized by $Fm3m$-KCl:Tl, 30°C, or simply by cub.-KCl:Tl. If the material be melted, it is found that TlCl volatilizes rapidly because the boiling point of TlCl is about the same as the melting point of KCl (776°C). Hence, much higher initial proportions of TlCl must be used to obtain bright luminescence emission, and it will be found that the product made by melting together and cooling 0.1 mole of KCl and 0.01 mole of TlCl is considerably more efficient than the product made by grinding together the optimum proportions of 0.1 mole of KCl and 0.003 mole of TlCl. A short exposure of KCl at room temperature to TlCl vapor does not afford perceptible luminescence emission, probably because the proportion of condensed TlCl is too low. When the TlCl content is about one tenth of that given above, the product emits chiefly in the ultraviolet.[162–166] (*Note:* These halide phosphors are hygroscopic and should be stored in desiccators or other moisture-free containers.)

2. To 1 mole (40.32 g) of MgO add an alcohol solution containing about 0.00033 mole (0.0758 g) of dissolved $SbCl_3$. Stir well while evaporating to dryness, grind and mix, transfer to a silica boat, and

place the uncovered boat and contents in a silica-tube combustion furnace. Heat to 1000°C for about an hour while passing dry nitrogen (N_2) saturated with carbon disulphide (CS_2) through the tube. Cool in the $N_2 + CS_2$ stream, and transfer to a good desiccator. The MgO is converted to MgS according to the reaction,

$$2MgO + CS_2 \rightleftharpoons 2MgS + CO_2 \qquad (28)$$

and a large portion of the Sb activator is lost by sublimation of $SbCl_3$ (the optimum proportion of Sb in the completed phosphor is of the order of 0.02 weight per cent). The resultant phosphor luminesces green–yellow, with a weak afterglow, under UV, x rays, and CR and may be symbolized by *Fm3m*-MgS:Sb, 1000°C, or simply by cub.-MgS:Sb.[145]

3. Intimately mix 1 mole (81.38 g) of ZnO and 0.006 mole (0.426 g) of MnO in a clean platinum crucible, and add *slowly* about 2.5 moles (50 g) of HF (about 110 cm^3 of 47 per cent hydrofluoric acid). There is about 25 per cent excess of HF over the amount necessary to convert the oxides to ZnF_2 and MnF_2. Evaporate to dryness in a good hood (the excess HF volatilizes), cover the crucible, and heat at about 800°C for 1 hr in a silicon carbide resistance-element electric furnace.[88] Remove and cool in air. The resultant phosphor luminesces orange with a short orange afterglow, under x rays and CR and may be symbolized *on a mole basis* by *P4/mnm*-$ZnF_2 \cdot 0.006MnF_2$, 800°C, or simply by tetr.-ZnF_2:Mn.[58]

4a. Intimately mix 100 g (1.026 mole) of *pure* ZnS [58] with 2 g (0.0342 mole) of NaCl and 0.0211 g (0.000157 mole) of $CuCl_2$, and place in a clean iron-free fused-silica, alumina, or Vycor crucible. Cover the crucible, place it in a larger refractory crucible where it is surrounded with finely divided ZnS, and place it in the electric furnace at room temperature. (The ZnS-surround material is preferably made by heating a mixture of ZnS and about 4 per cent NaCl to 1250°C.) Raise the temperature to 850°C in about one hour, remove the crucible to cool in the air, and wash the contents of the inner crucible several times with distilled water saturated with H_2S.[444] The resultant phosphor luminesces green, with a moderate green afterglow, under 3650-Å UV and under CR and may be symbolized *on a weight basis* by *F43m*-ZnS:Cu[Cl_2](0.01), [NaCl(2)], 850°C, or simply by cub.(or α^*)-ZnS:Cu.[34] (This is the **zinc-blende, blende,** or **sphalerite** form of zinc sulphide.) The **numbers in parentheses** indicate the *initial* weight per cent, and the **square brackets** [ ] indicate doubt as to the existence or proportion of bracketed ingredient in the final phosphor. In this case, the chlorides may react with the ZnS to form $ZnCl_2$, that is, $ZnS + 2NaCl = ZnCl_2 + Na_2S$, which melts at 262°C and boils at 732°C. Since the NaCl, in particular, provides a low-melting liquid phase of itself (melting point = 801°C) and by reaction during crystallization, it is called a **flux.** Any Na_2S formed by reaction is left behind but is largely removed (along with residual NaCl) by the final washing. The Cu, on the other hand, is permanently incorporated in the ZnS crystals, and its relative weight per cent is indicated in parentheses (the weight of the activator anion being ignored), the weight of the host crystal being set equal to 100. (On this basis the total of phosphor plus activator is slightly more than 100 per cent.)

4*b*. If the same ingredients be used as in 4*a* but the temperature of crystallization be raised above about 1020°C (for example, 1250°C), the phosphor crystals are found to have hexagonal, instead of cubic, structures, and the phosphors have long green afterglows after cessation of excitation. This phosphor may be symbolized by $C6mc$-ZnS:Cu[Cl_2] (0.01), [NaCl(2)], 1250°C, or simply by hex.(or β^*)-ZnS:Cu [58] (the **wurtzite** form of ZnS).

5. Pass a stream of H_2S into about 2 liters of an aqueous solution containing 1 mole (161.44 g) of $ZnSO_4$ and 0.02 mole (3.02 g) of $MnSO_4$ until no more precipitate is formed. The reaction is

$$ZnSO_4 + 0.02MnSO_4 + 1.02H_2S \rightleftharpoons$$

$$ZnS \downarrow + 0.02MnS \downarrow + 1.02H_2SO_4 \quad (29)$$

Wash the precipitate several times with water, and dry the precipitate at 110°C. Examination of the dried product shows that it has a feeble orange luminescence under UV; that is, the two sulphides have already formed some solid solution (pure ZnS, so prepared, shows no luminescence). If 2 per cent of NaCl be added and the resultant mixture be heated, as described in 4*a*, to about 1000°C, the resultant cooled phosphor has an efficient orange luminescence under UV, but an inefficient orange luminescence under CR. If, instead of the NaCl, about 2 per cent of an oxygen-containing compound (for example, $ZnSO_4$) be added before the heating at 1000°C, the resultant phosphor has a relatively inefficient orange luminescence under UV, but now has an efficient orange luminescence under CR. In all cases, the perceptible afterglow is quite short. These phosphors may be symbolized *on a mole basis* by $F\bar{4}3m$-ZnS·0.02MnS, [NaCl(2), or $ZnSO_4$(2)], 1000°C, or simply by cub.-ZnS:Mn.[153-154] Heating at temperatures higher than 1020°C produces the hexagonal structure, as in 4*b*, without noteworthy changes in properties.

6. Intimately mix 50 g $SrCO_3$, 0.019 g $Eu_2(SO_4)_3$, 0.017 g $SmCl_3$, 10 g S, and 40 g Se in a clean silica crucible. Heat at about 1100°C,

for about 10 min, in an atmosphere of nitrogen (N_2) which has been bubbled through CSeS. Remove and cool in a stream of cold (N_2 + CSeS). Break up and mix the cooled material with 2 g CaF_2 (or LiF), 10 g S, and 40 g Se, and reheat at about 950°C for 10 min in (N_2 + CSeS), cooling as before. The resultant phosphor luminesces yellow, under 3650-Å UV and is quite complex, since (1) the $SrCO_3$ is transformed into variable proportions of SrS, SrSe, and $SrSO_4$, and (2) the fluoride flux (and/or its reaction products) remain in the final product. This phosphor may be tentatively symbolized by $Fm3m$-(SrS: SrSe): $Pnma$-$SrSO_4$: Eu [$\frac{3}{2}SO_4$] (0.02): Sm [Cl_3] (0.02): $Fm3m$-[CaF_2($\approx$4)], 1100°C and 950°C, [N_2 + CSeS], or simply by cub.-(SrS:**SrSe**):[flux]:Sm:Eu, or by cub.-Sr(S:**Se**):$SrSO_4$:CaF_2:Sm:Eu.[58, 63]

Another procedure and formulation for the preparation of an excellent phosphor of this type has been devised by R. Ward.[177a] This procedure comprises wetting 10 g of $SrCl_2$ with an aqueous solution containing 0.0085 g of $EuCl_3$ and 0.021 g of $SmCl_3$, evaporating the water, and drying at 200°C, then grinding the dried product with 100 g of SrSe (from well-purified $SrSeO_3$ heated in a stream of pure dry ammonia at 835°C) and 4 g of pure CaS in a moisture-free container, and finally heating the ground dry mixture in pure dry oxygen-free nitrogen at 1050°C for 30 min. The yellow-luminescing product may be symbolized simply by cub.-(Sr:Ca)(S:**Se**):$SrCl_2$:Sm:Eu. (*Note:* Phosphors of this type should be kept in desiccators or other moisture-free containers to avoid decomposition and deterioration of the material.)

7. Intimately mix 1 mole (56.08 g) of CaO with 1 mole (232.0 g) of WO_3 in a clean silica crucible, heat in an electric furnace at about 1100°C for about an hour, and cool in air.[145-147,149] The resultant phosphor luminesces a pale violet under x rays, 2537-Å UV, and CR. The material *may* have a slight excess of tungsten [or Ca] (by selective loss of oxygen during heating) which performs the role of an activator. On this basis, the phosphor may be symbolized *tentatively* by $I4_1/a$-$CaWO_4$:[W], 1100°C, or simply by tetr.-$CaWO_4$:[W] [58] (sometimes called **scheelite,** after the natural mineral).

8. Fill a clean silica container with ZnO, cover with a silica lid, and place inside a larger crucible where it is surrounded with lampblack. Place in an electric furnace at about 1000°C for 1 hr, and then shut off the power, leaving the crucible to cool in the furnace. The lampblack (C) reacts with atmospheric oxygen to produce CO which reacts to reduce some of the ZnO according to ZnO + CO = Zn + CO_2. The resultant phosphor luminesces a pale blue–green, under 3650-Å UV, x rays, and CR and may be symbolized by $C6mc$-ZnO:[Zn], 1000°C, [CO], or simply by hex.-ZnO:[Zn].[58]

9*a*. Intimately mix 2 moles (162.76 g) of ZnO with 1 mole (60.06 g) of SiO_2 in a clean platinum or palladium crucible, and heat for about an hour at 1250°C in an electric furnace, cooling in air. The resultant phosphor luminesces a pale violet, under x rays and CR and may be symbolized by $R\bar{3}$-Zn_2SiO_4:[Si], 1250°C, or simply by rbhdl.(or α)-Zn_2SiO_4:[Si].[61,72,430a]   (Artificial **willemite**).

9*b*. Intimately mix 2 moles (162.76 g) of ZnO with 1.012 moles (60.781 g) of SiO_2 and 0.012 mole (0.851 g) of MnO in a clean platinum crucible. Cover the crucible, place it in an electric furnace at about 1250°C for about one hour, remove, and cool in air. The resultant phosphor luminesces green, under 2537-Å UV and CR and may be symbolized on a mole basis by $R\bar{3}$-$Zn_2SiO_4 \cdot 0.012MnSiO_3$, 1250°C, or simply by rbhdl.(or α)-Zn_2SiO_4:Mn (sometimes *called* **willemite**).[61,72]

9*c*. If the foregoing material be melted at about 1600°C and rapidly cooled by plunging small thin portions into cold water, the resultant phosphor luminesces yellow, under 2537-Å UV and CR and has a different crystal structure whose space group has not been determined. This phosphor may be symbolized simply by β-Zn_2SiO_4:Mn.[61]

10. To 500 ml of water in which are dissolved 0.1 mole (52.28 g) of $Ba(NO_3)_2$ and 0.0086 mole (2.85 g) of $Pb(NO_3)_2$ add, with stirring, 60 ml (10 per cent excess) of 2 molal ($2M$) Na_2SO_4 (that is, an aqueous solution containing 2 moles of Na_2SO_4 per liter of total volume). The reactions are

$$Me(NO_3)_2 + Na_2SO_4 \xrightarrow{\longleftarrow} MeSO_4 \downarrow + 2NaNO_3 \qquad (30)$$

where Me stands for Ba and Pb. Wash the precipitate several times with hot water, and dry at 110°C. Examination of the dried product shows that it has no luminescence under UV, but moderate pale-violet luminescence under x rays and CR. After heating the material to about 1150°C for 1 to 6 hr and cooling slowly, there is still no luminescence under UV, but there is a much stronger pale-violet luminescence, with very short afterglow, under x rays and CR. This phosphor may be symbolized *on a mole basis* by $Pnma$-$BaSO_4 \cdot 0.086PbSO_4$, 1150°C, or simply by rhomb.-$BaSO_4$:Pb.[210,246]

11. Intimately mix 1 mole (101.94 g) of Al_2O_3 with 0.005 mole (0.76 g) of Cr_2O_3, melt the mixture at about 2100°C in the oxidizing portion of a hydrogen–oxygen flame, and cool in an oxidizing atmosphere. The resultant phosphor is **artificial ruby** which luminesces red, under UV or CR, and may be symbolized by $R\bar{3}c$-$Al_2O_3 \cdot 0.005Cr_2O_3$, 2100°C, $[O_2]$, or simply by rbhdl.-Al_2O_3:Cr.[144-148,152,193-196]

Variations of the foregoing illustrative procedures and ingredients afford innumerable other phosphors with distinctive luminescences.[58,144-216, *et al.*] For example, variations of the method described in 9*b* for preparing rbhdl.-Zn_2SiO_4:Mn include (1) the hydrofluoric-acid process, and (2) the so-called carbonate process. The **hydrofluoric-acid process** is based on the hypothetical aqueous reaction,[145]

$$2ZnO + 2SiO_2 + 4HF \overset{\longrightarrow}{\longleftarrow} Zn_2SiO_4 + SiF_4 \uparrow + 2H_2O \qquad (31)$$

or

$$2ZnF_2 + 2SiO_2 \overset{\longrightarrow}{\longleftarrow} Zn_2SiO_4 + SiF_4 \uparrow \qquad (32)$$

The strong activity of the reaction forming the volatile ($\uparrow$) SiF_4 decreases the average particle sizes of the ingredients (MnO or MnF_2 may be included) and promotes intimate mixing. The Zn_2SiO_4 is not formed in solution but forms when the dried reaction products are heated at about 1000°C. This procedure, although in favor some time in the past, is not used at present because it is dangerous (toxic HF!), expensive, and does not yield uniform high-efficiency phosphors. The **carbonate process** is based on decomposition reactions of the type,[58,191,204]

$$2Zn(NO_3)_2 + SiO_2 + (NH_4)_2CO_3 \overset{\longrightarrow}{\longleftarrow}$$

$$2ZnCO_3 \downarrow + SiO_2 \downarrow + 2NH_4NO_3 \qquad (33)$$

followed by

$$2ZnCO_3 + SiO_2 \overset{heat}{\longrightarrow} Zn_2SiO_4 + CO_2 \uparrow \qquad (34)$$

In this case, the finely divided silica is suspended in the nitrate solution [$Mn(NO_3)_2$ may be included], and the carbonate is precipitated around the silica particles which are "swept down" out of suspension. As an alternative, the Zn and Mn may be precipitated as hydroxides or oxalates. The method allows uniform and intimate mixing of the ingredients, particularly when colloidal silica is used. In its final stages, the reaction reduces essentially to the solid-state oxide reaction (Fig. 6):

$$2ZnO + SiO_2 \overset{\longrightarrow}{\longleftarrow} Zn_2SiO_4 \qquad (35)$$

Many novel procedures have been proposed for the synthesis of phosphors, including explosive reactions of powdered elements (for example, ignition of powdered Zn + powdered S, or Se, or Te [160]) and electrolytic introduction of activators into previously crystallized material,[161] but these have not proved so successful as the indicated high-temperature solid-state-reaction methods for preparing phosphors from compounds.

There are some major differences between the "conventional" chemistry in fluids (including gases and liquids) and chemistry involving solid-state reactions. In the familiar chemistry of dilute solutions, ionic molecules readily dissociate into ions, for example,

$$KCl \rightleftharpoons K^+ + Cl^- \tag{36}$$

and

$$ZnSO_4 \rightleftharpoons Zn^{2+} + SO_4^{2-} \tag{37}$$

where the complex SO_4^{2-} ion is known as a **radical.** The degree of ionization and separation of the ions is proportional to the dielectric constant ϵ_D of the solvent medium and to the degree of dilution. In water ($\epsilon_D \approx 80$) the degree of ionization is generally higher than in phosphor solids whose dielectric constants are of the order of 6. Also, the ions in water are well separated by the intervening nonionized water molecules, as contrasted with the closely packed adjacent anions and cations in, for example, a crystal of KCl. The influence of water in affecting the energy levels of the valence electrons in compounds is indicated by the fact that solid *anhydrous* $CuSO_4$ is colorless, whereas solid $CuSO_4 \cdot 5H_2O$ and aqueous solutions containing Cu^{2+} ions have a blue color under white light. In anhydrous $CuSO_4$, the copper atoms are surrounded by the oxygen atoms of the sulphate radicals, and the solid does not have allowed electronic transitions with energy differences in the range from 1.7 to 3.1 ev (visible radiation). In anisotropically paramagnetic $CuSO_4 \cdot 5H_2O$, the copper atoms are surrounded by 4 water molecules and 2 oxygen atoms of the sulphate radical,[24, 39] and the solid has allowed electronic transitions of absorption in the red-to-green region (1.7 to 2.5 ev) of the visible spectrum.

In the formation of a solid precipitate by reaction of fluids, the highly mobile reactant atoms, ions, and molecules can, in effect, seek each other out to combine in exact stoichiometric proportions and reject excesses of one or more of the reactants. In the formation of a solid by solid-state reaction, however, (1) the reactant atoms, ions, and molecules usually diffuse very slowly through the relatively unyielding solids; (2) excesses of the reactants are often occluded or otherwise incorporated in the solid product as material in solid solution, or as separate solid phases (interspersed crystallites or vitreous inclusions); and (3) the high temperatures used in most solid-state reactions often promote attack on the container and promote decomposition and selective volatilization. The solids formed by solid-state reactions are, therefore, usually much less uniform in composition than those produced by low-temperature reactions in the fluid state. In general, some of the solid ingredients remain partially unreacted, and so conventional chemical equations are rarely carried to completion during reactions in the solid state.

It is essential to have a uniform initial mixture of fine particles for most solid-state reactions, because the reactants sometimes diffuse only one or two particle diameters (or less) away from their points of origin. For example, when a small lump of $CuCl_2$ is placed in the center of a batch of ZnS (with or without 2 weight per cent of $ZnCl_2$) and the batch is heated at 900°C for about an hour, the green luminescence occasioned by the Cu extends less than 2 mm into the surrounding blue-luminescing cub.-ZnS:[Zn]. The green–blue boundary is very sharply defined. Some reactants, such as solid SiO_2, have such low mobilities that their reactant partners, for example, ZnO and MnO, must diffuse to them (Fig. 6). The formation of a liquid (flux) phase, even a thin liquid layer on each particle, may assist greatly in producing a homogeneous distribution of certain reactants. For example, the presence of water vapor or certain halides promote rapid formation of several silicate- and sulphide-type phosphors.[58, 173, 221, 223, 225, 253] It is well known that the inclusion of some fluxes is beneficial in certain phosphors intended for excitation by low-energy photons (for example, ZnS:Mn prepared with NaCl flux and excited by UV). On the other hand, when high efficiency is required under excitation by high-energy particles, it is often necessary to avoid all fluxes, especially when the fluxes tend to become occluded as luminescence-inert material, or when they react with the other phosphor ingredients to form inert or vitreous phases.

It is noteworthy that successful phosphor research requires painstaking purification of all ingredients, and extraordinary carefulness and cleanliness during preparations, in order that the beneficial effects of tiny proportions of possible useful activators (*impurities*) may not be obscured by the presence of unknown contaminants.[58] In many cases, the final phosphor is the most sensitive indicator of contamination, often exceeding chemical and spectroscopic techniques in analytical sensitivity. The essential features of phosphor syntheses are (1) clean working conditions; (2) pure ingredients, generally prepared at low temperatures to preserve reactivity; (3) accurate proportioning and coprecipitation, cocomminution, or other thorough mixing of the ingredients, usually in as fine a state of subdivision as possible; and (4) reaction and crystallization of the ingredients at an optimum temperature and pressure, for an optimum time (and time cycle of temperature and pressure), and in an optimum atmosphere. These optima, which depend on the natures of the ingredients and on the desired properties of the resultant phosphor, must be determined empirically for each phosphor.

Changes in one of the parameters of phosphor preparation usually necessitate changes in the other parameters in order to obtain optimum characteristics of the resultant phosphors. For example, an increase in the temperature of reaction and crystallization usually allows or requires a decrease in the time of heating and permits the use of a smaller proportion of flux (if any). At crystallization temperatures which are near the boiling points of any of the phosphor ingredients, or their reaction products, the proportion of *initial* ingredient in the *final* phosphor generally decreases with increasing temperature and heating time. Chemical analyses have shown that all the Mn activator remains in zinc-silicate:Mn phosphors, at temperatures up to 1400°C and heating times of several hours, because none of the oxygen-containing ingredients or their reaction products volatilize appreciably at or below 1400°C. Also, Cu activator, is well retained in chloride-fluxed ZnS:Cu phosphors up to at least 1400°C, even though CuCl boils at 1366°C. On the other hand, Sb volatilizes rapidly as $SbCl_3$ (boiling point = 223°C) during the formation of cub.-MgS:Sb at 750° to 1000°C. Also, for the successful preparation of the hex.-ZnS:Pb:Cu phosphor, as reported by Fonda,[169] it is necessary to add a large excess of $PbSO_4$ to the initial halide-fluxed mixture because a variable amount (up to 90 weight per cent) is generally lost during heating at 1040°C ($PbCl_2$ boils at 950°C). A similar uncertainty of the proportions of the ingredients in the *final* phosphor is introduced when the ingredients are coprecipitated from solution. Here, the uncertainty is traceable to differences in the solubility products of the precipitates and to a selective decrease in the solute concentrations at the beginning and end of precipitation. It is customary to denote the activator and host-crystal proportion in the final phosphor as being the same as in the initial ingredient mix, but the foregoing examples serve as a warning that this is not always accurate.

The rate at which the hot phosphor is cooled is sometimes important, for example, hex.-ZnO:[Zn] should be cooled very slowly when prepared by method 8, and β-Zn_2SiO_4:Mn should be cooled very rapidly. For most phosphors, the rate of cooling is not critical, except that slow cooling favors the attainment of an equilibrium distribution of crystal imperfections. In the case of a cub.-Mg_2TiO_4:Mn phosphor, E. Tiede and F. A. Kröger report that cooling rates as slow as 4°C hr^{-1} are required to obtain maximum luminescence efficiency after heating the phosphor batch in oxygen at 1300°C, although equivalent results are obtained when the batch is cooled rapidly and then annealed in oxygen at 550°C for 48 hrs.[154] It appears that the Mn activator atoms in this phosphor should be highly oxidized, being perhaps largely Mn^{4+} in the most efficient samples, and the reaction rates and thermodynamic equilibria in this system are such that the proportion of Mn^{4+} is made largest by prolonged heating in an oxidizing atmosphere at a temperature near 550°C. Rapid cooling tends to (1) preserve the valence states prevalent

EXAMPLES OF THE INITIAL INGREDIENTS, CONDITIONS OF PREPARA

No.	Ingredients and Crystallization, Mole Proportions unless Otherwise Indicated	Temp., °C	Simplified Notati Main Structure—Fo
1	Grind together 100 g KCl + 2 g $SbCl_3$	30	Cub.-KCl:Sb
2	1KCl + 0.004 to 0.4TlCl	30–800	Cub.-KCl:Tl
3	$1ZnF_2$ + 0.006MnF_2 (see Nos. 57 and 58)	900	Tetr.-ZnF_2:Mn
4	$0.1ZnF_2$ + $0.9MgF_2$ + 0.006MnF_2	800	Tetr.-$9MgF_2 \cdot ZnF_2$:Mn
5	100 g ZnS + 2 g NaCl	940	Cub.-ZnS:[Zn]
6	100 g ZnS + 2 g NaCl + 0.016 g $AgNO_3$	940	Cub.-ZnS:Ag
7	100 g ZnS + 4 g NaCl + 2 g $BaCl_2$ + 0.016 g $AgNO_3$	1250	Hex.-ZnS:Ag
8	100 g ZnS + 2 g NaCl + 0.01 g $CuCl_2$	880	Cub.-ZnS:Cu
9	100 g ZnS + 2 g NaCl + 0.01 g $CuCl_2$	1220	Hex.-ZnS:Cu
10	100 g ZnS + 2 g NaCl + 0.032 g $AgNO_3$ + 0.02 g $CuCl_2$	1260	Hex.-ZnS:Ag:Cu
11	100 g ZnS + 2 g NaCl + 4.5 g $MnCl_2$	1000	Cub.-ZnS:Mn
12	100 g ZnS + 2 g NaCl + 5 g $PbSO_4$ + 0.0002 g $CuCl_2$	1200	Hex.-ZnS:Cu:Pb
13	100 g ZnS + 2 g $CaSO_4$ + 0.01 g $CuCl_2$ + 0.001 g $Fe_2(SO_4)_3$	1200	Hex.-ZnS:Cu:Fe
14	48 g ZnS + 52 g CdS + 2NaCl + 0.016 g $AgNO_3$	940	Hex.-ZnS(48)CdS(52):Ag
15	86 g ZnS + 14 g CdS + 1 g NaCl + 1 g NH_4Cl + 0.5 g $BaCl_2$ + 0.015 g $CuCl_2$	1250	Hex.-ZnS(86)CdS(14):Cu
16	100 g ZnSe + 2 g NaCl	780	Cub.-ZnSe:[Zn]
17	100 g ZnSe + 2 g NaCl + 0.008 g $AgNO_3$	780	Cub.-ZnSe:Ag
18	100 g ZnSe + 2 g NaCl + 0.01 g $CuCl_2$	780	Cub.-ZnSe:Cu
19	100 g ZnSe + 2 g NaCl + 4.5 g $MnCl_2$	780	Cub.-ZnSe:Mn
20	100 g SrS + 6 g CaF_2 + 6 g $SrSO_4$ + 0.04 g $SmCl_3$ + 0.04 g $Eu_2(SO_4)_3$	1000	Cub.-SrS:$SrSO_4$:CaF_2:S
21	5 g SrS + 95 g SrSe + 6 g CaF_2 + 6 g $SrSO_4$ + 0.04 g $SmCl_3$ + 0.04 g $Eu_2(SO_4)_3$	1000	Cub.-Sr(S:Se):$SrSO_4$:Ca
22	100 g SrS + 6 g LiF + 6 g $SrSO_4$ + 0.04 g $SmCl_3$ + 0.04 g $CeCl_3$	1000	Cub.-SrS:$SrSO_4$:LiF:Sm
23	100 g CaO + 6 g Na_2SO_4 + 2 g CaF_2 + 0.04 g $Bi_2O_2CO_3$ + 100 g S	950	Cub.-CaS:Na_2SO_4:CaF_2
24	$1CaO + 1WO_3$	1100	Tetr.-$CaWO_4$:[W]
25	$2MgO + 1WO_3$ (see Nos. 55 and 56)	1100	Monocl.-Mg_2WO_5:[W]
26	$2MgO + 1WO_3$ + 0.1 g U_3O_8	1100	Monocl.-Mg_2WO_5:U
27	1ZnO heated in CO	1000	Hex.-ZnO:[Zn]
28	$2BeO + 1SiO_2$	1400	Rbhdl.-Be_2SiO_4:[Be]
29	$2ZnO + 1SiO_2$ (+ optional $0.02TiSiO_4$)	1250	Rbhdl.-Zn_2SiO_4:[Si] (or
30	$2ZnO + 1.012SiO_2 + 0.012MnO$	1250	Rbhdl.-Zn_2SiO_4:Mn
31	$2ZnO + 1.012SiO_2 + 0.012MnO$ melt and quench	1600	β-Zn_2SiO_4:Mn
32	$8ZnO + 1BeO + 5SiO_2 + 0.25MnO$	1250	Rbhdl.-$Zn_8BeSi_5O_{19}$:Mn

Pdλ	1	2537	3650	4500	Emission Spectrum Wavelength, Å	Decay Type, L = long	Stim.	Quench	References
	x	x	x		2990 3850 5700	?			65
	x	x			4570 5400 6100	ϵ^{-at}			162–166
.00	x				5870	ϵ^{-10t} L			58
25	x				5900	$\epsilon^{-10t} \to t^{-n}$ L			58
.00	x	x	x		3400 4700	t^{-n}			58
.10	x	x	x		4550	t^{-n}			58
.60	x	x	x		4350	t^{-n}			58
.50	x	x	x	x	5280	t^{-n} L			58
210	x	x	x	x	5160	t^{-n} L			58
20	x	x	x	x	4400 5130	$t^{-n_1} \to t^{-n_2}$ L			58, 167
.10	x	x	x		5910	ϵ^{-at}	x		145, 153, 154
	x	x	x		5000	t^{-n} L	x		168, 169
	x	x	x		5160	t^{-n} L		x	58
70	x	x	x	x	5730	t^{-n}			58
40	x	x	x	x	5600	t^{-n} L			58
	x	x	x		6430	t^{-n}			170
	x	x	x		5700	t^{-n}			170
	x	x	x		6500	t^{-n}			170
	x	x	x		6450	?			170
	x	x	x	x	6300	t^{-n}	x		58, 171–180
	x	x	x	x	5700	t^{-n}	x		58, 171–180
	x	x	x		4850	t^{-n}	x		58, 171–180
	x	x	x	x	4580	t^{-n} L			145, 181–186
34	x	x			4300	ϵ^{-at}			58, 155
47	x	x			4930	ϵ^{-at}			58, 155, 187–190
	x	x	x		5130	ϵ^{-at}			58
84	x	x	x		3850 5050	ϵ^{-at}			58, 102, 354
	x				3050	ϵ^{-at}?			58
	x				4150	? L			58, 61, 154, 158, 159
.00*	x	x			5250	ϵ^{-80t}			58
76	x	x			5630	ϵ^{-80t}			58
80	x	x			5300	$\epsilon^{-at} \to t^{-n}$			58, 191

TABLE 5 (*Continu*

No.	Ingredients and Crystallization, Mole Proportions unless Otherwise Indicated	Temp., °C	Simplified Nota Main Structure—I
33	$8ZnO + 1BeO + 5SiO_2 + 0.45MnO$	1250	Rbhdl.-$Zn_8BeSi_5O_{19}$:M
34	$5ZnO + 3BeO + 1TiO_2 + 1ZrO_2 + 6SiO_2 + 0.06MnO$	1200	?-$Zn_5Be_3TiZrSi_6O_{24}$:M
35	$9ZnO + 1BeO + 1SnO_2 + 6SiO_2 + 0.054MnO$	1100	Rbhdl.-$Zn_9BeSnSi_6O_{24}$:
36	$2ZnO + 1.012GeO_2 + 0.012MnO$	1100	Rbhdl.-Zn_2GeO_4:Mn
37	$4MgO + 1GeO_2 + 0.01MnO$	1100	?-Mg_4GeO_6:Mn
38	$1CdO + 1.006SiO_2 + 0.006MnO$	1200	Orthorhombic-$CdSiO_3$:
39	$1ZnO + 1.006Al_2O_3 + 0.006MnO$	1600	Cub.-$ZnAl_2O_4$:Mn
40	$1ZnO + 1.006Ga_2O_3 + 0.006MnO$	1600	Cub.-$ZnGa_2O_4$:Mn
41	$3ZnO + 2.012B_2O_3 + 0.018MnO$	900	Triclinic-$Zn_3B_4O_9$:Mn
42	$2CdO + 1.006B_2O_3 + 0.012MnO$	800	?-$Cd_2B_2O_5$:Mn
43	$1Al_2O_3$	1600	Rbhdl.-Al_2O_3:[Al]
44	$1Al_2O_3 + 0.005Cr_2O_3$	2100	Rbhdl.-Al_2O_3:Cr
45	$2CdO + 1.02P_2O_5 + 0.02MnO$	1100	?-$Cd_2P_2O_7$:Mn
46	$2CaO + 1.03P_2O_5 + 0.0001Dy_2(C_2O_4)_3 \cdot 10H_2O$	1050	?-$Ca_2P_2O_7$:Dy
47	$3CaO + 2P_2O_5 + 0.1CePO_4$ (heat in H_2)	1250	γ-$Ca_3(PO_4)_2$:Ce
48	$1ZnSO_4 + 0.025MnSO_4$	900	Orthorhomb.-$ZnSO_4$:M
49	$1CdSO_4 + 0.035MnSO_4$	850	Orthorhomb.-$CdSO_4$:M
50	$6ZnO + 1V_2O_5$	800	?-$Zn_6V_2O_{11}$:[V]
51	$1BaO + 1SiO_2 + 0.001Eu_2(SO_4)_3$ (heat in H_2)	1000	?-$BaSiO_3$:Eu
52	$1SrO + 1SiO_2 + 0.001Eu_2(SO_4)_3$ (heat in H_2)	1000	?-$SrSiO_3$:Eu
53	$1SrO + 1BaO + 2SiO_2 + 0.002Eu_2(SO_4)_3$ (heat in H_2)	1000	?-$(Sr:Ba)SiO_3$:Eu
54	$1SrO + 1CaO + 2SiO_2 + 0.002Eu_2(SO_4)_3$ (heat in H_2)	1000	?-$(Sr:Ca)SiO_3$:Eu
55	$1CdO + 1WO_3$	1100	Monocl.-$CdWO_4$:[W]
56	$1CdO + 1WO_3 + 0.1$ g U_3O_8	1100	Monocl.-$CdWO_4$:U
57	$1ZnF_2 + 0.002CbF_5$	600	Tetr.-ZnF_2:Cb
58	$1MgF_2 + 0.006MnF_2$	800	Tetr.-MgF_2:Mn
59	$1BaSO_4 + 0.09PbSO_4$	1150	Rhomb.-$BaSO_4$:Pb
60	$1MgS + 0.0003SbCl_3$ (heat in $N_2 + CS_2$)	1000	Cub.-MgS:Sb
61	$3Ca_3(PO_4)_2 + 1CaF_2 + 0.05Sb_2O_3 + 1(NH_4)_2HPO_4$	1000	Hex.-$3Ca_3(PO_4)_2 \cdot CaF$
62	$3Ca_3(PO_4)_2 + 1CaF_2 + 0.05Sb_2O_3 + 70$ g $Mn_3(PO_4)_2 + 1(NH_4)_2HPO_4$	1000	Hex.-$3Ca_3(PO_4)_2 \cdot CaF$
63	$3Ca_3(PO_4)_2 + 1CaCl_2 + 350$ g $Mn_3(PO_4)_2 + 1(NH_4)_2HPO_4$	1000	Hex.-$3Ca_3(PO_4)_2 \cdot CaC$
64	$2MgO + 1TiO_2 + 0.002MnO$ (heat in O_2, cool *slowly*)	1300	Cub.-Mg_2TiO_4:Mn
65	$2CaO + 1MgO + 2SiO_2 + 0.05CeCl_3$ (heat in N_2)	1250	Tricl.-$Ca_2MgSi_2O_7$:Ce

Excited by					Emission Spectrum Wavelength, Å	Decay Type, L = long	Infrared		References
CR $\int Pd\lambda$	Photons, Å						Stim.	Quench	
	1	2537	3650	4500	3000 4000 5000 6000				
80	x	x			6100	$\epsilon^{-at} \longrightarrow t^{-n}$			58, 191
	x	x			4300 · · · 5500	$\epsilon^{-at} \longrightarrow t^{-n}$ L			61
90	x	x			5370	$\epsilon^{-at} \longrightarrow t^{-n}$ L			58, 665
70	x	x	x		5370	ϵ^{-at}			58, 192
	x	x	x		6420/6500 6260/6330	ϵ^{-at}			58, 192
54	x	x			5900	ϵ^{-at}			58
26	x		x		5130	ϵ^{-at}			58, 156, 157
24	x	x	x		5060	$\epsilon^{-at} \longrightarrow t^{-n}$			58, 156, 157
23	x	x			5500	ϵ^{-at}			58, 156, 157
72	x	x			6260	ϵ^{-at}			58
	x				3000	ϵ^{-at}?			58
	x	x	x		6927/6942	ϵ^{-at}			148, 193–196
44	x	x			6150	ϵ^{-at}			58
24	x		[x]		4800 · · · 5760	t^{-n}			58, 197
	x	x			3600	$\epsilon^{-10^4 t}$			198–200
55	x	x			6340	$\epsilon^{-at} \longrightarrow t^{-n}$ L			58, 145
46	x	x			5900	$\epsilon^{-at} \longrightarrow t^{-n}$ L			58, 145
	x	x	x		5650	?			202
	x	x	x		5070	ϵ^{-at}			58, 203
	x	x	x		4850	ϵ^{-at}			203, 375
	x	x	x		5200	ϵ^{-at}			58, 203
	x	x	x		4300	ϵ^{-at}			58, 203
25	x	x			4930	ϵ^{-at}			58
	x		x		6400	ϵ^{-at}			58, 189
100	x				4640	t^{-n}			58
9	x				5900	ϵ^{-at}			58
	x				3500	$\epsilon^{-10^6 t}$			244, 246
	x	x	x		5300	$\epsilon^{-10^6 t} \mp t^{-n}$			145
		x			4800	ϵ^{-at}			554
		x			4900 · · · 5800	ϵ^{-at}			554
		x			6000	ϵ^{-at}			554
			x	x	6550	$\epsilon^{-10^3 t}$			154
					3700	$\epsilon^{-10^6 t}$			679

only the 5760-Å band is excited by 3650-Å *UV*.

at high temperatures, (2) "freeze in" omission defects and other imperfections which are relatively abundant at high temperatures (eq. 21) and (3) afford in some cases an abnormally large density of traps (Fig. 9). Crystals with such abnormally high concentrations of imperfections are often unstable and tend to return to the state which would have obtained on slow cooling. High-melting phosphor crystals which have been cooled slowly enough to attain most nearly the equilibrium state corresponding to the temperature at which they are to be used, for example, room temperature, are generally most stable during operation. The greatest stability during operation is usually obtained with phosphor crystals which have been cooled slowly and which have simple compositions as close to stoichiometric proportions as possible. Complex compositions often do not undergo complete reaction and form heterogeneous products which are unstably intermediate to the vitreous and crystalline states. The β-Zn_2SiO_4:Mn phosphor has high stability, despite its rapid cooling, because the metastable crystals must be heated to about 1000°C before their atoms become sufficiently mobile to allow the crystalline transition to the stable rbhdl.(α)-form.[61,72]

Well over 10^5 samples of phosphors have been synthesized and tested during the past two decades of phosphor research in laboratories here and abroad. In this book, it is possible to cite only a few examples of the more interesting phosphors, some of which are listed in Table 5 with specific references to more complete descriptions of their syntheses and properties. Almost all the ZnS-type phosphors described herein were prepared from ingredients purified by the "acid process" (Appendix 1) to minimize the introduction of oxides which may be dissolved off laboratory glassware by alkaline solutions, and the silicate-type phosphors (including germanates, aluminates, borates, and stannates, *inter alia*) were prepared by the "carbonate process" to assure thorough and intimate mixing of the ingredients.[58,191,192,205]

The data given in Table 5 are adequate for preparing good phosphors, but not necessarily the best phosphors of each type. *There is no effective substitute for personal experience in synthesizing phosphors, and the results of different investigators often differ greatly according to their idiosyncratic techniques in preparing ostensibly identical phosphors.* This is often true even when the different investigators have the same ingredients and facilities. The differences are generally traceable to inadvertent contamination, incomplete mixing of ingredients or subsequent separation of ingredients as the result of differences in particle size and density, as well as to errors in computation or measurement. Variations in time, temperature, and atmosphere during crystallization also may produce discordant results. A furnace with a rotating circular hearth plate is useful in obtaining uniform heat treatment of several samples heated at one time,[88] but it is difficult to obtain identical time–temperature cycles during successive furnacings, especially when different sizes of crucibles and batches are used.

Most of the efficient phosphors, such as rhbdl.-Zn_2SiO_4:Mn, rhomb.-$BaSO_4$:Pb, and cub.(or hex.)-ZnS:Cu(or Ag, or Mn) are best crystallized in a neutral or slightly oxidizing atmosphere, but a few phosphors, such as γ-$Ca_3(PO_4)_2$:Ce and $SrSiO_3$:Eu (Table 5), are more efficient after being subjected to a reducing atmosphere during at least part of their crystallization time at high temperatures. In the latter cases, the activators Ce and Eu function best when incorporated in the host crystal as the (approximate) ions Ce^{3+} and Eu^{2+} instead of the more prevalent Ce^{4+} and Eu^{3+} ions which obtain under neutral or oxidizing conditions. During the high-temperature crystallization of efficient hex.-ZnS-type phosphors, there is a readily detectable evolution of SO_2 formed by interaction of the ZnS and the oxygen in the normal air atmosphere. Phosphors of this type crystallized in reducing atmospheres (H_2, H_2S) have very low efficiencies relative to those prepared in neutral (N_2) or slightly oxidizing atmospheres (SO_2, CO_2 [253, 255]). Also, it has been found that partial oxidation of pure ZnS (+ NaCl flux) at high temperature affords a green-emitting long-persistent phosphor with properties remarkably similar to those of hex.-ZnS:Cu,[58] and the deliberate inclusion of about 2 per cent of ZnO in the initial mix of a hex.-ZnS:Cu phosphor enhances the room-temperature persistence of the final product.[63] Similarly, oxygen-containing fluxes are often essential in the preparation of efficient long-persistent alkaline-earth sulphide (selenide) phosphors. Under conditions such as these, fluxes may be such potent and essential promoters of luminescence that they deserve to be called activators as well as (or instead of) fluxes.[*]

As an example of the various effects which may be obtained by alteration of the atmosphere used during crystallization, a hex.-ZnO phosphor prepared by burning zinc vapor in excess oxygen has its luminescence emission predominantly in a narrow ultraviolet band peaked near 3850 Å, whereas the hex.-ZnO:[Zn] phosphor prepared by heating ZnO in a reducing atmosphere has its luminescence emission predominantly in a broad visible band peaked near 5100 Å.[58] Contrarily, hex.-CdS emits in the infrared after being heated in a neutral or mildly oxidizing atmosphere but in the green after being heated in a reducing atmosphere.[154, 215]

With respect to pressure during crystallization, the more useful phosphors, such as hex.-ZnS:Ag and rhbdl.-Zn_2SiO_4:Mn, are best prepared at atmospheric or higher pressures, being less efficient when prepared by crystallization *in vacuo*. The infrared-emitting cub.-Cu_2O:[Cu] phosphor, however, is more efficient when crystallized *in vacuo* than at atmospheric pressure.[216]

References 58, 208–212, and 221 give many detailed examples of the purification, compounding, reaction, crystallization, and subsequent treatment of a number of practical phosphors. In commercial production a furnace may often be used for only one phosphor, but in research great care must be exercised to avoid cross contamination of different

---

[*] Fluxes can also provide *coactivator* anions to compensate activator cation charges.[755,757]

phosphors heated at the same time, or at different times, in the same furnace. Multiple containers and dense inert packing around the phosphor crucibles help to minimize extraneous contamination.

Some of the phosphors in Table 5 represent **families of phosphors** which are generated by gradual substitution of new phosphor-active ingredients, or by variation of the treatment or proportions of the indicated ingredients. For example, by increasing the ratio of CdS/ZnS in Nos. 14 and 15, the emission spectrum may be displaced gradually into the infrared, whereas a decrease in the CdS/ZnS ratio displaces the spectrum in the opposite direction (see Figs. 32–37). Outstanding examples of phosphor families are

(1) (Zn:Cd)(S:Se) with the range of activators and temperatures indicated in examples 5 to 19, (2) $(Zn:Be)_2(Si:Ge)O_4:Mn$, and (3) $(Zn:Cd)O:B_2O_3:Mn$, where the **colons** indicate variable proportions of the ingredients *as long as the total within any pair of parentheses is equal to one mole or stoichiometric proportion.*

In Table 5, the figures given under CR are (1) the relative rate of emission of luminescence photon energy (under CR excitation) at the indicated main peak of the emission spectrum, and (2) the foregoing peak value times the area under the entire luminescence emission spectrum $(= \int Pd\lambda = \int Pd\nu)$. The CR data were taken at 6 kv and about 1 μa cm^{-2}, and the reference standard, rbhdl.-$Zn_2SiO_4:Mn(0.3)$ has an efficiency of about 5 candles per watt in conventional unmetallized screens detected on one side [63] (efficiencies greater than 10 candles per watt are obtainable at higher voltages and lower current densities with metallized screens). The classification of phosphors into exponential-decay (ϵ^{-at}) and power-law-decay (t^{-n}) types is discussed in Chapter 5 along with the description of stimulation and quenching of long-persisting phosphorescences by infrared.

The *complete* **symbolization** of a phosphor should include designation of (1) crystal structure; (2) host crystal; (3) proportion, location, and effective valence of activator(s); (4) fluxes and other initial ingredients which influence the properties of the final phosphor (though they may or may not be present in the final product); and (5) conditions of crystallization and cooling. It has been customary, unfortunately, to use simplified designations which omit everything except the chemical identities of the host crystal and the activator(s). These simplified designations are often quite inadequate, because the properties of many phosphors, particularly the alkaline-earth(Ca,Sr,Ba)-sulphide phosphors, depend strongly on all the initial ingredients and on the detailed techniques of synthesis. Furthermore, there is a lack of standardization of symbolism in the literature. The phosphor symbolized by $C6mc$-ZnS:Cu or hex.-ZnS:Cu in this book, may be found symbolized by α-ZnS:Cu,

β*-ZnS:Cu, ZnSCu, ZnS/Cu, ZnS-Cu, ZnS(Cu), ZnS.Cu, and even ZnS·Cu in the literature.

Even if the same simplified symbolism were to be used throughout the literature, two phosphors designated merely by hex.-ZnS:Cu could have quite different properties, depending on the unstated purity of their ingredients, the kind and amount of flux, the proportion of activator, the size of batch, the size and kind of crucible, the atmosphere and temperature during crystallization, and the treatment (washing, grinding) after crystallization. In general, however, experienced phosphor researchers can approximate each other's products, *if* the products are available for direct comparison. Also, experienced phosphor researchers are able to supply missing necessary information in the cases of well-known phosphors, for example, hex.-ZnS:Ag is usually interpreted as $C6mc$-ZnS:Ag[NO$_3$](0.01), [NaCl(2)], 1100–1350°C, and rbhdl.-Zn$_2$SiO$_4$:Mn is usually interpreted as $R\bar{3}$-Zn$_2$SiO$_4$·0.01—0.05MnSiO$_3$, 1000–1300°C, with preparative procedures as described in examples 4*b* and 9*b*. Simplified formulas for phosphors will undoubtedly continue to be used, although they tax the memorizing powers of those who must interpret them. Fortunately, it is generally true that, when only the simplified formulas are used, it may be assumed that optimum proportions and conditions of synthesis were used to obtain maximum luminescence efficiency under a given **excitant** (for example, UV, x rays, or CR).

It is well to maintain a healthy skepticism toward results reported by a single investigator on phosphors, especially new or unusual phosphors, which are identified solely by simplified symbols, manufacturers' code numbers, or ambiguous terms such as "zinc sulphide phosphor," or "silicate phosphor." Experienced phosphor researchers know that even elaborate descriptions of the preparations of some phosphors do not always ensure successful reproduction of results by those "skilled in the art." An adequate description of the art of synthesizing uniformly superior phosphors must await a much larger book than the present small volume. For the present, reports on properties of phosphors should be weighed carefully according to the amount of specific information given about their preparation and measurement; with complete confidence reserved for results which have been confirmed by several *impartial* investigators.

Devising New Phosphors

It is not difficult for even a novice to devise and prepare merely new phosphors, but it is very difficult for even an expert to devise and prepare a "successful" new phosphor. By a "successful" new phosphor is meant a phosphor whose unambiguous preparation or properties shed new light on the fundamental constitutions and mechanisms of luminescent solids, or a phosphor which is superior to existing phosphors for practical uses. It is not possible, at present, to predict the properties of proposed new phosphors (which are not just slight modifications of known phosphors), and so the following summary of phosphor compositions is presented as an aid to those who wish to interpolate or extrapolate from the existing combinations.[58, 157]

From the chemical standpoint, the better phosphors comprise: (1) a well-crystallized colorless (or only lightly colored), high-melting host crystal containing mostly *singly valent* elements, such as those from periodic-system groups 1 (for example, Na, K, Rb), 2 (Be, Mg, Ca, Sr, Ba, Zn, Cd), 3B (B, Al, Ga), 4A (Zr, Hf, Th), 4B (Si, Ge), 5 (P, V), and/or 6A (Mo, W), combined with elements from groups 6B (O, S, Se, Te) and/or 7B (F, Cl, Br, I); and (2) a cocrystallized activator which may be formed by (*a*) a *presumed* selective decomposition of the host crystal, for example, $ZnO:[Zn]$, $ZnS:[Zn]$, $(ZnO:V_2O_5):[V]$, and $CaWO_4:[W]$, or (*b*) deliberate inclusion of small amounts of one or more *multivalent* impurity elements, for example, Cu, Ag,[217, 218] Au, Ce, Sm, Eu, Gd, Tb, Dy, Tl, Ti, Sn, Pb, Cb, Bi, Cr, Sc, Mn, Tc(Ma), P, and U, which are generally combined with nonmetallic elements (omitted in the simple notations). It is noteworthy that the sulphide(selenide)-type phosphors usually require fluxes to obtain best results, whereas the other phosphors are usually harmed rather than improved by fluxes. Although not all the indicated combinations afford phosphors which operate efficiently at or above room temperature, their possibilities are far from exhausted, and other elements will certainly be found useful in making phosphors.

The determination of optimum proportions and conditions of preparation of a proposed new phosphor is usually a lengthy task, because there are many independent parameters, such as (1) the proportions of the host-crystal ingredients (for example, the ratio of ZnO/SiO_2), (2) the proportion of activator(s) (for example, the ratio of $(ZnO + SiO_2)/MnO$), (3) the proportion and kind of flux(es), if required, (4) the atmosphere and temperature–time cycle during reaction, crystallization, and cooling. Thorough phosphor research requires that large numbers of samples be prepared and tested under various excitants and conditions of excitation to assay the utility of a proposed host crystal with or without activators. In general, if a stoichiometric host crystal gives negligible luminescence under all circumstances, the host crystal is of low utility even in nonstoichiometric proportions.

Optimum activator proportions usually range from about 0.0001 mole (for example, Ag, Cu, Sm, Eu, Bi) to about 0.01 mole (for example, Mn, Tl, Ti, Ce) per mole of host crystal, or about 0.01 to 1 weight per cent of activator. The occurrence of exceptions, such as $SrAl_2O_4:[AlF_3]$:

$Mn(0.05)$ [201] and $Ca_3(PO_4)_2$:$Bi(5)$,[154] make it advisable to investigate proposed new activator + host–crystal combinations in steps of 0, 0.0003, 0.001, 0.003, $\cdots$1, 3, 10, and 30 mole per cent, or weight per cent, to be assured that an optimum activator proportion is not missed. As is shown later, high activator proportions, for example, 3 weight per cent of Mn activator in rbhdl.-Zn_2SiO_4, are advantageous at very high excitation densities; hence, the optimum proportion depends on the conditions of excitation. At proportions above the maximum useful activator proportion, additional activator atoms decrease the efficiency of the phosphor by interacting detrimentally with other activator atoms which are already present and functioning, and increase the absorption of the phosphor for its own emission (activator compounds are usually highly colored).[61] If the additional activator atoms could be guided to specific spots in the crystal where the activator concentration is below the average, then the maximum activator proportion might be made considerably greater than that obtained by the purely statistical distribution which should be favored by very high crystallization temperature and rapid cooling. With the advent of relatively intense sources of neutrons and other particles which can produce transmutations of atoms, it may be possible to transmute part of the host-crystal atoms in some pure crystals into activator atoms. Here, the activator atoms induced in cold crystals may have quite different distributions among possible sites than are obtained by present methods of high-temperature reaction. A specific example, suggested by Professor J. Turkevich, is the transmutation of some of the stable $_{30}Zn^{64}$ in cub.- or hex.-ZnS into stable $_{29}Cu^{65}$ activator atoms by the reactions:

$$_{30}Zn^{64} + \text{neutron} \rightarrow {}_{30}Zn^{65} \xrightarrow{(250\ \text{days})} {}_{29}Cu^{65} + \text{positive beta particle}$$

With respect to fluxes, the proportions are rarely critical, although the kind, number, and proportion of fluxes are important in obtaining optimum performance from commercially useful phosphors.[58, 169–176, 205, 207–215] For example, detrimental crushing and grinding of ZnS-type phosphors is avoided when the ingredients are crystallized with about 6 per cent of NaCl at about 1250°C. The resultant batch contains crystals bonded together by water-soluble NaCl and Na_2S which can be readily dissolved away to leave unaggregated free-flowing phosphor crystals. The function of the fluxes appears to be chiefly to provide (1) a fluid phase for solvation and transport of ingredients, and (2) increased surface reactivity (lowered activation energy) and atomic mobility to facilitate crystal growth. In general, the flux materials should not dissolve and retain a large portion of the activator compounds, although some solvation may occur during crystallization at high temperature followed by a precipitating out of the activator compounds (and incorporation in the host crystal) as the temperature is lowered during cooling. The most popular fluxes are the water-soluble, low-melting-point alkali and alkaline-earth halides, borates, and sulphates, which are usually added in proportions of the order of 1 weight per cent of the batch prior to crystallization. Where possible, the residual fluxes are removed by water-washing the phosphor, except when the host crystal itself is water-soluble (in such cases, it is sometimes possible to use certain organic solvents; for example, methyl alcohol dissolves $MgSO_4$, and alcohol and glycerine dissolve some of the alkali halides).

When the optimum proportions and method of synthesis of a useful phosphor have been ascertained, it is often possible to relax somewhat the extreme precautions which are necessary for success in the laboratory. The intrinsic impurity tolerance of a phosphor is roughly proportional to its activator content; hence, some phosphors, such as rbhdl.-$Zn_2SiO_4 \cdot 0.01Mn$, can tolerate up to about 0.01 mole per cent of total metallic impurities, whereas other phosphors, such as cub.-$ZnS \cdot 0.0001Ag$, can tolerate only about 0.0001 mole per cent of total metallic impurities. The silicate phosphors used in "fluorescent" lamps are made successfully from commercially available USP (less pure than CP) ingredients under rather ordinary factory conditions, whereas the best sulphide phosphors used in cathode-ray tubes are made from specially purified LP ingredients in very clean surroundings and special apparatus.[58] One reason why large-scale production of phosphors is less rigorous than laboratory-scale research is that the surface exposed to contamination during processing increases as only the *square* of the radius of an assumedly spherical batch, while the volume increases as the *cube* of the radius of the batch (freshly crystallized batches of phosphors are usually examined under suitable sources of ultraviolet to detect and discard inhomogeneities, such as relatively nonluminescent "rinds" formed by reaction with the crucible or furnace atmosphere). Theoretically, it should be possible to prepare better phosphors on a large commercial scale than on a small experimental scale.

From both the physical and chemical standpoints, considerable improvement in the variety and utility of phosphors should result from the development and use of high-pressure furnaces for crystallizing phosphors.[65, 207, 212] By using applied pressures which are appreciable relative to the internal pressures of solids, it should be possible to (1) control selective decomposition and allow the use of higher temperatures for crystallizing more stable phosphors with controlled crystal sizes, (2) incorporate some of the more volatile elements as useful constituents in phosphors, (3) alter favorably the compositions of phosphors by controlling the partial pressures of constituent gases in the ambient atmosphere during crystallization, and (4) produce new crystal forms of phosphors.[58, 690]

Most of the past phosphor research has been aimed at satisfying physical requirements set by a few detectors, such as the human eye and other photosensitive devices, and by apparatuses catering to these detectors, for example, lamps, x-ray fluoroscopes, and cathode-ray tubes (CRT) used in television and radar. The research procedure has been largely empirical, aided by scientific intuition based on chemical experience and tempered by physical tests of the phosphors. It is likely that intuition will continue as a major factor for some time in determining the progress of phosphor research, because the present physical theories of the luminescence of solids are too qualitative to be practical guides to really new useful compositions.[34, 157]

CHAPTER 4

CONSTITUTIONS, STRUCTURES, AND ENERGY LEVELS OF PHOSPHORS

Sizes and Shapes of Phosphor Crystals and Particles

If all the phosphors described in the previous chapter were to be examined under a microscope, the sizes of their individual crystals would be found to range from less than 10^{-5} cm, for some silicate and aluminate phosphors, to over 1 cm, for carefully grown cub.-KCl:Tl, cub.-NaI:Tl, tetr.-CaWO$_4$:[W], monocl.-CdWO$_4$:[W], and rbhdl.-Al$_2$O$_3$:Cr. The latter five phosphors are rarities, in that they may be prepared as large transparent single crystals by skillfully cooling a melt.[162, 293, 677] Large batches of some of the other phosphors, such as tetr.-ZnF$_2$:Mn and rbhdl.-Zn$_2$SiO$_4$:Mn, may be melted and cooled to form moderately efficient phosphors, but the cooled melts are usually opaque aggregates of disoriented tiny crystals. In certain cases it is possible to grow thin acicular crystals from the vapor phase, for example, centimeter-long needles of hex.-CdS may be grown from CdS vapor at atmospheric pressure,[291] and needles of rbhdl.-Zn$_2$SiO$_4$:Mn have been grown up to 3 cm long by hydrothermal synthesis at about 1000°C and 1000-atmospheres steam pressure.[219, 220] Most of the efficient phosphors are made by solid-state reactions at temperatures below the melting points of their finely divided ingredients, however, and the sizes of their *separate* crystals generally range from less than 0.01 to about 100 microns.[58, 63, 678]

In addition to the wide range in the sizes of phosphor crystals, there are various degrees of aggregation of the individual crystals to form particles. **Phosphor particles** range from one crystal to an indefinitely large number of crystals which have become joined together during crystallization or which cohere even after previous dispersal. Separated crystals, which are placed in contact again, may cohere as a result of (1) differences in static charge, (2) attraction forces of the van der Waals type, and (3) intergrowth of their surfaces (especially at elevated temperatures, and in liquid media). In general, the degree of **aggregation,** that is, the number of crystals per particle, decreases as the average size of the crystals increases. Large crystals have lower surface–volume ratios than small crystals, and so there is less contact surface for cohesion

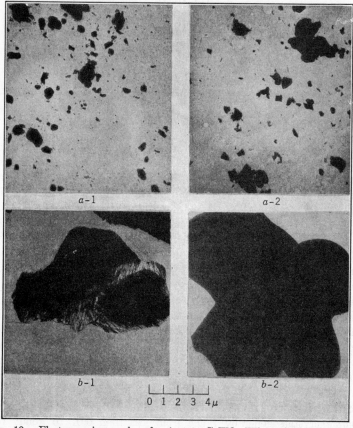

F<small>IG</small>. 10. Electron micrographs of *a*-1, tetr.-CaWO₄:[W], 1000 °C; *a*-2, rbhdl.-8ZnO·BeO·5SiO₂:Mn(1.4), 1250 °C; *b*-2, hex.-ZnS(48)CdS(52):Ag(0.01), 700 °C; and *b*-1, a particle of hex.-ZnS(48)CdS(52):Ag(0.01), 700 °C after being subjected to intense cathode rays at 50 kv and about 1 amp cm⁻².

in a given weight of large-crystal material than in the same weight of small-crystal material. Furthermore, the natural small-scale vicinal irregularities of crystal surfaces make it more difficult for the larger crystals to achieve uniform intimate contact with each other.

    Phosphor crystals vary in shape as well as in size. The individual crystals of a particular phosphor may exhibit shapes ranging from almost spherical or irregularly amorphous to rather definite polyhedra, such as straightsided fragments of cubes, octahedra, flat plates, and acicular prisms. Figure 10 gives several exemplary electron micrographs, showing the range of shapes and sizes of crystals and particles of various phosphors.[58, 678] Crystals of hex.-ZnS(48)CdS(52):Ag which have been skeletonized by bombardment with high-intensity cathode rays (CR), are shown in Fig. 10*b*-1. The skeletal striae are probably mostly ZnS left behind by selective volatilization of the CdS. The striae show that the internal structure of these crystals is rectilinear, even though the external shapes of the crystals may be partly curvilinear (Fig. 10*b*-2).

Internal Structures and Constitutions of Phosphor Crystals

This is the keystone subject in understanding the luminescence of solids. It is the focal point of common interest for both chemists and physicists who wish to know the "why" of phosphors. Our knowledge of the atomic interiors of phosphor crystals is derived *indirectly* from (1) "foresighted" information about chemical ingredients and conditions of synthesis, and (2) "hindsighted" information obtained by microchemical, spectroscopic, microscopic, and diffraction analysis, and by studies of the physical properties of the completed phosphor.[34] At present, our knowledge of the internal structures and constitutions of phosphors is far from satisfactory, as evidenced by the conflicting interpretations and hypotheses advanced by different investigators, or by the same investigator at different times.[58, 61–63, 65, 72–84, 110–129, 144–269, *et al*.] Some of the present views on the subject are introduced here, on a *tentative* basis, to aid in correlating several of the properties of phosphors described in the next chapter.

DEGREE OF CRYSTALLINITY OF PHOSPHORS. Many of the ingredients employed in making phosphors are purified by chemical methods, using solutions, and are eventually precipitated at a temperature near room temperature. Substances which are highly ionic in character, for example, KCl, tend to precipitate in externally apparent crystalline shapes [270a] and may be made luminescent without heating to high temperatures (Chapter 3 under Syntheses and Symbolism of Phosphors, example 1). Substances which are relatively nonionic, for example, ZnS and $BaSO_4$, tend to precipitate in rounded shapes with no well-defined angles and edges, and these substances (or materials incorporating large proportions of these substances) must be heated to temperatures of the order of 1000°C to make them efficiently luminescent.

If x-ray, electron, or neutron diffraction photographs were to be made of all the phosphors listed in Table 5, all the photographs would show rather sharp lines, indicating definite crystallinity (Fig. 11a). Some of the simpler phosphors, such as Nos. 5–11 and 30–33, have particularly high degrees of crystallinity and high luminescence efficiencies under excitation by *either* UV or CR. The more complex phosphors, such as Nos. 20–23 and 34, usually have greater degrees of amorphism and strain, as evidenced by more diffuse x-ray diffraction patterns and the appearance of diffraction lines of the fluxes as well as the host crystal. The relatively less perfect complex phosphors may have high luminescence efficiencies under UV excitation, but their efficiencies under CR excitation are usually quite low. In general, *a phosphor exhibits efficient luminescence under excitation by charged material particles, or high-energy photons ($h\nu \gg 10$ ev), only when it has a high degree of crystallinity.*

Figure 11a shows the principal lines in the x-ray diffraction patterns of some typical phosphors. These patterns merit study, especially with respect to the following features:

1. An *increase* in the spacing of the diffraction lines from the left-hand edge (that is, the center of the undiffracted primary x-ray beam) represents a *decrease* in the lattice spacing. This is shown in patterns 1 and 2 by the larger spacing and corresponding smaller length of the edge of the unit cell *a* of cub.-ZnS compared with cub.-NaCl (Note the line omissions and line-intensity reversals which characterize the change between the two cubic structures with different space groups and different ratios of atomic numbers of anions and cations).

2. The NaCl-flux patterns do not appear in diffraction photographs of the crystallized and washed ZnS phosphors. This is true even if as much as 20 weight per cent of NaCl flux is used.

3. The polymorphic and apparently enantiotropic transition from cub.- to hex.-ZnS is shown to be quite distinct in patterns 2 and 3. Individual crystals are always observed to be entirely one structure or the other, although in batches of crystals grown near the transition temperature of about 1020°C some individual crystals in a batch may be cubic while others are hexagonal. The distribution of structures near the transition temperature is influenced by the atmosphere, the size of the crystals, the flux, and the time of heating. It is reported that 0.01 per cent $CoSO_4$ promotes the cub. → hex. transition above 600°C.[272]

4. CdS may be precipitated in either the cubic or hexagonal form at 100°C,[270b] but exhibits only the hexagonal structure (pattern 4) at crystallization temperatures from about 600 to over 1400°C. There is an increasing tendency for (Zn:Cd)(S:Se) crystals to have the hexagonal structure, even at low crystallization temperatures (patterns 5 and 8), as the CdS and CdSe content is increased.[170]

5. ZnSe exhibits only the cubic structure, at least up to 1350°C (pattern 6). There is an increasing tendency for Zn(S:Se) crystals to have the cubic structure, even at high temperature, as the ZnSe content is increased.

6. Patterns 9 and 10 show the diffraction lines of two phosphors whose fluxes remain in the crystallized material. The extra lines in pattern 10 show particularly well the presence of residual fluxes. According to the *Strukturbericht*, Vol. 2, 1928–1932, pure crystallized SrS has $a = 5.999$ Å and CaS has $a = 5.667$ Å, and so pattern 9 represents a slightly expanded SrS structure, and pattern 10 represents a slightly contracted SrS structure (owing to some substitution of Ca for Sr). Many of the lines in patterns 8, 9, and 10 are abnormally weak and diffuse, denoting departures from crystalline regularity. Phosphors which exhibit such diffuse lines are generally inefficient under excitation by high-energy primary particles.[34, 170]

7. Patterns 11–14 show the diffraction lines of some representative phosphor host crystals which are dominated by oxygen, as contrasted with the sulphur-dominated host crystals represented in patterns 2–10. According to *Strukturbericht*, Vol. 4, 1936, the dimensions of the known unit cells of the fundamental members of this group of host crystals are:

DIMENSIONS OF UNIT CELLS IN Å (± 0.01 Å)

(6 simple molecules of A_2XO_4 per unit cell)

	Rhombohedral Basis		Hexagonal Basis		Density
	a_{rhomb}	α	a_{hex}	c_{hex}	g cm^{-3}
$R\bar{3}$-Zn_2SiO_4	8.62	107° 44′	13.92	9.33	4.241
$R\bar{3}$-Zn_2GeO_4	8.78	107° 50′	14.19	9.46	4.823
$R\bar{3}$-Be_2SiO_4	7.68	107° 59′	12.42	8.24	2.985
$R\bar{3}$-Be_2GeO_4	7.89	108° 6′	12.77	8.41	3.868

Zn_2SiO_4 and Zn_2GeO_4 form solid solutions in all proportions,[220] whereas, according to Fonda, Be_2SiO_4 forms solid solutions with Zn_2SiO_4 up to 30 mole per cent Be_2SiO_4 in the presence of 1 weight per cent of manganese silicate (activator), and up to 15 mole per cent Be_2SiO_4 in the presence of 4 weight per cent of manganese silicate.[221, 223] (In $R\bar{3}$-Zn_2SiO_4 each Zn and each Si atom is tetrahedrally coordinated by four O atoms, and each O atom is bonded to two Zn atoms and one Si atom. The Zn atoms are linked together through intervening O atoms, but the Si atoms are linked through O and Zn atoms.) The $R\bar{3} \rightarrow \beta$ transition is apparently monotropic, although the β form is stable for over 15 years at room temperature.

8. The most symmetrical (cubic) structures give the least complicated diffraction patterns, and the pattern complexity increases as the symmetry decreases and as the number of constituents in the crystal increases.

It should be noted that (1) all the seven crystal systems (Table 4) are represented among the phosphors in Table 5, and (2) those phosphors which crystallize in the systems with the highest degrees of symmetry (cub., hex., rbhdl., tetr.) generally have the highest luminescence efficiencies under excitation by CR. Sketches of some ideal high-symmetry

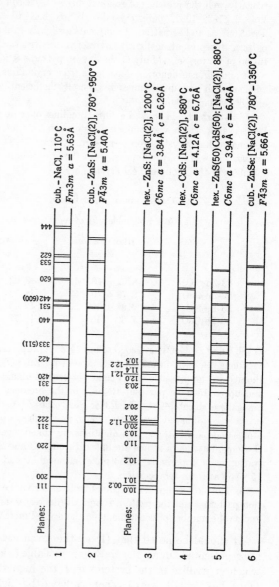

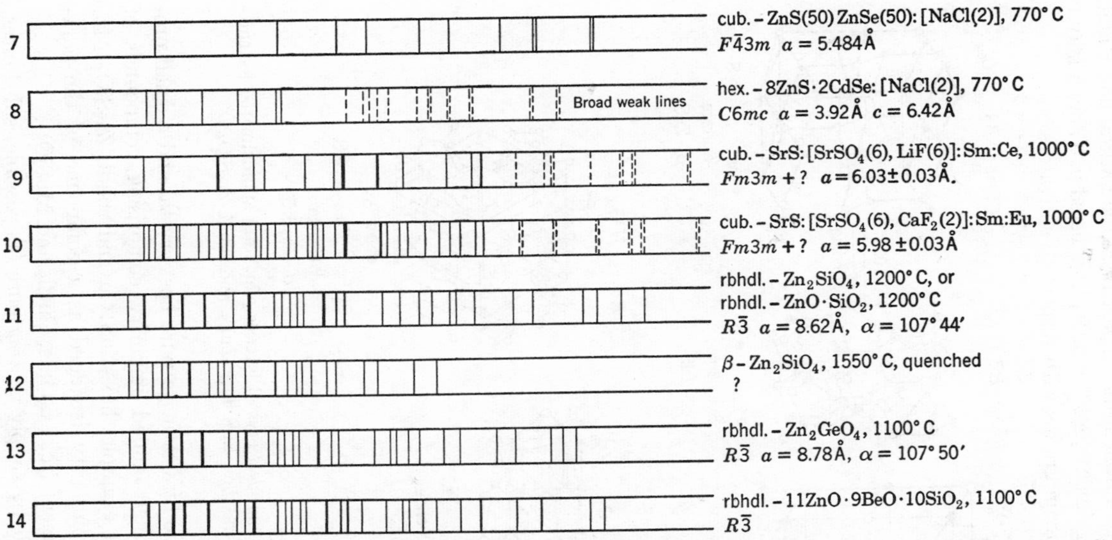

7   cub. – ZnS(50)ZnSe(50): [NaCl(2)], 770° C
$F\bar{4}3m$ $a = 5.484$ Å

8   hex. – 8ZnS·2CdSe: [NaCl(2)], 770° C
$C6mc$ $a = 3.92$ Å $c = 6.42$ Å
Broad weak lines

9   cub. – SrS:[SrSO$_4$(6), LiF(6)]:Sm:Ce, 1000° C
$Fm3m + ?$ $a = 6.03 \pm 0.03$ Å.

10   cub. – SrS:[SrSO$_4$(6), CaF$_2$(2)]:Sm:Eu, 1000° C
$Fm3m + ?$ $a = 5.98 \pm 0.03$ Å

11   rbhdl. – Zn$_2$SiO$_4$, 1200° C, or
rbhdl. – ZnO·SiO$_2$, 1200° C
$R\bar{3}$ $a = 8.62$ Å, $\alpha = 107° 44'$

12   β – Zn$_2$SiO$_4$, 1550° C, quenched
?

13   rbhdl. – Zn$_2$GeO$_4$, 1100° C
$R\bar{3}$ $a = 8.78$ Å, $\alpha = 107° 50'$

14   rbhdl. – 11ZnO·9BeO·10SiO$_2$, 1100° C
$R\bar{3}$

FIG. 11a. Principal lines in the x-ray diffraction patterns of some completed phosphors measured at room temperature. Copper-$K\alpha$ radiation, nickel filter.

crystal lattices, which are approximated by several outstanding phosphor host crystals, are shown in Fig. 11*b*.

CONSTITUTIONS OF PHOSPHORS. Apart from the natural uncertainty about purities of ingredients, all the phosphors listed in Table 5 may be separated into two main groups with respect to activators.

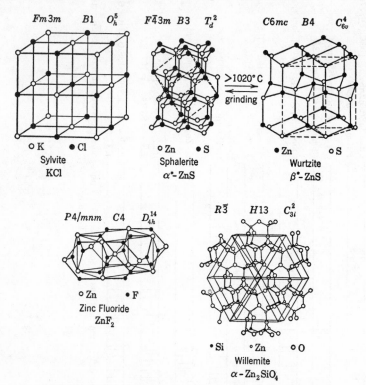

FIG. 11*b*.   Sketches of some *ideal* crystal lattices which are approximated by cub.-KCl, cub.-ZnS, hex.-ZnS, tetr.-ZnF$_2$, and rbhdl.-Zn$_2$SiO$_4$.

1. Phosphors made by heating a pure host crystal and a small proportion of a **consciously added** impurity comprising compounds of cationic elements, for example, hex.-ZnS:Cu and rbhdl.-Zn$_2$SiO$_4$:Mn, where the activator impurity is added as a compound of the paramagnetic cationic element (for example, CuCl$_2$ and MnSiO$_3$).

2. Phosphors made by simply heating pure (as possible) substances, such as CaWO$_4$, ZnO, and Zn$_2$SiO$_4$. These phosphors are sometimes *postulated* to be **"self-activated"** by selective volatilization of decomposition products during high-temperature crystallization. This is

assumed to produce a stoichiometric excess or structural displacement of one or more of the ingredient elements, although there is as yet no experimental confirmation of "self-activation."

When an impurity compound is consciously added, the nonmetallic (anion) constituent is often (but not always!) of minor importance. For example, practically identical hex.-$ZnS:Cu(0.01)$ phosphors may be made by using $CuCl$, $CuCl_2$, $CuBr_2$, CuS, $Cu(NO_3)_2$, or $CuCO_3$ as an initial ingredient for the introduction of the Cu activator. Also, practically identical rbhdl.-$Zn_2SiO_4:Mn(1)$ phosphors may be made by using MnO, MnO_2, $Mn(OH)_2$, $MnCO_3$, $Mn(NO_3)_2$, MnF_2, $MnBr_2$, $MnSiO_3$, or $MnSO_4$ as an initial ingredient for the introduction of the Mn activator (assuming about 1 per cent excess silica to combine with the manganese in all cases). Because the anionic constituent of the activator compound is often without particular effect, it has become customary to denote the metallic (cation) constituent of the added activator compound as the sole activator (phosphorogen), without indicating its state of chemical combination in the phosphor. That this is sometimes misleading, is exemplified by the cub.-$KCl:Sb$ phosphor, where the Sb must be added as the trichloride to obtain an efficient phosphor; also, the cub. (or hex.)-$ZnS:Mn(1)$ phosphors have quite different relative efficiencies, under UV and CR excitation, depending on whether the Mn is added as $MnCl_2$ or MnS, or as an oxygen-containing compound, such as $MnCO_3$, $Mn(NO_3)_2$, or $MnSO_4$ (improved CR response and diminished UV response are obtained with the latter compounds). These data re-emphasize the point that the simplified notations of phosphors not only are incomplete, but also tend to be misleading, because activators undoubtedly exist to some extent in the combined state, rather than in the indicated pure elemental state, in all phosphors. It is inconceivable that an atom (or ion) can be incorporated in a solid in such a manner that the atom or ion does not form bonds to some degree with its neighbors in the solid. To speak of free or elemental atoms in solids, even when such atoms are in stoichiometric excess, is definitely misleading. In this respect, it may be recalled that *chemical bonds are never completely ionic (heteropolar), or completely covalent (homopolar), but tend to fluctuate or exist as hybrids between the two extremes as indicated in eqs. 11 and 12.* The customary simplified notations of phosphors continue in use, largely for lack of more specific information about the natures and locations of the activators.

There are three general possibilities for the location of activator atoms (ions). These are: (1) *on* the host crystals in **surface sites** (including internal crevices); (2) *inside* the host crystal in **substitutional sites,** that is, the activator atom replacing a *regular* atom of the host crystal; and (3) *inside* the host crystal in **interstitial sites,** that is, the activator atom lodging between *regular* atoms of the host crystal. It is possible that activator atoms located within crystals but near their surfaces may have different effective valence states or bond-type distributions from those located well within the interiors of crystals. This difference may be caused by reactions between the surface layers and the atmosphere during heating, coupled with slow rates of diffusion of the reaction products into the interiors of the crystals. In general, however, the interiors of phosphor crystals are our chief concern.

The luminescence during excitation of most, if not all, of the phosphors discussed in this book is predominantly a *volume effect* which occurs throughout the excited portions of the solids. This is evidenced by (1) the observation that luminescence efficiency increases with increasing penetration of charged-material-particle excitants (for example, CR[34]), (2) the observation that the luminescence efficiencies of many phosphors remain high (sometimes increasing!) as their crystal sizes are diminished by chemical attack in solution,[102] and (3) the observation that individual crystals of phosphors lose efficiency uniformly on being ground.[242] In the latter case, the efficiency loss should be localized on the particular crystal faces or points which were mechanically affected, if the luminescence is predominantly a surface effect. In the case of a relatively surface-active phosphor, hex.-ZnO:[Zn],[58] no such spotty luminescence was observed on microscopic examination of ground crystals (which, incidentally, change their reflection color, under white light, from white to tan on grinding). Another specific example, showing that luminescence occurs chiefly in the volumes of phosphor crystals, is found in the rbhdl.-Zn_2SiO_4:Mn phosphor which can be synthesized with an average particle size ranging from less than 0.01 to over 3 microns without large change in luminescence efficiency under 4- to 10-kv cathode rays or 2537-Å ultraviolet.[678] That some phosphors are more surface-active than others, however, is indicated by their higher efficiencies under excitation by very low-velocity charged material particles which penetrate only a few atomic layers. Also, an excited ZnS:Ag phosphor has been reported to adsorb 50 per cent more of a commercial dye (Lanasol green, Ciba) than the same phosphor when unexcited, whereas irradiation reportedly reduces by about 90 per cent the amount of phenolphthalein adsorbed on CdS, and no difference was found in the amount of dye adsorbed on nonluminescent ZnS and $BaSO_4$ in the irradiated and unirradiated states.[60, 271] It is possible that excited surface atoms, or electrons trapped on surface sites, may play a prominent role in some afterglow phenomena, but the luminescence of most solids during and after excitation appears to be chiefly a volume effect.

Phosphors formed from host-crystal and activator ingredients which are **isostructural,** that is, which crystallize in the same space group and have common proportions and kinds of anions, for example, ZnF_2 + MnF_2 (both $P4/mnm$), $ZnAl_2O_4$ + $MnAl_2O_4$ (both $Fd3m$), and Al_2O_3 + Cr_2O_3 (both $R\bar{3}c$), undoubtedly have their activator cations located chiefly in substitutional sites, that is, Mn atoms substituted for Zn atoms in regular host-crystal sites, and Cr atoms substituted for Al atoms in regular host-crystal sites. It is convenient to use the words **cation** and **anion** to distinguish between relatively electropositive and electronegative atoms (ions) and radicals, remembering that *this does not mean that these atoms or radicals are completely ionic.* Isostructures usually obtain only when the radius ratios of the intersubstituted atoms (and radicals) are approximately equal, and the *radii of atoms depend strongly on their degree of ionization and coordination number, and on the nature (especially polarizing power) and spatial arrangement of their ligands (nearest bonded neighbors).* This is demonstrated, in part, in Table 6 which compares the radii of a number of elements combined in (1) the ionic state, assuming 6 coordination, as in the $Fm3m$ structure, and (2) the covalent state, assuming that the atoms are 4-coordinated, as in the $F\bar{4}3m$ or $C6mc$ structures of the ZnS type.[24]

In the tetr.-ZnF_2:Mn, cub.-$ZnAl_2O_4$:Mn, and rbhdl.-Al_2O_3:Cr phosphors, the sizes of the host-crystal atoms (Zn and Al) and the substitutionally located activator atoms (Mn and Cr) are within about 20 per cent of each other on an ionic basis and are probably nearer in size in the actual crystals, because the bonds are only partly ionic. The relative proportions of ionic and covalent bonding of a substitutional atom in a crystal, however, may be considerably different from those of the displaced regular host-crystal atom. There is a tendency to fill as nearly as possible the space normally occupied by the regular host-crystal atom, and this usually requires an increase or decrease in the *normal* size of the impurity atom (the *normal* sizes, shown in Table 6, are for atoms incorporated as host-crystal constituents, not as impurities). As shown in Table 6, cationic constituents, for example, Zn, Al, Mn, and Cr, *decrease* in size on becoming more ionic, whereas anionic constituents, for example, O and F, *increase* in size on becoming more ionic. In cases of substitutional location of activator atoms, the number and kind of coordinated anions is the same for the host-crystal and the activator atoms, but the bond angles (spatial orientations of the interatomic axes between ligands) are probably different because of local distortion due to the differences in normal sizes and electron configurations of the intersubstituted atoms. The ligand anions surrounding a substitutionally located activator atom experience a perturbation proportional to

TABLE 6

COMPARISON OF OCTAHEDRAL-IONIC AND TETRAHEDRAL-COVALENT CRYSTAL RADII (IN Å) OF SOME ELEMENTS ARRANGED ACCORDING TO THE PERIODIC SYSTEM

L. Pauling and M. L. Huggins [24]

IB	IIB	IIIB	IVB	VB	VIB	VIIB
	Be 1.06 Be^{2+} 0.31	B 0.88 B^{3+} 0.20	C 0.77 C^{4+} 0.15	N 0.70 N^{3-} 1.71	O 0.66 O^{2-} 1.40	F 0.64 F^- 1.36
	Mg 1.40 Mg^{2+} 0.65	Al 1.26 Al^{3+} 0.50	Si 1.17 Si^{4+} 0.41	P 1.10 P^{3-} 2.12	S 1.04 S^{2-} 1.84	Cl 0.99 Cl^- 1.81
Cu 1.35 Cu^+ 0.96	Zn 1.31 Zn^{2+} 0.74	Ga 1.26 Ga^{3+} 0.62	Ge 1.22 Ge^{4+} 0.53	As 1.18 As^{3-} 2.22	Se 1.14 Se^{2-} 1.98	Br 1.11 Br^- 1.95
Ag 1.53 Ag^+ 1.26	Cd 1.48 Cd^{2+} 0.97	In 1.44 In^{3+} 0.81	Sn 1.40 Sn^{4+} 0.71	Sb 1.36 Sb^{3-} 2.45	Te 1.32 Te^{2-} 2.21	I 1.28 I^- 2.16
Au 1.50 Au^+ 1.37	Hg 1.48 Hg^{2+} 1.10	Tl 1.47 Tl^{3+} 0.95	Pb 1.46 Pb^{4+} 0.84	Bi 1.46		

Further Crystal Radii of Ions (Based on $r_O{}^{2-} = 1.40$ Å)

IA	IIA	IIIA	IVA	VA	VIA	VIIA
Li^+ 0.60						
Na^+ 0.95						
K^+ 1.33	Ca^{2+} 0.99	Sc^{3+} 0.81	Ti^{4+} 0.68	V^{5+} 0.59	Cr^{3+} 0.64	Mn^{2+} 0.80
Rb^+ 1.48	Sr^{2+} 1.13	Y^{3+} 0.93	Zr^{4+} 0.80	Cb^{5+} 0.70	Mo^{6+} 0.62	
Cs^+ 1.69	Ba^{2+} 1.35	La^{3+} 1.15				

Also, Tl^+ 1.44, Mn^{3+} 0.62, Mn^{7+} 0.46, Fe^{2+} 0.75, Fe^{3+} 0.60, Co^{2+} 0.72, Ni^{2+} 0.70, Cr^{6+} 0.52, O^{6+} 0.09, S^{6+} 0.29, Se^{6+} 0.42, Te^{6+} 0.56, N^{5+} 0.11, P^{5+} 0.34, As^{5+} 0.47, Sb^{5+} 0.62, Bi^{5+} 0.74, Ti^{3+} 0.69, V^{3+} 0.66, and trivalent rare-earth ions 0.90 ± 0.05 Å.

the normal size and charge difference between the activator atom and the regular host-crystal atom. Because the nuclear charge Z of an anion is effectively diluted in being distributed among $Z + N$ electrons, anions are more polarizable (deformable) than cations with the same inert-gas configurations (and $Z - N$ electrons). Also, the polarizabilities α of ions increase with size according to the Born–Heisenberg relationship,

$$\alpha = A/(Z - \phi)^3 \tag{38}$$

where A is a constant for a series of ions with the same inert-gas structure, and ϕ has the values 0, 6, 13, 28, 46, corresponding to the structures of He, Ne, A, Kr, and Xe, respectively. (Incidentally, the contribution of an atom to the refraction of light by a substance is proportional to the polarizability of the atom.) [25]

Some activators, such as Mn and Cr, seem to occupy chiefly substitutional sites in their phosphor host crystals, although little is known about the kind and degree of local and long-range perturbations introduced by such foreign substitutional elements. Experimental substantiation of substitutional location is possible in cases where the host-crystal and activator substances (assuming common anions) form solid solutions which obey **Vegard's law;** that is, the structure remains unchanged, except that the average interatomic spacings vary linearly with the proportion of the solute (activator) substance.[39, 223, 224] Examples of the operation of Vegard's law are found in patterns 5 and 7 of Fig. 11a which shows representative materials chosen from the general systems hex.-ZnS:CdS and cub.-ZnS:ZnSe. These systems form solid solutions in all proportions, and their x-ray diffraction lines merely shift, without the appearance of new lines, as the proportions of the ingredients are changed. The lattice spacing increases with increasing CdS content in the first system, and the spacing increases with increasing ZnSe content in the second system. This is to be expected, because Cd is larger than Zn, and Se is larger than S (Table 6), as long as these elements intersubstitute in all proportions. In the system rbhdl.-Zn_2SiO_4:$Mn_2(SiO_4)$, the lattice expands with increasing Mn content, whereas in the system rhomb.-$CdSiO_3$:$Mn(SiO_3)$ the lattice contracts with increasing Mn content, because Mn is intermediate in size between Zn and Cd.[223, 224] The simple alterations in lattice spacings and the absence of additional lines in the x-ray diffraction patterns of these silicate phosphors, compared with the patterns of their pure host crystals, indicates that the Mn atoms substitute for the atoms of Zn or Cd in the host crystal. Therefore, when the formula for rbhdl.-Zn_2SiO_4:Mn(0.3) is written in the equivalent form rbhdl.-$Zn_2SiO_4 \cdot 0.012MnSiO_3$, it does *not* mean that the $MnSiO_3$ exists as such in the phosphor crystals, but rather as rbhdl.-Mn_2SiO_4 in solid solution (see the later discussion on deficiency structure). This re-emphasizes the fact that phosphor formulas have very limited significance with respect to the constitutions of phosphors.

For present purposes, a substitutional activator atom and its neighboring atoms will be called an **s center**. Figures 8b and 8c show two opposite types of local distortion which may exist in **s** centers. These distortions comprise a local departure from crystallinity and may (a) promote polymorphic transitions or (b) inhibit crystallization, even when the ingredients crystallize in the same space group.[272]

Phosphors formed from host-crystal and activator ingredients which crystallize in *different* space groups, for example, $Fm3m$-KCl + $Pm3m$-TlCl, and $F\bar{4}3m$-ZnS (or $C6mc$-ZnS) + $Fm3m$-Cu_2S (or $C6/mmc$-CuS) or $Pn3m$-Ag_2S (or $Fm3m$-MnS, $F\bar{4}3m$-MnS, or $C6mc$-MnS), may have their activator atoms in either substitutional or interstitial sites, depending on the radius ratios of the activator and host-crystal atoms and the degree of compactness of the host crystal. The close-packed predominantly ionic cub.-KCl:Tl phosphor apparently has its Tl activator atoms in substitutional sites, because (1) K^+ and Tl^+ are both monovalent (although there is a Tl^{3+}); (2) the ionic radii of K^+ and Tl^+ are within about 10 per cent of each other; (3) KCl and TlCl crystallize in the same point group; and (4) there is a simple correlation between the number of optical-dispersion electrons, calculated from the amplitude and half-width of the emission spectrum, and the chemically analyzable concentration of Tl.[273] The loose-packed (Fig. 10) semi-ionic [24] cub.-ZnS:Mn and hex.-ZnS:Mn phosphors apparently have their Mn activator atoms chiefly in substitutional sites, because the radii of bivalent Zn^{2+} and Mn^{2+} are nearly equal, and tristructural MnS crystallizes in the same structures as distructural ZnS.[39, 154] In this case, however, the fact that the host crystal has a very open structure, containing interstitial sites which are apparently equivalent in number and geometry to the regular crystal sites (Fig. 12), makes the demarcation between substitutional and interstitial sites less distinct. When the normal effective valencies and sizes of the activator and host-crystal atoms are quite different, and their solid compounds with common anions crystallize in different point groups, it is reasonable to expect that at least some of the activator atoms may lodge preferentially in interstitial rather than substitutional sites. The very different behaviors of ZnS-type phosphors with Cu and Ag activators, compared with phosphors with known substitutionally located activators, has led to the *postulate* that the Cu and Ag activators occupy chiefly interstitial sites in these phosphors.[34] Experimental verification of this postulate is made difficult by the very small optimum proportion of Cu or Ag (about 0.01 per cent, or about 1 per cent of the optimum proportion of substitutionally located Mn). Despite the lack of direct proof of activators in interstitial sites, except for some x-ray-diffraction evidence in the case of Cu-deficient semiconducting Cu_2Se,[274a] there are plausible reasons for assuming their existence, and an interstitial activator atom and its neighboring atoms may be designated as an **i center**.

As shown in Fig. 12, the lattice structure of cub.-ZnS may be visualized as two interpenetrating three-dimensional networks of edge-joined tetrahedra of atoms of (1) Zn, and (2) S, such that *ideally* every other Zn tetrahedron contains

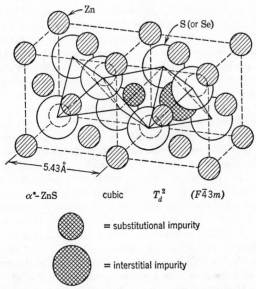

Zn

S (or Se)

5.43Å

α^*-ZnS   cubic T_d^2 $(F\bar{4}3m)$

⊞ = substitutional impurity

⊞ = interstitial impurity

FIG. 12. Sketch of two unit cells of the lattice of cub.-ZnS. The sizes and shapes of the Zn and S atoms (ions) depend on the unknown degree of homopolar and ionic binding (also, on the state of excitation of the atoms—see Table 23).

one S atom, and every other S tetrahedron contains one Zn atom. A similar alternating arrangement exists in hex.-ZnS, whose structure may be derived from the cubic form by a simple gliding (Fig. 5) or screw-rotation operation.[274b] These loose-packed low-coordinated structures are characteristic of many semi-covalent + semi-ionic phosphor host crystals (see Fig. 11b, where the close-packed predominantly ionic cub.-KCl is an exception). The interstices may harbor atoms dislodged from the host crystal or introduced as impurities. In Fig. 12, the small cross-hatched sphere represents an idealized impurity atom in a substitutional site, whereas the larger cross-hatched sphere represents an idealized impurity atom in an interstitial site (without indicating the inevitable distortion in either case). The structural similarity of the indicated interstitial and substitutional sites makes them geometrically equivalent, *but not crystallographically equivalent*, in crystalline ZnS. Statistical fluctuations in structure, especially during crystallization and cooling, result in some of the regularly located Zn atoms being lodged in interstitial sites in the cooled crystal (eqs. 19–21), leaving omission defects in the regular structure. These local dislocations upset the degree of perfection (increase the entropy) of the crystal without greatly changing its total energy, in the case of crystalline ZnS. When foreign atoms are introduced, however, an additional mixing entropy is involved, and the total energy of the crystal is increased by the "unnatural" perturbations in either substitutional or interstitial sites. It may be noted that the degree of local asymmetric distortion (Fig. 8d) which may be produced by an interstitial cation (for example, Cu^{2+}) in the ZnS structure should decrease to a minimum as the anion (for example, S^{2-}) necessary to balance the cationic charge is brought nearer until it occupies an adjoining interstitial site. Also, it is conceivable that in some cases an impurity atom may dislodge a regular host-crystal atom which may remain in an interstitial site adjacent the impurity atom. The varieties of structural irregularity which may be produced by interstitial imperfections appear to be considerably greater than those obtainable by simple substitution (without interstitial displacement of the host-crystal atom).

Considering next a predominantly ionic phosphor, for example, cub.-$KCl:Sb(Cl_3)$, the $SbCl_3$ molecules apparently wander at least part way into the KCl crystals by thermochemical diffusion, which is accelerated by the presence of oxygen or a film of adsorbed moisture, and is aided by the tendency for alkali halides to recrystallize even at room temperature (especially under pressure).[275-277] The alkali halides and antimony chlorides react to form compounds of the type K_2SbCl_6, which contain the antimony as Sb^{3+} and Sb^{5+} rather than the indicated Sb^{4+}.[278] According to Schleede, when $Pm3m$-CsCl is ground with $SbCl_3$ at room temperature no luminescence is obtained, but if the mixture be heated above about 400°C the CsCl host crystal assumes the $Fm3m$ structure (the same as for KCl) and luminesces with a yellow–green color just after rapid cooling, *even though the structure of the cooled host-crystal material reverts to the original Pm3m.* On standing, or on heating to about 250°C, the luminescent material reverts to the nonluminescent state without detectable change in structure as determined by x-ray diffraction.[65] From this it appears that the metastable $Fm3m$ structure is temporarily maintained in the neighborhood of some of the Sb-activator centers on cooling the material, and that these structurally metastable centers are gradually transformed into the general $Pm3m$ structure of the host crystal, whereupon the centers become nonluminescent under UV excitation. Although the locations of the Sb-activator atoms in the CsCl crystals have not been determined, these data clearly show that *the crystalline state and environment of an activator center are of major importance in determining its luminescence characteristics.*

Further examples of the influence of the structure and cation valency of the host crystal on the effective valence state of foreign elements have been reported by Selwood who finds that (1) when low concentrations of manganous nitrate $[Mn(NO_3)_2]$ are decomposed on gamma alumina (cub.-Al_2O_3), the resultant manganese oxide contains mainly cub.-Mn_2O_3, with Mn^{3+}, whereas at higher concentrations of $Mn(NO_3)_2$ more of the normal tetr.-MnO_2, with Mn^{4+}, is produced; (2) when the decomposition is carried out on rutile (tetr.-TiO_2), only tetr.-MnO_2 is obtained, regardless of concentration; and (3) when nickel oxide is formed on cub.-MgO only cub.-NiO, with Ni^{2+}, is formed, whereas the nickel oxide formed on cub.-Al_2O_3 contains some Ni^{3+}.[22] With such a strong influence of the substrate on the deposited substance, it is reasonable to assume that *in* phosphors the effective valence states and degree of ionic and covalent bonding of the activator atoms are strongly influenced by the composition and crystal structure of the host crystal, and by the location of the activator atoms if structurally different sites may be occupied in the host crystal.

In the relatively ionic alkaline-earth-sulphide-type phosphors (for example, Nos. 20–23 in Table 5), the host crystals are predominantly cubic (*Fm3m*), but the structures are complicated by considerable proportions of residual non-isostructural sulphate and fluoride fluxes (Fig. 11*a*). The residual fluxes may allow the formation of centers having structures which are quite different from the bulk of the host crystal. Present sparse information on the properties of these phosphors points to chiefly i centers as the active agents in luminescence, but the great complexity of these phosphors discourages further comment at this time. It is ironic that the alkaline-earth-sulphide phosphors were the first to be reported as having been made artificially (*circa* 1603 [145]), because they seem to be among the most difficult phosphors to prepare, use, and interpret. Extensive studies of these complex phosphors led Lenard to postulate that activator centers were localized colloid-like complexes, varying in size and composition within a given crystal, with different centers having different intrinsic durations of afterglow (*Zentren verschiedener Dauer*).[145, 267, 269] This general concept is close to the truth insofar as the *sizes* of centers are concerned. With respect to their *structure*, however, "quasicrystalline" should be substituted for "colloid-like" because it is improbable that there is a very small completely amorphous region bound in a highly regular crystal matrix. Also, the efficiency of energy transfer from the regular host crystal into the centers should decrease with increasing size and amorphicity of the center, and so phosphors with large completely amorphous centers should be practically nonluminescent under excitation by charged material particles which give up their energy chiefly to the greater proportion of host-crystal atoms. The fact that many phosphors do have fairly high luminescence efficiency ($\cong$ 10 per cent) under excitation by charged material particles, indicates that in at least those cases there is appreciable structural continuity between the regular host crystal and the activator center.

For our purposes, an impurity center may be pictured as a small region wherein there is a contest for structural dominance between the many united host-crystal atoms and the single impurity atom (or distinct group of atoms). An impurity atom strives to rearrange the host-crystal atoms about it to produce an appropriate configuration for its size, charge, bonding (directional) characteristics, and deformability. The ambient host-crystal atoms, on the other hand, strive to maintain their unperturbed configurations [groupings, orientations (symmetries), and spacings]. Thermal agitation, thermal expansion, and state of excitation may greatly affect this structural contest, especially when the impurity and host-crystal atoms are very dissimilar. For example, it has been shown that an Sb impurity atom in cub.-CsCl:Sb can coerce its immediate host-crystal-atom neighbors to remain in a configuration associated with a face-centered-cubic array, even when the vast majority of the host-crystal atoms revert to a body-centered-cubic array as the crystal is cooled from 700° to 300°K. Evidence obtained in these laboratories indicates that a Ti impurity atom in rbhdl.-Zn_2SiO_4:[Si] may act chiefly to perturb the SiO_4 tetrahedra at 300°K, whereas at 80°K the Ti atom may succeed in drawing about it 6 O atoms (as in titanate crystals) only to have these 6 O atoms wrested from it when the host-crystal structure prevails again on warming to 300°K. Among the wide variety of different combinations and proportions of host crystals and impurities, it is probable that some centers may change their configurations several times as the temperature (or intensity of excitation) is increased or decreased. Depending on the material and the conditions of operation, a transition from one configuration to another may be gradual or abrupt; also, a transition may be either reversible or irreversible within a given temperature range.

In a few cases, it is possible to associate luminescence with an intrinsic transition of an ion or radical.[65,75,116,144] For example, many of the rare-earth ions, Sm^{3+}, Eu^{3+}, Gd^{3+}, Tb^{3+}, and Dy^{3+}, exhibit similar line spectra of luminescence when incorporated in gases, liquids, and solids. Also, the uranyl radical (UO^{2+}) exhibits a distinctive molecular band spectrum of luminescence in liquid solution and in solids. In these cases, and in the cases of certain presumably pure organic crystals, such as naphthalene, it might be argued that the activator concept is unnecessary, because the active center (ion, distinct simple molecule, or radical) luminesces almost independently of its surroundings. The luminescence is not completely independent of the surroundings, however, because the efficiencies of the luminescent centers vary considerably in different environments, and spectral displacements are caused by environment even in the rare-earth emission spectra (see Fig. 13). Also, there are cases, such as the platinocyanide radical, $Pt(CN)^{2-}$, where the luminescence efficiency is greatly improved when the luminescent radical is in a solid rather than a liquid; and there are other cases, such as the Cu^+ and Mn^{2+} activator ions, which are nonluminescent in liquid solution or in their own pure compounds in solid form (at room temperature [154]), where appreciable luminescence efficiency is obtained only when the luminescence centers are suspended in a crystalline matrix of another material (that is, a host crystal). The well separated centers in the host crystal become more capable of transforming excitation energy into luminescence photons instead of heat [note, again, that compounds of most activator elements absorb strongly in the visible and near-visible region of the spectrum, for example, CuS and Cu_2S (black), Ag_2S (black), MnS (black), MnO (green), MnO_2 (black), PbO (yellow), BiS (gray), Ce_2O_3 (gray–green), and Cr_2O_3 (green)]. It is at least convenient to refer broadly to luminescence-active atoms, ions, radicals, or distinct molecules as **centers**, or as **activator centers** (or just **activators**), especially in *real* solids where the activator concept serves as a constant reminder of the special imperfections which seem to be essential for efficient luminescence, at least at operating temperature near or above room temperature. There is a tendency to cling to the activator concept, whether based on an impurity or on an imperfection in general, because it has utility in circumventing the selection rules which drastically restrict the possible radiative transitions in *ideal* crystals.

In contrast with the foregoing paragraph, the very existence of an activator has not been proved in any of the so-called "self-activated" phosphors, such as cub.-ZnS:[Zn], hex.-ZnS:[Zn], cub.-ZnSe:[Zn], hex.-CdS:[Cd], hex.-ZnO:[Zn], rbhdl.-Zn_2SiO_4:[Si], rbhdl.-Al_2O_3:[Al], and monocl.-Mg_2WO_5:[W], where the **brackets** indicate *doubt* as to the presence of the *presumed* activator. Most of these pure-substance phosphors are about as efficient as the best phosphors made from the same host crystals with consciously added activators. In the ZnS-type phosphors, it is sometimes assumed that an infinitesimal trace of [Zn] or [Cd] activator is left behind when the more volatile sulphur leaves at a faster rate during partial decomposition at the high temperatures of crystallization.

It is reported that the thermal dissociation constant K for the reaction $2ZnS \rightleftarrows 2Zn + S_2$ at temperatures from 800 to 1200°C is given by $\log_{10} K = \log_{10} P_{Zn}{}^2 P_{S_2} = 21.012 - 40{,}585T^{-1}$, where P_{Zn} and P_{S_2} are the equilibrium pressures of Zn and S_2 measured in atmospheres and T is expressed in °K (the corresponding expression for $2CdS \rightleftarrows 2Cd + S_2$ is $\log_{10} K = \log_{10} P_{Cd}{}^2 P_{S_2} = 20.486 - 33{,}970T^{-1}$).[661]

The residual [Zn], or [Cd], is then accredited with producing the intense visible, or red + infrared, emission band of these phosphors when excited at room temperature, but the same phosphors have emission bands in the ultraviolet, or green, region (a narrow group of bands adjacent the long-wave edge of the host-crystal absorption band) which appear during excitation at very low temperatures or during *intense* excitation at room temperature. Unless two types of [Zn] centers, for example, be postulated, or each [Zn] center has two greatly different characteristic frequencies of emission, one of these emissions (presumably the UV band) is left to be accounted for as a radiative transition by the unactivated host crystal. A possible explanation, without postulating two kinds of [Zn] centers (or even a stoichiometric excess of Zn), is that some of the regular Zn atoms may be dislocated into interstitial sites, where they may be designated as [Zn] *impurities*; then, (1) the [Zn] atoms may form the centers producing one emission band, and (2) a neighboring S atom (either in a normal or interstitial site) perturbed by the interstitial [Zn] may be the emitter producing another emission band. The possibilities of anions as emitters are manifest in certain tungstate-type phosphors where the shape and position of the emission spectrum are practically invariant whether Mg, Cd, or Zn be used as the cation. This makes plausible the assumption that the WO_4 radicals, or WO_4 radicals perturbed by excess or interstitial cations or [W], play the role of an activator.[58, 61, 155] Furthermore, the practically identical emission spectra of many oxygen-dominated crystals, without added activator, for example, hex.-ZnO:[Zn] (visible band) and monocl.-Mg_2WO_5:[W], indicate that their common O atoms or O-tetrahedra may be the chief sources of the photon emissions during luminescence.

One of the disturbing features of the concept of "self-activation" is that many of the postulated excess or interstitial cations, such as [Zn] (but not [W]), have only one known ionic valence (for example, $Zn \rightarrow Zn^{++}$ only, although there is some polarographic evidence for the existence of the univalent ion Zn^+ [666]). This is in contrast to the fact that consciously added useful activators are definitely and invariably multivalent (for example, $Cu \rightarrow Cu^+$ or Cu^{2+}, $Ag \rightarrow Ag^+$ or Ag^{2+}, and $Mn \rightarrow Mn^{2+}$ or Mn^{3+} etc.; compare Chapter 3 under Devising New Phosphors). The conventional ionic concept of valence should be modified in cases, such as [Zn], where there is a strong tendency to form partially covalent rather than completely ionic bonds, and the number of possible covalent bonds should be given weight in assessing the valence possibilities of an element.[12, 24] In the absence of specific information about the presence or absence of activators in "self-activated" phosphors, it is a matter of individual preference whether to denote such phosphors with or without presumed [activators]. The procedure followed here is to retain the activator concept, using the brackets as a reminder of the uncertainty of their existence.

There are considerably more data and indirect reasoning which could be presented regarding the microconstitutions of phosphors, but most of it is controversial. For example, when ZnS is crystallized at about 1100°C to form a blue-emitting phosphor and is then reheated with chlorides of Cu, Ag, and Mn, the characteristic luminescences of these activators are reported to appear in the following order under UV excitation: Cu > 330°C, Ag > 400°C, and Mn > 700°C. That the activators penetrate into the *volumes* of the ZnS crystals, even at these low temperatures, is shown by dissolving the final products in dilute hydrochloric acid and noting that the characteristic luminescence emission color produced by the activator (for example, green in the case of Cu) persists even when the crystals have been so reduced in size that they make only a slight turbidity in the acid solution.[102] Tiede and Riehl argue that the Ag and Cu atoms diffuse through and occupy the vacant interstitial sites at temperatures below that required for place exchange (*Platzwechsel*) with the regular host-crystal atoms, whereas the Mn atoms diffuse through the crystal only above the place-exchange temperature and occupy substitutional sites. Schleede, on the other hand, argues for more consideration of the role of crevices and faults in diffusion processes, and points out that the melting points of the activator chlorides (CuCl = 430°C, AgCl = 450°C, and $MnCl_2$ = 650°C) are close enough to the above phosphor-formation temperatures to vitiate possibly the postulated place-exchange idea.[65]

A place-exchange temperature T_x is a matter of degree, because the probability **P** per unit time of displacing a regularly located atom increases exponentially with increasing temperature, that is

$$\mathbf{P} = \nu_a \epsilon^{-E^*/kT} \tag{39}$$

where ν_a is the vibration frequency of the atom in its normal (equilibrium) state ($\nu_a \approx 10^{12}$ sec^{-1}), and E^* is the activation energy, that is, the height of the potential barrier between the original and the subsequent (displaced) equilibrium positions of the atom in the crystal.[29, 35, 36] According to eq. 39, the probability per unit time of place exchange between atoms is small but finite up to $T_x \approx E^*/30k$ and then increases rapidly at increasingly higher temperatures. Also, the probability of place exchange in a given period of time is proportional to the time, and so appreciable place exchange may be effected at temperatures below $T_x \approx E^*/30k$ by using long times. (Figure 90 shows that Mn does not replace Zn appreciably in precrystallized rbhdl.-Zn_2SiO_4 unless the host-crystal + activator-compound mixture is heated to above 700°C for several hours).

In addition to the influence of temperature on the diffusion of activators into crystals, there is the influence of crystal anisotropy which promotes different rates of diffusion in different crystallographic directions in nonisotropic crystals, and different degrees of adsorption and binding on different crystal faces. Particularly striking examples of the latter effect have been found in some luminescent single crystals of natural gypsum [monocl.$(C2/m)$-$CaSO_4 \cdot 2H_2O$ + unidentified activator] which appear as ◧◨ , where the light areas luminesce and the dark areas do not luminesce under excitation by UV.[102,145] Similar selective-inclusion effects have been produced by incorporating certain dyes in artificial water-containing crystals, for example, monoclinic lead acetate crystals grown from fluorescein-containing solutions appear as ◧◨ , and triclinic barium acetate crystals grown from fluorescein-containing solutions appear as ◨◧ during excitation.[146] These results provoke the question as to whether the distribution of activator atoms in phosphor crystals is purely statistical, or whether it is a more regular arrangement, perhaps in the pattern of a tenuous superlattice permeating the host crystal. It is known from microscopic examinations of phosphors that inhomogeneities of the ingredient mix, prior to solid-state reaction, often produce variations in composition of different crystals of the same batch, although the *individual* crystals may *appear* to have uniform luminescence efficiency. Also, it is known that the rate of inclusion of activator compounds, such as $MnCl_2$ during the growth of cub.-NaCl:Mn from aqueous solution,[226-228] is proportional to the concentration of the activator compound in the solution, and so there is a diminishing concentration of activator in the outer reaches of the host crystal unless the activator compound is maintained at a constant concentration in the solution. These gross observations are suggestive, but the atomic-scale distributions of activators, as well as their specific locations, coordinations, bond types, and magnetic moments, remain to be determined.

The obtention and interpretation of pertinent experimental information regarding the microconstitutions of phosphors are fundamental problems in research on the luminescence of solids. These problems are admittedly complex and as such may be more suited to the training and average temperament of chemists (who are required to develop prodigious memories), but their solutions require the cooperation of chemists, crystallographers, physicists, and spectroscopists. Anomalies, such as "self-activation" and the apparent equivalence of Cu and O in producing the long-persistent green luminescence of hex.-ZnS:Cu (or hex.-ZnS:[O]),[58,63] must be resolved by thorough and careful chemical preparations coupled with accurate determinations of composition, crystal structure, spectra, dia- or paramagnetism, and other physical properties.

It has been reported that as little as 1 g of cupric ion (Cu^{2+}) in 10^9 liters of solution accelerates the oxidation of Na_2SO_3 to Na_2SO_4;[25] hence, it is possible that there is a catalytic effect of copper activator in promoting oxidation of sulphide-type phosphors. This would account for the apparent equivalence of copper and oxygen for activation purposes and perhaps make the oxygen the real activator.

Energy States and Energy Levels of Phosphors

The possible energy states and energy levels of solids are of fundamental interest in the study of phosphors because they *designate* (1) the energies which the system can absorb from primary excitants, (2) the energy transformations and transmissions which can take place in and on the phosphor crystals, and (3) the energies of photons and other particles which the system can emit. The compositions and structures (that is, the **constitutions**) of solids, of course, *determine* their energy levels.

Hypothetically, one might start with isolated atoms having discrete energy levels (Fig. 2), calculate the bandlike energy levels they would have when combined as isolated simple molecules (Fig. 3), and then calculate the more complex energy levels and bands which they would have in (1) *ideal* solids (Fig. 4), and (2) *real* solids (Fig. 9). This has been done for a few idealized chemically simple solids, such as diamond and some of the alkali halides,[35-37] and some attempts have been made to calculate the energy levels introduced into ideal solids by impurities. It has just been shown, however, that much is lacking in our knowledge of the microconstitutions of *real* solids, and it is these unknown microconstitutions which determine to a large degree the energy levels (and, hence, the luminescences) of solids. *Practically*, therefore, one determines rather empirically which atoms will combine, then arranges for the reaction and crystallization of large numbers of these combinable atoms, and attempts to diagnose the energy levels of the resultant solids. The diagnoses are hindered by experimental difficulties, such as the general unavailability of large phosphor crystals and strong continuous-spectrum sources of UV and x-ray radiation, and by the overlapping of complex host-crystal energy bands with each other, as well as the overlapping of the energy levels of host crystals and their imperfections. Under these circumstances, it has become customary to use bits of experimental information to devise fragmentary *ad hoc* energy-level diagrams with the hope that they may be expanded and revised to become of general utility. At present, it appears necessary to use at least three types of diagrams to represent the energy and momentum transformations and transductions in solids, for example, (1) potential-energy versus configurational-coordinate diagrams, such as Fig. 3, which indicate the potential energy of the system as a function of generalized interatomic distance; (2) total-energy versus distance diagrams, such as Fig. 9, which indicate the variation of the total energy of the system on moving an electron along a one-dimensional row of atoms in the solid; and (3) momentum diagrams, which should indicate the local changes in momenta of electrons and crystal atoms, but convenient forms of these latter diagrams are yet to be invented. Even these various diagrams are hardly adequate to depict the energy transformations, transductions, and transmissions in a finite *real* crystal with its local variations in composition, structure, bond types and directivities, atomic polarizations, electrostatic charges, magnetic moments, and surface discontinuities. In any event, it is difficult to indicate the *dynamic* energy transitions in solids by means of *static* diagrams. A good deal of active imagination is and probably always will be necessary to take into account temporary changes in atomic configurations, atomic motions, interatomic spacings, spatial distributions and densities of electron "clouds" (including polarizations), spin- and spin-orbit interactions, and binding energies. Nonetheless, even two-dimensional energy-state diagrams have been useful in partly simplifying the task of visualizing the electronic transitions which give rise to observed spectral lines and bands of absorption and luminescence emission, and in helping to picture some of the possible processes of transmission and temporary storage of energy in three-dimensional solids.

The energy levels of solids may be considered as divided into five general categories for the discussion of luminescence. As shown in Table 7 (and Appendix 4), transitions between these levels range from transitions in atomic nuclei, which produce gamma-ray photons with energies from 10^4 to over 10^6 ev, to transitions between various states of atomic motions in solids, which produce mainly low-energy photons with energies mostly below 1 ev when the solid is maintained near room temperature (as is the case during the luminescences of most solids). Gamma-ray photons are emitted when atomic nuclei make radiative transitions from excited states to states of lower energy,[5-11] and x-ray photons are emitted when an extranuclear electron from an outlying shell makes a radiative transition to an inner shell from which an electron has been ejected (for example, by absorption of energy from a high-velocity charged material particle).[5,91,92] Both these emissions are determined primarily by the nuclear mass and charge of the emitting atom, and neither emission is particularly sensitive to the environment of the emitting nucleus or atom in the solid. This is so because at least the terminal (lower, ground-state) energy levels of these high-energy-difference radiative transitions are well shielded by outer "shells" from external fields. The visible and near-visible line spectra and narrow-band spectra produced by electronic transitions between the discrete energy levels corresponding to different electron configurations in in-completed shells are somewhat affected by the environment of the

TABLE 7

GENERAL CLASSIFICATIONS OF THE EMISSIONS FROM LUMINESCENT SOLIDS

Excluding Deceleration Radiation

	Type of Emission	Radiation Sources	Common Type of Spectra
Energy of the emitted photon →	Gamma-ray fluorescence	1. Transitions of nucleons in atomic nuclei	Gamma-ray line spectra
	X-ray fluorescence	2. Electronic transitions in inner completed shells of atoms	X-ray line spectra
	Conventional luminescence	3. Electronic transitions in inner incompleted shells of atoms 4. Electronic transitions in outer (valence) shells of atoms and molecules	Visible and near-visible line spectra Visible and near-visible band spectra
	Thermal radiation	5. Transitions of atoms and groups of atoms vibrating and rotating; also electrons as in 3 and 4	Infrared band spectra

atom,[65, 68, 144, 193–196] and the band spectra produced by electronic transitions of valence electrons in outer shells are strongly influenced by the environment of the emitting atom (compare the different emissions obtained from the same activator, for example, Mn, in different host crystals in Table 5). *These latter two emissions (from 3 and 4 in Table 7) comprise the conventional luminescence which is our chief topic.* The low-energy infrared photons are produced chiefly by atomic motions (thermal agitation) in phosphor crystals. These atomic motions produce heat, and provide conduction losses and thermal radiation which are the chief competitors of luminescence radiation as a means of disposing of excitation energy.

In practice, most of the heat incidental to luminescence in solids is dissipated by conduction from excited portions of phosphor crystals to unexcited portions of the crystals and to neighboring crystals and substrates which are at lower temperatures. For a hypothetical *completely isolated* luminescing crystal the absorbed energy would be disposed of as luminescence and thermal radiation. If a phosphor crystal, in equilibrium initially at $0°K$, were to be excited to luminescence for a given time while suspended immovably in an evacuated chamber whose walls were maintained at $0°K$, then the relative proportions of (1) the total excitation energy E_A absorbed by the crystal, and (2) the total energy radiated as luminescence radiation E_L and thermal radiation E_H would be a direct measure of the **luminescence efficiency** $\mathcal{E}$ of the phosphor, according to

$$\mathcal{E} = 100 E_L/E_A = 100 E_L/(E_L + E_H) \quad \text{in per cent} \quad (40)$$

This relationship assumes that no energy is expended in producing electron emission, or chemical or structural changes and that the crystal has returned to $0°K$ at the end of the measurement interval, that is, $E_A = E_L + E_H$. If another initial temperature T were to be chosen for the foregoing experiment, then the thermal radiation $E_{H(T)}$, normally emitted by the solid *in equilibrium at the temperature T*, would have to be subtracted from the measured total thermal radiation E'_H, and so eq. 40 would become

$$\mathcal{E} = 100 E_L/E_A = 100 E_L/[E_L + E'_H - E_{H(T)}] \quad (41)$$

where E_A does not include the equilibrium thermal radiation, $E_{H(T)}$, absorbed by the crystal during the measurement interval. The distinctions between luminescence and thermal radiation are described in the next chapter.

DISCRETE ENERGY LEVELS. In commencing a discussion of the energy levels involved in (conventional) luminescence, we may first enquire briefly into the electronic configurations of some of the elements which produce line- or narrow-band emission spectra when they are incorporated as activators in phosphor host crystals. A number of such elements and their neighbors are tabulated in Table 8, using data from recent sources.[5, 22, 68-70] Typical activator elements which produce sharp-line emission spectra in certain crystals are Cr (No. 44 in Table 5) and Sm (Fig. 13).[144] Both these atoms have unfilled inner shells, that is, the $3d$ shell in Cr, and the $4f$ and $5d$ shells in Sm. In Sm, the transitions which give rise to such sharp emission lines are believed to be semi-"forbidden" transitions between states having the same $4f^N$ configurations and the same n values of the individual electrons, but different resultant L and S quantum numbers as a result of changes in electron spins and redistribution of electrons among the different $4f$ levels. The transitions are forbidden by the selection rules for dipole emission of the free atoms, but the first-order selection rules are apparently circumvented sufficiently to allow the observed radiative transitions when the atoms are placed in the fields inside a crystal. (Only rare earths with $4f^2$ to $4f^{12}$ configurations show sharp lines; the elements Ce $(4f^1)$ and Yb $(4f^{13})$ emit diffuse bands in the ultraviolet.[68, 193, 198-200]) The crystal

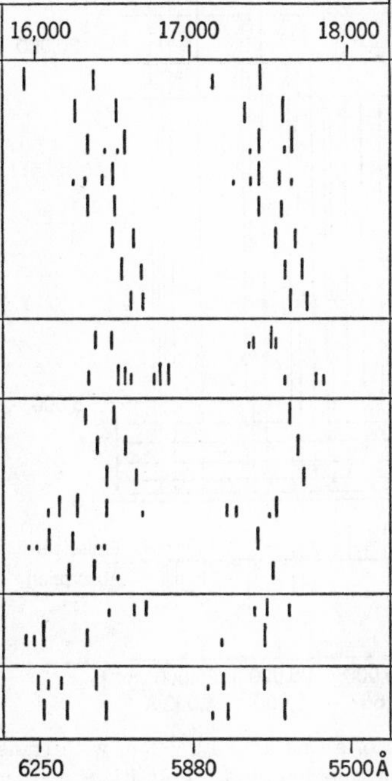

	Wave Number in cm^{-1}			Edge of Unit Cell, Å	Me-X Spacing, Å	Space Group
cm^{-1}	16,000	17,000	18,000			
cub.-MgO					4.21	2.10	$Fm3m$
cub.-CaO					4.80	2.38	$Fm3m$
cub.-SrO					5.15	2.59	$Fm3m$
cub.-BaO					5.50	2.77	$Fm3m$
cub.-MgS					5.19	2.52	$Fm3m$
cub.-CaS					5.69	2.80	$Fm3m$
cub.-SrS					6.01	3.00	$Fm3m$
cub.-BaS					6.37	3.18	$Fm3m$
cub.>ZnS hex.-					5.42 3.82-6.28	2.35	$F\bar{4}3m$ $C6mc$
hex.-BeO					2.69-4.37	1.64	$C6mc$
cub.-CaF$_2$					5.45	2.36	$Fm3m$
cub.-SrF$_2$					5.78	2.50	$Fm3m$
cub.-BaF$_2$					6.19	2.68	$Fm3m$
cub.-ZrO$_2$					5.10	2.21	$Fm3m$
cub.-CeO$_2$					5.40	2.34	$Fm3m$
cub.-ThO$_2$					5.57	2.42	$Fm3m$
tetr.-MgF$_2$					3.08-4.62	1.99	$P4/mnm$
tetr.-TiO$_2$					2.95-4.58	1.97	$P4/mnm$
rbhdl.-Al$_2$O$_3$					5.15*	1.84	$R\bar{3}c$
rbhdl.-Ga$_2$O$_3$					5.28*	1.99	$R\bar{3}c$

6250   5880   5500Å

FIG. 13. Main lines in the emission spectra of various host crystals with samarium activator. The wave number $\bar{\nu}$ is the number of waves per centimeter, that is, $\bar{\nu} = 1/\lambda = \nu/c$. (R. Tomaschek [65,144])

field also produces a Stark-effect splitting (degeneracy removal) of some of the energy levels, to an extent which depends on the J values of the levels and on the symmetry of the crystal. For example, the absorption transition of Eu^{3+} between an excited state with $J = 0$ and a lower state with $J = 2$ (Fig. 14) produces only 2 lines when the Eu^{3+} is in a high-symmetry field ($m3m$ symmetry in cubic EuF$_3$:BiF$_3$ mixed crystals) and 5 lines when the Eu^{3+} is in a low-symmetry field (mm symmetry in rhombic EuF$_3$). According to a group-theoretical analysis by Bethe, the intermediate-symmetry hexagonal or rhombohedral point

groups would produce 3 absorption lines, in this case, and this is confirmed experimentally.[144] It is interesting to note that the number of
lines predicted by the theory of degeneracy removal by crystal fields is
too few in the cases of Sm^{3+} and Eu^{3+} (Eu^{2+} gives diffuse bands [276])

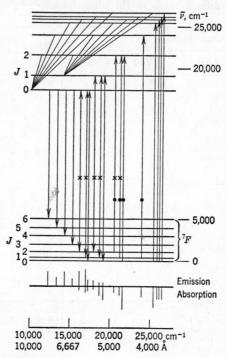

FIG. 14. Energy-level diagram of an isolated Eu^{3+} ion, with transitions correlated
with some major absorption and emission lines. Transitions marked $\times$ are strongly
temperature-dependent in absorption. Transitions marked $\cdot$ occur also as emissions
when the Eu^{3+} is incorporated in phosphors such as cub.-CaO:Eu. Diagonal lines
indicate transitions which occur during excitation of luminescence. (R. Tomaschek,
O. Deutschbein [65,144])

which are outstanding activators for infrared-stimulable phosphors (Nos.
20 and 21 in Table 5; note *band* emission spectra!). Contrarily, the
theory predicts too many lines for Pr^{3+}, Nd^{3+}, Er^{3+}, and Dy^{3+} (see
No. 46 in Table 5).[68]

With respect to the effect of crystal symmetry and composition on
the emission spectra of solids with rare-earth-ion activators, it should be
noted in Fig. 13 that the emission spectra of cub.-MgO:Sm and cub.-
CaF_2:Sm are different (this is true also when Eu activator is used instead

of Sm [65]) even though both host crystals crystallize in space group *Fm3m*. Also, it is reported by Deutschbein that the emission spectrum of Eu^{3+} in large uniaxial crystals of hex.-$Eu(BrO_3)_3 \cdot 9H_2O$ or hex.-Eu-ethyl-sulphate $\cdot 9H_2O$ is different when the crystals are viewed in different crystallographic directions, in that certain lines which are observed along one axis are not observed along another crystallographic axis.[65] The ordinary ray (electric vector perpendicular to the optic axis) viewed along the *c* axis contains only lines corresponding to $\Delta J = 2$, whereas the extraordinary ray (electric vector parallel to the optic axis) contains lines corresponding to $\Delta J = 1$. Taking the directions of polarization into consideration, the $\Delta J = 1$ transitions are tentatively ascribed to magnetic dipole radiation, and the $\Delta J = 2$ transitions are ascribed to electric dipole radiations. Although measurements are not available, it is to be expected that the luminescence efficiencies of anisotropic phosphor crystals should also differ to some degree along the different crystallographic axes, the anisotropy influencing both the absorption of primary excitation energy and the emission of luminescence radiation.

Examination of Fig. 13 shows that there are regular displacements of the emission lines of Sm in host-crystal sequences such as MgS-CaS-SrS-BaS. The displacements occur toward higher **wave numbers** (that is, shorter wavelengths) as the bonding energy decreases, and as the inter-atomic spacing (Me–X) and atomic number of the anion or cation increase. Closer examination of photographs of the actual spectra usually discloses many weaker satellite lines around the strong sharp lines, and these satellite lines have been correlated with atomic vibra-tions in the crystals. A heavier host-crystal element substituted in the same structure displaces the satellite lines toward longer wavelengths relative to the strong (parent) lines. This occurs as a result of the lower-ing of the crystal vibration frequency by the heavier element. The satellite lines often appear as relatively diffuse bands which become less diffuse as the temperature of the solid is lowered or the degree of crystal-linity of the solid is increased, or as the perturbation produced on the host crystal by the activator is decreased. As might be expected, the sharp emission lines produced by a rare-earth activator, such as Eu in a host crystal of $SrO \cdot B_2O_3$, become more diffuse and bandlike as the host crystal becomes more vitreous.[65, 144] In the emission spectrum of rbhdl.-Al_2O_3:Cr, there are three moderately strong diffuse bands, and one weak diffuse band, on the *short-wave* side of the two most intense lines near 6927 and 6947 Å (these diffuse bands are in addition to similar but weaker diffuse bands on the long-wave side of the sharp emission lines).[195] The short-wave diffuse emission bands are absent at 83°K but are prominent at 300 and 470°K. These satellite bands are believed to be due to superposition of atomic vibrations (between Cr and O atoms) and the electronic transitions causing the sharp-line radiations. The electronic transitions occur by effecting different groupings of electrons in the unfilled $3d$ levels of Cr.

Without going into details, it may be mentioned that the linelike emission spectrum of rbhdl.-Al_2O_3:Cr, and of Cr activator in a number of other oxygen-dominated host crystals, has been ascribed also to forbidden transitions, in this case in the Cr^{3+} ion which becomes luminescence-active only when surrounded by *four* oxygen atoms, as in $R\bar{3}c$-Ga_2O_3:Cr and $Fd3m$-Mg_2TiO_4:Cr.[194-196, 285] The chief emission band of cub.-Mg_2TiO_4:Cr extends from 7050 to 7600 Å with three main fine-structure peaks near 7080, 7140, and 7380 Å. Kröger reports a similarly shaped emission band extending from 6440 to 6830 Å for cub.-Mg_2TiO_4:Mn prepared in an oxidizing atmosphere and measured at 93°K.[154, 156] Similar red emission bands with fine structure are obtained from Mn incorporated in cub.-$ZnAl_2O_4$, cub.-$SrAl_2O_4$, cub.-$MgAl_2O_4$, and rbhdl.-Al_2O_3 (see also Mg_4GeO_6:Mn, No. 37 in Table 5), but the emission bands of these phosphors occur in the green region of the spectrum when the crystallizations are done in a reducing atmosphere. Kröger reasons from these data that Mn^{2+} ($3d^5$, $^6S_{5/2}$) is responsible for the green emission bands (without fine structure), and that Mn^{4+} ($3d^3$, $^4F_{3/2}$), which is isoelectronic with Cr^{3+} ($3d^3$, $^4F_{3/2}$—Table 8), is responsible for the red emission bands (with fine structure). If this be true, then the radiative transition in Mn^{2+} must occur between different energy levels from those in Mn^{4+}. This is so because eq. 7 shows that on going from Mn^{2+} to Mn^{4+} the increased effective nuclear charge per electron causes an increase in a given characteristic frequency, instead of the apparent decrease that is observed here. It is possible that the Mn ions may go into structurally different sites with different coordination numbers which could help produce the observed changes in effective valence.[243]

As additional information about phosphors emitting line spectra, it is reported that all diamonds excited at $-180°C$ emit a sharp line at 4156 Å and that this line decreases in sharpness and intensity as the temperature is raised until at 350°C it is no longer detectable.[55] The activator in this case has not been identified. Some diamonds at room temperature have blue luminescence emission with good intensity under excitation by 3650-Å UV, whereas others which have equal or superior gem quality do not give perceptible luminescence at room temperature.

From this brief account of some of the phosphors which give sharp-line spectra or bands with fine structure, it appears that the sharp lines arise from transitions between energy states with the same principal

TABLE 8

ELECTRONIC CONFIGURATIONS AND OTHER PROPERTIES OF CERTAIN TRANSITION-GROUP ATOMS AND IONS, AND THEIR NEIGHBORS IN THE PERIODIC CHART OF THE ELEMENTS [5,22,68]

At. No.	Symbol	Configurations of the Neutral Atoms. ( ) = No. in Filled Level									Ground State Term	Observed Oxidation States, Valencies	Effective Magnetic Moments of Ions in Bohr Magnetons ( ) = Theor.					
		3d (10)	4s (2)	4p (6)	4d (10)	4f (14)	5s (2)	5p (6)	5d (10)	6s (2)			Ion	Lowest Term	μm (eff.) (approx.)	Ion	Lowest Term	μm (eff) (approx.)
22	Ti	2	2								$^3F_2$	3+, 4+	Ti^{4+}	$3d^0\ ^1S_0$	0	Ti^{3+}	$3d^1\ ^2D_{3/2}$	1.8
23	V	3	2								$^4F_{3/2}$	2+, 3+, 4+, 5+	V^{5+}	$3d^0\ ^1S_0$	0	V^{2+}	$3d^3\ ^4F_{3/2}$	3.9
24	Cr	5	1								$^7S_3$	2+, 3+, 6+	Cr^{3+}	$3d^3\ ^4F_{3/2}$	3.9	Cr^{2+}	$3d^4\ ^5D_0$	4.9
25	Mn	5	2								$^6S_{5/2}$	2+, 3+, 4+, 6+, 7+	Mn^{4+}	$3d^3\ ^4F_{3/2}$	3.9	Mn^{2+}	$3d^5\ ^6S_{5/2}$	5.9
26	Fe	6	2								$^5D_4$	2+, 3+	Fe^{3+}	$3d^5\ ^6S_{5/2}$	5.9	Fe^{2+}	$3d^6\ ^5D_4$	5
27	Co	7	2								$^4F_{9/2}$	2+, 3+	Co^{3+}	$3d^6\ ^5D_4$	(4.9)	Co^{2+}	$3d^7\ ^4F_{9/2}$	4.5
28	Ni	8	2								$^3F_4$	2+, 3+	···		···	Ni^{2+}	$3d^8\ ^3F_4$	3
29	Cu	10	1								$^2S_{1/2}$	1+, 2+	Cu^{2+}	$3d^9\ ^2D_{5/2}$	2	Cu^{+}	$3d^{10}\ ^1S_0$	0
30	Zn	10	2								$^1S_0$	2+	Zn^{2+}	$3d^{10}\ ^1S_0$	0	···		···
···	···	10	···								···	···	···	···	···			
56	Ba	10	2	6	10	0	2	6	0	2	$^1S_0$	2+	Ba^{2+}	$4f^0\ ^1S_0$	0			
57	La	10	2	6	10	0	2	6	1	2	$^2D_{3/2}$	3+	La^{3+}	$4f^0\ ^1S_0$	0			
58	Ce	10	2	6	10	2	2	6	0	2	^3H	3+, 4+	Ce^{3+}	$4f^1\ ^2F_{5/2}$	2.5			
59	Pr	10	2	6	10	3	2	6	0	2	^4I	3+, 4 or 5+	Pr^{3+}	$4f^2\ ^3H_4$	3.5			
60	Nd	10	2	6	10	4	2	6	0	2	^5I	3+, (4+?)	Nd^{3+}	$4f^3\ ^4I_{9/2}$	3.6			
61	···	10	2	6	10	5	2	6	0	2	^6H	···	61^{3+}	$4f^4\ ^5I_4$	(2.7)			
62	Sm	10	2	6	10	6	2	6	0	2	$^7F_0$	2+, 3+	Sm^{3+}	$4f^5\ ^6H_{5/2}$	1.6			
63	Eu	10	2	6	10	7	2	6	0	2	$^8S^{\circ}_{7/2}$	2+, 3+	Eu^{3+}	$4f^6\ ^7F_0$	3.5			
64	Gd	10	2	6	10	7	2	6	1	2	$^9D^{\circ}_2$	3+	Gd^{3+}	$4f^7\ ^8S_{7/2}$	7.9			
65	Tb	10	2	6	10	8	2	6	1	2	^8H	3+, 4+	Tb^{3+}	$4f^8\ ^7F_6$	9.7			
66	Dy	10	2	6	10	10	2	6	0	2	^5I	3+	Dy^{3+}	$4f^9\ ^6H_{15/2}$	10.6			
67	Ho	10	2	6	10	11	2	6	0	2	^4I	3+	Ho^{3+}	$4f^{10}\ ^5I_8$	10.5			
68	Er	10	2	6	10	12	2	6	0	2	^3H	3+	Er^{3+}	$4f^{11}\ ^4I_{15/2}$	9.5			
69	Tm	10	2	6	10	13	2	6	0	2	$^2F^{\circ}_{7/2}$	3+	Tm^{3+}	$4f^{12}\ ^3H_6$	7.5			
70	Yb	10	2	6	10	14	2	6	0	2	$^1S_0$	2+, 3+	Yb^{3+}	$4f^{13}\ ^2F_{7/2}$	4.5			
71	Lu	10	2	6	10	14	2	6	1	2	$^2D_{3/2}$	3+	Lu^{3+}	$4f^{14}\ ^1S_0$	0			

$$\mu_{eff} = 2[S(S+1)]^{1/2}$$
$$\qquad\ = [N(N+2)]^{1/2}$$

where $2S+1$ is the multiplicity, and N is the number of electrons with unpaired spins.

quantum numbers, but different total angular momentum quantum numbers resulting from changes in the distributions of unpaired electrons in inner incompleted shells. When a highly charged incompleted-inner-shell ion, such as Eu^{3+}, Mn^{4+}, or Cr^{3+}, is combined in a solid, the electrons from adjacent atoms tend to neutralize (pair with) the unpaired spins but may not succeed in pairing all the spins; hence, line spectra or bands with fine structure are obtained. When less highly charged ions, such as Eu^{2+} or Mn^{2+}, are combined in a solid, the ligands apparently succeed in neutralizing practically all of the unpaired spins to the extent that the energy level of the unfilled shell drops well below that of the outermost valence shell. The spectra of the resultant phosphors are then relatively structureless bands arising from electronic transitions of valence electrons between states with different n's or l's. Measurements of the magnetic moments of phosphors (Table 8) should provide valuable information regarding the effective valences of their activators in the excited and unexcited states.[145, 279]

ENERGY BANDS. We shall next enquire into the energy levels involved in and on phosphors which exhibit band spectra without fine structure.[34-37, 65, 102, 154, 221-223, 280-308] The origin of energy bands in ideal crystals was sketched in connection with the energy-level diagram of Fig. 4, and some possible perturbations of these energy levels in real crystals were presented in Fig. 9. At this point, consideration of conservation of momentum is to be introduced into the discussion of energy transformations in solids.

The momentum p_m of a completely free material particle, for example, an electron, is equal to mv, and with this momentum is associated a diffraction wavelength λ_d, according to eq. 2. Rearranging eq. 2 gives

$$p_m = mv = h/\lambda_d = h\mathbf{K} \tag{42}$$

where $\mathbf{K}$ is the **wave number** of the *free* particle. From eq. 42, the equation for the kinetic energy of a completely free particle moving in a given direction may be written

$$E_k = mv^2/2 = h^2\mathbf{K}^2/2m \tag{43}$$

As long as the particle is truly free, E_k increases continuously as $\mathbf{K}^2$ (or v^2), and the curve of E_k vs. $\mathbf{K}$ is a continuous parabola as indicated by the dotted outer curve in Fig. 15a.

When a primary particle enters the periodic field of an ideal solid, the solid may take up momentum p_m from the particle in discrete amounts which are related to the lattice-identity translations x_τ and the wave-number vector κ of the crystal by

$$p_m = h/x_\tau = h\kappa/2\pi \tag{44}$$

Both p_m and κ are proportional to x_τ^{-1}, that is, to the **reciprocal lattice** which is derived in the case of orthogonal lattices by taking the reciprocals of the vectors describing the positions of the atoms in the idealized crystal lattice.[18, 50, 300] By choosing a particular lattice point as an origin, location vectors may be drawn from the origin to all the other lattice points. The shortest vectors in the reciprocal lattice are the points farthest from the origin in the original lattice, and *vice versa;* the directions of the vectors remaining unchanged. The reciprocal lattice defines a **momentum space** whose vectors represent the allowed quantized momenta which can be taken up by the idealized crystal.[732]

When the primary particle is inside the crystal and is influenced by the periodic crystal field, the wave-number vector of the influenced (bound) particle is denoted by **k**. As shown by the discontinuous solid curve overlaying the dotted parabola in Fig. 15a, there are marked gaps (forbidden regions) in the E_k versus **k** curve. These discontinuities occur when the particle is diffracted, as expressed by the Bragg relation,

$$2\pi n\mathbf{k}^{-1} = n\lambda_d = 2x_\tau \sin \theta \qquad (45)$$

where θ is the angle between the path of the particle and a lattice plane, and x_τ is the spacing between planes. The general wave-function of an electron moving through an ideal crystal is given by

$$\psi = \epsilon^{i(\mathbf{k},\,\mathbf{r})} f(\mathbf{r}) \qquad (46)$$

where **r** is the position vector of the electron, $f(\mathbf{r})$ is a periodic function determined by the structure of the crystal. Because the ideal lattice is periodic, the **reduced wave-number vector** $\mathbf{k}_r$ defined in a unit-cell interval of the reciprocal lattice with cell dimension a,

$$-\frac{\pi}{a} \leq \mathbf{k}_r \leq \frac{\pi}{a} \qquad (r = x,\, y,\, z \text{ directions}) \qquad (47)$$

suffices for all electron states in the entire ideal crystal. The quantity $\mathbf{k}_r$ is related to the momentum of the electron by

$$p_m = mv = h\mathbf{k}_r/2\pi \qquad (48)$$

and the **selection rule** governing optical transitions in solids is that *only those transitions are allowed for which* $\mathbf{k}_r$ *remains constant; that is, momentum must be conserved.* Since the momentum of an optical photon is negligible, this simply means that conservation of momentum during electronic transitions is accomplished chiefly by momentum interchanges between electrons and atoms, that is, between electrons and elastic waves **(phonons)** of atomic motions in the crystal. A truly free electron cannot absorb or emit a photon. This is so because, upon annihilation or creation of the photon, the **K** of the electron would be altered, and momentum would not be conserved. When the electron is bound to a third particle (the emitted photon being counted as one particle), however, the third particle can conserve the momentum of the system by

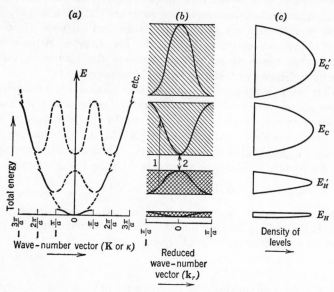

FIG. 15a–c.   (a) The dotted parabola shows the variation of the energy E of a completely free electron as a function of wave-number vector **K**. The solid lines show the variation of E as a function of κ when the electron is moving in a given direction in the periodic field in a crystal. (b) Horizontally expanded energy-level diagram based on (a), assuming the lower two bands to be normally filled and the upper two bands ($E_C \equiv E^*_C$) to be normally empty (insulator crystal). (c) Schematic of the density of the energy levels in each band, assuming no overlapping of bands (densities of different bands not to scale). The electron velocity is also a minimum at both the upper and lower edges of each band.

assuring that at least two particles share the total momentum both before and after absorption or emission of a photon. An electron in a crystal is not completely free, because even in the conduction band it is subject to the periodic field of the crystal.

Regarding the previously mentioned **phonons** (undulatory atomic motion in solids), phonon wavelengths are quantized as multiples of the identity translations x_τ, phonon momenta are determined by x_τ^{-1}, and κ is the number of phonon waves per unit distance in the direction of their propagation.   The energies of phonons are expressable in a form similar to that used for photons (eq. 1), that is,

$$E = h\nu_a = hv/\lambda \tag{49}$$

where v is the velocity of sound in the solid (about 10^6 cm sec^{-1}). The minimum phonon wavelength is about twice the minimum identity translation, that is $2x_{\tau_0}$, and the maximum phonon wavelength is the longest dimension of the crystal. According to Fig. 12, $2x_\tau$ is 10.86 $\times$ 10^{-8} cm along the major axes of cub.-ZnS, and so the *maximum* energy an individual phonon may have in an ideal crystal of cub.-ZnS is about 0.04 ev, and the corresponding *maximum* phonon frequency is about 10^{13} sec^{-1}. If the crystal were a cube measuring 10^{-3} cm (10 microns) along an edge, then the *minimum* energy of an individual phonon in the crystal would be about 4×10^{-6} ev, and the corresponding *minimum* phonon frequency would be about 10^9 sec^{-1}. In complex crystals, such as rbhdl.-Zn_2SiO_4, there are additional high-frequency near-optical vibrational modes which may be excited within the strongly bonded radical groups,[16, 28, 36, 300] for example, the SiO_4^{2-} group, but these optical modes are generally little excited at the temperatures at which phosphors operate efficiently. It is important to note that in the neighborhood of imperfections in *real* crystals phonons may have different local momenta than those allowed in *ideal* crystals. *In the neighborhood of imperfections, therefore, optical transitions may be allowed which are forbidden in the more perfect portions of a crystal.*

It should be noted that both phonons and photons may be created and annihilated, and both particles obey Bose–Einstein statistics which allow an unlimited number of noninteracting particles in a system to have a given energy (quantum) value. Except at very high temperatures, the indestructible interacting electrons in solids obey Fermi–Dirac statistics (Fig. 15*d–f*) which, in accordance with the exclusion principle, allow only one electron in a system to have a given set of the four quantum numbers n, l, m_μ, and s_μ.

With reference to Figs. 15*d–f*, an analysis by J. Bardeen [287] shows that the height of filling of energy levels in surface states on real (semiconductor) crystals is such that the Fermi level ζ is not necessarily midway between the two bands, as shown, but there may be a partial filling of the available levels to an extent depending on the nature of the material and the treatment of its surface. Furthermore, the presence of bound surface charges may alter the contours of the energy-level diagrams such that the bands are bent over a region extending about 100 Å deep into the solid. At equilibrium, then, the Fermi level in the interior of the semiconductor solid adjusts to the same height as that at the surface of the solid, whereas the energy levels in the region immediately below the surface of the semiconductor may be bent up or down, depending on whether the semiconductor conducts current predominantly by "free" electrons (N-type semiconductor with electron-donor impurities providing occupied levels just below the conduction band) or by positive holes (*P*-type semiconductor with electron-acceptor impurities providing unoccupied levels just above the topmost filled band). For a given semiconductor crystal, the type of conduction is strongly influenced by minute traces of certain impurities.

A pictorial representation of the selection rule governing optical transitions in *ideal* solids is shown in Fig. 15*b*, where the allowed energy bands contain plots of E vs. $\mathbf{k}_r$. These plots are expanded portions of the central portion (between $-\pi/a$ and π/a) of Fig. 15*a* and show the

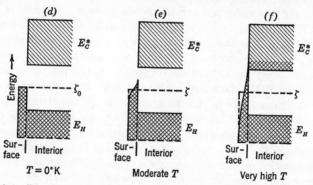

Fig. 15*d–f*. Distribution in energy of surface and interior electrons for an *ideal* near-insulator crystal (poor intrinsic semiconductor). The distribution follows the Fermi function, $f = [\epsilon^{-(E-\zeta)/kT} + 1]^{-1}$, which approaches $f = \epsilon^{-(E-\zeta)/kT}$ at very high temperatures. [$\zeta = (\partial F/\partial N)_U$, where F is the free energy of the N electrons which have a total energy U.] [18, 36]

different allowed energies associated with different values of the reduced wave-number vector in each band. Figure 15 is derived from a one-dimensional periodic structure but may be used as an approximation of a three-dimensional simple cubic structure.[18, 19, 36] The selection rule for optical transitions is expressed graphically by allowing only those transitions which begin and terminate on the E vs. $\mathbf{k}_r$ curves. Assuming the state of affairs shown in Fig. 15*b*, the excitation transition 1 must start and terminate as shown (an equivalent transition occurs an equal distance to the right of $\mathbf{k}_r = 0$); that is, it is not possible in the ideal crystal to raise an electron having an energy lying above or below E_{H_0} (in the uppermost E_H band) into the specific energy level denoted by the intersection of 1 with the E vs. $\mathbf{k}_r$ curve in E^*_C.

Assume that an electron has just been raised from the uppermost E_H band to a level above the lower edge of E^*_C. The lifetime of the excited state prior to radiation is of the order of 10^{-8} sec (limited by oscillator damping, Chap. 5),[2, 116] whereas energy dissipation as phonons can occur in times as short as 10^{-13} sec. Hence, the excited electron can soon lose energy, which appears as phonons (heat) in the crystal, and the electron drifts (in energy) down to the lower edge of E^*_C where it has a minimum kinetic energy ($\mathbf{k}_r = 0$) and the minimum potential energy allowable in E^*_C. The positive hole left in E_H is filled by electrons from higher energy levels, so that the positive hole moves to the *upper* edge of E_H where it, also, has a minimum kinetic energy and a minimum potential energy for the given band. A minimum-energy optical transition between the two bands (from the lower edge of E^*_C to the upper edge of E_H) then becomes possible if the excited electron and the positive hole happen to be sufficiently near each other in the crystal. Otherwise the uncombined electron and positive hole remain as stored excess potential energy in the crystal. If the uppermost filled band were the band now labeled E^*_C, then the minimum-energy allowed transition between the excited electron (in $E^*_{C'}$) and the positive hole (in E_C) would obtain only in the discontinuity regions near $\pm\pi/a$, that is, in the anomalous energy ranges where the particles are reflected from crystal planes. As shown in Fig. 15c, the density of levels in a simple band reaches its maximum value near the middle of the band and decreases to zero at the upper and lower edges of the band. Absorption and excitation transitions are, therefore, relatively infrequent from edge to edge of the different allowed bands. The maximum density of levels is proportional to the number of possible electrons in the band (this density increases according to the expanded degeneracy expression, $2(2l + 1) = 2, 6, 10, 14, \cdots$, for the s, p, d, f, $\cdots$ shells), and is inversely proportional to the width of the band (the $3d$ band of atomic Cu has 10 electrons, but is narrower than the $4s$ band which has a maximum of 2 electrons [19, 24, 36]). In a given solid, the density of levels generally increases with decreasing principal quantum number n, that is, the density of levels in, for example, the p bands, increases on going from $3p$ to $2p$ to $1p$ (highest density in the narrow x-ray bands). There is an inherent indeterminacy of the energy level of an excited electron or positive hole, and so a radiative transition between a broad band and a very narrow band is more probable for a wide range of energy values in the broad band than is the case for transitions between two broad bands (where small changes in E make large changes in $\mathbf{k}_r$ in both cases). If, for example, a positive hole is localized (trapped) on a discrete energy level, it can combine with electrons with a moderate range of $\mathbf{k}_r$ values from an upper broad band. The localized partner, being strongly bound to an atom can readily transmit to the crystal the moderate momentum differences which must be disposed of in order to permit creation of a photon and yet conserve momentum. This points to the utility of narrow or discrete energy levels which may be introduced by impurities and other imperfections in *real* crystals.

It is possible to determine the band widths and the densities of levels in the uppermost *occupied* bands of solids by measuring the spectra of their soft x-ray emission (fluorescence).[18, 35, 309] By bombarding the solid with sufficiently high-velocity primary electrons, secondary electrons may be ejected from the narrow inner (K, L) shells of the atoms or ions. Electrons from higher-lying occupied bands make radiative transitions to combine with the positive holes in the temporarily depleted K or L levels, producing emission spectra which may be interpreted in terms of the band widths and energy-level densities of the broader upper bands. Table 9 gives some results of measurements by O'Bryan and Skinner who measured the soft x-ray emission spectra of a number of solids.[309]

TABLE 9

BAND WIDTHS OF SOME OF THE UPPER FILLED BANDS OF SEVERAL SOLIDS, AS DETERMINED FROM SOFT X-RAY EMISSION SPECTRA

O'Bryan and Skinner

Substance	LiF	NaF	KF	MgF$_2$	LiBr	NaBr	KBr	RbBr	AgBr
Crystal structure	cub.	cub.	cub.	tetr.	cub.	cub.	cub.	cub.	cub.
Total band width, ev	18	16	15	19	>7	6.5	6.1	5.6	>7
Band width at half max. int., ev	2.7	2.0	1.7	3.1	1.2	0.75	0.55	0.45	1.1
Anion band	2p	2p	2p	2p	4p	4p	4p	4p	4p

Li$_2$O	BeO	B$_2$O$_3$	MgO	Al$_2$O$_3$	SiO$_2$	CaO	SrO	BaO	ZnO
cub.	hex.	?	cub.	rbhdl.	hex.	cub.	cub.	cub.	hex.
16	23	19	21	21	20	15	14	14	18
2.7	5.5	3.5	5.5	4.0	4.0	2.2	2.0	2.2	3.0
2p	2p	2p	2p	2p	2p	2p	2p	2p	2p

BeO	B$_2$O$_3$	MgO	Al$_2$O$_3$	ZnS	MnS	Cu$_2$S
hex.	?	cub.	rbhdl.	cub.	cub.	cub.
				hex.	hex.	
12	13	10	13	8	7.5	7.0
8	9	7	8	3.8	3.7	3.2
2s	2s	2s	2s	3s	3s	3s

Careful examination and interpretation of the soft x-ray emission spectra of solids gives information regarding the influence of crystal structure and type of bonding on the upper filled bands. The influence of crystal structure appears as a fine structure, although it is not so pronounced as in the absorption spectra.[18] Emission transitions from *anion* levels are observed as main bands which may have satellites representing radiative transitions from temporarily neutral atoms, the ratios between the main bands and the satellites giving a measure of the relative proportions of ions and neutral atoms in the solid (it being remembered that each atom *fluctuates* between the allowed extremes of being charged and neutral; eqs. 11–15). The relatively broad $2p$ bands of oxygen in BeO, MgO, Al_2O_3, SiO_2, B_2O_3, and ZnO indicate considerable interaction between the electrons of neighboring oxygens (overlapping of ψ's of electrons in adjoining oxygens), and increasing relative breadth (half-width/total width) of the band represents increasing covalent bonding. O'Bryan and Skinner find that the oxygens do not have more than one effective negative charge per atom; that is, O^-, but not O^{2-}, is found in the listed oxygen-dominated solids.

The normally unoccupied conduction bands in insulator crystals are generally broader than the uppermost filled bands, especially when the conduction bands overlap. The spacing between the lowest conduction band and the highest filled band of most phosphor *host crystals* is apparently of the order of 3 ev, because most useful phosphors emit luminescence photons having energies ranging from about 1 to 7 ev. Direct experimental information about the widths of conduction bands and the widths of the forbidden energy ranges between the broad allowed upper bands is a major lack in phosphor research. One of the obvious difficulties is that particular wavelengths in observed spectra of absorption and emission represent merely *differences* between energy levels, and so the absolute energy values of the initial and final levels involved in a transition cannot be determined from spectra alone.

In the absence of readily interpretable experimental information on the pertinent portions of the energy-level diagrams of phosphors, it has become customary to draw tentative diagrams which are adjusted to fit whatever fragmentary information is available. It is preferred to keep the present discussion on a broader plane, although it may be instructive to consider some specific phosphors, for example, cub.-ZnS with and without added activators, if only to demonstrate our need for more experimental information. If the *idealized* insulator host crystal of cub.-ZnS were completely ionic (this is a *half-truth* according to Pauling [24] and the data in Table 9),[283] then the $30 - 2$ electrons per Zn^{2+} ion and $16 + 2$ electrons per S^{2-} ion would be arranged in filled allowed bands ranging in energy from the topmost $3p^6$ (S^{2-}) band down to the $1s^2$ (Zn^{2+}) level. If there are no exciton levels below the lowest conduction band (excitons are postulated when absorption is observed without concomitant electric conduction [281]), then the first absorption band for a photon-excited electronic transition would correspond to the raising of electrons from the $3p$ (S^{2-}) band to the $4s$ (Zn^+) band. On a local scale, this corresponds to an S^{2-} ion losing an electron to become

$3p^5$ (S^-), and a Zn^{2+} ion gaining an electron to become $4s^1$ (Zn^+). The long-wave edge of the first absorption band, which is about 3500 Å (or 3.5 ev) for ZnS,[154, 280, 282] might correspond to the excitation transition denoted by 2 in Fig. 15*b*. This edge-to-edge transition would produce only feeble spectra because the density of states at the edges of the bands is very low, but with increasing energy of the primary particle there is a higher probability of transitions where electrons are raised from the denser portion of the filled band to the denser portion of the conduction band. As the energy of the primary particle is further increased, transitions *to* progressively higher conduction-band levels, and *from* progressively lower filled levels, become possible, so that the resultant spectra become quite complex. When the energy of the primary particle exceeds about 9650 ev, electrons may be raised from the lowest $1s^2$ levels of Zn into the conduction band (the lowest K level of Zn corresponds to an absorption near 1.3 Å).

Electrons in the conduction band are experimentally detectable because they may be made to move toward an anode when an electric field is applied to the crystal. When electrons in an insulator crystal are raised into the conduction band by heat (Fig. 15*f*) or by energy from primary particles, the insulator becomes a conductor whose conduction is proportional to the number, velocity, and lifetime of the "free" electrons in the normally unoccupied bands. If the conduction effect is produced by primary photons, it is known as **photoconduction,** and a solid which exhibits the effect to an appreciable degree is called a **photoconductor.** When excited electrons have been given sufficient kinetic energy to escape from the crystal, they are called **photoelectrons** if they were excited by primary photons and **secondary electrons** if they were excited by primary electrons or other charged material particles (primary electrons, of course, excite *internal* secondary electrons as well as the usual externally detectable secondary electrons[117, 119-134, 258, 288-294]). Conduction effects due to positive holes may be detected by measurements of the sign of the thermoelectric and Hall effects,[35-37] although only the *net* effect is determined when both electrons and positive holes participate.

Earlier in this chapter, it was indicated that cub.-ZnS:[Zn] may have displaced Zn atoms (and/or S atoms) in interstitial sites, cub.-ZnS:Cu and cub.-ZnS:Ag probably have Cu and Ag impurity atoms in interstitial sites, and cub.-ZnS:Mn has Mn impurity atoms predominantly in substitutional sites. From Table 5, it is seen that the peak wavelengths of the luminescence emission bands of these phosphors occur in the following order: cub.-ZnS:Ag, 4550 Å (2.73 ev), cub.-ZnS:[Zn], 4700 Å (2.64 ev), cub.-ZnS:Cu, 5280 Å (2.35 ev), and cub.-ZnS:Mn, 5910 Å (2.1 ev). *If* it were certain that all these emissions were the result of transitions from or very near the common lower edge of the host-crystal conduction band to an unoccupied level supplied by the activator, then a simple energy-level diagram, such as one based on the lower part of Fig. 9, could be drawn for each phosphor by merely adjusting the spacing between E^*_C and E_I in proportion to the electron-volt separations given in the foregoing sequence. This simple diagram would imply that photoconduction should always accompany luminescence, because the electrons would have to be raised from the normally filled activator levels E_I, or the normally filled host-crystal levels E_H, into the conduction band E^*_C, before a radiative transition could take place to E_I (if a positive hole is not produced by direct excitation from E_I, then a positive hole in a lower-lying E_H band may transfer to E_I *and be trapped* by the transition of an electron from E_I to E_H). This simple energy-level diagram ignores the possibilities that (1) electrons might fall from the conduction band to normally unoccupied discrete impurity levels, for example, E^*_I in Fig. 9, before making a radiative transition to E_I, and (2) electrons might be raised initially to the discrete unoccupied impurity levels where the electrons could not move freely through the crystal, but might make radiative transitions to lower levels. Nonetheless, such simple energy-level diagrams have been used frequently for predominantly i-center phosphors, such as cub.-ZnS:[Zn], cub.-ZnS:Cu, and cub.-ZnS:Ag, which apparently exhibit concurrent and proportional photoconduction and luminescence (that is, photoconduction parallels luminescence), and which have temperature-dependent predominantly nonexponential decays of afterglow emission after cessation of excitation.[35, 36, 74, 102, 119-123, 154, 158, 221, 222, 305] More complex energy-level diagrams, in which radiative transitions involve impurity-induced discrete excitation levels, are frequently used for predominantly s-center phosphors, such as cub.-ZnS:Mn and rbhdl.-Zn_2SiO_4:Mn, which apparently do not have a marked correlation between photoconduction and luminescence, and which have predominantly exponential afterglow decays whose rates are relatively independent of temperature.[113, 119, 154, 221, 222]

Figure 16a shows some *conjectural* simplified energy-level diagrams which are based on the foregoing fragmentary information. The relative locations and widths of the normally filled low-density activator bands are particularly uncertain. Some broadening of the normally filled activator bands may be caused by a small interaction of the sparsely distributed activator atoms, and by their nonuniform perturbations of the host crystal according to statistical variations in location and local constitution in the neighborhood of the different impurity atoms in a given host crystal.

Although the diagrams of Fig. 16a are quantitatively inadequate, they may be used to illustrate some of the possible energy changes which may occur in or on phosphor crystals.

1. Excitation. A primary particle **(excitant)**, of energy E, may give up some or all of its energy to raise electrons from occupied levels to unoccupied levels when (a) the energy separation between the two levels does not exceed the energy of the primary particle, and (b) momentum

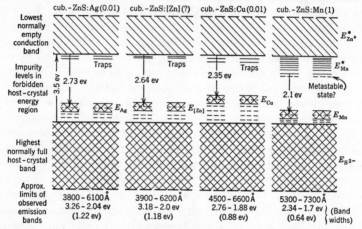

FIG. 16a. Conjectural partial energy-level diagrams of several phosphors based on the same host crystal, *assuming* radiative transitions can occur from (or very near) the lower edge of the conduction band. Dashed levels between the normally occupied activator and host-crystal bands indicate hypothetical levels possibly arising from local perturbation of the host crystal.

is conserved. In the more perfect portions of the host crystal, only those excitation transitions are allowed in which $\mathbf{k}_r$ remains constant (Fig. 15b), but in the neighborhood of imperfections (for example, dislocations, omissions, surface discontinuities, and impurities [35, 36, 286–288, 302]) the ideal host-crystal selection rule may be disobeyed, and there may be many additional allowed energy levels incorporated locally in the forbidden region of the ideal host-crystal energy-level diagram. Where these additional levels are normally occupied, the absorption spectrum of the real crystal exhibits a weak "tail" on the long-wave side of the first absorption band.[221] This "tail" may sometimes be identified as the **activator absorption band,** corresponding to transitions from E_{Ag} to $E^*_{Zn^+}$, or from E_{Mn} to E^*_{Mn} or $E^*_{Zn^+}$, as contrasted with the stronger **host-crystal absorption band** corresponding to transitions from $E_{S^=}$ to $E^*_{Zn^+}$. The first activator and host-crystal absorption bands often merge (see Fig. 24),[170] indicating that there is either an overlapping of the corresponding occupied bands, or the presence of additional occupied levels in the immediate neighborhood of the impurity. As previously mentioned, an excited electron in the conduction band tends to lose energy rapidly to the crystal (as phonons), until the energy of the electron has been reduced so that it occupies levels near the lower edge of the conduction band. At this point, the excited electron has relatively little kinetic energy, and so it may become trapped by dropping into an unoccupied level introduced by an un-ionized impurity (or other imperfection), giving up excess energy as phonons. Electrons in traps, such as those indicated in Fig. 16a, may be raised into the conduction band again by supplying sufficient energy, which may be absorbed from the crystal (phonons, heat) or from excitant particles.

2. Emission. According to these simple diagrams, luminescence emission occurs (except in cub.-ZnS:Mn) when an electron makes a radiative transition from a level near the lower edge of the conduction band to combine with a positive hole in the normally filled activator band, the energy difference between the two levels being emitted as a photon. In predominantly i-center phosphors, the excitation is often associated with an ionization internal to the crystal (complete ionization requires ejection of the electron from the crystal), and the emission represents a radiative transition which occurs during the recombination of an excited "free" electron and an atom (or ion) which has lost an electron. In predominantly s-center phosphors, such as cub.-ZnS:Mn and rbhdl.-Zn_2SiO_4:Mn, the excitation is generally associated with the raising of an electron to a higher level, *without* "internal ionization" (for example, $E_{Mn} \rightarrow E^*_{Mn}$), and the emission represents the radiative return to a level near the ground-state level. In some phosphors, such as cub.-ZnS:Mn(1), it is possible to have both processes taking place (see double emission band in Fig. 42), that is, both [Zn] i centers and Mn s centers participate, and so the energy-level diagram of the phosphor may be represented by a composite of the diagrams shown for cub.-ZnS:[Zn] and cub.-ZnS:Mn, and the phosphor may be denoted as cub.-ZnS:[Zn]:Mn. It may be noted that for a given phosphor, such as cub.-ZnS:Mn, part of the absorbed excitation energy may be used to raise electrons to the intermediate (localized) discrete levels, and part of the input energy may be used to raise electrons into the conduction band. The emission process, then, is a composite of radiative returns of (1) nonfreed localized excited electrons, and (2) freed electrons which may be detained in traps some distance away from their parent atoms. These distinct processes may often be distinguished by their different rates, as shown in Fig. 79c, noting that process 1 is relatively independent of the operating temperature of the phosphor, whereas process 2 is strongly temperature-dependent.

If the excitation and emission process occurs in times approximating the natural lifetimes of excited nonmetastable isolated atoms (about 10^{-8} sec for optical transitions), the process is called **fluorescence,** whereas longer-duration processes are called **phosphorescence.** Phosphorescence involves temporary storage of potential luminescence energy, for example, in the form of trapped excited electrons, or electrons in metastable states. **Metastable states** correspond to electrons in levels from which radiative transitions to lower unoccupied levels are "forbidden" by first-order selection rules (which may be upset to a degree depending on the perturbation experienced by the emitting atom in the solid). Electrons in metastable states may be raised to higher levels where radiative transitions to the given lower levels are allowed. In this respect, metastable states are similar to traps, except that the lower local levels of trapping centers may sometimes be completely occupied, and so a radiative transition is not possible until (1) the electron is ejected from the particular trap and travels to another center which has a lower-lying positive hole, or (2) a positive hole appears during the lifetime of the trapped electron in the trapping center, such as by an electronic transition from the filled activator level to a positive-hole vacancy in the underlying host-crystal band.

Under certain circumstances, it is conceivable that a positive hole initially localized in, say, a low-lying Ag level in Fig. 16a may be filled by an electron raised thermally from the topmost filled band, whereupon the positive hole is transferred to the filled band and is free to move through the crystal until trapped by an electron falling (in energy) from a localized normally filled impurity level in the forbidden region. This is essentially a mechanism for transferring positive holes from one localized energy level to an equivalent or different localized energy level in the forbidden region. Such a mechanism has been proposed by Schön and Klasens to allow positive holes to transfer from low-lying Ag or [Zn] levels to higher-lying Cu, Co, or Ni levels in zinc-sulphide phosphors containing pairs of activators, for example, [Zn] + Cu, Ag + Co, etc.[233] According to the simplified version of this hypothesis, the intensity of excitation, I and the intensity of luminescence emission L are related such that I/L increases linearly with $L^{-1/2}$, and the rate of increase should itself increase with increasing temperature of the phosphor during luminescence. Some experimental evidence has been advanced in support of this view, but anomalous results are also found, and further experimental work is necessary to promote confidence in the applicability of the proposed mechanism.

STOKES' LAW. Resonance radiation is not observed in phosphors emitting broad bands, because the energy levels occupied by the excited electron and its residual positive hole move closer together (by giving up energy as phonons) before a radiative transition occurs. In more general terms, this means that the energy difference between the ground- and excited-energy levels, at equilibrium a short time after excitation, is less than the energy difference at the instant of excitation, for example, $E_e < E_a$ in Fig. 3. As a consequence, the energy of the emitted photon, $h\nu_e$, is usually considerably less than that of the absorbed primary photon, $h\nu_a$. The statement that $h\nu_a \geq h\nu_e$ is known as **Stokes' law.** This is a restatement of the law of conservation of energy on a quantum basis.[34] It is at least theoretically possible to have $h\nu_a < h\nu_e$ by statistically abnormal phonon enrichment of the initial ground-state level and/or the excited-state level at the instants of excitation and emission, respectively, or by **superexcitation,** that is, raising an already excited electron to a still higher level.* A few apparent violations of Stokes' law have been reported, but with only very feeble intensities of emission;[195, 196] furthermore, the thermodynamic implications of Stokes' law have been the subject of considerable dispute.[295-299]

RADIATIONLESS TRANSITIONS. An excited "free" electron (in the conduction band) and its residual positive hole usually have appreciable kinetic energies and have momenta which differ at least in direction, and so their recombination is improbable until energy losses as phonons have reduced their kinetic energies and momenta to sufficiently low values. This is an added reason for the low probability of resonance and anti-Stokes' radiations. Imperfections facilitate recombinations by localizing (trapping) one recombination partner whose ligands can dissipate moderate momentum differences as phonons when the radiative recombination takes place. In other words, a small portion of the available free energy of an excited electron may be sacrificed as phonons to conserve momentum during a radiative transition. The larger the proportion of excitation energy which is transformed into phonons, the less will be the luminescence efficiency, and the greater will be the heating of the phosphor crystal during luminescence. Certain impurities, notably Fe, Co, and Ni which are known to "poison" many phosphors, apparently enhance radiationless transitions at the expense of radiative transitions, perhaps by introducing many closely spaced local unoccupied levels between the conduction band and the highest filled band of the host crystal. Surface discontinuities (Fig. 15d–f), which may produce levels in the forbidden region extending over 100 Å into the volume of the crystal [286-288] and relatively distorted or vitreous regions in the crystal are other possible sources of closely spaced local levels which may facili-

* Intense laser beams produce frequency conversion by non-linear interactions in solids.[764, 766]

tate the radiationless transition of excited electrons or excitons back to the ground-state level.

An alternative pictorial "explanation" of radiationless loss of large electronic excitation energies is based on the postulate that the ground-state and excited-state energy levels of an imperfection center closely approach each other in a plot of potential energy *versus* configurational coordinates (generalized internuclear distances, compare Figs. 3 and 16*b*).[61, 678] If in Fig. 3 E_1 and E_2 were drawn so that they came close together just above c^*_3, then excitation to a level above $a^*_3 - c^*_3$ would afford a high transition probability to the near-by general ground-state level from which the bulk of the excitation energy would be dissipated as phonons (heat) in returning to the normal ground-state level (for example, $a_2 - c_2$ in Fig. 3).

Other "explanations" of radiationless loss of energy include (1) collapse of exciton states by small increments corresponding to phonon energies (there is less than 0.04 ev per phonon in the *ideal* crystal), and (2) interaction of "free" electrons with each other, as in a gas (*Plasmawechselwirkung*), or interaction with atoms of the crystal (*Vielfachstösse*).[102, 302-307] According to the *Vielfachstösse* hypothesis, an electron in the conduction band of a crystal at 0°K does not give up energy to the crystal, but, as the temperature of the crystal is increased and the oscillating atoms interfere more with electron movement, the proportion of excitation energy which is transferred per unit time to the crystal increases rapidly until, above a rather sharply defined temperature (the **temperature break point** T_B), all the energy of an excited electron may be given up in a single radiationless transition involving simultaneous or rapid sequential production of perhaps 100 phonons. The *Vielfachstösse* hypothesis has been greatly elaborated to "explain," among other things, an observed decrease of luminescence quantum efficiency with increasing frequency (energy) of primary photons, by postulating that electrons which are raised over about 1 ev above the lower edge of the conduction band become particularly susceptible to radiationless-transition losses by multiple-phonon production. This behavior is explained more plausibly by the fact that the higher-energy photons are strongly absorbed by the host crystal, and so their penetrations are shallow, and luminescence is then produced in a smaller excited volume in and near the inefficient distorted surface layers of the phosphor crystals. Also, the location and width of the conduction band is determined by the host crystal, so that the *Vielfachstösse* hypothesis assumes that the T_B of ZnS-type phosphors (specified at 500°K) is independent of the activator. This phase of the theory is discredited by the observation that the T_B of a given phosphor host crystal sometimes differs markedly with different activators. For example, the approximate T_B values of some ZnS-type phosphors excited by 3650-Å ultraviolet are: cub.-ZnS:Ag (270°K), cub.-ZnS:Cu (440°K), hex.-ZnS:Ag (285°K), and hex.-ZnS:Cu (490°K), while cub.-ZnS:[Zn]:Mn exhibits two T_B values at about 270°K and 490°K, and hex.-ZnS:Ag:[O] exhibits two T_B values at about 290°K and 400°K.[58, 63, 154, 308] (*Cf.* Fig. 18a.)

PHOTOCONDUCTION. Measurement of the electric conduction of a solid gives a value for the current which is a measure of the charge displacement per unit time in the direction of the applied field. Such a current in nonmetallic solids may comprise (1) electrolytic conduction: transport of mobile ions; (2) electronic conduction: transport of "free" electrons; and (3) positive-hole conduction: transport of excess positive charge from ion to ion by transfer of electrons in the opposite direction. Care must be taken to identify and distinguish the magnitudes of these possible contributions to measured current values. Also, it should be kept in mind that a current value, for, say, pure electronic conduction does not give by itself the number of electrons in the conduction band at any instant. This is so because different electrons moving in a given solid may have different velocities and different lifetimes ("relaxation times") as "free" electrons between liberation and capture (by traps or by recombination with positive holes). The different lifetimes occur partly because of variations in velocity according to the height of the energy level in the conduction band, and partly because of variations in the distances between electron traps or other obstacles to movement in the direction of the applied field. Independent measurements of average velocity or of average lifetime and average distance between captures must be made in order to interpret conduction data in terms of number of charge carriers. In the case of measurement of the conduction of a mass of small phosphor crystals during or after (nonuniform) excitation, the conduction process may be quite complex. The charge carriers may be "free" electrons *and* positive holes (this current *adds* to the "free"-electron current) whose average velocities, lifetimes, and distances between captures may change during the course of the measurement. In the case of measurement of conduction after excitation, for example, these changes may occur because (1) a sufficient density of trapped electrons may build up in certain imperfect regions of the solid to provide local space charges which retard and deflect current flow and may be evidenced by polarization of the conduction cell, and (2) the first "free" electrons liberated from predominantly shallow traps may have velocities, lifetimes, and distances before capture which differ from those of electrons which are liberated later from predominantly deep traps. In cases where electrons are trapped very near their parent atoms (positive holes), some of these electrons may return to their points of origin by moving *against* the direction of the applied field. Hence, the number of "free" electrons estimated from conduction data need not correspond with the number of emitted luminescence photons, even if these photons were produced solely by returning trapped electrons. Furthermore, it is possible for an excited electron to be freed and trapped one or several times before making a radiative (or nonradiative!) recombination, so that the rate of change in the number of conduction electrons need not correspond with the rate of change of luminescence emission (compare Figs. 84 and 85).

The photoconductions of phosphors have not been investigated thoroughly enough to eliminate discrepancies reported by various researchers. It is certain that luminescence emission and photoconduction parallel each other during growth (during excitation) and decay (after excitation) for certain strongly photoconducting i-center phosphors, such as hex.-ZnS:Cu and hex.-9ZnS·CdS:Cu (see Figs. 116 and 117), whereas there are only a few scattered reports of weak photoconduction for predominantly s-center phosphors, such as cub.-ZnS:Mn and rbhdl.-Zn_2SiO_4:Mn.[58, 119–123, 154] In the predominantly s-center phosphors, the weak photoconduction may be attributed to a minor proportion of excited activator atoms whose excited electrons have wandered from their parent atoms, and this "internal ionization" may have a higher probability of occurring when the activator atoms are located in interstitial sites.[34, 63] Very high nonpolarizing photoconduction is obtained from acicular hex.-CdS crystals grown from CdS vapor.[291] These long crystals exhibit a sharp maximum of photoconduction when excited by photons with wavelengths near the long-wave edge of the first absorption band (near 5200 Å), the photoconduction decreasing rapidly to a low value on either side of the sharp spectral maximum (a similar effect is exhibited by hex.-ZnO [215b]). The low photoconduction produced by photons absorbed in the strong host-crystal absorption band may be attributed to shallow penetration of the excitant radiation and to the high resistance to motion of electrons liberated in the distorted surface layers. The low photoconduction produced by photons absorbed in the weak imperfection absorption band may be attributed to the small proportion of imperfections ([Cd]?) in the volume of the crystal. Maximum photoconduction is observed where the host-crystal and imperfection absorption bands overlap the most. The photoconduction of these long hex.-CdS crystals is orders of magnitude greater than that obtained from the usual tiny crystals of conventional phosphors (these are reported to give at most about 30 per cent of the internal photoeffect obtainable from metallic selenium under intermittent illumination [121]).

In contrast with the strong structureless photoconduction excitation band located on the absorption edge ($\approx$ 2.4-ev photons) of the relatively homopolar hex.-CdS, Ferguson reports oppositely peaked photoconduction and photoelectric-emission bands *with fine structure* overlapping the structureless host-crystal absorption band of predominantly ionic cub.-NaCl in the primary-photon energy range from 9.4 to 14 ev.[281] The fine structure indicates that in real crystals the simple energy-band pictures, such as those presented in Fig. 16a, may be complicated by additional excitation or trapping levels which overlap the given host-crystal bands.

In hypothetical *ideally perfect* single crystals at 0°K, a "free" electron in the conduction band should be able to travel through the entire length of the crystal without hindrance, but in *real* crystals operated at (for example) room temperature, thermal scattering and the presence of trapping imperfections, mosaic structure, and impurities often limit the mean free path of "free"-conduction electrons to less than 10^{-5} cm.[35, 36] Electric conduction in real crystals, then, may be a process of repeated trapping and release of excited "free" electrons, whereby electrons detained in deep traps tend to build up a space charge which produces externally evident polarization effects. Crystals with chiefly shallow traps (short afterglows), such as hex.-CdS or high-cadmium hex.-(Zn:Cd)S:Ag, should have higher photoconductions than similar crystals with many deep traps, because electrons in the very shallow traps may be raised back into the conduction band quickly by energy derived from thermal agitation in the crystal.

There are many conflicting reports regarding photoconductions of phosphors, for example, Bergmann and Ronge investigated 20 fine-crystal ZnS-type commercial phosphors under intermittent illumination and reported the highest photoconduction from the phosphors with little or no afterglow,[121] whereas other investigators report (1) little difference in the photoconductions of phosphors with different after-glows, for example, ZnS:Mn and ZnS:Ag,[119] or (2) large differences in photoconduction with different afterglows, for example, about 10 times as much photoconduction from ZnS:[Zn], or ZnS:Cu, as from ZnS:Mn.[154] Gudden and Pohl, and Van Heerden report that higher photoconduction is obtained by exciting ZnS-type crystal in their weak activator absorption bands than in their strong host-crystal absorption bands,[289-291] whereas Bergmann and Ronge report that certain hex.-(Zn:Cd)S:Ag phosphors have two spectral maxima (at 3500 and 3800 Å) for excitation of photoconduction *and* luminescence.[121] Also, Van Heerden reports little conduction excited in cub.-ZnS crystals by beta rays, whereas Frerichs and Warminsky report very high conduction excited in crystals of hex.-CdS (or CdSe, or CdTe) by beta or gamma rays.[290-291] It is probable that some inadequately reported factors, such as the purities of ingredients, details of preparations, crystal structures, and crystal sizes and orientations, may exert such strong effects on photoconduction of solids, that many of the reported results must be considered specific to the particular sample and experimental technique, for example, measurement of surface versus volume conduction, effects of adsorbed films of moisture, and elimination of polarization effects. (According to A. E. Hardy, polarization is observed in the measurement of photoconduction of ZnS-type phosphors; that is, the photoconduction cell retains the polarity of the applied potential difference after the power supply has been disconnected. Also, when the polarity of the applied potential difference across the cell is reversed, either during excitation or phosphorescence, some time elapses before a measurable current flows in the opposite direction.)

ENERGY TRANSMITTAL IN SOLIDS. It appears, at present, that even in phosphors with concurrent luminescence and photoconduction the activator centers and other crystal imperfections play predominant roles in trapping excited electrons and positive holes and in converting excitation energy into phonons as well as photons. In an efficient phosphor, most of the centers preferentially transform the excitation energy into luminescence photons with energies of the order of 3 ev, rather than into phonons with energies less than about 0.04 ev.

The host crystal provides a regular framework in which the activator centers are suspended and through which excitation energy may be transmitted to and from the centers. Primary excitation energy may be absorbed either by the activator centers directly, or by the larger population of host-crystal atoms, and energy may be transmitted through the host crystal as secondary internal excitation energy (or stimulation energy) in the form of excited electrons, positive holes, excitons, resonance transfers, imperfect collisions of the second kind, photons, and phonons. **Resonance transfers** and **collisions of the second kind** involve excitation, without ionization, of an atom which passes its excitation energy on to a neighboring atom or center when there is sufficient overlap of the ψ's of the atoms (centers). Resonance transfer should occur without energy degradation (for example, without energy loss as heat), whereas collisions of the second kind imply that the electronic excitation energy of one atom is transformed, in whole or in part, into increased kinetic energy of the interacting atoms. In a conventional collision of the second kind, all of the electronic excitation energy is converted into kinetic energy of the interacting atoms, that is, the excitation energy is completely degraded into phonons. Excitons, resonance transfers, and imperfect collisions of the second kind differ (perhaps trivially) in that the same bound excited electron moves through the crystal in exciton motion, whereas different bound excited electrons pass along the excitation energy in resonance transfers or imperfect collisions of the second kind. Long-range excitons may be produced in the more perfect portions of the host crystal, but usually not in the impurity centers, because there is generally a large difference between the energy required to raise an electron from a filled host-crystal band into an exciton level and the energy required to raise an electron from a (usually shallower) filled activator into the same level. The relative participations of these possible forms of energy transmittal, in specific phosphors under specific conditions of operation, are at present unknown, although electrons, photons, and ψ-overlap transfers appear to be the chief means of energy transmittal in most phosphors.

GENERALIZED ENERGY-LEVEL DIAGRAMS FOR PHOSPHOR CENTERS. Figures 16b and 16c are generalized and simplified energy-level diagrams which have been found useful in interpreting many of the luminescence phenomena that occur in solids.[678] The diagrams, which should be used together, are intended to portray the elemental energy transformations in a typical phosphor center, that is, in the perturbed region produced by a substitutional or interstitial impurity atom (Fig. 8).

Figure 16b, which is an extended solid-state version of Fig. 3, shows the variation of potential energy of a phosphor center (in only two states) as a function of a generalized configurational coordinate $\bar{x}$ which represents three-dimensional changes in average internuclear spacings and possible changes in the geometric arrangements of the atoms in the center.[34-37, 61, 74] A major new feature of this diagram is the near approach of the ground-state and excited-state curves at f. When the center is in energy level f, it is assumed to be in a configuration where there is a high probability that any excitation energy stored in the center or delivered to the center will be dissipated quickly as heat to the surrounding crystal. An unexcited center in equilibrium at, say, room temperature may be in the ground-state vibrational level E_a from which it can be raised to excited-state level E^{**}_b by absorbing energy equal to $E^{**}_b - E_a$. The excited center gives up some of this energy $E^{**}_b - E^*_c$ as heat in about 10^{-12} sec and comes to equilibrium in the excited-state level E^*_c. When the selection rules are favorable, the center may make a spontaneous radiative transition (in an average time, usually longer than 10^{-8} sec, determined by the nature of the center and the host crystal) by emitting the energy $E^*_c - E_d$ as a luminescence photon. At this point, the center still has an excess of vibrational energy above the initial equilibrium energy level, and so it returns to the original ground-state level by giving up the energy $E_d - E_a$ as further heat to the surrounding crystal. [When the selection rules for a radiative transition from E^*_c are unfavorable, that is, E^*_c is a **metastable state,** additional energy may have to be provided to raise the center into a higher excited state (not shown) from which a radiative transition is permitted. Under these circumstances, the center in state E^*_c functions as a **trap.**]

When the temperature of the solid is raised, the equilibrium levels E_a and E^*_c rise in proportion to the additional vibrational and rotational energy in the center. When the center is in the indicated excited state, the probability of being raised thermally into the crossover level at f is equal to $\nu_a \epsilon^{-\Delta E^*/kT}$, where ν_a is the frequency of vibration of the system ($\approx 10^{12}$ sec^{-1}), and $\Delta E^* = E^*_f - E^*_e$ is the activation energy. The diagram in Fig. 16b indicates how a phosphor center may function as (1) an *activator*, by providing a highly probable radiative transition

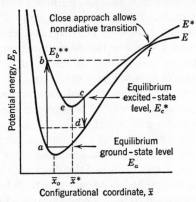

Fig. 16b. Generalized energy-level diagram of a typical activator center as a function of averaged interatomic configuration for the ground state and *one* (lowest) excited state. In many centers there are several different excited-state levels for each filled ground-state impurity level (of which there may be several).

from an excited-state level well below f, (2) a *trap*, by requiring an additional activation energy to raise the excited center into a state (not shown) from which a radiative transition is probable, and (3) a *poison* (or *killer*, or *quencher*), by having the excited-state equilibrium level sufficiently near or above f so that radiationless transitions predominate. According to the diagram, a center may operate *predominantly* as an activator and/or a trap at low temperatures and become *increasingly* a poison center as the operating temperature of the solid is raised to increase the probability of radiationless crossover. The diagram emphasizes, also, that the average internuclear spacings and atomic configurations in a center change during the luminescence process.

Figure 16c, which is an amended version of Fig. 9b, shows some of the higher-energy levels of both the surrounding host crystal and the impurity center as a function of distance along (through) a row of atoms passing through the impurity atom. Specifically, one may imagine this picture to obtain for a center formed by a substitutional Mn activator atom in hex.-ZnS:Mn or rbhdl.-Zn_2SiO_4:Mn. The diagram is drawn for a specific configurational coordinate $\bar{x}$ (Fig. 16b), and so one must

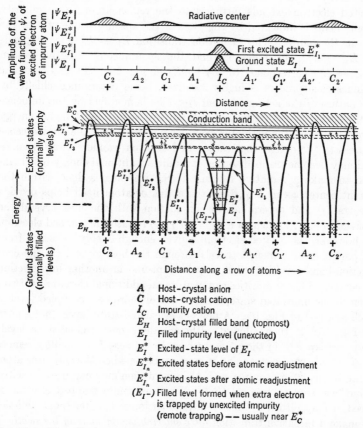

A — Host-crystal anion
C — Host-crystal cation
I_C — Impurity cation
E_H — Host-crystal filled band (topmost)
E_I — Filled impurity level (unexcited)
E_I^* — Excited-state level of E_I
$E_{I_n}^{**}$ — Excited states before atomic readjustment
$E_{I_n}^*$ — Excited states after atomic readjustment
(E_I-) — Filled level formed when extra electron is trapped by unexcited impurity (remote trapping) — — usually near E_C^*

FIG. 16c. Generalized energy-level diagram of a typical activator (or trapping) center as a function of distance along a row of atoms passing through the mid-point of the center. The breadths of the excited-state levels $E^*_{I_n}$ increase as n increases, and so the upper levels may overlap each other and E^*_C. [If the upper drawings were for *static* atoms (no thermal agitation) with $|\psi|$ measured along a *line* through the atomic *nuclei*, then $|\psi|^2$ would be zero in each nucleus and have a maximum on each side of the nucleus—see Table 23.]

imagine that the atomic configurations, internuclear spacings, potential barriers, and energy levels change after electronic transitions in the center. The impurity atom I_C is shown as introducing an additional occupied ground-state level (or group of levels) E_I and a series of excited-state levels $E^{**}_{I_n}$ (before internuclear readjustment) and $E^*_{I_n}$ (after internuclear readjustment) extending up into the conduction band. As the energy of the excited state increases, the wave function ψ of the excited electron extends farther into the region around the impurity atom. This is indicated by the plots of the absolute value of ψ at the top of the figure. The excitation transition $E_I \rightarrow E^{**}_{I_1}$ corresponds to $E_a \rightarrow E^{**}_b$ in Fig. 16b, and the spontaneous radiative return $E^*_{I_1} \rightarrow E^*_I$ corresponds to $E^*_c \rightarrow E_d$ in Fig. 16b. In this case, the luminescence process is highly localized, being determined chiefly by the nature of the activator atom I_C which is modified by the influence of the host crystal. When the excitation proceeds to much higher energy levels, however, the activator atom loses control over the luminescence process to the extent that the excited electron wanders away from I_C and may become trapped so that it requires additional energy for liberation to make a radiative return. If the trapping is done within the parent center, as in $E^*_{I_2}$, then the activator atom itself helps provide the trap(s) and determines in conjunction with the host crystal the activation energy, such as $E^{**}_{I_2} - E^*_{I_2}$, to release the trapped electron. If, however, the excited electron is given enough energy to escape from its parent center, then it travels through the host crystal (in the conduction-band levels) until it may be captured in another imperfection or center. Some impurities can capture an additional electron (or two), even in the unexcited state, to form one or more new occupied level(s) such as $(E_{I}-)$ in Fig. 16c. A new set of excited-state levels (not shown in Fig. 16c) is then allowed, although these new excited-state levels may overlap E^*_C in those cases where E_I- is near E^*_C (shallow remote trap). This type of trapping is most probable when the activator atom is multivalent, for example, a Cu^{2+} impurity ion may capture an excited "free" electron to become Cu^+. When a remote trapped electron is freed, it may return to its own or another ionized center (positive hole) to make a radiative transition *after entering one or more of the relatively discrete excited-state levels associated with the center.* Accordingly, an excited *"free"* electron which belongs to the host crystal at large must become *bound* to the localized center before it can make a radiative transition. This mechanism is much more probable than the direct radiative transition of "free" electrons from the conduction band (as in Fig. 16a).

There is another distinctive process whereby an excited electron trapped at E_I- might make a radiative transition. In this process, a positive hole may come to the trapping center through the filled band E_H and approach I_C where an electron may drop from E_I into the positive hole. The trap then becomes a normal excited center with a positive hole localized at E^*_I and an excited (formerly trapped) electron at $E^*_{I_1}$ or some higher level. A radiative transition of the excited electron to E^*_I then returns the system to the ground-state level E_I. By an extension of this process, it may be possible under certain conditions to transfer excitation energy, in the form of a localized positive hole, from one center to another. If there be two different centers; one with a positive hole $E^*_I(1)$, the other in the ground state $E_I(2)$, then an electron might be raised thermally from E_H into $E^*_I(1)$, and the positive hole in E_H could wander through the normally filled band to the other center, where an electron could drop from $E_I(2)$ to provide a new localized positive hole $E^*_I(2)$. This process tends to transfer positive holes from energy levels near E_H to higher levels (positive holes naturally rise!).

The energy-level diagrams shown in Figs. 16b and 16c have been found useful for generalized descriptions of the luminescence of solids, but quantitatively accurate diagrams for specific phosphors are not yet available. For a given phosphor center, these diagrams are determined by the chemical species, charge, bonding characteristics, and crystallographic location of the activator, coupled with the chemical and structural nature of the host crystal. These component factors are themselves complex, and so their combination and mutual perturbation to form a real phosphor center is necessarily complicated. Also, a specific diagram may represent a given center, which is called a *system*, whereas a phosphor crystal is an *ensemble* of systems having statistical deviations from the mean. Despite the apparent difficulty of calculating energy-level diagrams in advance of making new phosphors, or of determining the diagrams from measurements on the completed phosphors, such diagrams offer considerable promise as a graphic means for correlating the rapidly increasing empirical data on luminescent solids.

CHAPTER 5

LUMINESCENCE OF PHOSPHORS

Definitions (Luminescence versus Thermal Radiation)

Strictly speaking, luminescence should be limited to the emission of *visible* radiation (1.7-ev to 3.1-ev photons), but this traditional *subjective* limitation is ignored in the following objective discussion, largely because the human eye is but one of many photosensitive devices now being used to detect luminescence. Some of the awkward consequences of limiting luminescence to the vaguely defined visible region of the spectrum become obvious when a phosphor is considered, such as tetr.-$CaWO_4$:[W] (No. 24, Table 5), whose emission band extends well above and below the short-wave limit of vision near 4000 Å. If the ultraviolet part of the emission were to be ignored as not satisfying the original visible-radiation concept of luminescence, it would be very difficult to describe the characteristics of the phosphor in a general way, unless a new word (radiescence?) were to be invented and introduced to cover the gamut of observed visible and near-visible emissions of phosphors, that is, emissions of photons with energies ranging from about 1 ev (12,400 Å) to about 10 ev (1,240 Å). For lack of a more suitable established word, **luminescence** is here used in the broad sense that it is *independent* of the similarly derived words *luminance* and *luminosity* which pertain to the human eye as the sole detector (compare Appendix 3 and Fig. 120).[58]

Broadly and objectively speaking, therefore, **luminescence** affords emission of radiation (of subatomic origin) in excess of thermal radiation; that is, luminescence affords photon emission in excess of the photon emission produced *entirely* by thermal agitation (see Appendix 4). Luminescence is generally excited by primary photons or charged material particles having individual energies ranging from about 2 ev to over 10^6 ev and affords emitted photons with energies in excess of about 1 ev.[34, 63] When luminescence is excited by energy liberated during chemical reactions, the liberated energy per emitting atom or molecule usually exceeds 1 ev. These excitation energies are hundreds to millions of times greater than the energies of individual phonons in solids. A single phonon can increase the energy of an electron or atom in a solid by at most a few hundredths of an electron volt, whereas the individual primary particles generally used to excite luminescence can occasion energy increases up to the total amount of energy carried by the primary particle (except for rest-mass energy), that is, tens to millions of electron volts.

The generation of electromagnetic waves by antennas, as in radio and television, is a third distinct radiative process in addition to those that produce luminescence and thermal radiation. Common radio-frequency antennas are made of highly conducting materials whose nearly free electrons are readily displaced *as a group* by applied alternating electric fields. The frequency of the photons radiated from such antennas is determined chiefly by the frequency at which the driving circuits are operated. These radio-frequency photons have energies ranging from about 10^{-9} to 10^{-2} ev which are considerably less than the 10^{-1} ev corresponding to the peak wavelength of emission from a black body at room temperature (Figs. 17 and 120). Although the radio-frequency process produces radiation in excess of thermal radiation, it is *not* included in the definition of luminescence because the radio-frequency process is *not* localized on an atomic scale and does *not* depend on the composition of the antenna.

In order to obtain barely visible emission of thermal radiation from a solid, the temperature of the solid must be raised above about 900°K to obtain an appreciable probability of getting 1.7-ev (or greater) electronic excitations by the cumulative action of phonons. The thermal radiations of solids generally approximate the radiation of the perfect **black body** which has complete absorptivity at all frequencies and, by Kirchhoff's law, emissive power at all frequencies.[5, 28] (Note that the term "black body," like "luminescence," started as a concept restricted to the visible portion of the spectrum and has been extended to cover the entire electromagnetic spectrum.) A perfect black body in equilibrium in a vacuum has a **thermal emissive power** P_ν at frequency ν, which increases exponentially with absolute temperature T, according to **Planck's radiation law:**

$$P_\nu = 8\pi h\nu^3 c^{-3}(\epsilon^{h\nu/kT} - 1)^{-1} \text{ erg cm}^{-2} \text{ sec}^{-1} \tag{50}$$

The thermal radiations from real solids *cannot exceed* the emission which would be obtained from a perfect black body at the given temperature and frequency. Most real solids emit somewhat less thermal radiation than the maximum which is expressed by eq. 50.[105] The peak wavelength λ_{pk} of the broad emission band of black-body radiation is inversely proportional to T, according to **Wien's displacement law:**

$$\lambda_{pk} = 0.29T^{-1} \text{ cm} = 2.9 \times 10^7 T^{-1} \text{ Å} \tag{51}$$

The **total thermal emissive power** P_T of a perfect black body increases as T^4, according to the **Stefan–Boltzmann law:**

$$P_T = 2\pi^5 k^4 T^4/15c^2 h^3 = 5.7 \times 10^{-5} T^4 \text{ erg cm}^{-2} \text{sec}^{-1} \tag{52}$$

At room temperature (about 300°K), λ_{pk} is in the far infrared at 97,000 Å, and P_T is only 4.6×10^5 erg cm^{-2} sec^{-1}, so that there is an inappreciable amount of radiation in the visible region between 4000 and 7000 Å. At the temperature of an incandescent-lamp filament (about 2800°K), λ_{pk} is about 10,000 Å, and P_T is 3.5×10^9 erg cm^{-2} sec^{-1}, so that there is appreciable emission in the visible part of the spectrum. Thus far, no solid material has been developed to endure prolonged operation above about 4000°K, and so most of the energy emitted

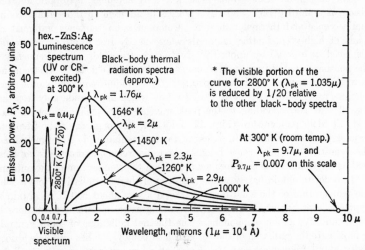

FIG. 17. The room-temperature luminescence emission spectrum of a typical phosphor [hex.-ZnS:Ag(0.015), 1250°C], and the thermal-radiation spectra of a black body at various temperatures. By varying the intensity of excitation, the relative *height* of the luminescence spectrum may be varied from zero to values much higher than the indicated black-body spectra. The black-body spectra are drawn to scale relative to each other.

from practical incandescent solids lies in the infrared, and their efficiencies of light production are usually less than about 5 per cent.[259]

In practice, the luminescence emissions and thermal radiations of phosphors are readily distinguishable because there are large differences between their spectral and temperature characteristics (Figs. 17 and 18). Thermal radiations from solids are generally broad bands (with fine structure evident in some selective radiators, such as the Welsbach $ThO_2:CeO_2$ gas-mantle coatings), and their incandescence efficiencies increase rapidly and constantly with increasing temperature up to the melting point of the solid. Phosphors have relatively narrow emission bands, even on an energy basis, and usually have pronounced temperature break points, T_B, above which luminescence efficiency decreases rapidly to vanishingly low values.   In the temperature range between about 700 and 1000°K there is an overlapping of feeble luminescence and feeble incandescence, but most phosphors are operated at temperatures below 500°K, where there is negligible thermal radiation in the spectral

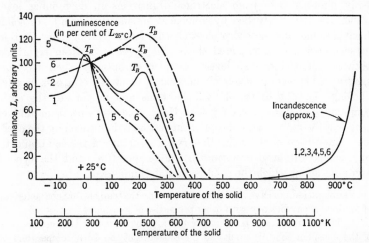

Fig. 18a.   The temperature dependence of luminescence emission (under steady excitation) and incandescence output of several phosphors. The luminescence emission curves are set equal at 25°C and, as in Fig. 17, may be raised or lowered by varying the intensity of excitation.   Increasing the excitation density sometimes decreases the temperature break point T_B,[211, 311] and some phosphors exhibit hysteresis; that is, the L vs. T curves for increasing T do not overlap the curves for decreasing T.[221]

Phosphor	Cryst. Temp.	Constant Excitation by
1. hex.-ZnS:Ag(0.015),[NaCl(2)]	1240°C	3650 Å
2. hex.-ZnS:Cu(0.003),[NaCl(2)]	1240°C	3650 Å
3. cub.-ZnS:Cu(0.003),[NaCl(2)]	660°C	3650 Å
4. ? -ZnS:Mn($\approx$2),[NaCl(2)]	$\approx$1000°C*	3650 Å
5. rbhdl.-Zn$_2$SiO$_4$:Mn(0.3)	1250°C	2537 Å
6. β-Zn$_2$SiO$_4$:Mn(0.3)	1550°C, quench	2537 Å

* Curve 4 from P. Brauer, PB19206.

region of luminescence emission from the phosphor.   For this reason, luminescence has often been referred to as "cold light."   If room temperature were much higher than 300°K, it would be necessary to cool almost all phosphors in order to use them for producing efficient visible and near-visible luminescence.   Also, when it is feasible to operate phosphors at temperatures well below room temperature, the number of available useful phosphors is greatly increased.[155]   These observations apply to the *chief topic of this book*, that is, (conventional) **luminescence,** wherein the emitted photons have energies ranging from about 1 to 10 ev. [The low efficiency of x-ray fluorescence, $h\nu > 100$ ev (Table 7 and Appendix 4), is practically uninfluenced by temperature, because x-ray photons arise from electronic transitions to deep inner levels which are little perturbed by influences outside the atom.] In solids, (conventional) luminescence photons arise from electron redistributions in shallow inner incompleted shells (*line spectra*) or from transitions between allowed energy levels of valence (optical) electrons (*band spectra*). The energy levels of the valence electrons of atoms in solids "belong" in part to the individual atoms and in part to the entire community of atoms in the crystal. Hence, the band-spectrum luminescences of the more useful phosphors are strongly influenced by the chemical composition of the solid, the crystal structure of the solid, and the physical conditions (for example, the temperature, and the type of excitant) under which the solid is made to luminesce.

Various attempts have been made to correlate the temperature characteristics of luminescence efficiency with other properties of phosphors.

1. The existence of a temperature break point T_B below the melting point of the phosphor solid, is similar to the existence of the **Curie temperature** T_C in ferromagnetic (ferrimagnetic?) ferrites of the type $MeO \cdot ZnO \cdot 2Fe_2O_3$ (Me = Mn, Ni, Cu, Mg).[21] These ferrites are strongly ferromagnetic at temperatures below T_C and rapidly decrease in magnetizability (electron-spin alignment) to weak paramagnetism at temperatures immediately above T_C [T_C ranges from about 200 to 650°K, whereas the melting points of the ferrites (and most useful phosphors) lie above about 1600°K]. The Curie temperature has been correlated with the exchange energy E_R of electrons in incomplete shells, according to [18, 36]

$$E_R \approx kT_C \tag{53}$$

From eq. 53, the exchange energy is of the order of a few hundredths of a volt, which allows the exchange bonds to be broken by thermal agitation at relatively low temperatures. The analogy of T_B to T_C is suggestive that exchange forces may play an important part in luminescence, but the weak magnetizability of phosphors, and the strong dependence of T_B (but not T_C) on traces of impurities makes the analogy tenuous.

2. In Möglich and Rompe's *Vielfachstösse* hypothesis, an attempt is made to associate T_B with the **Debye characteristic temperature** T_θ of the host crystal,

$$T_\theta = h\nu_m/k \tag{54}$$

where ν_m is the maximum frequency [306] ($\approx 10^{13}$ sec^{-1}) of phonons in the crystal (ν_m is determined by the minimum identity translation and by the elastic constants of the crystal [5, 25, 28]). Published values of T_θ are in the range from 100 to 1000°K (AgBr = 144°K, KBr = 177°K, KCl = 227°K, NaCl = 281°K, CaF$_2$ = 474°K, and FeS$_2$ = 645°K).[36] Schön reports that the luminescence efficiencies of hex.-ZnS:[Zn], hex.-ZnS(75)CdS(25):Cu, and hex.-ZnS(50)CdS(50):[Zn] *decrease* on going from 90 to 300°K when the phosphors are excited by primary photons absorbed in the host-crystal absorption band, and *increase* on going from 90 to 300°K when the phosphors are excited by primary photons absorbed in the longer-wave activator absorption band [307] (compare Fig. 18c). Schön concludes that the results in the host-crystal absorp-

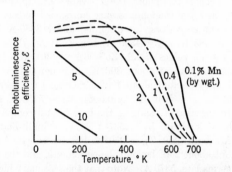

FIG. 18b. Relative photoluminescence efficiencies (that is, light output at constant excitation input) of some rbhdl.-Zn_2SiO_4:Mn phosphors, excited by 2537-Å UV, as a function of operating temperature and proportion of activator. (G. R. Fonda and C. Zener [154,260])

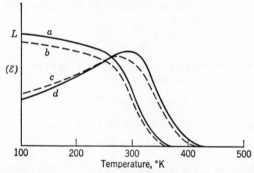

FIG. 18c. Curves showing the different variation of relative photoluminescence output L or efficiency ε obtained when a ZnS:[Zn] phosphor is excited by radiation (1) on the short-wave side of the long-wave edge of its strong host-crystal absorption band * (curves a and b) and (2) on the long-wave side of the strong absorption edge (curves c and d).

ZnS:[Zn] phosphor excited by:

a. 2537-Å UV
b. 3132-Å UV
c. 3653-Å UV
d. 4043-Å Violet light

* *See Fig. 24b.* (G. F. J. Garlick and A. F. Gibson [238])

tion region support the *Vielfachstösse* hypothesis, whereas the contrary results in the activator absorption region are due to "other causes." This is not a completely convincing support of the *Vielfachstösse* hypothesis, because it is known that the T_B's of these and many other phosphors often depend very strongly on small proportions of activators (compare curves 1 and 2 in Fig. 18), whether the phosphor be excited by UV, x rays, or CR.[58, 170] (In the case of some presumably i-center phosphors, the T_B values are apparently related to the peak wavelengths λ_{pk} of the luminescence emissions bands of the same host crystal with different activators, for example, under 3650-Å UV excitation, hex.-ZnS:Ag, $T_B = 285°K$, $\lambda_{pk} = 4350$ Å; hex.-ZnS:Cu, $T_B = 490°K$, $\lambda_{pk} = 5160$ Å, whereas the T_B values of cub.-ZnS(60)ZnSe(40) with [Zn], Ag, or Cu activator are apparently all at about 300°K, although the material with Cu has a secondary bulge extending about 100° higher than the other two, and the λ_{pk} values are: [Zn], $\lambda_{pk} = 5640$ Å; Ag(0.005), $\lambda_{pk} = 5150$ Å, and Cu(0.005), $\lambda_{pk} = 5700$ Å. In the case of changing the host-crystal compositions of some s-center phosphors excited by 2537-Å UV, rbhdl.-Zn$_2$SiO$_4$:Mn and rbhdl.-Zn$_2$GeO$_4$:Mn both have $T_B \approx 250°K$, but the silicate has a secondary bulge extending out about 200° beyond the germanate, and the λ_{pk} values are: silicate, 5250 Å; germanate, 5370 Å. Note that the secondary high-temperature extensions of the ε vs. *T* curve correspond to the *larger* λ_{pk} value in the case of the i-center phosphors, and to the *smaller* λ_{pk} value in the case of the s-center phosphors).[58, 63]

According to quantum theory, the relationship between the number of phonons N_p and the temperature *T* of an ideal solid is given by [18]

$$N_p \propto \int_0^{\nu_m} \frac{\nu^2 d\nu}{\epsilon^{h\nu/kT} - 1} \tag{55}$$

where ν is the phonon frequency, and ν_m is defined as in eq. 54. At low temperatures relative to T_θ, $kT \ll h\nu_m$, and

$$N_p \propto T^3 \quad \text{when} \quad T \ll T_\theta \tag{56}$$

whereas, at high temperatures relative to T_θ, $h\nu/kT \ll 1$, and

$$N_p \propto T \quad \text{when} \quad T \gg T_\theta \tag{57}$$

If we set $h\nu/kT = x$, the heat capacity C_V of a crystal at constant volume is given by the Debye relation,[5, 25, 36]

$$C_V = 3R \left[\frac{12T^3}{T_\theta^3} \int_0^{T_\theta/T} \frac{x^3 \, dx}{\epsilon^x - 1} - \frac{T_\theta}{T} \frac{3}{\epsilon^{T_\theta/T} - 1} \right] \tag{58}$$

wherefrom,

$$C_V \propto aT^3 \quad \text{when} \quad T \ll T_\theta \tag{59}$$

and

$$C_V = 3R \quad \text{when} \quad T \gg T_\theta \tag{60}$$

R being the **universal gas constant** = 8.314 $\times$ 10^7 erg deg^{-1} g atom^{-1}, or about

2 cal deg^{-1} for each gram atom in the solid. Note that Boltzmann's constant per molecule is

$$k = R/N_A = 8.314 \times 10^7/6.023 \times 10^{23} = 1.38 \times 10^{-16} \text{ erg deg}^{-1}$$

$$= 8.62 \times 10^{-5} \text{ ev deg}^{-1} \tag{61}$$

Classically, each atom in an ideal solid vibrates independently about its equilibrium position and has 3 degrees of freedom in both kinetic and potential energy. With each degree of freedom is associated a maximum energy $\frac{1}{2}kT$, or a total of $\frac{6}{2}kT = 3kT$ per atom. For N_A atoms, this amounts to a total vibrational energy per gram atom,

$$E_{\text{vib}} = 3N_A kT = 3RT \tag{62}$$

The classical **heat capacity** C_V is the rate of change of vibrational energy with changing temperature at constant volume of the solid, that is,

$$C_V = (\partial E_{\text{vib}}/\partial T)_V = 3R = 5.96 \text{ cal deg}^{-1} \text{ g atom}^{-1} \tag{63}$$

In other words, a gram atom of an ideal classical solid absorbs (or loses) 5.96 cal ($= 2.49 \times 10^8$ ergs $= 1.56 \times 10^{20}$ ev) for each degree rise (or fall) in the temperature of the solid. The more accurate quantum theory of specific heats introduces the concept of a maximum vibrational frequency ν_m, which is imposed when the atoms do not vibrate independently, but move so that they follow the patterns of the possible longitudinal and transverse waves in the crystal. With ν_m is associated a temperature T_θ below which not all the possible vibrational modes in the solid are active, and so C_V increases in proportion to T^3 as additional modes are brought into action to store more vibrational energy in the crystal. At temperatures well above T_θ, even the highest-frequency mode ν_m is active, and so C_V becomes independent of T because no further vibrational modes can be brought into operation by increasing the temperature of the ideal solid. At very high temperatures a small electronic contribution to C_V is occasionally observed. C_V values higher than $3R$ may be obtained by exciting atomic or molecular rotations, as well as vibrations, but rotational motions seldom make large contributions to the specific heat of solids.

A plot of vibrational frequency versus wave-number vector (or reciprocal wavelength) of the vibrational waves in a phosphor host crystal exhibits (1) low-frequency acoustical branches, and (2) higher-frequency optical branches.[16, 28, 35, 36, 300] The **acoustical vibration modes** arise from cooperative motion of near-by particles (atoms or ions) regardless of species, in producing vibrational waves, whereas the **optical vibration modes** arise from oppositely disposed vibrations of different species of particles. By different species of particles is meant particles with different masses and possibly different charges. The separation, in frequency, between the acoustic and optical branches vanishes for identical atoms equally bonded to each other and generally increases with increasing bonding of different atoms into segregated groups, as in complex crystals containing distinct radicals. In general, the optical vibration modes are best excited in polar crystals wherein the electric vector of incident radiation may cause oppositely disposed displacements of cations and anions. High-frequency (near-infrared) optical vibration modes are particularly pronounced in complex structures containing radicals, as in rhomb.-$ZnSO_4$ and rbhdl.-Zn_2SiO_4 where the highest-frequency optical vibration modes are associated with vibrations within the SO_4^{2-} and SiO_4^{2-} radicals. In polar crystals, the fluctuating charge displacements arising from optical vibrations of cations versus anions adds considerably to the scattering of "free" electrons moving through the crystals.

The Debye temperature T_θ of a solid may represent a partial measure of the average resistance of the atoms to being displaced thermally to an extent that they might interfere with the transport of excitation energy or with the probability of radiative transitions, but it is not clear to what extent these separate links in the luminescence chain may be affected in different phosphors, especially when different activator atoms may be located in either interstitial or substitutional sites to disturb the periodicity which is required for the simple concept of phonons existing as elastic waves.

3. Another approach to the explanation of T_B has been made by von Hippel, Seitz, and Mott, who suggested that the potential-energy versus configurational-coordinate curves (compare Fig. 16b) of the excited and ground states of phosphors may approach each other very closely at moderately large interatomic spacings, which might be achieved by thermal agitation when the system is in the excited state.[35,74] According to this picture, sufficient thermal agitation would raise the excited system into a state where the two levels approach closely enough to allow a very low-energy transition (infrared emission?) to an elevated level of the ground state. After the transition, the excitation energy is dissipated largely as heat, as the system returns down a "ladder" of closely spaced vibrational levels to the original ground-state level. On this basis, it is proposed that the luminescence efficiency ε might follow a relation of the type,

$$\varepsilon = 1/[1 + (\nu_a/\mathbf{P})\epsilon^{-E^*/kT}] \tag{64}$$

where ν_a is the vibrational frequency of the atoms (about 10^{12} sec^{-1}), $\mathbf{P}$ is the probability per unit time that an excited electron will make a *radiative* transition, and E^* is the thermal activation energy required to raise the system from the equilibrium radiative excited state to a higher level where the nonradiative transition to the ground state occurs ($E^* = f - e$ in Fig. 16b). Equation 64 gives a curve of constant efficiency from $T = 0°K$ up to T_B (which may be adjusted, by judicious selection of ν_a, $\mathbf{P}$, and E^*, to fit the particular experimental data), but the equation does not account for those cases where the luminescence efficiency decreases on both sides of T_B (note curves 1, 2, 3, and 4 in Fig. 18a).

4. An attempt has been made to correlate the temperature break point T_B with the peak wavelength λ_{pk} of the luminescence emission band of some i-center phosphors, by assuming that T_B represents the temperature above which an appreciable number of electrons may be raised thermally into the activator band from the topmost filled host-crystal band (for example, from $E_{S^{2-}}$ to E_{Ag}

in Fig. 16a).[58] If this process were probable, it would tend to fill the normally receptive activator levels, thereby preventing radiative transitions of excited electrons which would normally return from still higher levels. It might appear that the thermal raising process would require an enormous temperature, because the (apparent) *optically* determined energy separations between the middle of the activator band and the top of the highest filled host-crystal band are about 1 ev (for example, these are $3.5 - 2.73 = 0.77$ ev for cub.-ZnS:Ag, and $3.5 - 2.35 = 1.15$ ev for cub.-ZnS:Cu); or else the process would require a considerable overlap of the bands or a "ladder" of closely spaced levels between the bands. These energy separations are estimated on the basis of photon-excited transitions, however, and it is shown later that the corresponding thermally excited transitions may proceed at much lower values of kT than appear to be necessary from spectral data; for example, Klasens [233] reports that the *thermally* determined separation between the top of the highest filled host-crystal band and the activator band of cub.-ZnS:Ag(0.01) is 0.37 ev, compared with the afore-mentioned *optical* separation of 0.77 ev. On this hypothetical basis, the ratio of the T_B values for the two cub.-ZnS phosphors with different activators should be proportional to the ratio of the cited band separations. that is,

$$T_{B(Ag)}/T_{B(Cu)} \approx 0.77/1.15 = 0.61 \tag{65}$$

which is in fair agreement with the observed ratio,

$$T_{B(Ag)}/T_{B(Cu)} \approx 270°/440° = 0.59 \tag{66}$$

Similar calculations for the hex.-ZnS:Ag and hex.-ZnS:Cu phosphors give a ratio based on the energy-band separations $0.66/1.11 = 0.60$, whereas the observed $T_{B(Ag)}/T_{B(Cu)}$ ratio is $285°/490° = 0.58$. These agreements may be only coincidence, but the results suggest that other host crystals, which operate well with two different i-center activators, should be investigated for confirmation or denial of the underlying hypothesis. It may be noted that this hypothesis does not pretend to describe the shape of the ε vs. T curve above or below T_B, nor does it allow ready calculation of the T_B's of s-center phosphors, whose radiative-transition levels have a spectroscopically indeterminate spacing relative to the host-crystal levels (at least at present). Analagous reasoning to the foregoing has been presented also by Klasens, with emphasis on the role of migration of positive holes.[233]

5. Another approach to the problem has been made by Szigeti and Nagy, who report that the electric conductance g (determined by dielectric losses of the *unexcited* phosphor at 20 megacycles), of rbhdl.-Zn_2SiO_4:Mn may be correlated with the variation of the luminescence efficiency ε under excitation by 2537-Å UV over the temperature range from 300 to 650°K.[310] They report that ε and g remain constant up to about 445°K, whereafter ε decreases and g increases such that straight lines are obtained when log ε and log g are plotted against T^{-1} (this reported variation of ε vs. T does not agree with curve 5 of Fig. 18, which shows ε steadily decreasing with T from 150°K, with only a slight "bump" near 445°K). Szigeti and Nagy deduce that the 2537-Å photons are absorbed

by (a) radiative and (b) nonradiative centers, such that a constant number N_0 of photons is emitted at temperatures below 445°K where there is a maximum number of radiative centers (and g = constant = g_0), and the total number of photons N_T emitted at temperature T is given by

$$N_T = N_0[cg_0/c(g_0 + ag_T)] \tag{67}$$

where c is a constant, a is the absorption coefficient for the nonradiative centers (if the absorption coefficient for the radiative centers is assumed to be 1), and g_T is the conductance of the unexcited phosphor at temperature T (above 445°K). The corresponding total conductance is given as

$$g = g_0 + g_T = g_0 + A\epsilon^{-E^*/2kT} \tag{68}$$

where $g_0 = 0.2 \times 10^{-6}$ ohm^{-1} cm^{-1}, $A = 5.4 \times 10^{-3}$ ohm^{-1} cm^{-1}, and the thermal activation energy, E^*, is 0.7 ev [this is presumably the energy required to raise an electron from E^*_{Mn} into E^*_{Zn} (compare Fig. 16a), thereby transforming the radiative center into a nonradiative center]. These authors claim that similar correlations between the variation of ε and g as functions of T were found for other phosphors, but judgment may be reserved until more complete information is made available regarding the experimental techniques and the identities of the phosphors.

PHOSPHORS, PIGMENTS, AND SCOTOPHORS. Solids may be divided roughly into the three subject classes with respect to some of the externally evident changes produced when the solid is struck by primary particles, including photons and charged material particles. Primary particles may be *reflected* or *transmitted* by a crystal. When a particle is transmitted, it may (1) pass unchanged through the crystal, or (2) be scattered with only slight loss of energy to the crystal, or (3) lose all or part of its kinetic energy by absorption processes in the crystal. The absorbed energy may produce heat, electron emission, structural or chemical changes, and/or luminescence. Materials which naturally reflect some photons of light and absorb other photons of light (**selective reflection,** or **selective absorption,** of photons with different wavelengths), are called **pigments** or **dyes** when the absorbed photon energy does not produce appreciable luminescence, that is, the intensity of the reflected light is equal to or less than that of the incident light at every wavelength. The spectral characteristics of the reflected light are determined jointly by the primary radiation and the pigment. Many pigments suffer permanent changes in selective reflection during exposure, especially under strong actinic radiations such as occur in sunlight, and the production of these induced discolorations is known as **solarization.** Some practically colorless solids, such as cub.-KCl, may be made to

tenebresce, that is, darken and bleach under suitable irradiations. The darkening effects may be produced by primary x rays or cathode rays, and the bleaching may be accomplished by heat and/or by irradiating the solid with photons with wavelengths lying within the new visible or invisible absorption band produced by the primary radiations. Solids which exhibit such reversible induced absorptions **(tenebrescence)** have been called **scotophors** (dark bearers), as a companion name for phosphors (light bearers).[58, 63]

Luminescent materials in general **(luminophors)** may exhibit the properties of pigments and/or scotophors, but they have the additional property of being able to convert part of the absorbed primary energy into emitted luminescence radiations whose spectral characteristics are determined almost entirely by the luminophor. As examples, a red-reflecting pigment (for example, hex.-HgS) absorbs blue light which is converted into heat, whereas some red-emitting phosphors—for example, hex.-ZnS(20)CdS(80):Ag(0.01), [NaCl(2)], 950°C—absorb blue light and convert a considerable part of the absorbed energy into red luminescence emission which adds to the red light reflected by the phosphor.[312] Other red-emitting phosphors, such as Mg_4GeO_6:Mn and $Cd_2B_2O_5$:Mn, are colorless under white light; that is, they do not exhibit selective absorption in the *visible* region of the spectrum. An efficient pigment or scotophor converts absorbed primary energy into heat or chemical change, usually without appreciable luminescence, whereas an efficient luminophor (or phosphor) converts an appreciable part of the absorbed energy into luminescence photons. Some materials, of course, may exhibit simultaneously the properties of phosphors *and* pigments or scotophors; for example, a cub.-KCl scotophor has a weak UV + blue cathodoluminescence emission, and a hex.-ZnS:Ag(0.5) phosphor emits blue light which it also absorbs (as a pigment), the phosphor being gray by reflected white light.[58, 61] Furthermore, a material which is an efficient phosphor at a low temperature may become merely a pigment at higher temperatures where the ground- and excited-state energy levels approach each other sufficiently to allow nonradiative transitions (near approach of c to f in Fig. 16b).

Definitions. As shown in Table 10, the generic term *luminescence* is subclassified according to (1) a prefix denoting the means used for excitation **(excitant),** and (2) separate terms denoting the duration of afterglow **(persistence)** after cessation of excitation. Combinations of these classifications are often convenient, for example, **cathodophosphorescence** is phosphorescence after excitation by cathode rays (CR). The generic term *luminophor* is subclassified into **fluorophors,** or **fluors** (fluorescent materials) and **phosphors** (phosphorescent materials).

TABLE 10

SOME OF THE TERMINOLOGY OF LUMINESCENCE AND RELATED PHENOMENA

Simple Absorption and Reflection

Dyes and **pigments** convert incident radiation (photons) into (*a*) selectively reflected radiation, and (*b*) internal heat.

Tenebrescence

Scotophors are similar to dyes and pigments, except that at least part of their selective absorption of radiation is nonintrinsic, for example, new absorption bands may be *induced* by treatment with x rays or material particles.[58,63,66,67]

Luminescence [Physical (Electronic) Action]

Luminophors, or **lumophors**, convert part of the energy of absorbed photons or material particles into emitted radiation in excess of thermal radiation.
Fluorophors, or **fluors**, exhibit only fluorescence.
Phosphors exhibit phosphorescence (with or without fluorescence).

Designation	*Means Used for Excitation* (**Excitant**)
Photoluminescence	Low-energy photons (visible light, UV)
Roentgenoluminescence	High-energy photons (x rays, gamma rays)
Cathodoluminescence (Electroluminescence)	Cathode rays (CR), beta rays
Ionoluminescence (Radioluminescence *)	Alpha particles, ions
Triboluminescence	Mechanical disruption of crystals

Designation	*Duration of Detectable Afterglow* (**Persistence**)
Fluorescence	Shorter than about 10^{-8} sec for optical photons
Phosphorescence	Longer than about 10^{-8} sec for optical photons
	Effect of Irradiation or Heating During Phosphorescence
Stimulation (*Ausleuchtung*)	Phosphorescence intensity *increased* during irradiation or heating
Quenching (*Tilgung*)	Phosphorescence intensity *decreased* during irradiation or heating

Luminescence (Chemical Action)

Designation	*Excitant*
Chemiluminescence	Energy from chemical reactions
Bioluminescence	Energy from biochemical reactions

Designations Which Are Not Recommended

Candoluminescence (Non-black-body emissions observed at very high temperatures)
Thermoluminescence (Phosphorescence obtained at various temperatures)

---

* Radioluminescence has been used to describe luminescence excited by any or all radioactive-disintegration products. In the case of radium, the alpha-particle excitation predominates.

The demarcation, at 10^{-8} sec, between fluorescence and phosphorescence requires an explanation. This demarcation is based on the experimental observation that the **natural lifetime of the excited state** τ_F of an isolated nonmetastable atom is about 10^{-9} to 10^{-8} sec between the instant of excitation and the instant of emission of an **optical luminescence photon** (that is, a photon with an energy of the order of 2 ev).[5] The natural lifetime of the excited state, considering the atom as a radiating dipole, is determined by interaction of the fields of the dipole and the emitted photon (oscillator damping), and this interaction effect depends on the energy of the photon.[2]

To obtain a rough picture of the radiative process for a *free* simple atom, first note the differences in size and shape of the electron clouds ($\psi\psi^*$ distributions) of a hydrogen atom in the ground state ($1s$) and in two excited states ($2s$ and $2p$) as shown in Table 23. Next, visualize the H* atom in the $2p$ excited state with the positive nucleus and single negative electron forming a dipole at any instant. When the excited atom makes a radiative transition from the $2p$ to the $1s$ state, the *average* length of the dipole decreases from about 2.2 Å to about 0.53 Å. At the same time there is created (emitted) a photon of energy 10.2 ev, with frequency 1.465×10^{15} sec^{-1}, and wavelength 1215.7 Å. During the brief interval of about 10^{-8} sec that the atom and its photon "offspring" are very near each other, the oscillatory electric field of the emerging photon interacts with the field of the radiating dipole such that the oscillations of the dipole are damped while it is collapsing from the $2p$ to the $1s$ state. [The transition $2s \rightarrow 1s$ is forbidden (not observed) because it would correspond to $L = 0 \rightarrow L = 0$.]

Classically, the rate of emission of electromagnetic energy from a radiating dipole, considered as a simple harmonic oscillator, is proportional to the product of the fourth power of the frequency ν of the emitted photons and the square of the dipole moment p_d of the oscillator; so the rate of emission of photons is proportional to $\nu^3 p_d{}^2$, and the natural lifetime τ_F of the excited state is proportional to $\nu^{-3} p_d{}^{-2}$. Furthermore, the natural lifetimes of normal excited states of isolated atoms or ions are related to the natural line breadths $\Delta\nu$ of the emission lines, such that the line breadth $\Delta\nu_{1/2}$ at half the peak intensity equals the total radiative-transition probability **P** per unit time,

$$\mathbf{P} = \Delta\nu_{1/2} \tag{69}$$

Remembering that $E = h\nu$ (eq. 1), and applying the indeterminacy principle for energy and time, we have

$$\Delta E \Delta t \geq \hbar \equiv h/2\pi \tag{70}$$

so the energy E^* of the excited state is defined with an indeterminacy

$$\overline{\Delta E}^* \geq \hbar \tau_F{}^{-1} \tag{71}$$

where τ_F is the lifetime for a radiative transition of an isolated nonmetastable atom. Equation 71 relates the width $\overline{\Delta E}^*$ of the energy level of an excited atom to the lifetime τ_F of the excited state, and $\overline{\Delta E}^*$ is about 10^{-7} ev for the emission of a (conventional) luminescence photon. (This width is trivial compared with the approximately 1-ev widths of the emission and energy bands of the more useful phosphors. In the case of x-ray fluorescence emissions, however, the observed line widths are of the order of 3 ev, and so the natural lifetimes of the excited states are of the order of 10^{-15} sec.[5]) The distinction made in Table 10 between fluorescence and phosphorescence is, therefore, less arbitrary than it may at first appear, because the natural lifetime of the excited state of an isolated atom is related simply to the energy of the emitted photon and Planck's

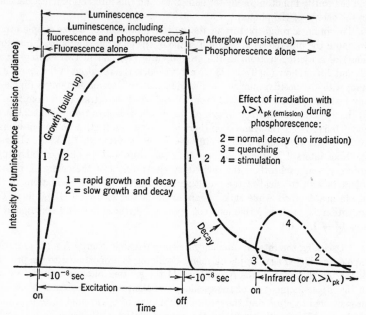

FIG. 19. Terminology of luminescence–process and intensity–time relationships during growth and decay of conventional luminescence emission.

fundamental action constant. For (conventional) luminescence emissions, all persistences longer than about 10^{-8} sec are called phosphorescence to indicate an abnormal delay, such as may be caused by a metastable state, or by electron trapping in solids (treated under Luminescence as a Function of Time in this chapter). Any pure fluorescence of phosphors would be observed as luminescence emission during the first 10^{-8} sec of excitation (Fig. 19); thereafter, the emission comprises both fluorescence (*if observed*) and phosphorescence until 10^{-8} sec after cessation of excitation, whereafter only phosphorescence is observed. *Fluorescence, then, is a limiting case of phosphorescence.*[678]

There is considerable confusion in the nomenclature of luminescence. The term *radioluminescence*, for example, is patently ambiguous, and the term *fluorescence* has been used variously to describe (1) luminescence emissions with afterglows shorter than about 10^{-8} sec (the definition which is preferred here), (2) emissions with afterglows which are short relative to the 0.1-sec persistence of human vision (the original *subjective* basis for distinguishing between fluorescence and phosphorescence), (3) afterglows which decay exponentially (irrespective of duration), (4) luminescence during excitation (irrespective of duration), and (5) transitions between *singlet* states of organic molecules in glassy media.[77] Usage 3 would require that the afterglows of phosphors which exhibit concomitant exponential and power-law decays (such as Nos. 32–35 in Table 5), be called fluorescence *and* phosphorescence, while usage No. 4 fails to recognize that phosphorescence emission may be an important component of the total luminescence emission during excitation times lasting longer than 10^{-8} sec. Usage 5 is similar to 3, except that the terms fluorescence and phosphorescence, which have been used for centuries to describe the *durations* of the afterglows of chiefly inorganic luminescent materials, are appropriated and defined in terms of a difficultly identifiable mechanism *without regard for duration of afterglow* in the field of organic luminescent substances and materials. Only definition 1 preserves the well-established original sense of the terms fluorescence and phosphorescence to denote a demarcation in the time of detectable persistence of luminescence emission after cessation of excitation. Definition 1 simply changes the demarcation time from the *subjective* 0.1-sec persistence of human vision to the *objective* natural lifetime of the excited states of isolated nonmetastable atoms or ions, where the natural lifetime is inversely related to the separation of the energy levels of the ground and excited states. In attempts to classify luminescence with respect to temperature, the term **thermoluminescence** has been used to denote phosphorescence obtained at various temperatures (the prefix *thermo* confusingly implies luminescence excited by heat!),[102, 119, 120, 256–258] and **candoluminescence** has been used to denote the emission of non-black-body radiations at high temperatures (these appear to be selective thermal radiations rather than luminescence[74]).[150–151] These misleading terms are not recommended, although **thermostimulation** and **photostimulation** (and **thermoquenching** and **photoquenching**) may be used to distinguish between the stimulation (or quenching) of phosphorescence by heat and by photons. On this basis, **phosphorescence,** *when unqualified,* usually means phosphorescence which proceeds spontaneously or is thermostimulated at room temperature (about 300°K).

In addition to the foregoing terms, the following less prominent terms have been used as indicated: **Crystalloluminescence** is obtained during crystallization, for example, a blue luminescence emission is observed when a saturated aqueous solution of sodium chloride is shaken with hydrochloric acid solution,[313, 314] **galvanoluminescence** is obtained during electrolysis,[81] for example, a luminescence emission is observed at an aluminum anode during the electrolysis of a solution of $MgBr_2 \cdot 2(C_2H_5)_2O$, or of citric acid;[315, 316] and **electroluminescence** is obtained during electric discharges in gases, for example, the luminescence emissions obtained from neon and mercury vapor lamps, and from lightning.[317, 320] The excitation processes in some of these cases are quite complex and, in crystalloluminescence and galvanoluminescence, may involve the combined effects of chemical action, photons, and charged material particles. Basically considered, all primary luminescence excitations should be attributable to one or more of the following excitants: (1) photons, (2) moving charged material particles, and (3) energy liberated during chemical or crystallographic changes. It is possible that only 1 and 2 may be found to be the actual excitants in most luminescences. Among terms which are little used are the following which are *not recommended;* "fluorochemistry," "ultraluminescence," and "infraluminescence." [318]

A satisfactory description of experiments on the luminescence of a solid generally requires an explicit description of the solid, the excitant, and the conditions of excitation and measurement (including specification of the times and temperatures involved, the atmosphere in which the phosphor is tested, and a statement as to whether luminescence or phosphorescence is being measured). In general, experimental results are understood as having been obtained at room temperature, unless otherwise specified. Unless otherwise specified, also, the results herein were obtained with the materials in a vacuum or at atmospheric pressure, that is, at external pressures which are negligible relative to the internal pressures of solids. The luminescence of solids is practically unaffected by uniform *hydrostatic* pressures up to the internal pressure of the particular solid. Relatively stable phosphors, such as rbhdl.-Zn_2SiO_4:Mn and hex.-ZnS:Cu, may be compressed in oil (or even molten plastic suspensions) at pressures of the order of 10^9 dynes cm^{-2} (10^6 g cm^{-2}, or 10^3 atmospheres) without appreciable loss of luminescence efficiency, although the phosphorescence of a hex.-ZnS:Cu phosphor was reportedly decreased by 50 per cent at 20,000 atmospheres.[319] Nonuniform *mechanical* pressure, for example, grinding, is especially injurious to phosphors and should be avoided whenever possible.[58, 75, 102, 145, 240–242]

Excitation Processes

There is, at present, a paucity of reliable data on the absorption and excitation characteristics of phosphors. The chief obstacles in the way of obtaining these fundamental data are: (1) the difficulty of generating and measuring photons and material particles with *all* energies in the range from about 1 to 10^6 ev, and (2) the difficulty of determining absorption coefficients of highly scattering layers of tiny phosphor crystals. Phosphor research would be aided greatly by the development of (1) an intense photon source giving a known power output at all frequencies from about 10^{13} (cycles) $\sec^{-1}$ ($h\nu = 0.04$ ev) to about 10^{20} $\sec^{-1}$ ($h\nu = 4 \times 10^5$ ev), and (2) methods for producing large (centimeter-size) single crystals of all phosphors without adversely affecting their luminescence characteristics. It is possible to generate a quasi-continuous spectrum of photons by electronic circuits "broadcasting" ("generating") at frequencies up to about 10^{10} $\sec^{-1}$, and it is possible to generate a continuous spectrum of photons by thermal radiations of solids in the frequency range from about 10^{12} to 10^{15} $\sec^{-1}$, but from 10^{10} to 10^{12} $\sec^{-1}$ and above about 10^{15} $\sec^{-1}$ the available photon sources are spectrally discontinuous. In fact, it is relatively easy to devise electronic circuits to make electrons in solids oscillate in unison and produce coherent (inphase) radiations at frequencies below about 10^{10} $\sec^{-1}$, since these frequencies are far below the characteristic fre-

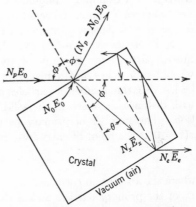

Fig. 20. Schematic of a single simplified phosphor crystal being excited by a beam of primary particles. In an actual phosphor screen, the crystals have wide ranges of random shapes, sizes, orientations, and degrees of optical contact, as well as random voids between the phosphor particles.

quencies of the atoms and atomic groups in solids used to construct such circuits. As the frequency is raised above 10^{10} $\sec^{-1}$, however, it becomes increasingly difficult to produce coherent radiations efficiently. At frequencies above about 10^{14} $\sec^{-1}$, the allowed electronic transitions in solids are quite discontinuous, and the probability of an electronic transition at any instant is determined chiefly by the local (subatomic) conditions in the solid. Thermal radiation and luminescence radiation, * therefore, are "noise" in the sense that the emitted photons are incoherent (that is, the photons are emitted at random instants of time).[58]

The absorption of primary excitant particles (photons or charged material particles) by a phosphor crystal occurs according to the following outline:

1. REFLECTION VERSUS TRANSMISSION. Consider a beam of N_p primary particles per unit time, with identical ("monochromatic")

* Except for stimulated (laser) radiation, which is coherent.[764-9]

energy E_0 striking an assumed flat face of a crystal at angle ϕ to the surface normal (Fig. 20). Of the N_p primary particles, $N_p - N_0$ will be reflected, and N_0 will be admitted by the crystal. For a given crystal, the proportion of admitted particles, N_0/N_p generally increases with increasing E_0 and decreasing $|\phi|$. (With an actual phosphor screen, comprising many different sizes, shapes, and orientations of crystals, the average crystal shape and size and the screen texture influence N_0/N_p. For example, a photon reflected from a single-crystal screen is lost, whereas a photon reflected from a single crystal in a multicrystal screen may be deflected into and absorbed by a neighboring crystal.)

2. REFRACTION, AND GENERAL EXPONENTIAL LOSSES. The N_0 admitted particles are transmitted and refracted at angle θ in the solid, where θ approaches ϕ with increasing E_0 ($\mu \rightarrow 1$, Fig. 21) and decreasing $|\phi|$. Also, the number of particles in the beam decreases exponentially with distance traversed in the solid, because of scattering and absorption. The number of primary particles N_x in the beam at distance x($x = 0$ at the point of entry into the crystal) is given by

$$N_x = N_0 \epsilon^{-ax} \tag{72}$$

where the absorption coefficient a is a function of the nature and energy E_0 of the primary particle and of the nature of the solid. In the case of primary material particles, a may vary with x because the energy of the particle varies with x. In the case of primary photons, a may be as large as 10^6 cm^{-1} when all the host-crystal atoms take part in the absorption of photons with a given energy.

3. ENERGY LOSSES OF PRIMARY PHOTONS IN SOLIDS. Low-energy photons generally give up *all* their energy in one absorption act with a *bound* electron, but the general absorption coefficient is the sum of

$$a = a_{\text{excitation}} + a_{\text{ionization}} + a_{\text{scattering}} + a_{\text{pair formation}} \tag{73}$$

$$\leftarrow \text{ ultraviolet } \rightarrow | \leftarrow \text{ x rays } \rightarrow | \leftarrow \text{ gamma rays } \rightarrow$$

where the lower line indicates the regions of photon energies in which the different energy-loss processes predominate (see Appendix 4). For most phosphor excitations by photons, there is negligible energy loss by scattering, and the absorption coefficient is a function of the number N_σ of atoms per unit volume and the **characteristic frequencies** (that is, the allowed energy-absorption transitions) ν_c of the atoms in the solid

$$a = f(N_\sigma, \nu_c) \tag{74}$$

Figure 21 shows a simplified schematic of the variation of index of refraction μ with photon wavelength for a hypothetical solid which is transparent in the visible region of the electromagnetic spectrum.[107]

The index of refraction of a nonisotropic (noncubic) crystal depends on the frequency, the direction of propagation, and the direction of polarization ("vibration direction" of electric vector) of the photon in the crystal, as well as on the individual and mutually induced polarizabilities of the atoms in the crystal. According to Bragg, the mutually induced polarization of adjacent atoms acted on by light is inversely proportional to the cube of their distance from each other; hence, the preferred orientations of certain closely bound radical groups (for example, NO_3^-, CO_3^{2-}, and SiO_4^{2-}) in crystals tends to make such crystals birefrigent.[47]

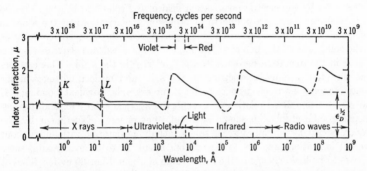

FIG. 21. Simplified schematic of the variation of index of refraction of a colorless insulator over a wide range of photon wavelengths.

Wherever there is a marked change in μ, there is strong reflection and absorption of photons, indicating a characteristic frequency with high **oscillator strength** (high density of absorbing atoms or molecules). In the dotted anomalous regions, where $\nu_{photon} = \nu_c$, the amplitude of electronic oscillation is a maximum. At very long wavelengths, $\mu = \epsilon_D^{1/2}$, where ϵ_D is the low-frequency dielectric constant of the solid.

4. ENERGY LOSSES OF CHARGED MATERIAL PARTICLES IN SOLIDS. Material particles, *especially those having a net electrostatic charge*, generally give up their energy *bitwise* by successive inelastic collisions with free or bound electrons.   The *average* partial (bitwise) loss of energy, $\overline{\Delta E}$, by material particles appears to be relatively independent of ν_c, being almost directly proportional to the total number of electrons per unit volume,

$$\overline{\Delta E} \propto N_{\sigma_1} Z_1 + N_{\sigma_2} Z_2 + \cdots N_{\sigma_i} Z_i \tag{75}$$

where the subscripts 1, 2, $\cdots i$ denote the different species of atoms in the solid. Primary electrons apparently lose an *average* of 10 to 30 ev per inelastic collision, almost independent of the nature of the solid, or of the energy of the primary electron [from about 100 ev up to 3×10^6

ev $(5m_0{}^-c^2)$, whereafter deceleration radiation (*Bremsstrahlung*) becomes important].[2, 34, 321-323]   At each individual inelastic collision, however, there is a possibility that a material particle may lose all or any part of its kinetic energy.[733]

It has been found experimentally that the average expenditure of primary energy per ion pair produced in gases is about 30 ev.[3] This value is little dependent on the energy of the (high-energy) primary charged material particle and the nature of the gas. When an atom is placed in a solid, however, the electrostatic potential energy between the nucleus and a valence electron is decreased by the electric polarization field of the solid and by the increased nucleus–electron distance r due to deformation (polarization) of the atom in the solid. Each of these decreases is proportional to the dielectric constant ϵ_D of the solid, and so the potential binding energy of a valence electron which is $Z_{\text{eff}}e^2/r$ in the free atom becomes *ideally* $Z_{\text{eff}}e^2/r\epsilon_D{}^2$ in the solid. This assumes that the atom retains its electrons in the solid and neglects changes in the kinetic energy of the electron caused by, for example, covalent bonding. If one uses the simplified relation 30 ev/$\epsilon_D{}^2$, the average energy expenditure per ion pair produced in solids should be about 10 ev or less. Experimentally, the average energy loss for all causes in solids, including excitation with and without internal ionization, appears to range from about 10 to 30 ev. Again, there appears to be little change in the mean value of about 15 or 20 ev for different energies of the primary particles or for different solids.[321-323]

It might be expected that $\Delta\bar{E}$ should be proportional to the energy difference (forbidden region) between the upper edge of the topmost filled band and the exciton levels or the lower edge of the conduction band, because these are the smallest energy expenditures for exciting (ionizing) the host-crystal atoms. This does not seem to be so, because the cathodoluminescence efficiencies of phosphors having forbidden regions ranging in width from about 10 ev to 2 ev do not generally increase as the energy difference decreases. The results indicate that the energy of charged material particles is used chiefly to raise electrons from the dense middle portions of the topmost filled band(s) well up into the high-density regions of the conduction bands. The width of the forbidden region ΔE_f may, however, be important in determining the lowest voltage electrons which can excite detectable luminescence. For example, hex.-ZnO:[Zn] ($\Delta E_f \approx 3$ ev) is readily excited by 5-volt primary electrons, whereas tetr.-CaWO$_4$:[W] and tetr.-ZnF$_2$:Mn ($\Delta E_f > 4.5$ ev) require about 10-volt primary electrons (on this basis, freshly cleaved hex.-CdS:[Cd] ($\Delta E_f \approx 2$ ev) should be excitable with about 3.5-volt electrons).

5. Penetrations and Energy Losses of Primary Particles. Figure 22 shows the variation in relative number, N_x/N_0; energy,

hv_x/hv_0 and V_x/V_0; and power, P_x/P_0 (imparted to the solid), for primary photons and electrons traversing a crystal up to the **penetration**

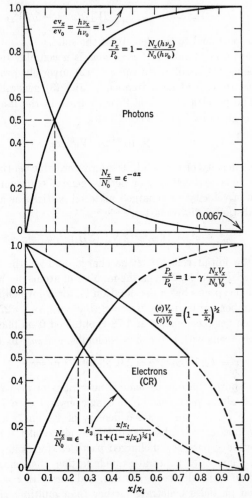

FIG. 22. Variation in relative number of particles, N_x/N_0, energy per particle, hv_x/hv_0 and V_x/V_0, and beam power (imparted to the solid), P_x/P_0, for beams of photons and electrons penetrating a solid. $\gamma \approx 1$ for small values of x/x_l, and $\gamma > 1$ for large values of x/x_l; $k_3 \approx 31.7$.

"limit" x_l, beyond which point negligible power is transmitted. The two sets of curves show that the two excitants differ in that a primary photon maintains $hv_x = hv_0$ until it is absorbed (photon disappears), whereas a primary electron loses kinetic energy $[E_k = (e)V]$, according to the **Thomson–Whiddington law,**[58, 91, 92]

$$V_0{}^2 - V_x{}^2 = bx \quad (i.e., \quad dV_x/dx = -b/2V_x) \tag{76}$$

where b is proportional to the density of the solid.

The penetration "limit" x_l for *photons* is a complex function of ν_0 and ν_c; x_l being (1) small when $\nu_0 = \nu_c$, (2) inversely proportional to the oscillator strength at each frequency, and (3) generally larger for larger ν_0. The penetration "limit" for *electrons* in the range from 10^3 to 10^5 volts is given by Terrill's equation,

$$x_l = 2.5 \times 10^{-12} \sigma^{-1} V_0{}^2 \text{ cm} \tag{77}$$

where σ is the density of the solid, in grams per cubic centimeter.[91]

Equation 76 is a special case of Bethe's general equation (eq. 78) for the variation of velocity v of charged material particles as a function of penetration distance x into matter,[34, 324, 733]

$$v_0{}^4 - v^4 = 16\pi e^4 z^2 (\Sigma N_{\sigma_i})(\Sigma B_i) x / m_0{}^- m_p \tag{78}$$

where v_0 is the initial velocity, z the charge, and m_p the mass of the primary material particle, ΣN_{σ_i} is the number of atoms per unit volume, and each atomic species has a deceleration factor B_i, which is related to its atomic number, for example, $B_{Be} \approx 8$, $B_O \approx 13$, $B_S \approx 22$, $B_{Zn} \approx 33$. The correlation between eqs. 76 and 78 is obtained from line 7, Table 1, in that, for electrons with $v \ll c$, e in coulombs, and $m_0{}^-$ in kg,

$$v = (2eV/m_0{}^-)^{1/2} = 5.9 \times 10^5 V^{1/2} \text{ m sec}^{-1} \tag{79}$$

Figure 23 shows the *approximate* penetrations of primary photons, electrons, and alpha particles into a single crystal of a hypothetical colorless phosphor with an assumed density of 4 g cm^{-3} and a strong characteristic host-crystal absorption in the ultraviolet. For a given energy per primary particle, uncharged particles (for example, photons and neutrons) generally penetrate much farther than singly charged particles (for example, electrons, protons, and ions with $z = 1$), and the singly charged particles penetrate farther than multiply charged particles (for example, alpha particles, and other ions with $z \geq 2$). At very high energies, the greatest penetrations are obtained with particles having the highest mass.[2]

Also, at high primary-particle energies, energy loss by **deceleration radiation** (*Bremsstrahlung*)[2] becomes important, in addition to the energy losses by electronic excitations and ionizations produced as the charged material particles penetrate into solids. Deceleration radiation is produced when an electron passes close enough to an atomic nucleus to be deflected ("accelerated") and loses some energy as a photon. From the preceding discussions, it is seen that high-energy particles, including photons and charged material particles, may produce a considerable variety of secondary, tertiary, etc., particles having a large range of energies in the solid.

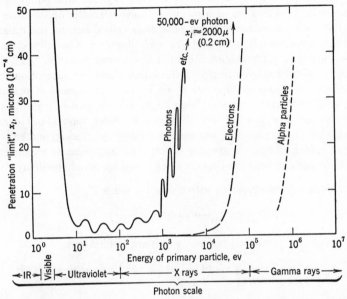

FIG. 23. The penetration "limit" x_l, as a function of the energy of the primary particle, for photons, electrons, and alpha particles penetrating an "average" optically transparent solid with an assumed density of 4 g cm^{-3}.

6. SCREEN TEXTURE AND SCATTERING. It should be kept in mind that conventional phosphor screens, which are usually about 2 to 10 average crystal diameters thick, comprise randomly oriented tiny crystals of irregular size and shape, whereby the screens have unavoidable variations in **texture** (number, sizes, and shapes of particles as a function of location on the plane of the screen). Primary particles striking these screens at different places encounter different ϕ's (Fig. 20), and have different probabilities of (1) being reflected backwards without transmission through any crystal; (2) being reflected backwards with a variable amount of transmission through one or more crystals; (3) passing through the screen without striking any particles (incomplete coverage of the substrate); (4) passing through the screen by multiple reflec-

tions, without transmission through any crystal; and (5) passing through the screen by variable combinations of refraction and reflection, whereby a primary particle may travel a distance x *in the crystals*, ranging from almost 0 up to many times the thickness of the screen. The statistical indeterminacy of x makes it practically impossible to determine quantitatively the absorption coefficients of fine-crystal phosphors. It is not possible, generally, to immerse the fine crystals in a liquid of the same index of refraction as a means to prevent scattering, because (1) the indices of refraction of many phosphors extend beyond the range of present immersion liquids, and (2) such liquids are usually strongly absorbing toward ultraviolet photons and material particles. The chief hope for obtaining quantitative information about the absorption characteristics of phosphors appears to lie in the development of practical universal methods for growing large single crystals, that is, crystals with at least one face having both dimensions greater than about 4 mm. Meanwhile, most phosphor screens are strongly scattering, and so their *transmissions of light* (neglecting absorption) approximate exponential relations such as the following ones developed by Mukhopadhyay: [324, 325]

For isotropic (cubic) crystals, with refractive index μ',

$$L = L_0 \epsilon^{-4\pi^2 x_s x_c (\mu' - \mu)^2 \left(\frac{\pi}{8} - \frac{\pi^2}{36}\right) \lambda^{-2}} \tag{80}$$

For anisotropic (uniaxial) crystals, which are birefringent,

$$L = L_0 \epsilon^{-4\pi^2 x_s x_c \left[(\mu_0 - \mu)^2 \left(\frac{\pi}{8} - \frac{\pi^2}{36}\right) + (\mu_e - \mu_0)^2 \left(\frac{\pi}{30} - \frac{\pi^2}{324}\right) + (\mu_e - \mu_0)(\mu_0 - \mu) \left(\frac{\pi}{12} - \frac{\pi^2}{54}\right) \right] \lambda^{-2}} \tag{81}$$

where L is the **intensity of transmitted light** (number of photons per square centimeter per second), of wavelength λ, encountering the Nth layer of crystals of diameter x_c; x_s is the thickness of the screen (Nx_c); μ is the index of refraction of the medium surrounding the crystals (air or vacuum); μ_0 is the refractive index for the ordinary vibration, and μ_e is the refractive index for the extraordinary vibration in doubly refracting crystals, where $\mu_e > \mu_0 > \mu$. From eqs. 80 and 81, it is seen that increasing transmission is obtained with increasing μ and λ, and with decreasing x_c, x_s, and μ' (or μ_0 and μ_e), a maximum value being obtained when $\mu = \mu'$. It is necessary to go to either very large or vanishingly small crystals to obviate scattering, and the large crystals are preferable, since vanishingly small crystals tend to be more distorted by greater susceptibility to surface tension. The amplitude A of light *scattered* from an assemblage of N_c crystals is proportional to λ^{-2}, and the *intensity of scattered light* L at a point a distance x from a scattering crystal of volume V_c is proportional to λ^{-4}, as given by the **Rayleigh scattering formula,**[107]

$$L = L_0 \left(\frac{\mu' - \mu}{\mu}\right)^2 (1 + \cos^2 \phi_s) N_c \pi V_c / \lambda^4 x \qquad (82)$$

where $L_0 = A^2$ is the intensity of the incident light; μ' and μ are the optical densities (indices of refraction) of the crystal and the ambient medium, respectively; and ϕ_s is the angle between the path of the incident beam and the direction in which scattering is measured.

7. ABSORPTION SPECTRA OF PHOSPHORS. Figure 24a shows part of the *approximate* relative room-temperature absorption spectra of some typical phosphors whose crystals average about 1 to 10 microns in diameter (see Appendix 4). According to Mollwo, the absorption coefficients $a = x^{-1} \log_e(L_0/L)$ of transparent layers of oxides and sulphides of zinc and cadmium (produced by treating evaporated and condensed metal layers with oxygen or sulphur) have approximately the values given in Table 11.[282] With respect to the effect of variation

TABLE 11

	Absorption Edge, Å	Absorption Coefficient a, 10^4mm^{-1}						
		2000 Å	2500 Å	3000 Å	3500 Å	4000 Å	4500 Å	5000 Å
Cub.-ZnS $F\bar{4}3m$	3500	11.5	6	3	>0			
Hex.-ZnO $C6mc$	3850	5	3.5	2.7	2.7	0		
Cub.-CdS $F\bar{4}3m$	5120	5	4.7	2.6	2.1	1.7	1.0	0.2
Cub.-CdO $Fm3m$	5800	3.3	2.6	1.9	1.6	1.3	0.8	0.4

of operating temperature, the absorption and excitation spectra of hex.-ZnO:[Zn], ZnS:[Zn], ZnS:Ag, and ZnS:Cu phosphors shift to longer wavelengths at the rate of 0.4 to 1 Å °K^{-1} without large change in shape. Figure 24b shows the excitation spectrum of a ZnS:[Zn]

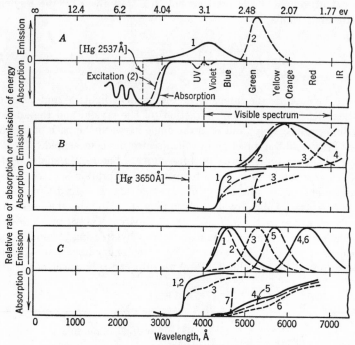

Fɪɢ. 24*a*. Relative absorption and emission spectra of some typical oxygen-dominated (silicate) and sulphur-dominated (sulphide) phosphors. The cub.-ZnS:[Zn] phosphor, especially when prepared in a reducing atmosphere, has a narrow edge-emission band located over the edge of the absorption band near 3500 Å. The edge-emission band appears at 90°K under excitation by UV,[154] and at 300°K under excitation by high-intensity CR.[354]

A	1. rbhdl.-Zn_2SiO_4:[Si], 1200°C	λ_{pk} = 4100 Å
	2. rbhdl.-Zn_2SiO_4:Mn(0.3), 1200°C	λ_{pk} = 5250 Å (2.36 ev)
B	1. hex.-ZnS·CdS:[Zn:Cd],[NaCl(2)], 780°C	λ_{pk} = 6000 Å
	2. hex.-ZnS·CdS:Ag(0.005),[NaCl(2)], 780°C	λ_{pk} = 5770 Å
	3. hex.-ZnS·CdS:Cu(0.005),[NaCl(2)], 780°C	λ_{pk} > 7100 Å
	4. hex.-CdS:[Cd], 800–1300°C	
C	1. cub.-ZnS:[Zn],[NaCl(2)], 780°C	λ_{pk} = 4650 Å
	2. cub.-ZnS:Ag(0.005),[NaCl(2)], 780°C	λ_{pk} = 4590 Å
	3. cub.-ZnS:Cu(0.005),[NaCl(2)], 780°C	λ_{pk} = 5290 Å
	4. cub.-ZnSe:[Zn],[NaCl(2)], 780°C	λ_{pk} = 6450 Å
	5. cub.-ZnSe:Ag(0.005),[NaCl(2)], 780°C	λ_{pk} = 5700 Å
	6. cub.-ZnSe:Cu(0.005),[NaCl(2)], 780°C	λ_{pk} = 6450 Å
	7. cub.-ZnSe:Mn(1),[NaCl(2)], 780°C	λ_{pk} = 6450 Å

phosphor at two different temperatures, as determined by Garlick and Gibson.[238]

The strong host-crystal absorption of rbhdl.-Zn_2SiO_4:Mn occurs at wavelengths below about 3000 Å (below 2500 Å, according to Kröger [154]),[96, 326, 328, 329] whereas the host-crystal absorption of hex.-ZnS·CdS (equimolecular proportions) occurs at wavelengths below about 4500 Å.[170] The fine structure of this strong absorption has not been determined with certainty; such a determination would have to be made on very thin sections of large crystals at low temperatures. In both cases, in Fig. 24a, the host-crystal absorption spectrum overlaps

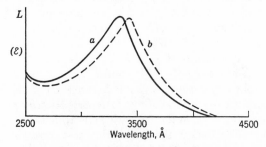

FIG. 24b. Excitation spectrum of a ZnS:[Zn] phosphor at 90°K (a) and 230°K (b).
(G. F. J. Garlick and A. F. Gibson)

the emission spectrum of the heated "pure" host crystal (without added activator). When small proportions of one or more activator impurities, for example, Mn, Ag, Cu, are incorporated in these host crystals, one or more additional weaker absorption bands usually appear on the long-wave side of the strong host-crystal absorption. As shown in Fig. 24a, the absorption (excitation) spectrum of the rbhdl.-Zn_2SiO_4:Mn phosphor is separated by over 1 ev from its emission spectrum, whereas these spectra overlap in the case of the sulphide phosphors with or without the Ag or Cu activators. In this respect, it has been reported that there is a 0.6 to 1.5-ev separation between the absorption and emission bands of alkali halides with Mn activator, whereas these bands adjoin each other when the same host crystals contain Ag, Cu, or Pb activator.[330, 331]

The absorption spectrum of a phosphor may be divided into several major regions: (1) the strong short-wave **host-crystal-absorption region,** (2) the intermediate **absorption-edge region,** where the strong host-crystal absorption merges into (3) the weak long-wave **imperfection-absorption region,** and (4) the weak longer-wave *induced* **trapped-electron absorption region.** With reference to Figs. 9 and 16c, the host-crystal absorption corresponds to transitions of electrons bound to host-crystal atoms at the instant of absorption, that is, transitions from normally filled levels of the host crystal (for example, E_H) to higher unoccupied levels (for example, E^{**}_E, E^{**}_C), whereas the imperfection absorption corresponds to transitions of electrons bound to (a) impurity atoms, that is, transitions from E_I to E^{**}_I, E^{**}_E, or E^{**}_C, or (b) host-crystal atoms near an imperfection, for example, transitions from E_H to E^{**}_0. The trapped-electron absorption corresponds to transitions of trapped electrons, for example, transitions from $E^*_{I_3}$ to $E^{**}_{I_3}$ or from (E_I-) to E^*_C. The absorption coefficient (a in eq. 72) in each region is proportional to the product of (1) the number N_B of bound electrons which can make a transition *from* a given state, and (2) the number N_S of unoccupied states *to* which the transition can be made. In the transition $E_H \rightarrow E^{**}_C$, both N_B and N_S are large; hence, $N_B N_S$ and a are large. In the transition $E_H \rightarrow E^{**}_0$, N_B is large, but N_S is small, and in the transitions $(E_I-) \rightarrow E^{**}_C$ and $E^*_{I_3} \rightarrow E^{**}_{I_3}$ or E^{**}_C, N_B is small while N_S is large; hence, $N_B N_S$ and a are small in these cases.

It is interesting to note that Frerichs reports stronger photoconduction, at 1 to 5 kv cm^{-1}, in the 50-Å-wide absorption-edge region (near 5200 Å) of hex.-CdS than in the neighboring host-crystal and imperfection-absorption regions.[291] This indicates that the filled bands E_H and E_I overlap in hex.-CdS. The *moderate* total photoconduction excited by radiation absorbed in the longer-wave host-crystal-absorption region is not limited by the product $N_B N_S$ (both N_B and N_S are large), but presumably by the high resistivity of the faulty less crystalline surface layers in which most of the short-wave primary radiation is absorbed; that is, the conduction band is relatively discontinuous throughout the more distorted surface layers, or "free" electrons become trapped and build up space charge. The *moderate* photoconduction excited by radiation absorbed in the imperfection-absorption region is limited by the low density of imperfections (low N_B) in the highly crystalline volume of the hex.-CdS crystal.

As the temperature of hex.-CdS is increased, the absorption edge shifts to longer wavelengths at the rate of about 1 Å °C^{-1}, and the photoconduction decreases steadily to a negligibly small value at 300°C. Frerichs also substantiates Hardy's report that infrared irradiation quenches the photoconduction of hex.-(Zn:Cd)S.[123] In the case of hex.-CdS, the quenching region extends from 8000 to 13,000 Å. Apparently, quenching radiations and increased temperature both increase the probability of recombination of trapped electrons and positive holes without long-range motion of the − or + charges.

8. EXCITATION SPECTRA OF PHOSPHORS. It is, of course, axiomatic that a phosphor crystal must absorb those photons which excite it, but it does not follow that all absorbed photons excite photoluminescence, even when the energy of the primary photon is greater than that of the emitted photon. In many cases, there is a fairly strong absorption commencing at a certain (ill-defined) wavelength λ_0, but excitation of detectable photoluminescence does not occur until the wavelength of the primary photon is made considerably smaller than λ_0. This reportedly occurs in cub.-KCl:Ag(0.003) which emits a band centered around 2720 Å and absorbs photons with wavelengths below 2300 Å, but requires primary photons with wavelengths below 2060 Å for the production of appreciable photoluminescence.[330] In this case, the energy of absorbed primary photons is converted into heat, or something other than photoluminescence, in the spectral region between the long-wave edge of the excitation band at 2060 Å and the long-wave edge of the absorption band at about 2300 Å. The material functions as a *pigment* in the spectral region from 2060 to 2300 Å and functions increasingly as a *phosphor* in the region immediately below 2060 Å.

The excitation spectra of phosphors, therefore, do not necessarily coincide with their absorption spectra, although luminescence is generally excited by photons whose energies lie in or near the region of strong host-crystal absorption. The rbhdl.-Zn_2SiO_4:[Si] phosphor, for example, is excited by 1-Å x rays but is not excited by 2537-Å UV, whereas the rbhdl.-Zn_2SiO_4:Mn(1) phosphor is weakly excited by 1-Å x rays, strongly excited by 2537-Å UV, and feebly excited by 3650-Å UV. The hex.-$ZnS \cdot CdS$ phosphors in Fig. 24a are excited by 1-Å, 2537-Å, and 3650-Å photons and are not excited appreciably by primary photons with wavelengths on the long-wave side of the strong host-crystal absorption edge, even though these phosphors are moderately absorbent in this imperfection-absorption region. An exception to the statement that phosphors are generally excited by photons absorbed in the host-crystal absorption region is reported by Kröger.[154] Kröger reports that cub.-Mg_2TiO_4:Mn has good luminescence with a quantum efficiency of about 55 per cent when excited by 3250-Å photons in the imperfection (activator)-absorption region but is practically unexcited by primary photons with wavelengths below 2800 Å, that is, in the host-crystal absorption region.

There are reports of weak self-excitation, that is, a phosphorescing layer exciting an adjacent layer of the same (unexcited) phosphor,[102,145] but experiments in these laboratories have shown that such excitation is negligible when the phosphor has a single simple emission band. In our experiments, phosphors (for example, (ZnS(80)CdSe(20):[Zn],

[NaCl(2)], 780°C) which have outstanding overlaps of absorption and emission spectra near 5461 Å,[170] were exposed to the 5461-Å line from an H4 mercury-vapor lamp, and an attempt was made to photograph luminescence radiation from the phosphor with a small grating spectrograph. No evidence of an emission spectrum was obtained; the spectrograph plates showed only the reflected 5461-Å primary photons.

In general, absorbed monochromatic radiation with wavelengths shorter than the host-crystal absorption edge will excite luminescence, although the efficiency of excitation often falls to very low values in the soft x-ray region. When excitation occurs, and when the phosphor has a single (simple) emission band, the *entire* luminescence emission band is produced, regardless of the wavelength of the primary radiation. For example, a rbhdl.-Zn_2SiO_4:Mn(0.3) phosphor at 300°K emits practically the same band (Fig. 52) under excitation by 1-Å photons, 2537-Å photons, 12-v CR, and 120,000-v CR. *The shapes and locations of single (simple) emission bands of phosphors are practically independent of the energy of the individual excitant particles* (it being assumed that the intensity of excitation is insufficient to increase the temperature of the phosphor crystals appreciably). For primary particles with energies much less than that corresponding to the long-wave absorption edge of the host crystal, there is an increased probability that the energy of the absorbed particle will be converted into heat rather than used to *excite* luminescence (the heat may thermostimulate phosphorescence if there are trapped excited electrons left over from previous excitations). In terms of energy-level diagrams, such as Figs. 9 and 16, it appears that strong luminescence is obtained when the primary particle has sufficient energy to raise electrons from the normally filled host-crystal bands (E_H, or lower levels) to exciton (E^*_E) or conduction (E^*_C) levels. When luminescence is excited by primary photons with energies lying *between* the long-wave side of the strong host-crystal absorption band and the short-wave side of the luminescence emission band, it is interpreted as corresponding to excitation transitions between normally filled activator levels (E_I) and higher unfilled levels (E^{**}_I, E^{**}_E, E^{**}_C). Such luminescences are generally rather weak, except in s-center phosphors where the luminescence action appears to be localized in the activator atoms (for example, Mn atoms in rbhdl.-Zn_2SiO_4:Mn).

The constitution of the host crystal and the nature of the activator center are both important in determining the excitation spectra of phosphors.

A. As examples, for phosphors whose anions are predominantly *oxygen:*

1. Effect of activator: Rbhdl.-Zn_2SiO_4:[Si], 1250°C and rbhdl.-Zn_2SiO_4:Mn, 1250°C are both excited by 50-kv copper-target x rays, but only the phosphor with Mn activator is excited by 2537-Å photons. In some cases, a second activator renders a phosphor photoluminescent under primary photons which do not excite the material containing only one activator (see Figs. 101 and 102).

2. Effect of host crystal: Rbhdl.-$Zn_2SiO_4 \cdot 0.01$Mn is only feebly excited by 3650-Å photons, whereas rbhdl.-$Zn_2GeO_4 \cdot 0.01$Mn is strongly excited by 3650-Å photons, and rbhdl.-$9ZnO \cdot BeO \cdot 6SiO_2 \cdot 0.05$Mn is also well excited by 3650-Å photons [note that the Mn proportion is held constant relative to the total cation $(Zn + Be)$ proportions]. Rhomb.-Mg_2SiO_4:Mn is only feebly excited by 2537-Å UV, whereas Mg_2GeO_4:Mn is strongly excited by both 2537-Å and 3650-Å UV. Also, cub.-$ZnAl_2O_4$:Mn and cub.-$ZnGa_2O_4$:Mn are both excited by x rays, but only the cub.-$ZnGa_2O_4$:Mn is excited to blue–green luminescence emission by 2537-Å UV.

B. As examples, for phosphors whose anions are predominantly *sulphur or selenium:*

1. Effect of activator: Hex.-ZnS:[Zn], hex.-ZnS:Ag(0.01), hex.-ZnS:Cu(0.01), and hex.-$ZnS \cdot 0.01$Mn(S) are all well-excited by 3650-Å photons, but only the latter two phosphors, with Cu and Mn activator, are well excited by 4358-Å photons.

2. Effect of host crystal: Longer wavelengths (in the absorption-edge region) may be used to excite cub.-ZnS, with or without Cu, Ag, or (to a lesser extent) Mn activator, with increasing degree of substitution of Se for S in the ZnS host crystal. Similar results obtain for increasing degree of substitution of Cd for Zn in hex.-ZnS with or without Ag or Cu activator.

C. As examples, cross-comparing oxygen- and sulphur-dominated phosphors:

The luminescence efficiency of hex.-ZnS:Cu is much higher than that of tetr.-$CaWO_4$:[W] under 2700–4500-Å photons, whereas tetr.-$CaWO_4$:[W] is much more efficient that hex.-ZnS:Cu under 2000–2600-Å photons.[326] Dr. D. H. Tomboulian [327] reports practically no luminescence response from hex.-ZnS:Ag, hex.-$ZnS \cdot CdS$:Ag, and hex.-ZnO:[Zn] under 150-Å and 1000-Å photons, but good response from rhomb.-$BaSO_4$:Pb and tetr.-$CaWO_4$:[W] under the same primary radiations. At very high primary-photon energies, it is important to have elements of high atomic number in the phosphor, in order to obtain appreciable absorption of the primary radiation. Moderate luminescence is obtained from hex.-ZnS:Ag, hex.-ZnO:[Zn], and rbhdl.-Zn_2SiO_4:Mn under hard x rays or soft gamma rays (≈ 500 kv), but higher luminescence efficiency is usually obtained from the heavy-element-containing hex.-$6ZnS \cdot 4CdS:Ag$, tetr.-$CaWO_4:[W]$, tetr.-$CaWO_4:Pb$, and rhomb.-$BaSO_4:Pb$.

Primary photons with different energies often excite luminescence (*during* excitation) and phosphorescence (*after* excitation) emission of different degrees, depending on the energy of the excitant photon. Certain phosphors which exhibit luminescence and appreciable *visible* phosphorescence under one excitant may exhibit only luminescence, with very little visible phosphorescence, under another excitant. Whenever phosphorescence is observed after excitation, however, luminescence is always observed during the excitation (assumed longer than 0.1-sec duration). As examples:

(1) $Ba_3Si_2O_8 \cdot 0.3Pb$ and $Ca_2P_2O_7 \cdot 0.001Dy$ are both excited to blue–white luminescence and long-duration visible phosphorescence ($\gg 0.1$ sec) by copper-target x rays (50-kv), but $Ba_3Si_2O_8 \cdot 0.3Pb$ is excited to blue–white luminescence and only short-duration phosphorescence (< 0.1 sec) by 2537-Å photons, and $Ca_2P_2O_7 \cdot 0.001Dy$ is excited to only orange–yellow luminescence and short-duration phosphorescence by 3650-Å (no appreciable excitation by 2537-Å photons).

(2) Rbhdl.-$Zn_2SiO_4 \cdot 0.01Mn$, $Cd_2B_2O_5 \cdot 0.01Mn$, orthorhomb.-$CdSiO_3 \cdot 0.01Mn$, and certain rbhdl.-$(Zn:Be)_2SiO_4:Mn$ phosphors are excited to luminescence and long-duration phosphorescence ($\gg 0.1$ sec) by copper-target x rays and 2537-Å photons but are excited to only luminescence and short-duration phosphorescence (< 0.1 sec) by 3650-Å photons.

(3) Rbhdl.-$Zn_2GeO_4 \cdot 0.01Mn$, cub.-$ZnS \cdot 0.01MnS$, hex.-$ZnS:Ag(0.01)$, and hex.-$ZnS:Cu(0.01)$ are all excited to luminescence and long-duration phosphorescence by 2537-Å and 3650-Å photons, but the intensity of phosphorescence is much greater after excitation by the 3650-Å photons. Also, cub.-$KCl \cdot 0.03TlCl$ is excited to luminescence without perceptible phosphorescence emission by 2537-Å photons, whereas excitation by photons with wavelengths below 2300 Å affords luminescence emission followed by phosphorescence emission which may be detected for several hours.[165]

(4) Some cub.-$ZnS \cdot 0.01MnS$ and hex.-$ZnS:Cu(0.01)$ phosphors exhibit more intense phosphorescence after excitation by 4047-Å photons than after excitation by an equal or greater intensity of 3650-Å photons.

In the first two cases, 1 and 2, the occurrence of luminescence and the absence of long-duration phosphorescence after excitation by the longer-wavelength primary photons is attributed to highly localized transitions, for example, between E_I and $E^*_{I_1}$ in Fig. 16c. These localized transitions generally afford **exponential decays** of phosphorescence,

$L \propto \epsilon^{-at}$, where the average lifetime of the excited state, $\bar{\tau} = a^{-1}$, rarely exceeds 0.1 sec (obtained for tetr.-ZnF_2:Mn), and so ϵ^{-at} decays are rarely visible longer than about 0.5 sec. At the shorter exciting wavelengths, transitions may be made to the exciton or conduction bands, for example, from E_H or E_I to E^*_E or E^*_C in Fig. 16c, which allow the excitation energy to wander farther afield in the crystal. Such nonlocalized excitations permit remote trapping of electrons and positive holes, and generally afford **power-law decays** of phosphorescence, $L \propto t^{-n}$ either alone, or superimposed on exponential decays. These t^{-n} decays may last for days, even at room temperature, especially when the traps are quite deep (optical trap depths greater than about 1 ev). In the last two cases, 3 and 4, the greater intensity of phosphorescence after exposure to the longer-wave (lower-frequency) primary photons is attributed to their greater penetration (Fig. 23) and lesser **energy deficit**:

$$\Delta h\nu = h\nu_{absorbed} - h\nu_{emitted} \qquad (83)$$

The quantity $h\nu_{emitted}$ remains constant, being determined by the phosphor, and so $\Delta h\nu$ increases as $h\nu_{absorbed}$ increases, that is, the energy wasted as heat, $E_h = \Delta h\nu$, increases as the frequency of the primary photon increases, as long as only *one* photon is emitted per primary photon absorbed. The deeper penetration and lower energy deficit of long-wave primary photons both tend to reduce the heat residue *per unit excited volume* in the phosphor, thereby allowing more excited electrons to be detained in shallow (low-binding-energy) traps, from which they may be ejected later by thermal agitation, to contribute to phosphorescence emission.

The excitation of intense *short-term* phosphorescence by primary material particles is generally less efficient than by long-wave ultraviolet photons. This is understandable in that (1) the $\overline{\Delta E}$ absorbed from charged material particles (eq. 75) is usually larger than 10 ev, whereas the $h\nu$ values of the commonly used mercury radiations are 2537 Å = 4.9 ev and 3650 Å = 3.4 ev, so that the E_h value is considerably greater in the case of the material-particle excitants; and (2) except for extremely high energies, the penetrations of the commonly used material-particle excitants are considerably less than those of the common UV excitants (Fig. 23).[34, 63] As an example, during the first few seconds of phosphorescence, the emission intensity of hex.-ZnS:Cu is much less after excitation by 10^4-ev CR or 10^6-ev alpha particles than after excitation by 3650-Å or 4358-Å photons. This is true, even though the power input of material particles per unit area be made the same or be made arbitrarily greater than the photon power input during excitation. *Long-term* phosphorescence, which involves deep traps, appears to be less affected by differences in the heat residues left by different excitants. For example, some phosphors with only very deep traps ($\approx$ 1-ev optical trap depth), such as cub.-$Sr(S:Se):SrSO_4:CaF_2:Sm:Eu$ and cub.-$SrS:SrSO_4:LiF:Sm:Ce$, are readily excited to a high degree of potential phosphorescence (stimulable by infrared) by beta particles, alpha particles, and gamma rays, as well as by ultraviolet photons. Similarly, the intensity of phosphorescence emission of rbhdl.-$Zn_2SiO_4:Mn$ with low Mn-content is higher after excitation by copper-target x rays than by ultraviolet.

9. EXCITATION DENSITIES PRODUCED BY PHOTONS AND BY CHARGED MATERIAL PARTICLES. Phosphors may be excited to different degrees of luminescence by photons and charged material particles, not only because these two types of particles are intrinsically different, but also because they differ in their ranges of available energies per particle and **particle densities,** or **excitation intensities** (numbers of particles passing unit area in unit time). That is, in addition to the marked differences in penetration, and the all-at-once versus bitwise loss of energy by photons and material particles, respectively, there are considerable differences in the individual energies and total beam intensities available with the two general types of particles. The spectrum of available primary photons with energies greater than about 4 ev is quite discontinuous, with the highest-energy (per particle) x-ray source affording photons with maximum energies of about 10^5 ev, and higher-energy photons being available from only a scattered few radioactive gamma-ray sources.[90, 91, 103, 104] Photons are uncharged, and so they must be focused into beams by lenses or other optical media whose absorption characteristics limit their use to a relatively small low-energy portion of the entire radiation spectrum. The absorption and low refraction of high-energy photons by matter make it difficult to produce beams of x-ray photons or gamma-ray photons except by inefficient collimation, or by mirrors receiving radiation at grazing incidence.[397] It is a relatively "easy" matter, however, to produce, focus and deflect beams of *charged* material particles with almost any energy from less than 1 ev to over 3×10^8 ev in modern high-vacuum devices such as cathode-ray tubes (CRT), cyclotrons, betatrons, and synchrotrons. Furthermore, the power intensities (watts cm^{-2}) of beams of charged material particles may be made enormously greater than the intensities of photon beams. Photon beams from unmodulated sources afford less than about 10^{19} 4-ev photons per square centimeter per second, whereas modern projection CRT can deliver a CR beam with about 10^{20} 25-kv primary electrons per square centimeter per second,[34] with higher intensities available at higher voltages. The 4-ev photons penetrate about 10^{-3}

cm in an "average" phosphor, and afford an *average* of about 10^{22} absorbed photons (excitation "bits") per cubic centimeter per second in the penetrated portion of the phosphor crystal. The 25-kv primary electrons penetrate about 10^{-4} cm and, assuming an *average* bitwise energy loss of 30 ev per inelastic collision in the solid, this CR beam affords an *average* of about 10^{26} excitation "bits" per cubic centimeter per second in the penetrated portion of the phosphor crystal. In both cases, the average density of excitation "bits" in the solid is much higher at the point of entry of the beam and much lower at the penetration "limit" (see the exponential power-loss curves in Fig. 22).

Experiments with hex.-ZnO:[Zn] and hex.-9ZnS·1CdS:Cu(0.0073) phosphors excited by low intensities of 3650-Å UV and 10-kv CR, have shown that the excitations are practically additive; that is, the luminescence output (during excitation) produced by both excitants simultaneously is equal to the sum of the luminescence outputs excited by the excitants separately. Departures from additivity may be expected, of course, at very high excitation densities approaching saturation of the luminescence mechanism.

The excitation process in **triboluminescence,** that is, luminescence resulting from the fracturing of crystals, is believed to be essentially due to electron bombardment. As the crystal planes separate along the fracture, there is usually an unbalance of charge on the two new faces, and, since the distance between the faces is quite small in the initial stages of the fracture, the high electric field gradient causes electrons to leave the negative face and strike the positive face **(triboelectric effect).** The resultant electron bombardment excites cathodoluminescence in the phosphor and may excite luminescence in an intervening gas, for example, atmospheric nitrogen.[332, 333] An inrushing gas in the fracture may add to the excitation process by frictional ionization and subsequent attraction to the charged faces (compare with the faint-blue [nitrogen] glow observed when a roll of friction tape is unrolled in the dark). Some phosphors, notably tetr.-ZnF$_2$:Mn, cub.-ZnS:Mn, and Ca$_2$P$_2$O$_7$:Dy, display remarkable triboluminescence when lightly ground or agitated in a hard container. A friable cake of unground Ca$_2$P$_2$O$_7$:Dy phosphor may be made to luminesce and phosphoresce by simply scratching it with a fingernail.

Triboluminescence is a consequence of the triboelectric effect when there is sufficient electric discharge and the surfaces separated or the intervening gas are cathodoluminescent. Triboelectricity and triboluminescence are observable also when certain liquid or solid particles are rolled or shaken in vessels of different electron affinities and electric polarizabilities. Under these conditions, the foreign particles become electrically charged relative to the immediate wall of the container. [This **contact electrification** is somewhat analogous to that exhibited in colloidal dispersions, where the substance having the greater dielectric constant charges positively (loses electrons) with respect to the substance of lower dielectric constant (Coehn's law).] [324] When a charged foreign particle is moved to a new location on the oppositely charged surface, the mechanical energy disrupts the initial electrostatic equilibrium, and a triboelectric discharge occurs during this disruption and re-establishment of potential difference. A familiar example of this type of triboelectrically produced cathodoluminescence is found in the light emitted during the shaking of drops of mercury in clean fused-silica containers. Here, the mercury vapor is excited to luminescence by the discharge between the mercury drops and the wall of the insulator vessel. By coating the wall with phosphors which are excited by UV radiation, the color of the emitted light may be varied and the visual efficiency increased. [332] Under suitable conditions, the phosphor coating (or agitated phosphor particles in a suitable container) may be excited directly by the electron discharge.

Although the excitation processes which occasion **chemiluminescence** and **bioluminescence** have not been determined in detail, the excitation energy in both cases is derived from energy transformations during chemical reactions. When these energy transformations afford a new minimum of potential energy, by rearrangements of electron configurations and internuclear distances (see Fig. 3), some of the excess energy may be used to raise an electron of one of the reactants (or a ligand of the reactant) to an excited state from which a radiative transition is possible. [34, 81] One of the most spectacular chemiluminescence reactions is the oxidation of 3-aminophthalhydrazide (luminol) in alkaline solution. [334] The quantum efficiency of this outstanding chemiluminescence reaction is about 0.3 per cent, a value which is indicative of the generally low luminescence efficiencies of chemiluminescence and bioluminescence. [334-340]

Energy Transformations and Storage Processes

It may be restated, for emphasis, that phosphors are generally structure- and impurity-sensitive crystalline inorganic materials which luminesce appreciably, that is, emit detectable amounts of nonthermal radiation (comprising photons with energies in excess of about 1 ev), when excited by primary photons or (charged) material particles having energies in excess of about 2 ev. Also, a beam of primary particles penetrating a phosphor crystal decreases in particle density and total energy approximately as ϵ^{-ax}, by absorption and scattering, and the absorption of a quantum ("bit") of energy (> 2 ev) from the primary beam by an atom raises one of the atom's bound electrons to a higher energy state (simple excitation), or the electron may be freed from the atom ("internal ionization"). For excitation of (conventional) luminescence, the useful part of the total energy of a primary particle is its *free energy.* The **"free energy"** is the energy in excess of the particle's self energy ($m_0 c^2$ for material particles, zero for photons, compare Table 1) and is referred to simply as the **energy** of the particle. Only in gamma-ray fluorescence does the self-energy of the particles appear to play an appreciable part in the luminescence action.

If attention be focused on a particular atom in a solid which has just absorbed several electron volts of (free) energy from a primary particle, the absorbed energy may be converted wholly or partially, *at that spot in the crystal,* into (1) a luminescence photon, and/or (2) one or more phonons. If the absorbed energy is not converted entirely into a photon or phonons at the point of absorption, the absorbed energy (or residual energy after 1 and 2) may be transmitted through the crystal as (3) an excited "free" electron and its residual positive hole, (4) an exciton, (5) resonance transfers, or (6) imperfect collisions of the second kind (4, 5, and 6 are probably indistinguishable in practice and may be grouped as ψ-overlap energy transfers). The energy transported from the point of absorption by any of these means may be absorbed at some other place in the crystal; whereupon, in all cases except 2, there is again a probability of local conversion into a luminescence photon, or into phonons, as well as a probability of the energy being wholly or partially retransmitted and reabsorbed repetitively. Conversion of excitation energy into phonons represents an irreversible loss insofar as the *direct* production of luminescence is concerned and, in fact, usually decreases the general probability of obtaining further luminescence. The major competition during energy transformations and transmissions in phosphors is that between conversion of excitation energy into (1) *photons* with energies greater than about 1 ev, and (2) *phonons* with energies less than about 0.04 ev. Because an atom in a crystal perturbs its neighbors upon absorption or emission of energy, part of the excitation energy is necessarily consumed as increased atomic motion. Increasing atomic motion offers increasing interference with subatomic radiative-transition processes and with energy transport through crystals. Ideally, then, luminescence should be favored most at the absolute zero of temperature where atomic motion is least, and purely electronic processes should proceed with a low probability of excitation energy being degraded into heat or chemical change. This is evidenced by the observation that some solid activator compounds, which are nonluminescent at room temperature, become luminescent at very low temperatures. As examples, it is reported that (1) rose-colored tricl.-$MnSiO_3$ is nonluminescent at room temperature but exhibits a red luminescence emis-

TABLE 12

SOME VISIBLE PHOTOLUMINESCENCE EMISSIONS OBSERVED AT ROOM TEMPERATURE AND AT LIQUID-AIR TEMPERATURE

Solid Substance	25°C (298°K), Excitation by 3650 Å	25°C (298°K), Excitation by 2537 Å	−180°C (93°K), Excitation by 3650 Å	−180°C (93°K), Excitation by 2537 Å	Remarks
1[a] cp Cu_2Cl_2	....	....	Faint red	....	Becomes green, then inert, on warming from 90°K.
2 cp CuI_2	Violet-purple	....	Violet	....	Violet $\lambda_{pk} \approx 4280$ Å; Red $\lambda_{pk} \approx 6400$ Å; *λ_{pk} at 5400 Å shifts to 5700 Å on warming.
3[b] cp $AgCl$	....	....	Bright blue-green *	Blue-green	
4 cp Ag_2SO_4	....	....	Bright yel.-orange	Very bright blue	
5[c] cp $MnBr_2 \cdot 4H_2O$	Red	Faint red	Orange	Faint orange	
6 cp Mn borate	....	....	Faint white	Faint red	
7 cp $MnSO_4 \cdot 4H_2O$	Faint white	....	Faint red	....	
8 cp $Mn(C_2H_3O_2)_2 \cdot 4H_2O$	Faint red	....	Bright orange	....	
9[d] cp Bi_2O_3	....	....	Pink	....	
10 cp Bi borate	....	....	Blue	....	
11 cp $BiONO_3$	....	....	Bright yel.-orange	....	
12[e] cp TiO_2	....	....	Tan-orange	....	
13 cp TiF_4	....	....	....	....	
14[f] cp ZrO_2	Pale blue	Pale blue *	Pale green-yel.	Green	*Phosphoresces for several seconds. Has glow curve.
15 cp $ZrSO_4$	Pale green-white	....	Green-white *	....	*Green phosphorescence at 90°K
16[g] cp ThO_2	Pale blue	....	Pale blue	....	No change
17 cp $ThSO_4$	Blue	....	Bright blue	....	*Green phosph., at 90°K, appears to be exponential, e^{-1} value ≈ 1 sec.
18 cp $Th(NO_3)_4$ (Bakers Anal.)	....	....	Green *	....	*Green phosph. as for No. 17.
19 cp $Th(NO_3)_4$ (Kahlbaum)	Blue-white *	....	Bright blue-white †	Faint blue-white †	*White phosph. at 300°K; † green phosph., at 90°K, as for No. 17.
20 Hex.-CdS (sublimed in H_2 at 1000°C)	Green	Green	Orange	Orange	Luminescence-pure CdS
21 Hex.-CdS (sublimed in H_2 at 1000°C)	Red	Red	Green	Green	Luminescence-pure CdS
22 Li_2SiO_3	Faint green	Faint green	Faint green	Faint green	No change.

CHEMICALLY PURE (CP) SOLID SUBSTANCES WHICH HAD NO VISIBLE PHOTOLUMINESCENCE EMISSIONS AT 90°K AND 300°K UNDER 3650-Å OR 2537-Å UV

(a) $CuF_2 \cdot 2H_2O$, $CuCl_2$, $CuBr_2$, $CuCO_3$, $CuSO_4$, Cu borate, $Cu(NO_3)_2 \cdot 3H_2O$

(b) Ag_2O, $AgNO_3$, Ag_2CO_3

(c) MnF_2, $MnCl_2 \cdot 5H_2O$, MnO_2, $MnCO_3$, $MnSiO_3$ (black), $Mn_3(PO_4)_2 \cdot 7H_2O$, MnS, $MnC_2O_4 \cdot 4H_2O$

(d) $Bi(NO_3)_3$

(e) $Ti(NO_3)_4$

(g) $ThCl_4$

Misc.: $PbSe$, $PbSO_4$, Cr_2O_3, $NaCl$, KCl, $BaCl_2$, $SrCl_2 \cdot 6H_2O$, $SrAl_2O_4$, Li_2SO_4, Li_2CO_3, ZnS or CdS (precipitated and dried at 100°C), H_3BO_3.

sion band extending from 6000 to over 7500 Å ($\lambda_{pk} \approx$ 6650 Å) at 90°K; (2) hex.-$MnCl_2$ is nonluminescent at room temperature but exhibits a narrow emission band near 6325 Å at 20–80°K; and (3) cub.-Cu_2Cl_2 is nonluminescent at room temperature but exhibits a red luminescence emission near 80°K and a green luminescence emission at about 140°K (the color change may accompany a structural transition occurring between the two temperatures); [154, 347] and (4) various compounds of Ce^{3+}, for example, $CePO_4$, CeF_3, $Ce_2(CO_3)_3$, and $Ce_2(SO_4)_3$, exhibit narrow overlapping-doublet emission bands peaked near 3500 Å when excited at 90°K (these bands are called **edge emission bands,** because the emission spectrum overlaps the long-wave edge of the absorption band of the solid). [154, 199] Some additional data, obtained in these laboratories, are given in Table 12. As is discussed later, however, some phosphors, such as $Cd_3(BO_3)_2$:Mn and hex.-ZnS:Ag, exhibit much lower photoluminescence efficiency at 150 than at 300°K. [58, 63]

In those cases where all the absorbed excitation energy is converted into photons and phonons at the point of absorption, that is, the excited electrons remain attached to their parent atoms, and metastable states are absent, the radiative return to the ground state approximates a spontaneous monomolecular process ($L \propto \epsilon^{-at}$) without altering the electric conductivity of the crystal. This procedure apparently obtains in ʻcertain predominantly exponential-decay phosphors, such as tetr.-ZnF_2:Mn and rbhdl.-Zn_2SiO_4:Mn, excited by photons with sufficiently low energy ($<$ 10 ev) to be absorbed by the activator centers rather than by atoms of the host crystal. [34] Wandering of the excited state, by excitons or resonance transfer, is improbable in these cases, because the energy levels of the activator and host-crystal atoms are mismatched. In other cases, however, the occurrence of strong conduction paralleling nonexponential decay of phosphorescence and the ability to freeze-in potential luminescence energy by exciting the phosphor at a very low temperature indicate that ionization and remote electron trapping play important roles in the luminescence process. This procedure apparently obtains in certain long-phosphorescing sulphide-type phosphors, such as cub.-Sr(S:Se):$SrSO_4$:CaF_2:Sm:Eu and hex.-ZnS:Cu, which exhibit photoconduction and power-law decays of phosphorescence ($L \propto t^{-n}$) after excitation by low-energy photons.

When the energies of the primary particles are high ($\gg$ 10 ev), then the more numerous host-crystal atoms absorb most of the input energy in all cases, and the absorbed energy must be transported to the activator centers in "bits" exceeding about 2 ev in order to obtain efficient luminescence. Under these circumstances, there may be either an exponential decay of phosphorescence without appreciable concomitant electric conduction or a power-law decay with conduction, or both, depending on the nature of the phosphor. When the site of energy absorption is remote from the site of photon emission, the absorbed energy must be transmitted to the remote activator centers as excited electrons, positive holes, or ψ-overlap transfers. The excitation energy delivered to the activator center may then produce simple excitation (no conduction observed) or internal ionization (conduction observed), according to the magnitude of the transmitted energy "bit" and the nature of the activator center. When power-law decay is observed without appreciable conduction, as in some luminescent glasses, the excitation energy is absorbed efficiently (insofar as luminescence is concerned) only by the activator centers, and the trapped electrons remain in the centers. Energy transfer through distorted, vitreous, or inhomogeneous structures is inefficient because the energy-level diagram varies, with possible discontinuities, throughout the distorted host material, and the distortions introduce local "forbidden" levels which promote nonradiative transitions.

Temporary storage of (free) energy may occur in or on phosphor crystals by trapping of (1) excited quasifree electrons, (2) positive holes, and (3) excitons. Trapping is most probable in the neighborhood of crystal imperfections, although there is the possibility that an excited electron, for example, may "dig its own hole" by providing its own imperfection, for example, by producing a **polaron** which is a polarization-type perturbation in the host crystal (compare Fig. 8d).[35, 341, 686] The evidence for trapping in some phosphor crystals is quite conclusive. When phosphors such as Nos. 5–19, 23, 35, and 46 of Table 5 are excited at about 80°K, their luminescence emissions decrease abruptly when the excitation ceases, even though most of these phosphors have intense phosphorescence emissions at room temperature (300°K). That there is still some stored excitation energy in the crystals may be demonstrated by allowing the previously excited cold phosphor to warm slowly, whereupon thermostimulated phosphorescence emission is obtained. As shown in Fig. 25, the thermostimulated emission is a function of temperature, and the resultant curves of phosphorescence output as a function of temperature of the phosphor are called **glow curves.**[58, 221] The different component bands of the glow curves in Fig. 25 are interpreted as being evidence for different trapping states, the higher-temperature bands corresponding to greater trap depths [stronger binding in (to) the trap]. The phosphor operating temperature T at the peak of each glow-curve band for very slowly rising temperature is related to the corresponding thermal trap depth (thermal activation energy) E^* by the *approximation* $E^* \approx 30kT$. This approximation comes from $\nu_a \epsilon^{-E^*/kT}$ which expresses the probability per unit time of thermal emptying of traps wherein electrons make ν_a attempts to escape per second. If the **attempt frequency** ν_a is of the order of the highest vibration frequency of the atoms in the crystal, that is, about 10^{13} sec^{-1} (eq. 39), then, for $E^* = 30kT$,

$$\nu_a \epsilon^{-E^*/kT} \approx 1 \text{ sec}^{-1}$$

and so the average lifetime of an electron in a trap under the given conditions is about one second, or N electrons have a mean escape rate of N per second. From the foregoing equation, or the equivalent $E^* \approx kT \log_\epsilon \nu_a = 2.3kT \log_{10} \nu_a$, it may be calculated that $E^* \approx 25kT$ when $\nu_a = 10^{11}$ sec^{-1}, $E^* \approx 21kT$ when $\nu_a = 10^9$ sec^{-1}, and $E^* \approx 16kT$ when $\nu_a = 10^7$ sec^{-1}. Garlick has reported that ν_a is apparently of the order of 10^9 sec^{-1} for trapped electrons in several zinc-sulphide-type phosphors,[237, 238] and Williams reports $\nu_a \approx 10^{10}$ sec^{-1} for a tetr.-ZnF$_2$: Mn phosphor.[262] According to these reports, the attempt frequency ν_a is considerably less than the atomic-vibration (phonon) frequency ν_{at}. If ν_a is proportional to ν_{at}, then ν_a should vary slowly with change in temperature of the solid, because ν_{at} is determined largely by the atomic masses and interatomic binding energies, and has an upper limit ν_m which is determined by the minimum identity translation distance of the crystal (eqs. 54 and 55).

According to Fig. 25, the long visible phosphorescence emission of hex.-ZnS:Cu is associated with moderately deep traps which are occasioned *by the Cu activator* (Fig. 25*d*). Figure 25*b* shows that cub.-ZnS:Cu has mainly shallow traps, in keeping with the less intense shorter-persistent phosphorescence of cub.-ZnS:Cu relative to hex.-ZnS:Cu (see Fig. 80*b*).[63] Figures 25*a* and 25*c* show the pronounced effect which (1) infinitesimal traces of so-called **killers (poisons),** such as nickel compounds, and (2) variations of the composition of the host crystal have on the glow curves of this system of phosphors.[119, 120] Figure 25*e* shows that the deeper (higher-temperature) traps are filled first during weak excitation at low temperatures, and Fig. 25*f* shows that the shallower traps do not remain filled during excitation at elevated temperatures (compare curves 1 and 3). Figure 25*g* shows that there is conduction current i which parallels the thermally released luminescence emission L of a previously excited phosphor [an intensive study of trapping in cub.- and hex.-ZnS:Cu(0 to 0.3) phosphors is being done by R. H. Bube in these laboratories].

Glow-curve bands lying below about 290°K indicate relatively shallow traps which empty rapidly at room temperature, whereas glow-curve bands lying well above 290°K indicate deeper traps which empty

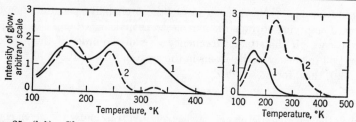

FIG. 25a (*left*). Glow curves for a hex.-ZnS:Cu phosphor made with and without a trace of Ni.

1. hex.-ZnS:Cu(0.01), 1100°C
2. hex.-ZnS:Cu(0.01):Ni(0.0005), 1100°C

FIG. 25b (*right*). Glow curves for ZnS:Cu(0.01) phosphors crystallized at different temperatures.

1. cub.-ZnS:Cu, 700°C
2. hex.-ZnS:Cu, 1100°C

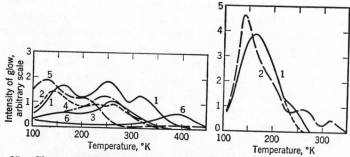

FIG. 25c. Glow curves for hex.-ZnS:CdS:Cu(0.01) phosphors crystallized at 1100°C. Curves 4, 5, and 6 are multiplied by 5, 25, and 25, respectively.

Curve	1	2	3	4	5	6	
ZnS	100	95	85	70	55	45	Weight
CdS	0	5	15	30	45	55	per cent

FIG. 25d. Glow curves for hex.-ZnS:[Zn] and the same phosphor reheated at 400°C with sufficient Cu salt to give 0.01% Cu.

1. hex.-ZnS:[Zn], 1100°C
2. No. 1 mixed with 0.01 weight per cent Cu and reheated at 400°C

FIG. 25. Glow curves (thermostimulated phosphorescence emission as a function phosphors. (G. F. J. Garlick and A. F. Wells,[58,221]

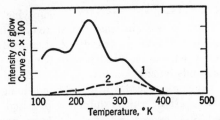

FIG. 25e. Glow curves for a hex.-ZnS:Cu phosphor after strong and weak excitation by 3650-Å UV. The area under curve 1 is actually 1300 times the area under curve 2.

1. Strongly excited at 90°K
2. Weakly excited at 90°K

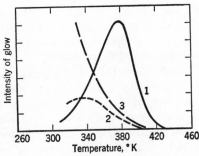

FIG. 25f. Glow curves for a hex.-ZnS:Cu phosphor after various excitations by 3650-Å UV.

1. Strongly excited at 370°K and then cooled to 295°K
2. Weakly excited at 295°K
3. Strongly excited at 295°K (curve reduced in height)

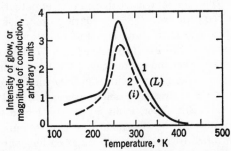

FIG. 25g. (1) Glow curve, and (2) concomitant curve of electric conduction current versus temperature for a previously cooled and photoexcited hex.-9ZnS·CdS:Cu phosphor layer clamped between oppositely charged electrodes.

of slowly increasing temperature after excitation by UV) of some long-persistent also M. H. F. Wilkins and G. F. J. Garlick [238a]) slowly at room temperature. The rate of emptying decreases with increasing trap depth (or with increasing peak temperature of the component glow-curve band) and increases with the temperature of the phosphor. Phosphors which exhibit short persistence at room temperature have glow-curve bands lying below 290°K, those which exhibit intense long persistence have pronounced glow-curve bands lying between about 290°K and 450°K, and those which exhibit abrupt initial decays followed by long weak phosphorescence emissions at room temperature have glow-curve bands lying above about 450°K. Some phosphors apparently have only very shallow traps and very deep traps, with few traps in the intermediate region. For example, the highly photoluminescent and cathodoluminescent cub.-MgS:Sb phosphor has a very rapid initial decrease in afterglow [down to 1 per cent of its luminescence emission (during excitation) in about 10^{-6} sec after intense excitation], but thereafter there is a long feeble afterglow which may be correlated with two glow-curve bands peaked near 400 and 500°K.

In phosphors Nos. 20–22 in Table 5, the phosphorescence decreases rapidly to vanishingly low values even at room temperature but becomes intensely thermostimulated above 500°K. This indicates a *thermal* trap depth kT of a few tenths of an electron volt, for example 0.65 ev.[221, 263] The same phosphors may be strongly photostimulated by 10,000-Å photons, and some photostimulation is obtained out to about 14,000 Å. This indicates an *optical* trap depth hc/λ of about 0.9 ev minimum, or about 1.3 ev for maximum photostimulation. (The difference between the thermal and optical trap depths is discussed in the next paragraph.) As an example of these deep-trap phosphors, Fig. 26 shows the intensity-time and spectral characteristics of a cub.-Sr(S:Se):SrSO$_4$:CaF$_2$:Sm: Eu phosphor which exhibits remarkable storage of potential phosphorescence at room temperature.[63, 221, 342, 343] During excitation by blue light ($\lambda \approx 4500$ Å) this phosphor emits a yellow radiation in the form of a simple *band* (compare with Fig. 14!) peaked near 5750 Å and acquires an induced stimulation spectrum peaked near 10,000 Å. In one experiment, when the blue excitation light was turned off, the luminescence decreased in about a second to a weak phosphorescence below about 10^{-3} mL (compare Appendix 2). After about 20 sec waiting in the dark after cessation of excitation, the phosphor was photostimulated by 8000- to 30,000-Å infrared from a 26-watt tungsten-filament lamp 16 cm distant (radiation passed through a Corning 254 filter). The photostimulated emission comprised the same yellow band as before, giving an initial luminance of 5 mL, and decaying to 1 mL in about 40 sec under the infrared irradiation. Higher intensities of infrared afford greater initial luminances of photostimulated phosphorescence, followed by more rapid decreases of luminance with time during stimulation. Some of the phosphors of this type can retain over half their initial stored potential phosphorescence for six months at room temperature and for even longer times at lower temperatures.[58, 171–180, 221, 342–344]

The infrared-stimulable phosphor, cub.-$SrS:SrSO_4:CaF_2:Sm:Eu$, has an **optical trap depth** (minimum) of about 0.9 ev ($\approx$ 14,000 Å), whereas its **thermal trap depth** is about 0.7 ev (glow-curve maximum near 550°K; trap depth determined from plot of log L vs. $1/T$, where L is the

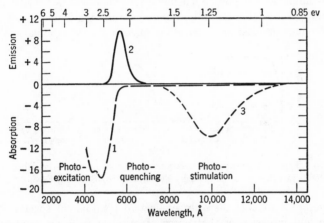

FIG. 26a. The excitation (1), emission (2), and induced stimulation (3) bands of a cub.-$Sr(S:Se):SrSO_4:CaF_2:Sm:Eu$ phosphor.[58]

thermostimulated phosphorescence emission). The disparity between the two activation energies required to release electrons from the same traps represents the difference between the mechanism of excitation (stimulation) by absorption of photon energy and by absorption of phonon energy. The occasional *photons* release electrons from traps by excitation *at the instant the stimulating photon is absorbed*, and require a relatively large energy to raise electrons from a low average "ground-state" level (in the trap) to a higher level (for example, from a_1 to a^*_3 in Fig. 3, or from a to b in Fig. 16b). The omnipresent *phonons*, on the other hand, produce ν_a *attempted* releases per second and may release electrons from traps *at the most propitious times*, that is, when the trapping energy is weakest and a relatively small energy is required to raise the trapped electron from a high "ground state" level (in the trap) to a free excited-state level (for example, from $E^*_{I_2}$ to $E^{**}_{I_2}$ in Fig. 16c).

As shown in Fig. 26a, there is a photoquenching region between the emission band and the induced photostimulation band of the given phosphor. Sometimes there are two or more photoquenching bands which may be produced independently of the photostimulation band (Frerichs, PB56304 [291]). Irradiation of the previously excited phosphor with photons in the quenching region results in a conversion of potential phosphorescence-emission energy directly into heat. This effect occurs in the presence (Fig. 26b) or absence of photostimulating photons and is discussed under Stimulation and Quenching of Phosphorescence in this chapter. For the present, it should be noted that irradiation of a photostimulable *and* photoquenchable phosphor with a broad band of primary photons (for example, unfiltered tungsten-filament radiation) may produce simultaneous excitation, quenching, and stimulation. The

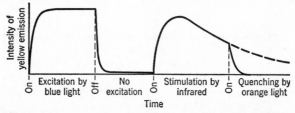

FIG. 26b. Typical intensity–time relationships for an infrared-stimulable, orange-light-quenchable phosphor such as cub.-Sr(S:Se):SrSO$_4$:CaF$_2$:Sm:Eu.

relative proportions of these three effects depend on the spectral distribution and intensity of the source of primary radiation, and on the spectral sensitivity curve of the phosphor at the given temperature. Where the photoquenching and photostimulation regions overlap,[145, 221] as they do in phosphors Nos. 20–23 in Table 5, it is important to avoid the overlap region when making experimental studies of photoquenching versus photostimulation, or of photostimulation versus thermostimulation.

More complicated generalized configurational-coordinate versus energy diagrams, such as the one shown in Fig. 27, are sometimes invoked in describing luminescence phenomena.[34–36, 221, 263, 267] The almost unlimited plasticity and multiplicity of such diagrams makes it possible to "explain" practically any set of experimental observations on phosphors. For example, the phosphor center "described" by Fig. 27 may exhibit fluorescence (emission from *D* to *E*) after excitation from *A* to *B*, and phosphorescence (with intermediate trapping in *F*) after excitation from *A* to *C*. Infrared, of energy *G* − *F*, would stimulate phosphorescence, whereas orange light, of energy *H* − *F*, would cause quenching by allowing a radiationless return to the ground-state level along line *HEA* (excitation energy given up as phonons). If phosphorescence is excited by longer wavelengths than is the case for fluorescence, then it is quite easy to draw line *CFE$_2$** so that *C* lies below *B*, and, if heat causes thermostimulation rather than quenching, it is a simple matter to redraw the central part of the diagram so that J lies below I. Similarly, the relative levels of D and F may be maneuvered to fit the facts. It is possible, also, to "allow" or "forbid" electron transitions from one level to another, at the real or apparent intersection of lines, or elsewhere, by invoking *ad hoc* selection rules, for example, postulating the same or different multiplicities for the different levels. Descriptions of this type are artistically stimulating, and some may incorporate verity, but they have yet to achieve practical significance. When it is possible to label the various abstract

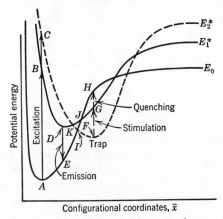

Fig. 27. Hypothetical schematic of some possible energy changes of a system (center) during luminescence.

levels with specific values related to the compositions and structures of the phosphor crystals, then these energy-state diagrams may become useful in phosphor research.

Emission Processes (Spectral Characteristics)

After a 2-ev or larger quantum of excitation energy has been absorbed somewhere in a phosphor crystal, and has survived transformation and storage processes without being reduced below about 1 ev, it may be transformed into a luminescence photon which can emerge from the crystal if the photon is not reabsorbed and converted into heat (see Appendix 4). Since the emitted radiation from phosphors is of practical importance, many measurements have been made on the intensities and spectral distributions of the luminescence photons which *escape* from thin screens made of myriads of tiny phosphor crystals. In contrast with the great difficulties encountered in measuring the absorption spectra of phosphors, measurement of their emission spectra is relatively easy.[93-98] For this reason, thousands of spectral-distribution curves of phosphor emission spectra are available (though not all are published) for study and interpretation in terms of the compositions and structures of their corresponding materials. These data do not afford complete information on the internal processes of emission, because only those photons are detected which have a high probability of escaping from the crystals. It is possible, for example, that in some cases there may be appreciable *internal* emission of ultraviolet, or infrared, which is undetected because the host crystals are strongly refractive and absorbent in the given spectral region. Similarly, if a phosphor screen selectively absorbs its own luminescence emission, the absorption probability is greatest for photons emitted from crystals lying farthest from the luminescence detector (for example the human eye, or other photosensitive device).

Most of the useful phosphors, as shown in Table 5, emit spectral *bands* which denote the presence of broad energy bands in the solid. The broadenings arise from (1) the extension of the ψ's of the host-crystal valence electrons over an entire single host crystal (giant molecule), coupled with the operation of the exclusion principle which requires all the electrons in such a giant molecule to have different quantum numbers; (2) the statistical distributions and differences of location and, hence, of energy, of impurities and other imperfections in a phosphor crystal, and the corresponding statistical distributions of perturbations which such imperfections cause in the energy bands of the host crystal; and (3) the perturbations caused by thermal agitation, and by atomic displacements accompanying absorption and emission of energy during luminescence. The spectral distribution of emission of a phosphor is, therefore, a function of (1) the composition, structure, and intrinsic degree of perfection of the phosphor host crystal; (2) the kinds and amounts of cocrystallized impurities, and their bond types, effective valencies, coordination numbers, and locations in the solid; and (3) the temperature of the phosphor during luminescence. This is exemplified by the data in Figs. 28–47.

It should be noted that most of the emission bands in Figs. 28–46 are quite simple in shape, especially when only one activator predominates.[58, 61,63,72,157,231, 312,349,352] There are some major discrepancies in the literature regarding the locations and structures of the emission spectra of such common phosphors as hex.-ZnS:Cu and hex.-ZnS:Ag.[145–147,254,265,268,350,351] The discrepancies arise in part from the use of photographic-measurement techniques, without adequate calibration of the spectral sensitivity of the emulsion, and in part from inadequate care in synthesis; that is, many of the phosphors cited in the literature contain unknown contaminants and incompletely reacted ingredients. This does not mean that all well-prepared phosphors with dominant single activators should have single simple emission bands; in fact, some phosphors, such as rbhdl.-Al_2O_3:Cr, Mg_2GeO_4:Mn, Mg_4GeO_6:Mn (Table 5), and $ThSiO_4$:Eu (Fig. 31) have very complex emission spectra comprising lines and narrow bands. Most of the useful phosphors, however, have broad emission bands which are practically devoid of fine structure at room temperature.[93,349]

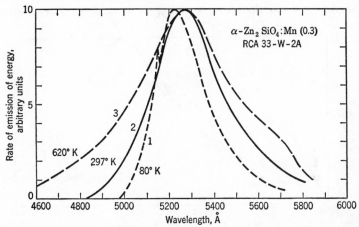

FIG. 28a. Relative spectral-distribution curves of the emission of a rbhdl.-Zn_2SiO_4:Mn(0.3) oxygen-dominated phosphor excited by 2537-Å ultraviolet at three different operating temperatures. Peaks not drawn to scale. The data were obtained with a grating spectrograph and microdensitometer and are corrected except for the (small) variation in spectral sensitivity of the emulsion. (R. E. Shrader)

	λ_{pk} Å	Relative Peak Output
1.	$\begin{cases} 5154 \text{ (line)} \\ 5227 \text{ (band)} \end{cases}$	150
2.	5275	100 (std.)
3.	5275	30

Figures 28–30 show the various effects of temperature (of the luminescing material) on the spectral-distribution curves of the emission bands of several phosphors.[158,352] In general, the emission bands tend to become narrower and sometimes develop lines as the temperature of the luminescing material is decreased. For example, rbhdl.-Zn_2SiO_4:Mn, which has a smooth band at room temperature, exhibits some line structure at 80 and 20°K.[347,348] For the most part, however, the emission bands are at least 100 Å wide at temperatures as low as 80°K. This corresponds to an energy breadth in excess of about 0.05 ev, or a value over seven times as large as that of kT ($= 0.007$ ev at 80°K). Furthermore, the broadening produced by a given change in temperature is

considerably greater than that calculable directly from the change in kT. The difference in half-widths of the emission bands of rbhdl.-Zn_2SiO_4:Mn at 80 and 300°K is about 0.07 ev, whereas the corresponding change in kT is only about 0.019 ev. This indicates that there may be a leverage factor, perhaps due to rapidly varying curves of potential-energy versus configurational-coordinates representing the excited and ground states

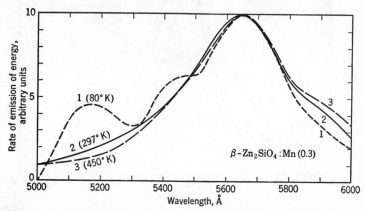

Fig. 28b. Relative spectral-distribution curves of the emission of a β-Zn_2SiO_4:Mn(0.3) oxygen-dominated phosphor excited by 2537-Å ultraviolet at three different temperatures (not shown in curve 1 are two very weak *lines* at approximately 5345 and 5380 Å). Peaks not drawn to scale. Data obtained as in Fig. 28a. (R. E. Shrader)

	Relative Peak Output (Approx.)
1. (80°K)	50
2. (297°K)	50
3. (450°K)	30
.. (475°K)	15
.. (500°K)	3.5

of the system (Figs. 3 and 16b). The locations and widths of the *line* emission spectra of some luminophors, such as rbhdl.-Al_2O_3:Cr, and of Sm, Pr, and Eu in various host crystals (Figs. 13 and 14), are relatively unaffected by moderate temperature changes (especially decreases) because the principal emission transitions occur between shielded inner energy levels of atoms which have incomplete inner shells.[65, 144, 194-196] Intermediate cases are provided by phosphors, such as $Ca_2P_2O_7$:Dy, Mg_4GeO_6:Mn,[58] Mg_2TiO_4:Mn, Mg_2TiO_4:Cr,[154, 156] and $ThSiO_4$:Eu, which have very narrow emission bands or line structure superimposed on a band background.

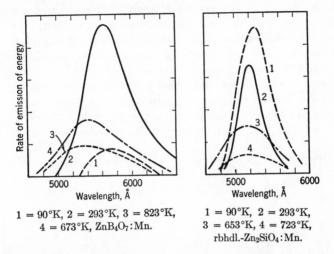

1 = 90°K, 2 = 293°K, 3 = 823°K,
4 = 673°K, ZnB_4O_7:Mn.

1 = 90°K, 2 = 293°K,
3 = 653°K, 4 = 723°K,
rbhdl.-Zn_2SiO_4:Mn.

FIG. 28c. Relative spectral-distribution curves of the emissions of two oxygen-dominated phosphors excited by cathode rays at several different temperatures. Peaks drawn to scale. (E. A. Ab [158])

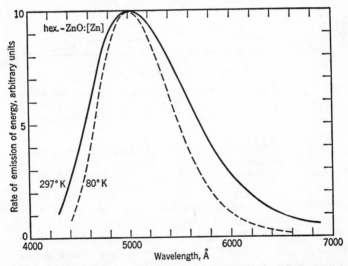

FIG. 28d. Spectral-distribution curves of the emission of a hex.-ZnO:[Zn] oxygen-dominated phosphor excited by 2537-Å ultraviolet at two different temperatures. Peaks not drawn to scale. (See also Fig. 50.) (F. J. Studer and L. Gaus [352])

Where *line*-emission spectra are observed, it is sometimes possible to identify the emitting atom (ion) and the energy levels which participate in the radiative transitions. As shown in Figs. 13 and 14, this has been done for some line-emitting luminophors, such as alkaline-earth oxides and sulphides with certain rare-earth activators, such as Sm and Eu. In these cases, the emission lines have been attributed to transitions between energy levels with different J values (due to spin changes and electron regroupings) in the incomplete $4f$ shells of the Sm^{3+} and Eu^{3+} ions.[144] Line-emitting luminophors of this type some-

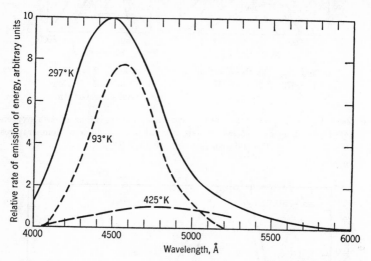

Fig. 29a. Relative spectral-distribution curves of the photoluminescence emission of a sulphur-dominated hex.-ZnS:[Zn] phosphor at three different operating temperatures. Data obtained as in Fig. 28a. (R. E. Shrader)

hex.-ZnS:[Zn], [NaCl(2)], 1250°C.
Excitation by 3650-Å UV from well-filtered CH-4 lamp 30-cm distant.
Resolution (band pass) = 5 Å.
Curves drawn to scale.

times exhibit resonance radiation; that is, some of the lines occur both in absorption and emission (Fig. 14). Some resonance lines are apparent in the spectrum of rbhdl.-Al_2O_3:Cr whose principal doublet lines are electronic transitions of Cr^{3+}. In addition to these lines, there is a weak band structure comprising many diffuse lines which become sharper at low temperatures. At low temperatures, the narrow bands lying on the long-wave side of the doublet become more intense in emission, whereas those on the short-wave side become more intense in absorption.

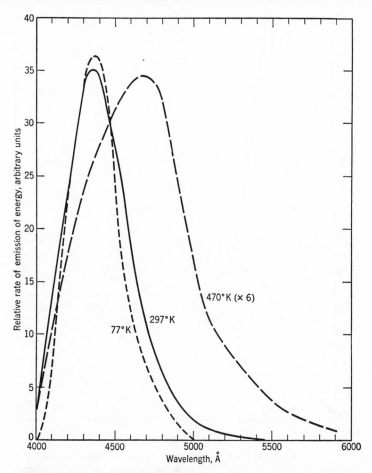

FIG. 29b. Relative spectral-distribution curves of the photoluminescence emission of a sulphur-dominated hex.-ZnS:Ag phosphor at three different temperatures. Data obtained as in Fig. 28a. (R. E. Shrader)

hex.-ZnS:Ag(0.015), [NaCl(4),BaCl$_2$(2)], 1260°C.
RCA 33-Z-20A.
Excitation by 3650-Å UV from CH-4 lamp as in Fig. 29a.
Resolution = 5 Å.
Curves drawn to scale, except the 470°K-curve ordinates should be divided by 6.

The band structure is explained as combinations of the electronic transitions of the Cr^{3+} ion with the vibrational frequencies of the host crystal (or the coupled Cr–O vibrations).[196]

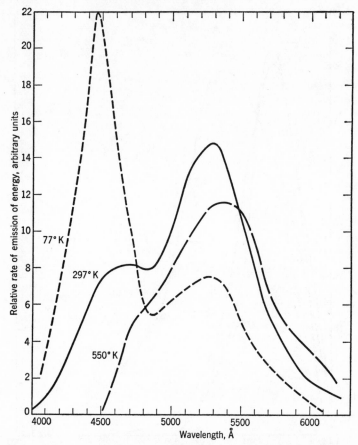

FIG. 29c. Relative spectral-distribution curves of the photoluminescence emission of a sulphur-dominated hex.-ZnS:Cu phosphor at three different temperatures. Data obtained as in Fig. 28a. (R. E. Shrader)

hex.-ZnS:Cu(0.01), [NaCl(2)], 1250°C.
Excitation by 3650-Å UV from CH-4 lamp as in Fig. 29a.
Resolution = 5 Å.
Curves drawn to scale.

When *band*-emission spectra are obtained, the absorption and emission spectra are quite different, and it has been difficult to identify the emitting atoms. With reference to Fig. 31, for example, the emission spectrum of crystallized $ThSiO_4$ (curve 1) is a feeble broad band peaked near the violet (short-wave limit near 3000 Å). A similar ultraviolet-plus-visible emission band is obtained when the Th is replaced by Zr,

Zn, Mg, Ca, Sr, Ba, or Cd, leading to the conclusion that the emission is associated in some way with the SiO_4 group ([Si]) or, more generally, with the Si–O bond.[72] In all the silicates which exhibit the characteristic UV + blue emission band (compare Fig. 51), the Si atoms are tetrahedrally surrounded by four O atoms. Distinct SiO_4 radical groups exist in only the orthosilicates (for example, rbhdl.-Zn_2SiO_4, rhomb.-Mg_2SiO_4, and tetr.-$ZrSiO_4$), whereas the SiO_4 groups are joined together

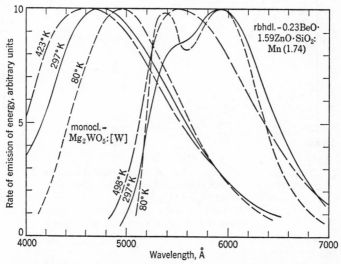

Fig. 30. Relative spectral-distribution curves of the emissions of two oxygen-dominated phosphors excited by 2537-Å ultraviolet at several different temperatures. Peaks not drawn to scale. (F. J. Studer and L. Gaus)

in chains or rings in other UV + blue-emitting silicates (for example, monocl.-$CaSiO_3$, monocl.-$CaMg(SiO_3)_2$, and monocl.-$MgSiO_3$). The Si–O bond is only about 50 per cent ionic, according to Pauling,[24] and it is possible that the excitation transition may involve electron transfer from one of the four O atoms, surrounding each Si, in an SiO_4 group, to the central Si atom. The resultant electron deficiency (positive hole) may then be exchanged among the tetrahedrally arranged O atoms, until the excited electron on the Si atom makes a radiative return to one of the ligand O atoms. In this case, the chief emitting atom would be an oxygen, although the entire SiO_4 group should be involved in the emission process. This host-crystal luminescence process was outlined without recourse to the activator concept, but curves 2 and 3 in Fig. 31 afford indirect evidence for the existence of an imperfection or impurity as a vital element in the luminescence process. Curve 2 shows that 0.2

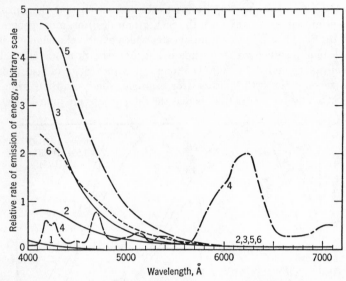

FIG. 31. Tracings from original recording-spectrophotometer curves of the luminescence emission spectra of ThSiO$_4$ crystallized (1) without added activator, (2) with
0.2 weight per cent of Cu (as CuO·SiO$_2$), (3) with 1 weight per cent of Ti (as
TiO$_2$·SiO$_2$), and (4) with 1 weight per cent of Eu (as Eu$_2$O$_3$·SiO$_2$). Curves 5 and 6
show the intensification action of Ti in some other silicates.

1. ThSiO$_4$:[Si], 1450°C, 1 hr   Excitation by 6-kv 25-μa cm^{-2}
2. ThSiO$_4$:Cu(0.2), 1450°C, 1 hr   CR.
3. ThSiO$_4$:Ti(1), 1450°C, 1 hr   Curves 1, 2, 3, 4 taken at same
4. ThSiO$_4$:Eu(1), 1450°C, 1 hr   gain and drawn to scale.
5. CaSiO$_3$:Ti(1), 1150°C, 1 hr
6. 7.5ZnO·BeO·5.4SiO$_2$:Ti(1), 1200°C, 1 hr

Band pass of spectroradiometer = 47 Å at 4000 Å, 157 Å at 5500 Å, and 342 Å
at 7000 Å (1-mm slits).[93]

Relative *visible* luminescence emissions, measured with an eye-corrected photocell
at 6 kv and 2 μa cm^{-2},

$$
\left.
\begin{array}{l}
1 = 1 \\
2 = 2 \\
3 = 4 \\
4 = 8 \\
5 = 11 \\
6 = 7
\end{array}
\right\} \text{based on rbhdl.-Zn}_2\text{SiO}_4\text{:Mn(0.3) as 100.}
$$

weight per cent of copper impurity *intensifies* the feeble emission band
of the "pure" silicate, and curve 3 shows that 1 weight per cent of titanium impurity *intensifies* the original band even more. The intensification
represents an increase in cathodoluminescence efficiency, and the efficiency increase is definitely associated with impurities. Extrapolating

to zero impurity or imperfection, therefore, it may be reasoned that a *perfect* crystal of $ThSiO_4$ should exhibit no luminescence (at least at room temperature).   On this basis, the assumption of self-activation, probably by interstitially displaced or stoichiometric-excess [Si] (or [Th]?), seems justified (similar reasoning applies to certain other "pure" phosphors, such as the luminescent tungstates which have luminescence-active WO_4 radicals [34, 58]).   The emission spectrum of $ThSiO_4$: Eu, curve 4 of Fig. 31, evidently represents a transition-type phosphor which exhibits both *line* emission (presumably from Eu^{3+}) and *band* emission (band near 6200 Å, presumably due to Eu^{2+}—compare with the Eu^{2+} band, at 6300 Å, of phosphor No. 20, Table 5).   Note that, according to curve 4 in Fig. 31, *the recording spectroradiometer used to measure almost all the emission spectra shown in this book, is capable of detecting "fine" structure down to the normal 1-mm-slit band passes shown.* By narrowing the slits of the instrument, it is possible to resolve the sodium *D* lines which are spaced only 4 Å apart.[93]   Although it is experimentally evident that the *bands* of $ThSiO_4$:Eu and cub.-SrS:$SrSO_4$: CaF_2:Sm:Eu are associated with the Eu (probably Eu^{2+}),[171–180, 221] it is not certain whether the Eu activator, or a ligand of the Eu (for example, O or S) is the chief emitting atom.   The fact that the Eu impurity does not just alter the efficiency of the "pure" host-crystal emission band, in these cases, indicates that the Eu is performing a role other than merely influencing the probability of an otherwise forbidden host-crystal transition (as in the case of the silicates with Cu [346b] or Ti [61, 154, 158, 159, 429, 430]).   The anomalously high electronic polarization of oxygen ion O^{2-} in the field of an adjacent titanium ion Ti^{4+} [64c] undoubtedly plays an important part in the intensification action of Ti activator in silicate phosphors; also, Ti prefers to be 6-coordinated with O, as compared with the normal 4 coordination of Si with O,[43] so traces of Ti produce a structural perturbation which may well upset the normal host-crystal selection rules in the immediate vicinity of a Ti impurity atom (ion).

Most of the useful phosphors have broad emission bands which are affected by temperature, as indicated in Figs. 28–30 (compare Fig. 50b). From these figures, it may be seen that increasing temperature can destroy line structure (Fig. 28a) and

1. Broaden the emission band without displacing its peak (visible band of hex.-ZnO:[Zn], rbhdl.-Zn_2SiO_4:Mn excited by ultraviolet *above* 297°K, and cub.-ZnS:Mn [154]).

2. Broaden the emission band and displace the peak to *longer* wavelengths (rbhdl.-Zn_2SiO_4:Mn excited by *ultraviolet* in the range *below* 297°K).

3. Broaden the emission band and displace the peak to *shorter* wavelengths (ZnB_4O_7:Mn and rbhdl.-Zn_2SiO_4:Mn excited by *electrons*, and monocl.-Mg_2WO_5:[W], excited by ultraviolet).

4. Broaden the emission band, leaving the foot of the short-wave side relatively unchanged, and displace the peak to longer wavelengths (hex.-ZnS:[Zn] and hex.-ZnS:Ag). Note that the emission band of hex.-ZnS:[Zn] displaces first to *shorter* and then longer wavelengths.

5. Slightly broaden the *edge*-emission band, and displace the entire band to longer wavelengths (UV band of hex.-ZnO:[Zn], Fig. 50b, and green emission band of hex.-CdS:[Cd] [154]).

6. Affect different component bands of a complex emission spectrum differently (note β-Zn_2SiO_4:Mn in Fig. 28b, hex.-ZnS:Cu in Fig. 29c, and the zinc–beryllium–silicate:Mn phosphor in Fig. 30).

The curves in Figs. 28c and 29a–c are drawn to scale to show differences in efficiency as well as differences in band shape and location, whereas the curves in Figs. 28a, 28b, 28d, and 30 are drawn with their peaks arbitrarily set equal to show spectral variations without showing variations in efficiency. According to Fig. 18, the luminescence efficiencies of different bands (associated with different activators) are affected differently by changes in the temperature of the luminescing phosphor. It is possible that the divergent results reported by Ab and Shrader for rbhdl.-Zn_2SiO_4:Mn may be attributed to the difference between excitation by high-energy electrons (predominantly host-crystal absorption) and excitation by low-energy photons (predominantly activator-band absorption). Apart from this apparent discrepancy, the temperature dependence of the emission spectra of predominantly i-center sulphide phosphors, with Ag or Cu activator, appears to be quite different from that of the predominantly s-center phosphors (for example, those with Mn activator).

EMISSION SPECTRA OF SOME PHOSPHORS WITH S(SE)-DOMINATED HOST CRYSTALS. It is instructive to examine the *room-temperature* emission spectra of a versatile general family of phosphors, such as cub.(or hex.)-(Zn:Cd)(S:Se), with and without the added activators Ag, Cu, and Mn.[170] This family provides many efficient photo- and cathodoluminescent phosphors at room temperature and readily indicates some of the major effects of variations in the compositions and structures of the host crystal, and in the nature and proportion of the activator(s). The spectral-distribution and relative luminescence-efficiency data of some representative phosphors of this family, shown in Figs. 32–47, were obtained at room temperature with the automatic recording spectroradiometer [93] (data on the visible absorption spectra of most of these phosphors are given in Ref. 170). Only one emission band is shown in most cases for both photoluminescence and cathodoluminescence, because the locations and shapes of the *single* simple emission bands are practically identical under the two different excitants. The cathodoluminescence data were taken at 6 kv and about 1 μa cm^{-2} (stationary unmodulated CR beam), the *approximate* relative peak-output values for all the figures being given with reference to a value of 100 for a standardized rbhdl.-Zn$_2$SiO$_4$:Mn(0.3) (RCA 33-W-2A) phosphor which has a cathodoluminescence efficiency of about 5 candles per watt (one side) at low excitation densities in conventional unaluminized CRT screens, or over 10 candles per watt for both sides of the screen.[58, 63, 72] All these data, for cathodoluminescence *and* photoluminescence, were obtained by measurements on the excited sides of the *same* thick phosphor coatings which were applied to clean Pyrex glass by settling the phosphor particles from a suspension in redistilled amyl acetate and baking in air for an hour at 275°C.[58] The photoluminescence data were obtained under excitation by radiation from a 100-watt-EH-4 mercury-vapor lamp (with Corning 5840 and 5970 filters) placed about 20 cm distant from the phosphor coating, and the approximate relative peak-output values are given with reference to a standardized hex.-9ZnS·CdS:Cu(0.0073), [NaCl(1), BaCl$_2$(0.5), NH$_4$Cl(1)], 1250°C (RCA 33-Z-21A) phosphor as a reference for comparison.[58, 214] By relative **peak output** is meant the relative height of the emission band (curve) at its highest point under the given excitation. The peaks of all the bands in Figs. 32–43 are arbitrarily drawn equal to 10 to show mainly differences in the locations and shapes of the bands. When the bands are replotted after multiplying their ordinates by the relative peak-output value for each band, the areas under the replotted curves are proportional to the relative luminescence efficiencies of the phosphors *under the given conditions of excitation.* That is, the area under each curve (as now drawn), when multiplied by the corresponding peak-output value is a relative measure of the total energy output in the form of luminescence photons. The curves may be converted from a linear wavelength scale to a linear frequency scale by applying the equation,

$$P_{\nu_2} = P_{\lambda_2}(\lambda_2{}^2/c) \tag{84}$$

where P_{ν_2} is the ordinate to be plotted at frequency ν_2 (corresponding to λ_2), when the ordinate which obtains at λ_2 is P_{λ_2}. The ordinate P is proportional to $P_\lambda\Delta\lambda$ (or $P_\nu\Delta\nu$) when $\Delta\lambda$ (or $\Delta\nu$) are very small. Hence, P is a measure of the rate of emission of energy as photons, that is, the radiance, and may be converted into a measure of the relative number of photons emitted per unit time by dividing each ordinate value by the corresponding value of hc/λ (or $h\nu$).

In the host-crystal system (Zn:Cd)(S:Se), ZnSe exhibits only the cubic structure ($F\bar{4}3m$) up to at least 1350°C, CdS exhibits only the hexagonal structure ($C6mc$) at all useful crystallization temperatures, and ZnS is cubic ($F\bar{4}3m$) up to about 1020°C above which it transforms into the hexagonal structure ($C6mc$). The *ideal* lattices representing these geometric arrangements are shown in Fig. 11, and Fig. 12 shows

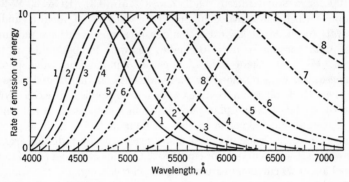

FIG. 32. Spectral-distribution curves of the emissions of selected members of the phosphor family (Zn:Cd)S, [NaCl(2)], 780°C.

Curve		1	2	3	4	5	6	7	8
Mole Proportions	ZnS	10	9.5	9	8	7	6	5	4
	CdS	..	0.5	1	2	3	4	5	6
Peak Output	6-kv CR	157	110	74	58	53	46	23	14
Excited by	3650-Å UV	53	23	21	50	35	30	24	11

All made with 2 per cent NaCl and *no* added activator at 780°C.

an enlargement of two unit cells of the $F\bar{4}3m$ structure. In this general phosphor system, compositions which are predominantly CdS (or CdSe) tend to be hexagonal, those which are predominantly ZnSe tend to be cubic, and those which are predominantly ZnS may be either cubic or hexagonal, depending on the temperature, pressure, and flux or impurities present during crystallization. Figures 32 and 35 show the room-temperature cathodoluminescence *band* spectra of selected members of the phosphor family (Zn:Cd)S:[Zn:Cd] produced by reaction and crystallization at 780 and 1200°C. The single (simple) emission bands, starting from either cub.-ZnS:[Zn], 780°C, or hex.-ZnS:[Zn], 1200°C, are displaced *gradually* toward longer wavelengths by increasing the proportion of CdS in the host crystal until the infrared-emitting hex.-CdS:[Cd] is obtained at either crystallization temperature. This gradual shift indicates that the tentatively postulated [Zn] and [Cd], if any, are not the emitting atoms. If they were, a double-band spectrum would be obtained when both [Zn] and [Cd] were present in the intermediate compositions. Apparently, therefore, the emitting atom in this family is the sulphur atom; that is, the radiative transition terminates in an energy band associated with sulphur in the crystal, which may be locally perturbed by [Zn] and [Cd] to provide for conservation of momentum during radiative transitions. The decrease in luminescence efficiency

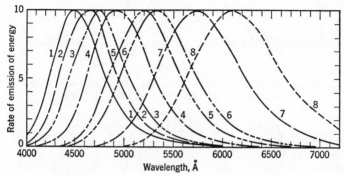

FIG. 33.   Spectral-distribution curves of the emissions of selected members of the phosphor family (Zn:Cd)S:Ag(0.005), [NaCl(2)], 780°C.

Curve		1	2	3	4	5	6	7	8
Mole Proportions	ZnS	10	9.5	9	8	7	6	5	4
	CdS	..	0.5	1	2	3	4	5	6
Peak Output	6-kv CR	104	90	104	109	120	107	88	78
Excited by	3650-Å UV	45	27	36	51	55	53	66	56

All made with 2 per cent NaCl and 0.005 per cent Ag at 780°C.

with increasing Cd content may occur as a result of the lower polarizing power of the larger Cd atom (ion) relative to that of Zn.

Some members of the cub.- and hex.-(Zn:Cd)S:[Zn:Cd] phosphor family, with or without Ag activator, have detectable narrow edge-emission bands in addition to the longer-wave broad emission bands shown in Figs. 29a–c, 32, 33, 35, and 36. The edge-emission bands are generally more pronounced when the host crystal is ZnS or CdS alone and are favored when the phosphors are crystallized in reducing atmospheres. When the phosphors are excited by UV at room temperature, the edge-emission bands are usually negligible. When a suitable UV-excited phosphor is operated at 90°K, however, Kröger has shown that the edge-emission band is readily detectable.[154] On the other hand, when the phosphor is excited *at room temperature* by weak or moderate intensities of CR, the edge-emission band is negligible, but under CR of sufficient intensity to saturate the longer-wave broad emission band

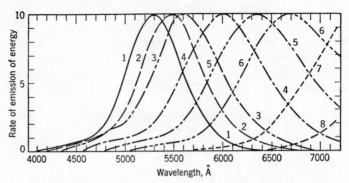

FIG. 34. Spectral-distribution curves of the emissions of selected members of the phosphor family (Zn:Cd)S:Cu(0.005), [NaCl(2)], 780°C.

Curve		1	2	3	4	5	6	7	8
Mole Proportions	ZnS	10	9.5	9	8	7	6	5	4
	CdS	..	0.5	1	2	3	4	5	6
Peak Output	6-kv CR	87	68	55	49	45	39	10 *	2 *
Excited by	3650-Å UV	56	30	28	39	40	46.5	22 *	7 *

All made with 2 per cent NaCl and 0.005 per cent Cu at 780°C.

* Output at 7100 Å.

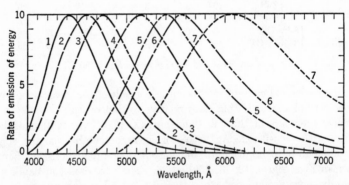

FIG. 35. Spectral-distribution curves of the emissions of selected members of the phosphor family (Zn:Cd)S, [NaCl(2)], 1200°C.

Curve		1	2	3	4	5	6	7
Mole Proportions	ZnS	10	9.5	9	8	7	6	5
	CdS	..	0.5	1	2	3	4	5
Peak Output	6-kv CR	307	123	84	68	58	51	21
Excited by	3650-Å UV	53	12	15	58	56	47	9

All hexagonal structures, formed at 1200°C. 2 per cent NaCl flux, and *no* added activator.

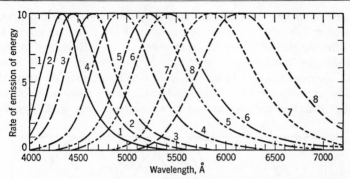

Fig. 36. Spectral-distribution curves of the emissions of selected members of the phosphor family (Zn:Cd)S:Ag(0.01), [NaCl(2)], 1200°C.

Curve		1	2	3	4	5	6	7	8
Mole Proportions	ZnS	10	9.5	9	8	7	6	5	4
	CdS	..	0.5	1	2	3	4	5	6
Peak Output {6-kv CR		415	170	161	147	143	124	64	55
Excited by {3650-Å UV		217	148	114	109	110	99	60	47

All hexagonal structures, formed at 1200°C. 2 per cent NaCl flux, 0.01 per cent Ag.

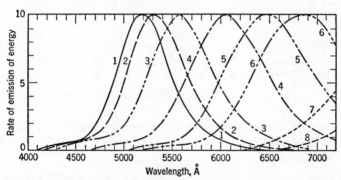

Fig. 37. Spectral-distribution curves of the emissions of selected members of the phosphor family (Zn:Cd)S:Cu(0.01), [NaCl(2)], 1200°C.

Curve		1	2	3	4	5	6	7	8
Mole Proportions	ZnS	10	9.5	9	8	7	6	5	4
	CdS	..	0.5	1	2	3	4	5	6
Peak Output {6-kv CR		145	94	71	56	41	30	13 *	5 *
Excited by {3650-Å UV		67	56	39	35	35	32	16 *	7 *

All hexagonal structures, formed at 1200°C. 2 per cent NaCl flux, 0.01 per cent Cu.

* = output at 7100 Å.

the edge-emission band appears and becomes increasingly prominent. This effect has been found by S. Lasof and F. H. Nicoll to obtain for hex.-ZnS:Ag(0.015) where the persistence of the CR-excited narrow èdge-emission band (peaked near 3450 Å) is of the order of 10^{-6} sec. S. M. Thomsen has found a similar effect with hex.-CdS:[Cd], where the broad orange-to-infrared emission band predominates at low CR current densities and the narrow green edge-emission band predomi-

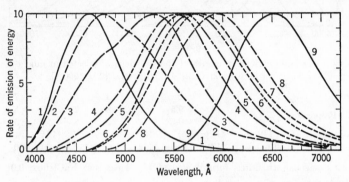

FIG. 38. Spectral-distribution curves of the emissions of selected members of the phosphor family Zn(S:Se), [NaCl(2)], 780°C.

Curve		1	2	3	4	5	6	7	8	9
Mole Proportions	ZnS	10	9.5	9	8	7	6	5	3	..
	ZnSe	..	0.5	1	2	3	4	5	7	10
Peak Output	6-kv CR	157	90	66	63	61	52	44	28	13
Excited by	3650-Å UV	53		20	57	56	49			8

All cubic structures, formed at 780°C. 2 per cent NaCl flux, *no* added activator.

nates at high current densities. This two-band behavior of the ZnS-type phosphors is analogous to that found for hex.-ZnO:[Zn] (see Fig. 50).

According to Kröger, the following results are obtained for ZnS-type phosphors excited by UV at 90°K:[154] For cub.-ZnS:[Zn], [KCl(2)], 900°C (in H_2S), the absorption edge is situated at 3410 Å (Fig. 24), with an edge-emission band extending from about 3370 to over 3600 Å, and for hex.-ZnS:[Zn], 1150°C, the absorption edge is situated at 3350 Å, with an edge-emission band extending from about 3310 to over 3600 Å (peak near 3400 Å).

The UV edge-emission band of cub.(or hex.)-ZnS:[Zn] is prominent (at low operating temperatures) after the material has been heated (crystallized) at about 1000°C in a reducing atmosphere (H_2 or H_2S) and is practically undetectable after the material has been heated at about 1000°C in an atmosphere of SO_2, CO_2, HCl, or air. It is after heating in the latter atmospheres that the normal blue emission band of cub.(or hex.)-ZnS:[Zn] appears most strongly; the blue emission band having very low efficiency after the material is heated in a reducing atmosphere.

For hex.-CdS:[Cd], the absorption edge is situated at 5170 Å, with an edge-emission band extending from about 5100 to 5600 Å (double-peak centered about 5170 Å).

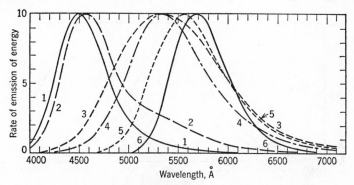

Fɪɢ. 39. Spectral-distribution curves of the emissions of selected members of the phosphor family Zn(S:Se):Ag(0.005), [NaCl(2)], 780°C.

1. cub.-ZnS:Ag(0.005)	780°C
2. cub.-9ZnS·ZnSe:Ag(0.005)	780°C
3. cub.-7ZnS·3ZnSe:Ag(0.005)	780°C
4. cub.-5ZnS·5ZnSe:Ag(0.005)	780°C
5. cub.-3ZnS·7ZnSe:Ag(0.005)	780°C
6. cub.-ZnSe:Ag(0.005)	780°C

Peak Outputs Excited by:

	6-kv CR	3650-Å UV
1.	104	45
2.	126	33
3.	61	53
4.	54	57
5.	38	56
6.	5.4	..

All these edge-emission bands are reported to exhibit fine structure, with wave-number differences of 360 cm^{-1} for the zinc sulphides, and 320 cm^{-1} for the cadmium sulphide. The absorption edge and the edge-emission band shift gradually to longer wavelengths with increasing Cd content, but fine structure disappears in the range from 9ZnS·CdS to ZnS·9CdS. Kröger ascribes these edge-emission bands to radiative transitions from the lower edge of the conduction band to the upper edge of the topmost normally filled host-crystal band (for example, from $E^*_{Zn}+$ to $E_{S^2}-$ in Fig. 16a), and explains the fine structure as arising from the coupling of this electronic transition with the vibrational levels of the crystal. It seems more probable that the edge-emission radiative transitions occur from high semilocalized excited-state levels, which may overlap the conduction band, of atoms near imperfections (for example, from a high E^*_{In} level to E^*_H at a point near A_2 or $A_{2'}$ in Fig. 16c).

Starting with cub.(or hex.)-ZnS:[Zn], and adding increasing proportions of **silver** (Ag) activator, the phosphor emission band *shifts*

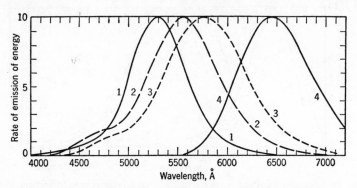

Fig. 40. Spectral-distribution curves of the emissions of selected members of the phosphor family Zn(S:Se):Cu(0.005), [NaCl(2)], 780°C.

1.	cub.-ZnS:Cu(0.005)	780°C
2.	cub.-8ZnS·2ZnSe:Cu(0.005)	780°C
3.	cub.-6ZnS·4ZnSe:Cu(0.005)	780°C
4.	cub.-ZnSe:Cu(0.005)	780°C

Peak Output Excited by:

	6-kv CR	3650-Å UV
1.	87	56
2.	70	61
3.	60	58
4.	40	45

toward *shorter* wavelengths (Fig. 44), and the luminescence efficiency *increases* up to about 0.03 weight per cent of Ag (larger proportions evidently do not go into solid solution, as evidenced by gray to black colors under white light).[61] This behavior indicates that Ag may act largely as an enhancing agent, much as the Ti activator does in rbhdl.-Zn₂SiO₄:Ti or ThSiO₄:Ti (Fig. 31). When an increasing proportion of Cd is substituted for Zn in the host crystal, the emission band is shifted gradually toward longer wavelengths (Figs. 32, 33, 35, and 36); whereas, when Se is substituted in increasing proportion for S in the host crystal, the emission band exhibits uneven displacements and broadenings, with some indications that the Se produces a new band (Figs. 38

and 39). Apart from some irregularities, the luminescence efficiency generally decreases with increasing proportion of Cd and/or Se.[170] As shown in Fig. 47, Au acts as a "poison" rather than an activator.[61]

Starting with cub.(or hex.)-ZnS:[Zn], and adding increasing proportions of **copper** (Cu) activator up to Cu(0.01), the original [Zn]

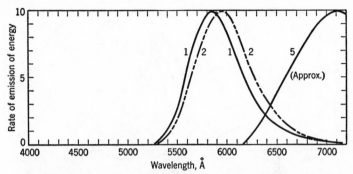

FIG. 41. Spectral-distribution curves of the emissions of selected members of the phosphor family (Zn:Cd)S:Mn(1), [NaCl(2)], 1050°C.

1. ZnS:Mn(1)	1050°C	
2. 7ZnS·3CdS:Mn(1)	1050°C	
3. 5ZnS·5CdS:Mn(1)	1050°C (curve not obtained)	
4. 3ZnS·7CdS:Mn(1)	1050°C (curve not obtained)	
5. CdS:Mn(1)	1050°C	

Peak Outputs Excited by:

	6-kv CR	3650-Å UV
1.	12	103
2.	2	13
3.	0.5	0.5
4.	..	..
5.	1	3

emission band *remains fixed* in position and *decreases* in efficiency as a *new* (Cu) *emission* band appears at *longer* wavelengths (Figs. 45a–c). The cathodoluminescence efficiency of the Cu band increases up to about 0.01 weight per cent Cu but has a *lower* efficiency than the original [Zn] band (again, increasing the activator proportion above the optimum gives increasingly dark products which strongly absorb their own emissions). Figures 45a–c show, however, that the multiband emissions may have different relative efficiencies under CR and UV. Figures 45b–c show also that the simple suppression of the [Zn] band and raising of the Cu band up to Cu(0.01) changes radically above Cu(0.01). In the cubic form, Cu(0.03) apparently intensifies the previously suppressed

[Zn] band (see Fig. 31) and in the hexagonal form, Cu(0.01) produces an intense photoluminescence emission band which is *intermediate* to the [Zn] and Cu bands (this may be the [Zn] band intensified and shifted to *longer* wavelengths by Cu, just as it is intensified and shifted to *shorter* wavelengths by Ag, Fig. 44). When an increasing proportion of Cd is

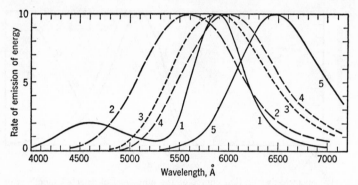

FIG. 42. Spectral-distribution curves of the emissions of selected members of the phosphor family Zn(S:Se):Mn(1), [NaCl(2)], 780°C.

1. cub.-ZnS:Mn(1)	780°C
2. cub.-8ZnS·2ZnSe:Mn(1)	780°C
3. cub.-5ZnS·5ZnSe:Mn(1)	780°C
4. cub.-3ZnS·7ZnSe:Mn(1)	780°C
5. cub.-ZnSe:Mn(1)	780°C

	Peak Outputs Excited by:	
	6-kv CR	3650-Å UV
1.	0.35	7.3
2.	2.1	6.7
3.	17.8	38
4.	11.5	22.8
5.	3.1	7.8

substituted for Zn in the host crystal, with optimum Cu(≈ 0.01) content, the Cu emission band is shifted gradually toward longer wavelengths (Figs. 34 and 37), and a similar gradual shift occurs when Se is substituted in increasing proportion for S (Fig. 40; compare with Fig. 39). As in the case of the system with [Zn] or Ag, the luminescence efficiency of the Cu band generally decreases with increasing proportion of Cd and/or Se.[170]

Starting with cub.(or hex.)-ZnS:[Zn], and adding increasing proportions of **manganese** (Mn) activator, the original [Zn] emission band *remains fixed* in position and *decreases* in efficiency as a *new* (Mn) *emission band* appears at longer wavelengths (Figs. 41 and 42). The luminescence efficiency of the Mn band increases up to about 1 to 2 weight per cent of Mn but has a *lower* efficiency than the original [Zn] band (MnS produces several new activator absorption (excitation) bands peaked near 3900, 4300, 4650, and 5000 Å,[154] whereas Ag and Cu produce

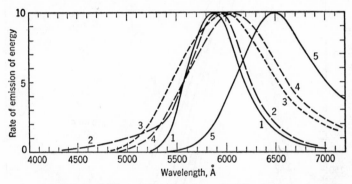

FIG. 43. Spectral-distribution curves of the cathodoluminescence emissions of selected members of the phosphor family Zn(S:Se):Mn(1), [NaCl(2)], 1200°C.

1.	hex.-ZnS:Mn(1)		1200°C
2.	↑	8ZnS·2ZnSe:Mn(1)	1200°C
3.		5ZnS·5ZnSe:Mn(1)	1200°C
4.	↓	3ZnS·7ZnSe:Mn(1)	1200°C
5.	cub.-ZnSe:Mn(1)		1200°C

Peak Outputs Excited by:

	6-kv CR	3650-Å UV
1.	3.3	42
2.	1	17
3.	6.7	23.5
4.	4.7	13.5
5.	1	4.6

single absorption bands near the absorption edge of the host crystal[170]). When an increasing proportion of Cd is substituted for Zn in the hexagonal host crystal, with optimum Mn content, the Mn emission band shifts very slightly toward *longer* wavelengths while the efficiency *decreases* to vanishingly low values until, near 100 per cent Cd substitution, the feeble deep-red emission band of hex.-CdS:Mn appears (Fig. 41). When an increasing proportion of Se is substituted for S in the cubic host crystal, with optimum Mn content, the Mn band shifts first to *shorter* wavelengths, broadening greatly, and then shifts to *longer* wavelengths while going through a maximum in efficiency (Fig. 42).

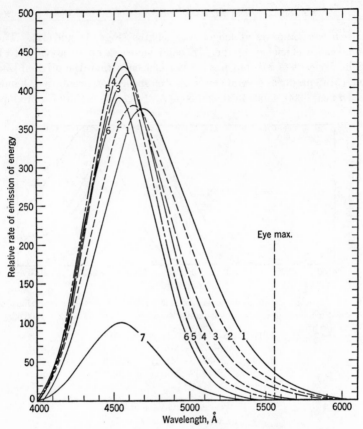

FIG. 44. Spectral-distribution curves of the cathodoluminescence emissions of selected members of the phosphor family cub.-ZnS:Ag(0 to 0.512%), [NaCl(2)], 940°C.

		Reflected Color	Luminescence Emission Color
1. cub.-ZnS:[Zn]		White	Light[3] blue
2. cub.-ZnS:0.002% Ag		White	Light[2] blue
3. cub.-ZnS:0.005	Ag	White	Light[1] blue
4. cub.-ZnS:0.008	Ag	White	Light blue
5. cub.-ZnS:0.032	Ag	White	Blue
6. cub.-ZnS:0.128	Ag	Tan–gray	Deep blue
7. cub.-ZnS:0.512	Ag	Dark gray	Very deep blue

All heated at 940°C for 2 hr.

Relative Visible Luminescence Output

1.	100%	5.	34.0
2.	71.5	6.	25.6
3.	52.2	7.	11.0
4.	43.6		

The following incidental data relating to Figs. 32–46 are of interest:

1. The emission band of cub.(or hex.)-ZnS:Mn lies on the *long-wave* side of the emission of cub.(or hex.)-ZnS:[Zn], whereas the emission band of hex.-CdS:Mn lies on the *short-wave* side of the emission of hex.-CdS:[Cd]. This indicates that the location of the Mn emission band does not depend directly on the energy levels of the host crystal (Fig. 16a).

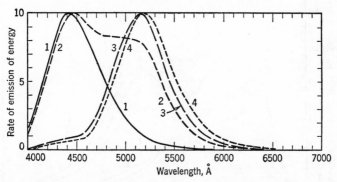

Fig. 45a. Spectral-distribution curves of the cathodoluminescence emissions of selected members of the phosphor family hex.-ZnS:Cu(0 to 0.01%), [NaCl(2)], 1200°C. (Note that phosphor No. 2, in particular, should be denoted by hex.-ZnS:[Zn]:Cu instead of the conventional designation.)

		Peak Output Excited by 6-kv 1-μa cm^{-2} CR
1.	hex.-ZnS:[Zn]	307 (λ_{pk} = 4430 Å)
2.	hex.-ZnS:Cu(0.001)	121 (λ_{pk} = 4470 Å)
3.	hex.-ZnS:Cu(0.005)	158 (λ_{pk} = 5160 Å)
4.	hex.-ZnS:Cu(0.01)	145 (λ_{pk} = 5190 Å)

All made with 2 per cent NaCl at 1200°C.

2. The emission bands of cub.-ZnSe:[Zn], cub.-ZnSe:Cu, and cub.-ZnSe:Mn are practically identical except for efficiencies. The Mn apparently lowers and the Cu enhances the efficiency of the material made without added activator.

3. In the case of the (Zn:Cd)S phosphors with [Zn], Ag, or Cu activator, (a) there is little overlap of the host-crystal absorption band and the emission band, and (b) the incorporation of as little as 0.005 weight per cent of Ag or Cu produces a large increase in the activator absorption band which overlaps the emission band (Fig. 24 [170]). In the case of the Zn(S:Se) phosphors with [Zn], Ag, or Cu, (a) there is a large overlap of the host-crystal absorption band and the emission band, and (b) the incorporation of Ag or Cu activator produces little change in

the long-wave "tail" of the absorption band (Fig. 24 [170]). In terms of the energy-level diagrams of Fig. 16a, it is possible that the normally filled activator levels $E_{[Zn]}$, E_{Cu}, and E_{Mn} are overlapped by the topmost filled host-crystal level (E_{Se}) in the cub.-ZnSe phosphors, and so the

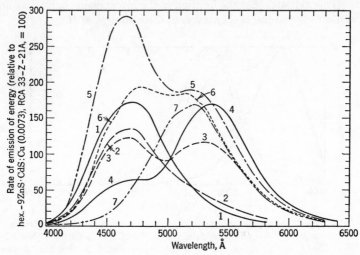

FIG. 45b. Relative spectral-distribution curves of the photoluminescence emissions of the phosphor family cub.-ZnS:Cu(0 to 0.3%), [NaCl(2)], 950°C (see note to Fig. 45a).

1. cub.-ZnS:[Zn]
2. cub.-ZnS:Cu(0.0003)
3. cub.-ZnS:Cu(0.003)
4. cub.-ZnS:Cu(0.01)
5. cub.-ZnS:Cu(0.03)
6. cub.-ZnS:Cu(0.1)
7. cub.-ZnS:Cu(0.3)

All made with 2 per cent NaCl at 950°C.
Excitation by 3650-Å UV from CH-4 lamp, 30-cm distant.
Curves drawn to scale.

room-temperature radiative transitions of the cub.-ZnSe phosphors with [Zn], Cu, and Mn activators terminate in E_{Se}. This casts doubt on the general validity of the conjectural diagrams in Fig. 16a, however, because the diagrams make it appear that the normally filled E_{Ag} level would be the first to be overlapped by an expanding (or rising) host-crystal band ($E_S \rightarrow E_{Se}$; compare Fig. 24).

4. Most of the single (simple) emission bands, when replotted on a linear energy (frequency or wave-number) scale, according to eq. 84,

may be represented by the probability formula,

$$P = a\epsilon^{-b_1\nu^2} \quad (\text{or } P = a\epsilon^{-b_2\bar{\nu}^2}) \tag{85}$$

where P is the radiance, that is, the rate of emission of luminescence energy per unit time, at frequency ν (or wave number $\bar{\nu}$), and a and b are constants for a given band.[349,704]

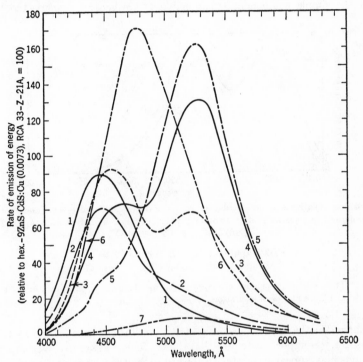

FIG. 45c. Relative spectral-distribution curves of the photoluminescence emissions of the phosphor family hex.-ZnS:Cu(0 to 0.3%), [NaCl(2)], 1250°C (see note to Fig. 45a).

1. hex.-ZnS:[Zn]
2. hex.-ZnS:Cu(0.0003)
3. hex.-ZnS:Cu(0.003)
4. hex.-ZnS:Cu(0.01)
5. hex.-ZnS:Cu(0.03)
6. hex.-ZnS:Cu(0.1)
7. hex.-ZnS:Cu(0.3)

All made with 2 per cent NaCl at 1250°C.
Excitation by 3650-Å UV from CH-4 lamp, 30-cm distant.
Curves drawn to scale.

5. Complex *band*-emission spectra, such as those shown in Figs. 45, 46, and curve 1 of Fig. 42, are generally resolvable into simple component bands which may be represented by eq. 85.

6. Figures 45a–46 show that the different component bands of multi-band emission spectra are selectively emphasized by different types and intensities of excitation.[72]. The locations, that is, the peak wavelengths, of the simple component emission bands are generally unaffected by

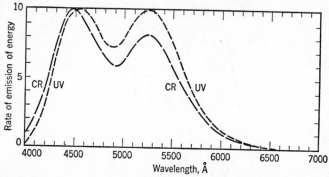

FIG. 46.   Relative spectral-distribution curves of the cathodoluminescence and photo-luminescence emissions of a two-band phosphor, hex.-$9.5ZnS \cdot 0.5CdS : Ag(0.02) : Cu(0.005)$, [NaCl(2)], 1200°C.

Peak output under 6-kv 2-μa cm^{-2} CR $= 63$ *
Peak output under 3650-Å UV $= 47$ *

* The peak-output values are comparable for the same type of excitation, but not for different types of excitation.

changes in the type of excitant at low excitation densities.[345] In this respect, it is important to note that the *resultant* peaks of two overlapping bands may appear to be shifted toward each other, whereas the individual peaks are really unchanged in spectral location. In general, increasing the excitation density in the phosphor volume *penetrated* by the excitant decreases the output of the long-wave emission bands relative to that of the short-wave emission bands in complex band-emission spectra.[58,72] This behavior is opposite to the effect of increasing the temperatures of the luminescing phosphors. For example, with (Zn:Cd)S: Ag:Cu phosphors which have their activator proportions adjusted to give two pronounced emission bands (short wave = Ag, long wave = Cu), increasing the phosphor temperature *increases* the relative height of the Cu band, whereas increasing the intensity of excitation *decreases* the relative height of the Cu band (note Figs. 48 and 50 [354]).

If a new host-crystal atom, which substitutionally replaces one of the host-crystal atoms of a phosphor, locates as a *nonligand* (that is, not a nearest neighbor) of an emitting atom, the (initially) foreign atom may

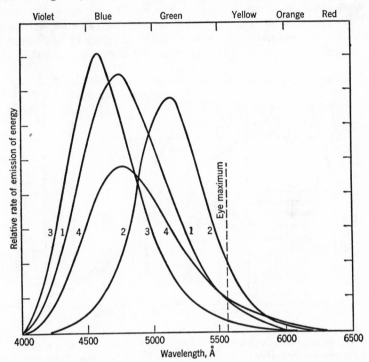

FIG. 47. Relative spectral-distribution curves of cub.-ZnS with several different activators. Excitation by cathode rays. Curves drawn to scale.

Curve	Phosphor	Emission Color
1.	cub.-ZnS:[Zn]	Light blue
2.	cub.-ZnS:Cu(0.01)	Green
3.	cub.-ZnS:Ag(0.008)	Blue
4.	cub.-ZnS:Au(0.002)	Blue–green

All crystallized with 2 weight per cent NaCl flux at 950°C.

be expected to exert an *indirect* effect on the spectral location of the emission. The emission band of the solid, then, should exhibit a *gradual displacement of the original band* with increasing proportion of next-nearest-neighbor foreign substituent. If, however, a new host-crystal atom locates as a *ligand* (combined nearest neighbor) of an emitting atom, the foreign atom may be expected to exert a *direct* effect on the spectral location of the emission. The emission band of the solid, in this case, may exhibit a *new band* which increases in relative intensity (at the expense of the original band) with increasing proportion of nearest-neighbor (ligand) foreign substituent. In both cases, increasing substitution of a foreign host-crystal element may give an increasing effect up to a proportion beyond which the efficiency of luminescence decreases rapidly as the result of structural distortions produced by

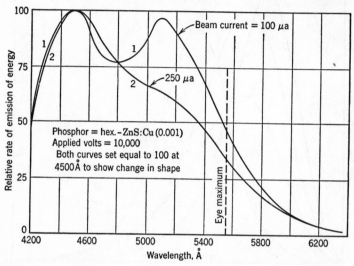

Fig. 48. Spectral-distribution curves of hex.-ZnS:[Zn]:Cu(0.001) (compare Fig. 45) under excitation by cathode rays, showing the influence of changing the excitation density (intensity).

exceeding the solubility of the foreign atoms in the original host crystal, or as the result of obtaining sufficient interaction between the foreign atoms to produce structural transitions.

By referring to Figs. 11 and 12, and visualizing Cd substituting for Zn, and Se substituting for S, in the (Zn:Cd)(S:Se) crystals, some of the data in Figs. 32–46 may be tentatively interpreted as follows:

1. The *gradual and regular* emission-band shift and efficiency variation occasioned by increasing proportions of the larger Cd atom (Table 6) in hex.-(Zn:Cd)S with [Zn], Ag, and Cu (Figs. 32–37), indicates that the cationic Cd exerts an *indirect* effect on the emitting atoms; that is, increasing the Cd content increases the lattice spacing without having individual Cd atoms directly influence the atoms responsible for the luminescence emission.

2. The *irregular* emission-band shift and efficiency variation occasioned by increasing proportions of Cd in hex.-(Zn:Cd)S:Mn (Fig. 41) is not so readily interpretable in terms of the propinquity of the Cd atoms and the emitting atoms. From data on the ease of saturation of the luminescence of this system with increasing excitation density, it appears that only about 1 per cent of the incorporated Mn atoms operate efficiently as activators, the remainder being chiefly host-crystal ingredients. It is apparent from Fig. 41 that the simultaneous presence of both Zn and Cd is detrimental, whereas with either cation alone moderate luminescence efficiency is obtained. This indicates that the active Mn atoms may be ligands of the regular Zn and Cd atoms and may be made inactive in the nonuniform field produced when both Zn and Cd are present at 300°K. The active Mn atoms may be either substituted for regular Zn atoms or located interstitially (Fig. 12). When these phosphors are cooled to 90°K, only the yellow–orange emission band of ZnS:Mn appears, and this band is shifted slightly to *longer* wavelengths (compare with the similar behavior of other s-center phosphors in Figs. 28–30). The hex.-7ZnS·3CdS:Mn and hex.-5ZnS· 5CdS:Mn phosphors which are practically nonluminescent at 300°K under 3650-Å UV, exhibit the same orange emission color and almost as bright photoluminescence as hex.-ZnS:Mn at 90°K. The hex.-3ZnS· 7CdS:Mn phosphor, which also exhibits only the orange emission of hex.-ZnS:Mn at 90°K, is stronger at 90 than at 300°K (but still feeble), and the room-temperature red emission color of hex.-CdS:Mn disappears; that is, it apparently shifts into the infrared at 90°K.

3. The *gradual and regular* emission-band shift and efficiency variation occasioned by increasing proportions of the larger Se atom (Table 6) in cub.-Zn(S:Se):Cu (Fig. 40) indicates that the anionic Se exerts an *indirect* effect on the emitting atoms.

4. The *irregular* emission-band shift and efficiency variation occasioned by increasing proportions of Se in cub.-Zn(S:Se) with [Zn], Ag, or Mn (Figs. 38, 39, 42, and 43) indicates that the Se atoms become *directly* involved in the luminescence emission process. The great broadening and anomalous shift of the emission band in Figs. 42 and 43 indicates that the Se is a ligand of the active Mn atoms, so that these atoms are located either in substitutional (Zn) sites or inside normally vacant S (Se) tetrahedra. In either case, the Mn atoms are surrounded by four S (Se) atoms, and the broadening and anomalous shift are observed to occur only when both S and Se are present to produce a nonuniform field around the active Mn atoms.

It may be profitable to review some of the aspects of the preparation of ZnS-type phosphors at this point.

1. Effect of Fluxes. In order to obtain appreciable luminescence efficiency of simple ZnS-host-crystal phosphors, it is usual to have a halide-type flux, for example, NaCl, present during crystallization at temperatures below about 1200°C, whereas efficient CdS-containing phosphors may be made without any flux.[312] It has been found that NaCl flux produces the same effect whether the NaCl is admixed with the host-crystal ingredients prior to crystallization or introduced as a vapor from an NaCl-containing surround during the crystallization at high temperatures. Chemical tests have not disclosed, with certainty, any appreciable residue of NaCl in the completed *acid-cleaned* sulphide phosphors, although about 10^{-4} mole Cl^- per mole of ZnS has been reportedly found on water-washed products.[156c] There is a possibility that the flux not only promotes crystallization, but also helps to produce a small dissociation of the ZnS.[253] The lower bonding energy of CdS may allow sufficient dissociation to produce the luminescence obtained in the absence of flux. In this respect, it should be noted that both green- and red-emitting cadmium sulphide (at room temperature!) may be produced by evaporating pure unfluxed CdS in a stream of hydrogen through a gradual temperature gradient near 950°C at atmospheric pressure (compare Ref. 221). Some of the cadmium sulphide crystals which luminesce a weak green at 300°K luminesce a strong orange at 90°K, and some of the CdS crystals which luminesce a deep red at 300°K luminesce a vivid green at 90°K.

2. Effect of Atmosphere during Crystallization. After pure ZnS (fluxed or unfluxed) is crystallized (a) in a strongly reducing atmosphere, for example, hydrogen or hydrogen sulphide, chiefly the UV edge-emission band appears; (b) in neutral or mildly oxidizing atmospheres (N_2, SO_2, CO_2, HCl), chiefly the blue [Zn] emission band appears, whereas (c) when the high-temperature crystallization is carried out in a stronger oxidizing atmosphere, for example, air or nitrogen containing a few per cent of oxygen, a green emission band sometimes appears which has all the characteristics of the band produced by normally introduced Cu activator.[58, 154, 253] If Cu activator is incorporated during crystallization in a reducing atmosphere, the blue emission band predominates, whereas, when the crystallization is done in an oxidizing atmosphere, the normal green emission band predominates.[255] This indicates that Cu^{2+} influences the green long-persistent emission band, and Cu^+ (Ag^+?) influences the blue emission band. A strong reducing atmosphere during high-temperature crystallization of hex.-ZnS:Ag is detrimental to phosphor efficiency,[253] presumably because the Ag^{2+} (or Ag^+) is reduced to a luminescence-inert effective valence state.

3. Effect of Temperature, Crystal Structure, and Rate of Cooling.
Rothschild reports that NaCl- or $BaCl_2$-fluxed cub.-ZnS:Cu(0.0075),
800°C, emits predominantly the green band, whereas the same phosphor
crystallized at 1150°C [hex.-ZnS:Cu(0.0075)] emits predominantly the
blue band.[255] Furthermore, rapid cooling reportedly favors the green
band, whereas slow cooling favors the blue band.   Also, the initially
suppressed blue emission band (Fig. 45) reappears at Cu proportions
greater than 0.01 weight per cent (even when about 1 weight per cent

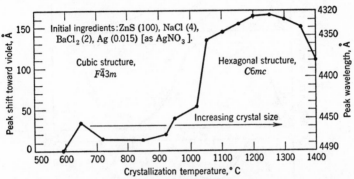

F_IG. 49.   Increasing temperature of crystallization of ZnS:Ag(0.015), [NaCl(4)],
[$BaCl_2(2)$] causes a marked shift in peak wavelength of emission during the change
from the $F\bar{4}3m$ to the $C6mc$ structure near 1020°C.   Heating time, at the indicated
top temperature (samples placed in cool furnace), varied from about 1 hr at 600°C
to about 10 min at 1400°C.

of Mn is present!), when the phosphor is crystallized at 950–1050°C.
Experiments in these laboratories (Figs. 45b, 45c, 49a) have confirmed
and added to these observations, for example:

1. Hex.-ZnS:Cu(0.01), [NaCl(2)] luminesces a strong *green* with an
intense long *green* phosphorescence emission when prepared by heating
to 1400°C in a dense surround of ZnS which had been previously crystallized with about 6 per cent of NaCl, whereas a phosphor made from the
same initial ingredients luminesces a strong *blue* with a short weak
pale-*blue* phosphorescence emission when prepared by heating to 1400°C
in a relatively porous surround of pure ZnS.   Both samples, when
spectrographically analyzed, had the same proportion of copper, thus
ruling out the possibility that the Cu had volatilized from the sample
which luminesced blue.

2. Hex.-ZnS:Cu(0.001), [NaCl(2)] heated to 1240°C and immediately quenched in cold water luminesces a strong *green*, whereas the
same initial ingredients heated to 1240°C and allowed to cool overnight

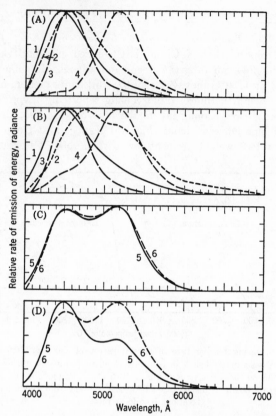

FIG. 49a. A comparison of the emission spectra of some ZnS-type phosphors, showing the effect of type of surround during high-temperature crystallization (compare Nos. 3 and 4 in *A* and *B*—both contain the full 0.01 weight per cent of Cu after crystallization) and the effect of slow and fast cooling in the absence of surrounds (see Nos. 5 and 6 in *C* and especially *D*). The same samples were tested in *A* and *B*, and in *C* and *D*.

	Relative Peak Outputs			Relative Peak Outputs
(A) 6-kv 1-μa cm⁻² CR excitation		**(C) 6-kv 1-μa cm⁻² CR excitation**		
1. hex.-ZnS:[Zn], [NaCl(20)], 1400°C	24	5. hex.-ZnS:Cu(0.001), 1240°C, slow cool		77
2. hex.-ZnS:[Zn] surround from above No. 3	3	6. hex.-ZnS:Cu(0.001), 1240°C, quenched		45
3. hex.-ZnS:Cu(0.01), ZnS surround, 1400°C	32	hex.-9ZnS·CdS:Cu(0.0073)		55
4. hex.-ZnS:Cu(0.01), ZnS [NaCl(20)] surround, 1400°C	107	rbhdl.-Zn₂SiO₄:Mn(0.3)		100 *
hex.-9ZnS·CdS:Cu(0.0073)	55	Nos. 5 and 6 had [NaCl(2)] flux and covers, but no surrounds.		
rbhdl.-Zn₂SiO₄:Mn(0.3)	100 *			
Nos. 3 and 4 had [NaCl(2)] flux.				
(B) 3650-Å UV excitation		**(D) 3650-A UV excitation**		
1. See (A)	20	5. hex.-ZnS:Cu(0.001), slow cool		69
2. See (A)	12	6. hex.-ZnS:Cu(0.001), quenched		79
3. See (A)	140	hex.-9ZnS·CdS:Cu(0.0073)		100 *
4. See (A)	175	Same as (C).		
hex.-9ZnS·CdS:Cu(0.0073)	100 *			
* Denotes reference standard.				

in the furnace luminesces *green–blue*. Both samples were heated in covered fused-silica crucibles without any surrounds. Referring to Fig. 45a, a rapid cooling favors the long-wave emission band, and slow cooling favors the short-wave emission band of this phosphor. Both samples had long green phosphorescence emissions of approximately equal intensity.

These results indicate that the Cu-containing phosphors may have been more affected by the mildly reducing SiC-element furnace atmosphere in the case of (a) the 1400°C material packed in porous ZnS, and (b) the slowly cooled 1240°C material, which was above 1000°C for a much longer time than the quenched sample. There is a need for further experimental work to clarify the actions of oxidizing and reducing atmospheres in the formation of ZnS-type phosphors.

The transition from the cubic to the hexagonal form of ZnS phosphors is generally marked by a displacement of the emission spectrum toward shorter wavelengths. This is quite evident for ZnS:[Zn], ZnS:Ag, and ZnS:Cu, but the emission band of ZnS:Mn is practically unaltered on crystallizing above and below the transition temperature near 1020°C. Figure 49 shows a plot of the peak wavelength of the emission bands of a series of ZnS:Ag phosphors crystallized at various temperatures from 580 to 1400°C. It is apparent that there is a pronounced shift of the emission band toward shorter wavelengths on going from 1000 to 1050°C.

From the experimental data which have been presented on the many variations in luminescence that may be obtained within the (Zn:Cd) (S:Se) system of phosphors,[221] it is not immediately apparent how even this one class of familiar materials may be interpreted in terms of quantitative conventional energy-level diagrams without encountering incompatibilities. For the present, then, it seems prudent to describe these and other phosphors largely on a phenomenological basis, without demanding that occasional (tentative) correlations and interpretations in one general system of phosphors be valid for all other general systems. Much of the following description of phosphors will be presented in the form of contrasts as well as comparisons among different interesting and useful phosphor systems.

EMISSION SPECTRA OF SOME PHOSPHORS WITH O-DOMINATED HOST CRYSTALS. The general family of phosphors which has oxygen-dominated host crystals is at least as ramified and versatile as the general family of sulphide(selenide)-type phosphors. From the standpoint of emission spectra, it may be recalled that the *line*-emission spectrum of Sm activator in cubic alkaline-earth oxides and sulphides occurs at *longer* wavelengths when the *smaller* and *lower*-atomic-number oxygen dominates the host crystals (Fig. 13). Also, when Nos. 20 and 21 in Table 5 are compared, it is evident that the *band* emission spectrum of Eu activator in cubic ŠrS and Sr(S:Se) occurs at *longer* wavelengths when the *smaller* sulphur dominates the host crystal.

Figure 50 shows the emission spectra obtainable from hex.-ZnO: [Zn].[354] It is not known, at present, whether the two emission bands are to be attributed to (1) two different activators (for example, [Zn$^+$] and

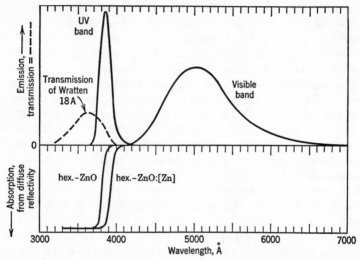

FIG. 50a. Absorption (from diffuse reflectivity) and emission spectra of some hex.-ZnO:[Zn] phosphors.[354]

[Zn^{++}] or [Zn] and [O]), or (2) transitions from low- and high-lying excited-state levels ($E^*_{I_n}$ in Fig. 16c), or (3) a given species of activator in structurally dissimilar sites, or (4) the reported difference in bonding along and away from the *c* axis of hex.-ZnO.[50a] When the oxide is made by burning zinc in an excess of oxygen, only the UV absorption-edge-emission band appears during cathodoluminescence at room temperature, but, when the zinc oxide is heated in a reducing atmosphere, the visible emission band appears alone at low excitation densities and the UV emission band appears only at very high excitation densities.[354] As shown in Fig. 50b (and contrary to the cited report [354]), the output of the visible band decreases more rapidly than that of the UV band as the temperature of the luminescing phosphor is increased, and the UV band shifts to longer wavelengths with increasing temperature (compare Fig. 28d). The shift of the edge-emission band of hex.-ZnO:[Zn] from about 3740 Å at 90°K to longer wavelengths with increasing temperature, at about 1 Å °C^{-1} above 300°K, is paralleled by a shift of the absorption edge in the same direction, just as the edge-emission band and absorption edge of hex.-CdS:[Cd] shift from about 4800 Å at 90°K to about 5200 Å at 300°K.[154, 215b]

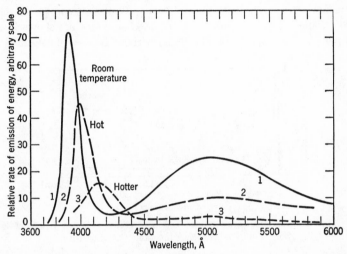

FIG. 50b.   Temperature dependence of the cathodoluminescence emission spectrum of hex.-ZnO:[Zn], sample A-1-11-11, excited by a 25-kv 100-μa ≈0.1-mm-diameter CR beam scanning an area 7 × 7 cm.   (F. H. Nicoll[354])

Approximate Temperature
of the Phosphor Screen
1.   300°K
2.   ↓
3.   >500°K

Aluminized screen with 1 mg cm^{-2} of phosphor.   Screen heated by CR beam.

When Fig. 50 is compared with curve 1 in Fig. 35, it is evident that the band emission spectra of the hexagonal oxide and sulphide of zinc occur at a *longer* wavelength when the *smaller* oxygen dominates the host crystal (the absorption edge and the absorption-edge emission band of the oxide also occur at *longer* wavelengths than the corresponding spectra of the sulphide).   The long-wave emission band of hex.-CdS:[Cd] lies in the near infrared (edge emission band near 5200 Å), and the long-wave emission band of hex.-CdO:[Cd] probably lies in the far infrared.   In contrast with the trend shown in these examples, the emission spectrum shifts to *shorter* wavelengths when the *smaller* S is substituted for Se in cub.-ZnSe:[Zn]. The change from a hexagonal to a cubic structure cannot account for this anomaly, because the emission bands of cub.- and hex.-ZnS:[Zn] are very close to each other (Figs. 32 and 35).

Figures 50 and 51 present, in part, the genesis of the emission bands of some members of the important silicate (germanate) phosphors.[58, 63, 72, 191, 192] The cathodoluminescence emission spectra of SiO_2, 1300°C, and several silicates without added activator are shown in the top part of Fig. 51. These phosphors without Mn are not excited by

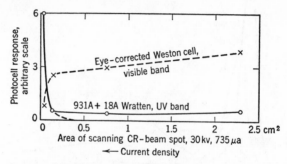

FIG. 50c.   Relative intensities of the UV and visible emission bands of hex.-ZnO:[Zn], A-1-11-11, as a function of CR intensity.

photons with wavelengths longer than 2500 Å, but they are excited to readily detectable though inefficient luminescence by x rays and cathode rays. The doublet structure of the emission band of SiO_2, 1300°C, may be caused by asymmetry of the SiO_4 tetrahedra in the anisotropic (hexagonal or rhombohedral) crystals of silica; that is, there may be different charge distributions and bonding energies for Si–O along different crystallographic directions, just as in the reported case of hex.ZnO.[50a]

It is noteworthy that the emission spectrum of rbhdl.-$3ZnO \cdot SiO_2$:[Si] peaks between the two bands of the emission spectrum of SiO_2 as well as between the two bands of the emission spectrum of hex.-ZnO:[Zn]. On going to the orthoproportion rbhdl.-Zn_2SiO_4:[Si], the emission band broadens in both directions, without shifting, but on going to rbhdl.-$ZnO \cdot 2SiO_2$:[Si], which has 400 per cent excess silica over orthoproportions, the emission band shifts to a band peaked near the short-wave peak of the emission spectrum of SiO_2 with at least one additional band peaked at still shorter wavelengths. The efficiencies of these phosphors are greatly improved by the incorporation of TiO_2 (compare Fig. 31). At the left of the second part of Fig. 51, the cathodoluminescence emission bands of rbhdl.-Zn_2SiO_4:[Si], rbhdl.-Zn_2GeO_4:[Ge], β-Zn_2SiO_4:[Si],

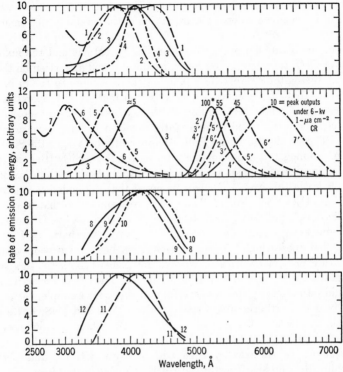

FIG. 51. Cathodoluminescence emission spectra of some oxygen-dominated phosphors of the silicate and germanate class.

1. SiO_2, 1300°C, 1 hr
2. rbhdl.-$ZnO \cdot 2SiO_2$, 1250°C, 1 hr
3. rbhdl.-$2ZnO \cdot SiO_2$, 1250°C, 1 hr
4. rbhdl.-$3ZnO \cdot SiO_2$, 1250°C, 1 hr
5. rbhdl.-$2ZnO \cdot GeO_2$, 1200°C, 1 hr
6. β-$2ZnO \cdot SiO_2$, 1600°C, 10 min, quenched
7. rbhdl.-$2BeO \cdot SiO_2$, 1400°C, 1 hr
3.′ rbhdl.-$2ZnO \cdot SiO_2 \cdot 0.012MnSiO_3$, 1250°C, 1 hr
5.′ rbhdl.-$2ZnO \cdot GeO_2 \cdot 0.012MnSiO_3$, 1100°C, 1 hr
6.′ β-$2ZnO \cdot SiO_2 \cdot 0.012MnSiO_3$, 1600°C, 10 min, quenched
7.′ rbhdl.-$3ZnO \cdot BeO \cdot 2.2SiO_2 \cdot 0.384MnSiO_3$, 1200°C, 1 hr
8. $CaO \cdot SiO_2 \cdot 0.005Re$, 1200°C, 7 hr, quenched
9. $SrO \cdot SiO_2 \cdot 0.006Re$, 1550°C, 1 hr, quenched
10. $BaO \cdot SiO_2 \cdot 0.006Re$, 1500°C, 4 hr, quenched
11. rhomb.-$2MgO \cdot SiO_2$, 1600°C, 20 min
12. tetr.-$ZrO_2 \cdot SiO_2 \cdot 0.005V$, 1600°C, 40 min

and rbhdl.-Be_2SiO_4:[Be] are shown located at successively *shorter* wavelengths in the given order. The expansion of the crystal lattice on going from rbhdl.-Zn_2SiO_4 to rbhdl.-Zn_2GeO_4 or β-Zn_2SiO_4 apparently decreases the attraction energy between the Zn cations and the anionic SiO_4 or GeO_4 tetrahedra. By lessening the external attraction on the O atoms, these atoms direct their bonding energies more to the Si or Ge cores, so that the tetrahedra probably contract by the operation of increased Si–O and Ge–O binding which increases the energy separation between the ground and excited states involved in luminescence emission. The smaller Be atom (Table 6) apparently outweighs the Si atom in determining the emission spectrum of rbhdl.-Be_2SiO_4:[Be], so that the luminescence emission is probably localized mostly within the BeO_4 tetrahedra in this case.[72] It may be noted that the emission spectrum of rbhdl.-Al_2O_3:[Al] (No. 43, Table 5) is very similar to that of rbhdl.-Be_2SiO_4:[Be], indicating again the important role of the O atoms in the luminescence-emission process. The two lower parts of Fig. 51 show that the cathodoluminescence emission bands of a wide variety of silicate phosphors (with or without intensifier-type activators such as Ag, Cu, Re,[346b] Ti, and V) are quite similar. The role of the activators is mainly to intensify the host-crystal emissions, for example, the emission bands of orthorhombic $CdSiO_3$, $CdSiO_3$:Cu, and $CdSiO_3$:Re are practically identical, being peaked near 4100 Å. These emission spectra were obtained by bombarding the phosphors with about 10-kv CR in a demountable cathode-ray tube with a silica window, and photographing the emission spectra with a Hilger E1 quartz-prism spectrograph. The exposure times were 10 to 20 min. Densitometer curves were obtained with an automatic recording Leeds and Northrup microdensitometer; then the curves were corrected for dispersion, and the antilog − 1 of the photographic-density ordinate values were plotted in Fig. 51. Some interesting cathodoluminescence emission spectra, which are not shown, include:

(1) $ZrO_2 \cdot ThO_2 \cdot 2SiO_2$:Ag, 1200°C, 60 min, emits a band extending from about 4500 to below 2900 Å, with a broad peak near 3100 Å; (2) rbhdl.-Ga_2O_3, 1000°C, 40 min, emits a band extending from about 4700 to below 3200 Å, with a peak near 3700 Å; (3) cub.-$ZnGa_2O_4 \cdot 0.005Cr$, 1600°C, 30 min, emits a weak band from 4000 to 5500 Å with a peak near 4500 Å and a strong band with faint structure from 5500 to above 7500 Å, with a peak near 6800 Å; (4) rbhdl.-$Al_2O_3 \cdot 0.009Cr$, 1600°, 30 min, emits a band extending from about 5800 to above 7500 Å, with strong lines at the peak near 6930 Å; (5) $CdO \cdot Ga_2O_3 \cdot 0.005Cr$, 1600°C, 30 min, emits a structureless band extending from 3200 to above 5100 Å, with a peak near 4100 Å; (6) $ZnO \cdot B_2O_3$, 800°C, 60 min, and $3ZnO \cdot B_2O_3$, 900°C, 60 min, both emit bands extending from about 5000 to below 2900 Å, with broad

peaks near 3450 Å; (7) $ZnO \cdot MgO \cdot 2SiO_2$, 1200°C, 60 min, emits a band extending from about 5000 to about 2900 Å, with a peak near 4100 Å; (8) $2BeO \cdot GeO_2 \cdot 0.01Mn$, 1400°C, 60 min, emits one narrow band extending from about 5600 to 6700 Å, with a peak near 6200 Å, and a broader band extending from about 4000 to below 2900 Å, with the peak apparently below 2900 Å; (9) hex.-$ZnO \cdot BeO$, 1300°C, 60 min, emits a doublet band from about 5100 to below 2900 Å, with peaks near 3000 and 4000 Å, and very low output at 3500 Å; and (10) hex.-ZnO, 1250°C, 60 min, in silicon-carbide resistance-furnace air atmosphere [initial ingredient was UV-emitting hex.-ZnO (no visible emission band as in Fig. 50a) made by burning zinc vapor in excess oxygen], emits a doublet band extending from about 4300 to above 6200 Å, with apparent peaks near 5200 and 5650 Å (orange–yellow emission color).

In the right side of the second portion of Fig. 51 is shown the same sequence of host crystals that is shown at the left, except that a rbhdl.-Zn:Be-silicate:Mn phosphor is substituted for rbhdl.-Be_2SiO_4:Mn because the latter is very inefficient (weak orange–red cathodoluminescence emission). Examination of the figure discloses that the band displacements are in the *opposite* order in the two sequences, and the emission band of rbhdl.-$(ZnO:SiO_2)$:Mn is practically unaffected by large variations in the ZnO/SiO_2 ratio, whereas this ratio has a large effect on the emission spectra of the phosphors made without Mn activator. As shown in Fig. 52, incorporation of increasing proportions of Mn activator in rbhdl.-Zn_2SiO_4:[Si] steadily reduces the luminescence efficiency of the [Si] band and produces a *new* emission band peaked near 5250 Å (compare with the similar action of Mn and Cu in ZnS, as indicated in Figs. 42 and 45).[61] The new band increases in efficiency up to about 1 weight per cent of Mn (peak output = 100), whereupon the [Si] band has completely disappeared. Higher proportions of Mn lower the luminescence efficiency of the 5250-Å band and shift the band slightly to longer wavelengths, as shown in Fig. 53. Although the host crystal is the nonstoichiometric $ZnO \cdot SiO_2$ in Fig. 52 and the stoichiometric (true compound [71b]) $2ZnO \cdot SiO_2$ in Fig. 53, the efficiency and emission spectrum of the zinc-silicate:Mn phosphor are practically the same in both cases (see Fig. 54).[72] When Be is substituted in increasing proportion for Zn in rbhdl.-Zn_2SiO_4:Mn(0.3), the efficiency of the 5250-Å band decreases steadily while a *new* band appears at about 6300 Å.[34, 93, 243] When an appreciable proportion of Be is incorporated in this phosphor, the addition of more Mn activator increases the efficiency of the 6300-Å band at the expense of the 5250-Å band (Fig. 55). To complete the cycle, it is shown in Fig. 56 that incorporation of Group IVA dioxides, for example, TiO_2, ZrO_2, HfO_2, and ThO_2, in rbhdl.-$(Zn:Be)_2SiO_4$:Mn apparently restores the original [Si] emission band (compare Fig. 56

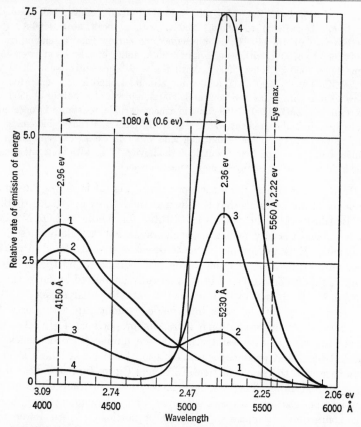

FIG. 52. Relative cathodoluminescence emission spectra of some rbhdl.-ZnO·SiO₂ host crystals made at 1200°C, 1 hr with increasing amounts of manganese activator (0, 0.0006, 0.006 and 0.015 weight per cent of Mn). At about 0.5 weight per cent Mn, the output of the UV + blue band is too low to be detected, whereas the output of the green band is 100. Incorporation of Cu or Ti intensifies the UV + blue emission band in the presence or absence of Mn.[346b, 429-430a] (Measurements made on the bombarded sides of thick samples.)

1. rbhdl.-ZnO·SiO₂:[Si], 1200°C, 1 hr
2. rbhdl.-ZnO·SiO₂·0.000012MnO, 1200°C, 1 hr
3. rbhdl.-ZnO·SiO₂·0.00012MnO, 1200°C, 1 hr
4. rbhdl.-ZnO·SiO₂·0.0003MnO, 1200°C, 1 hr

with Figs. 31 and 52).[34, 61, 157, 430a] (No trace of the [Si] emission band is found in the spectrum of rbhdl.-4ZnO·2BeO·CeO₂·4SiO₂:Mn).

The origins of the various emission bands in these oxygen-dominated phosphors, as well as in the sulphur-dominated phosphors, are still speculative. The visible emission band of hex.-ZnO:[Zn] is definitely promoted by the reduction process during preparation, but it is not certain whether to ascribe the band to centers involving traces of residual

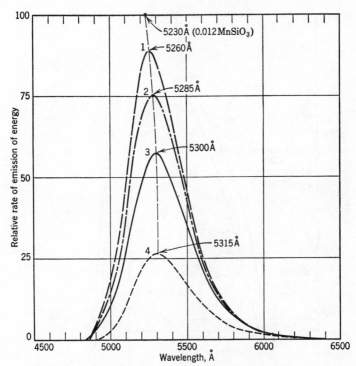

FIG. 53. Relative spectral-distribution curves of cathodoluminescence emission (bombarded side) of rbhdl.-Zn$_2$SiO$_4$ crystallized at 1200°C with 0.6, 1.2, 2.4, and 4.8 weight per cent of Mn activator [that is, Mn(2N), (4N), (8N), and (16N)].

1. rbhdl.-Zn$_2$SiO$_4$·0.024MnSiO$_3$, 1200°C, 1 hr
2. rbhdl.-Zn$_2$SiO$_4$·0.048MnSiO$_3$, 1200°C, 1 hr
3. rbhdl.-Zn$_2$SiO$_4$·0.096MnSiO$_3$, 1200°C, 1 hr
4. rbhdl.-Zn$_2$SiO$_4$·0.192MnSiO$_3$, 1200°C, 1 hr

sulphur or sulphide, to vacant O sites, or to excess Zn in regular or interstitial sites. The UV emission band of hex.-ZnO:[Zn] is partially overlapped by the absorption spectrum, and so the pigment-type absorption process must be practically saturated before the UV emission band appears.[354] Judging from the narrowness of the UV emission band, compared with the broad energy band of oxygen in ZnO (Table 9), the UV emission may involve radiative transitions from near the lower edge of the lowest conduction band ($4s$, Zn^+) to the upper edge of the topmost filled band ($2p$, O^{2-}). The doublet emission band of SiO_2, 1300°C, has not been identified with the constitution of this crystal, partly because

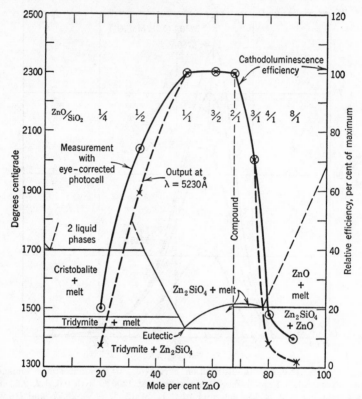

FIG. 54. Phase diagram of the system $ZnO:SiO_2$ (E. N. Bunting) [71b] with super-imposed plot of the relative cathodoluminescence efficiencies (bombarded side) of a range of $ZnO:SiO_2$ host crystals crystallized at 1250°C, 1 hr, with a constant ratio of 0.006 mole of Mn activator per mole of Zn.[72]

the detailed structures of silica crystals have not yet been determined,[39] but the similar doublet structure indicated in the emission spectra of rbhdl.-$ZnO \cdot 2SiO_2$:[Si], rbhdl.-Be_2SiO_4:[Be], and rbhdl.-Al_2O_3:[Al] suggests that the anisotropic uniaxiality of these rhombohedral crystals may produce the observed effect by different binding energies of the oxygens coordinated in the two different crystallographic directions. The 5250-Å (peak) emission band of rbhdl.-Zn_2SiO_4:Mn(0.3) is generally ascribed to transitions within Mn atoms (ions) substituted for Zn atoms (ions) in regular lattice sites.[154]

An attempt has been made by Butler to associate the absorption (excitation) spectrum with the $3d^5 \rightarrow 3d^4 4p$ transition of *free* Mn^{2+}, and the emission spectrum with the $3d^4 4p \rightarrow 3d^4 4s$ transition of Mn^{2+} by assuming that the energy

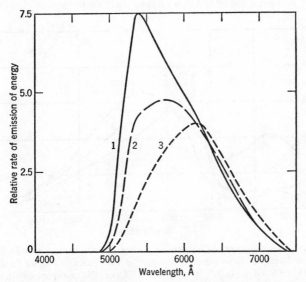

FIG. 55.   Relative spectral-distribution curves of the cathodoluminescence emissions (bombarded side) of three selected members of the phosphor family rbhdl.-$2ZnO \cdot BeO \cdot 3SiO_2$:Mn, 1200°C, 1 hr.

1. rbhdl.-$2ZnO \cdot BeO \cdot 3SiO_2 \cdot 0.018MnO$
2. rbhdl.-$2ZnO \cdot BeO \cdot 3SiO_2 \cdot 0.036MnO$
3. rbhdl.-$2ZnO \cdot BeO \cdot 3SiO_2 \cdot 0.072MnO$

All crystallized at 1200°C, 1 hr.

differences between these states become altered in the crystal to coincide with the observed data [Butler attempts also to correlate the emission spectrum of rbhdl.-Zn_2SiO_4:[Si] with the $3d^9 4p \rightarrow 3d^9 4s$ transition of *free* Zn^{2+}, but the fact that high-temperature-crystallized Zn_2SiO_4, $ThSiO_4$, $CdSiO_3$, $ZrSiO_4$, Mg_2SiO_4, $CaSiO_3$, $SrSiO_3$ and $BaSiO_3$ (Fig. 51) all have very similar emission spectra [58,72] vitiates this hypothesis].[247] Other investigators seek to correlate the observed emission band of rbhdl.-Zn_2SiO_4:Mn(1) with the known transitions of *free* atomic Mn (for example, the $3d^5 4s 4p - 3d^6 4p$ transition is about 5800 Å [349]). The existence of either entirely Mn^0 or entirely Mn^{2+} in the crystal is unlikely, because the binding is probably intermediate covalent and ionic. Still other investigators postulate that the 5250-Å emission band arises from Mn^{2+} replacing

Zn^{2+} in 4-fold coordination (by O), and that a Zn^{2+} or Mn^{2+} may replace a Si^{4+} with an accompanying interstitial Zn^{2+} or Mn^{2+} to maintain electrical neutrality.[243] This hypothesis is extended to account for the 6300-Å emission band of rbhdl.-$(Zn:Be)_2SiO_4:Mn$ phosphors as being caused by Mn^{2+} in 6-fold coordination, or in interstitial sites or flaws where it has "higher kinetic energy" than in the 4-fold coordinations. Schulman cites the presumed substitution of Mn^{2+} for Mg^{2+} in rhomb.-$Mg_2SiO_4:Mn$ as an example of the production of red emission

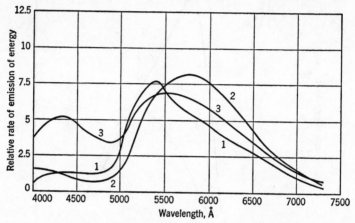

FIG. 56. Relative spectral-distribution curves of cathodoluminescence emission (bombarded side) of three selected members of the general phosphor family ZnO: $BeO:TiO_2:ZrO_2:SiO_2:MnO$. The group IVA oxides incorporated in silicates intensify the UV + blue emission band of the "pure" silicate.[61, 430a]

   1. $4ZnO \cdot 2BeO \cdot TiO_2 \cdot 4SiO_2 : Mn(N)$, 1150°C, 1 hr
   2. $4ZnO \cdot 2BeO \cdot ZrO_2 \cdot 4SiO_2 : Mn(N)$, 1250°C, 1 hr
   3. $5ZnO \cdot 3BeO \cdot ZrO_2 \cdot TiO_2 \cdot 6SiO_2 : Mn(N)$, 1200°C, 1 hr

centers by placing Mn^{2+} in 6-fold coordination with oxygen. Kröger and Linwood and Weyl evidently agree with this opinion to the extent of believing that dipositive Mn^{2+} causes the red emission, but Kröger asserts that the red emission *with fine structure* obtained from $Mg_2TiO_4:Mn$, $MgAl_2O_4:Mn$, $ZnAl_2O_4:Mn$, and $Al_2O_3:Mn$ is due to tetrapositive Mn^{4+} and requires an oxidizing atmosphere during crystallization.[156] Just such an emission spectrum with fine structure is displayed by $Mg_2GeO_4:Mn$ and $4MgO \cdot GeO_2:Mn$ (Fig. 57) after being crystallized in the normal air + CO atmosphere of a silicon-carbide-element resistance furnace.[58,262] It has been observed that the short-wave half (below about 6350 Å) of this emission spectrum disappears at 90°K, when the phosphor is excited by 3650-Å UV, indicating that there is a coupling between electronic transitions in the Mn and vibrations of the Mn relative to the surrounding host-crystal atoms (compare, rbhdl.-$Al_2O_3:Cr$). The similarity of the spectra of $Mg_2GeO_4:Mn$ and $Mg_2TiO_4:Mn$ make doubtful the various assertions regarding the emissions being caused by Mn in a definite simple valency or definite simple state of coordination. [The two emission bands of Mg_2SiO_4:Mn are located in the same red and infrared region of the spectrum (Fig. 58), but show no fine structure.[58,221]] On the basis of the data in Fig. 58, for example, it might be argued that the red emission from alumina and aluminates with Mn activator emanates from tripositive Mn^{3+}, because this emission is similar to that of Cr^{3+} in the same host crystals, and it is more likely that Mn^{3+} would replace Al^{3+} to occupy the same

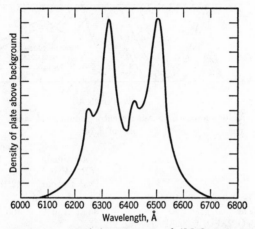

FIG. 57. Photoluminescence emission spectrum of $4MgO \cdot GeO_2$:0.01Mn, 1100°C, 1 hr. Excitation by emission from an RP-12 lamp (calcium-phosphate:Ce phosphor with an emission band peaked near 3500 Å; compare No. 47, Table 5). Measurement with grating spectrograph and densitometer. A fifth band sometimes appears in the center of the emission spectrum, and the short-wave half of the spectrum is absent at 90°K.

4-coordinated sites presumably occupied by Cr^{3+} (the fine structure of the red emission in Fig. 58 is not resolved because the spectroradiometer slits had to be opened wide to make a photoelectrically actuated recording of these weak emissions [93]). The strong green cathodoluminescence emission from many aluminates with Mn activator is not excited by 3650-Å UV and is only weakly excited by 2537-Å UV, whereas the red emission is selectively excited by 3650-Å radiation. Note that the red cathodoluminescence emission of the ZnBe-aluminate:Mn phosphors in Fig. 58 does not become pronounced until an abnormally large proportion of Mn activator is incorporated in the host crystals. In the absence of other data, this behavior might be interpreted alternatively as "evidence" that Mn atoms in low proportions in the host crystals tend to occupy perhaps only the 4-coordinated sites, whereas higher proportions force some Mn atoms to occupy the 6-coordinated sites in the spinel structure (cub.-$Fd3m$). The spinel structure has eight 4-coordinated sites and sixteen 6-coordinated sites per unit cell containing 32 oxygen atoms in cubic array.[39,285]

Another approach to the identity of the emitter in phosphors with Mn activator is based on the report that the emission bands of rbhdl.-$(Zn:Be)_2SiO_4:Mn$ phosphors approximate the absorption bands of $KMnO_4$ and $Zn(MnO_4)_2$, thereby leading to the assumption that the active emitting centers are MnO_4^- or MnO_4^{2-} groups.[310] It is further assumed that the cations do not influence the locations of the emission bands (reportedly comprising four overlapping component bands peaked at 2.35, 2.27, 2.17, and 2.05 ev), and that

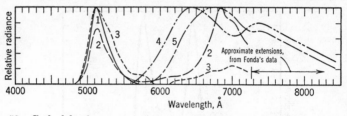

FIG. 58. Cathodoluminescence emission spectra (bombarded side), at 10 kv and $\approx 1 \ \mu a \ cm^{-2}$, of:

	Approximate Relative Peak Output	Spectroradiometer Resolving Power at	
		5250 Å	7000 Å
1. cub.-$2ZnO \cdot BeO \cdot 3Al_2O_3 \cdot 0.018MnO$, 1600°C, 1 hr	35 at 5150 Å	100 Å	260 Å
2. cub.-$ZnO \cdot BeO \cdot 2Al_2O_3 \cdot 0.009MnO \cdot 0.0015Cr_2O_3$, 1600°C, 1 hr	12 at 5150 Å	67 Å	175 Å
3. cub.-$ZnO \cdot BeO \cdot 2Al_2O_3 \cdot 0.12MnO$, 1600°C, 1 hr	1 at 5150 Å	133 Å	340 Å
4. rhomb.-$2MgO \cdot SiO_2 \cdot 0.012MnSiO_3$, 1600°C, 1 hr	10 at 6420 Å	133 Å	340 Å
5. monocl.-$MgO \cdot SiO_2 \cdot 0.006MnO$, 1600°C, 1 hr	10 at 6740 Å	133 Å	340 Å

only about 10^{-4} of the total Mn present in the phosphor exists in the presumed luminescence-active Mn^{7+} or Mn^{6+} state, the remainder being luminescence-inert Mn^{2+}. Yet another hypothesis concerning the role of Mn activator is based on the concept of exchange or resonance as applied to solids (R. Nagy [221]). Here it is assumed that chains of ...Zn–O–Si–O... atoms in, for example, zinc silicate may be considered as conjugated bond structures with alternating and exchanging high and low densities of bonding electrons on going along the chain, and it is further assumed that manganese atoms introduced into such resonance chains modify the resonance to allow emission of photons with longer wavelengths than the emission in the absence of manganese.

These inconclusive speculations are recorded here as a fair sample of the present state of phosphor literature. There is a need for better experimental evidence, rather than further speculation about the sources of these different emissions. More thorough studies of the spectra and magnetic properties of excited and unexcited phosphors at low temperatures may afford clues to the bond types, effective valencies, and coordination numbers of the emitting atoms and the symmetries of the perturbed crystal fields surrounding the emitting atoms. From the information available on the properties of the rhombohedral (α) and β-zinc-silicate:Mn phosphors, it appears that Mn atoms (ions to some variable degree) are in s (Zn) sites in both forms. Also, the 4 tetrahedrally arranged O ligand atoms of each luminescence-active Mn effectively shield the central Mn atom from the first-order effects of change in host-crystal structure on going from uniaxial (α = rhombohedral) to biaxial (β = triclinic, monoclinic, or rhombic). The appearance of several component lines and bands common to both the α and β forms (Figs. 28a–b) indicates that the 3d (inner) and 4s (valence) electron energy levels of Mn are near each other in both structures. The different relative intensities of component lines and bands observed with change in crystal structure and temperature are interpreted as being caused by changes in the relative heights of the 3d and 4s energy levels and by selective enhancement of transition probabilities involving electrons in these levels. The selective effects arise from changes in the magnitude and symmetry of the force fields involved in the structural contest between the Mn impurity and its host-crystal environment (note that the 3d electrons are bound chiefly to the Mn atoms, whereas the 4s electrons are shared with the ligand O atoms).

In addition to the multiband emissions already discussed, curves 3, 4, and 5 in Fig. 58 all show *two* emission bands associated with a single added activator in a given host crystal (compare cub.-KCl:Tl and $Ca_2P_2O_7$:Dy in Table 5). Fonda [221] has reported more fully on further examples of phosphors which exhibit a similar effect, as shown in Table 13. In all the cases of two-band emissions in the table and in Fig. 58, the short-wave emission band predominates at low concentrations of the added activator, whereas the long-wave emission band predominates at high concentrations of the added activator. Furthermore, the long-wave emission band usually predominates during the later stages of phosphorescence and is excited by lower-energy (longerwavelength) primary photons.

Continuing the general description of the emission spectra of phosphors with oxygen-dominated host crystals, the peak wavelength (λ_{pk}) of the emission band of rbhdl.-Zn_2SiO_4:Mn(0.3) *shifts* from 5250 to 5370 Å as an increasing proportion of Ge is substituted for Si in the host crystal (the λ_{pk} of rbhdl.-Zn_4SiGeO_8:Mn(0.3) is 5300 Å). This might be interpreted as evidence that a shift to longer wavelengths generally

TABLE 13

SOME PHOSPHORS WITH ONE ADDED ACTIVATOR AND TWO EMISSION BANDS

Phosphor	Excited by, Å	Peak Wavelengths of Emission Bands, Å
Hex.-ZnS:Pb(<5), 1150°C in nitrogen	2500–4400	4900 and 5800
Cub.-CaO:Pb(0.05–1), 1000°C	2500–4000 and CR	3700 and 6400
CaO.SiO$_2$:Mn(0.15–3.9), 1180°C in steam	CR	5500 and 6200
2MgO.SiO$_2$:Mn(0.15–6.2), 1180°C in steam	CR	6400 and 7400
MgO.SiO$_2$:Mn(0.15–6.2), 1180°C in steam	CR	6600 and 7400
Cub.-RbI:Tl (0.1–10), 450°C, 15 hr	2537 and CR	4350 and 4800
CaO.SiO$_2$:Pb(0.3–6):[CaF$_2$], 1100°C	2537 Å	3350

occurs when the lattice expands and the Mn atoms have more room, as in the less dense β-Zn$_2$SiO$_4$:Mn and perhaps in rbhdl.-Zn$_2$SiO$_4$:Mn with high Mn content (Fig. 53), but the contrary shift from $\lambda_{pk} = 5150$ Å for cub.-ZnAl$_2$O$_4$:Mn to $\lambda_{pk} = 5060$ Å for cub.-ZnGa$_2$O$_4$:Mn [Table 5; compare rhomb.-ZnSO$_4$:Mn (6340 Å) and rhomb.-CdSO$_4$:Mn (5900 Å)] indicates that this simple interpretation may not be justified.[157] The fact that the spectra simply *shift* in these cases indicates that Si, Ge, Al, and Ga are not nearest-neighbor ligands of the emitting atoms, and so the Mn atoms are the chief emitters rather than the O atoms. The appearance of a new emission band when Be is substituted for Zn in rbhdl.-Zn$_2$SiO$_4$:Mn is an indication that the small and strongly polarizing Be atoms become immediate ligands of one of the O atoms around the Mn atoms, thereby altering the force-field symmetry of the center and enhancing (producing) the characteristic emission band peaked near 6300 Å. When small amounts of rhomb.-Mg$_2$SiO$_4$:Mn are cocrystallized with rbhdl.-Zn$_2$SiO$_4$:Mn, the characteristic long-persistent red and infrared emission of rhomb.-Mg$_2$SiO$_4$:Mn (Fig. 58) appears immediately, indicating that a separate phase is formed.

When the crystal structure of phosphors with Mn activator and two oxides in the host crystal remains unchanged with large changes in the proportions of the two oxides, for example, ZnO/SiO$_2$ and ZnO/Al$_2$O$_3$, then the location and shape of the emission spectrum is little affected by the change in host-crystal proportions (compare curves 2′, 3′, and 4′ in Fig. 51). In the case of rbhdl.-Zn_2SiO_4:Mn, for example, a large excess of SiO_2 over orthoproportion continues to form the $R\bar{3}$ structure by simply leaving out the missing ZnO. It is found that a higher optimum proportion of MnO may be incorporated in these excess silica phosphors because there are more available vacant ZnO sites.[72] Hence, these crystallized materials with excess silica form **deficiency structures,** being much the same as if some ZnO were to be extracted from $R\bar{3}$-Zn_2SiO_4 without otherwise altering the $R\bar{3}$ structure.

In connection with spectral invariance on altering host-crystal proportions, it is notable that the emission spectrum of tetr.-$(Zn:Mg)F_2$:Mn is unchanged by variations of the ZnF_2/MgF_2 ratio.[58] When changes in the host-crystal structure are accompanied by changes in crystal structure, however, a *new* emission band appears to correspond to each different structure; for example (with Mn activator), the compound $ZnO \cdot B_2O_3$ emits one band, and the compound $5ZnO \cdot 2B_2O_3$ [220] emits another band. Such is the case when the ZnO/B_2O_3 or CdO/B_2O_3 ratio is altered in the borate phosphors shown in Fig. 59, and a similar effect is observed when the CaO/SiO_2 ratio is altered in $CaO:SiO_2$:Mn phosphors, or the MgO/SiO_2 ratio is altered in $MgO:SiO_2$:Mn phosphors (Fig. 58).[58,72] Chemically speaking, the characteristic emission bands may often be identified with certain stoichiometric proportions;[71b] for example, some typical host-crystal proportions and corresponding peak-wavelengths are: $2MgO \cdot SiO_2$:Mn (6420 Å), $MgO \cdot SiO_2$:Mn (6740 Å) (both have a second emission band at 7400 Å, Fig. 58 [221]); $ZnO \cdot B_2O_3$:Mn (5420 Å), $5ZnO \cdot 2B_2O_3$:Mn (5980 Å); $CdO \cdot 2B_2O_3$:Mn (5380 Å), and $2CdO \cdot B_2O_3$:Mn (6260 Å). Partial substitution of Cd for Zn in rbhdl.-Zn_2SiO_4:Mn (Fig. 60) apparently has much the same "poisoning" effect as when Cd is substituted in part for Zn in cub.(or hex.)-ZnS:Mn (Fig. 41). In this case, there is a stoichiometric and structural incompatibility between rbhdl.-Zn_2SiO_4:Mn and rhomb.-$CdSiO_3$:Mn, and the luminescence efficiency of the system decreases to very low values for phosphors in the composition range from about 30 to 90 mole per cent of Cd.

By way of further comparison of the effects of host-crystal variations in O- and S-dominated phosphors, D. Pearlman *et al.* reportedly obtain a *green–yellow* emission band peaked near 5250 Å when cub.-SrS:SrO:Bi(0.01) [$SrCl_2(12)$], 1000°C (in N_2) contains 3 per cent or less of SrO, but obtain only a *blue* emission band peaked near 4400 Å when the SrO content is 4 to 100 per cent.[172] Here, the limit of solubility of SrO in cub.-SrS is 3 per cent. With the appearance of a new phase in the *material* at SrO proportions above 3 per cent, therefore, the cub.-SrO:Bi emission arises and the cub.-SrS:Bi emission subsides.

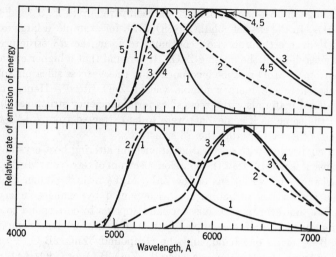

FIG. 59. Cathodoluminescence emission spectra (bombarded side) of some borate phosphors with manganese activator.

(Top) 1. $ZnO \cdot B_2O_3 \cdot 0.006MnO$, 900°C, 1 hr
        2. $3ZnO \cdot 2B_2O_3 \cdot 0.018MnO$, 900°C, 1 hr
        3. $2ZnO \cdot B_2O_3 \cdot 0.012MnO$, 900°C, 1 hr
        4. $3ZnO \cdot B_2O_3 \cdot 0.018MnO$, 900°C, 1 hr
        5. $3ZnO \cdot B_2O_3 \cdot Al_2O_3 \cdot 0.018MnO$, 1600°C, 1 hr

Peak Outputs under
6-kv 1-μa cm^{-2} CR

  1. 10
  2. 10
  3. 15
  4. 15
  5. 20

(Bottom) 1. $CdO \cdot 2B_2O_3 \cdot 0.006MnO$, 800°C, 1 hr
        2. $CdO \cdot B_2O_3 \cdot 0.006MnO$, 800°C, 1 hr
        3. $3CdO \cdot 2B_2O_3 \cdot 0.018MnO$, 800°C, 1 hr
        4. $2CdO \cdot B_2O_3 \cdot 0.012MnO$, 800°C, 1 hr

Peak Outputs under
6-kv 1-μa cm^{-2} CR

  1. 4
  2. 5
  3. 15
  4. 30

With respect to deficiency structure, R. Ward *et al.* (private communication) report that the limit of solubility of Sm_2S_3 in cub.-SrS is 35 per cent. With Sm_2S_3 proportions below 35 per cent, the cub.-SrS structure is maintained, although the average interatomic distance decreases and cation omission sites appear because 3 S atoms go into the structure for every 2 Sm atoms (a similar effect is reported by Ward for Ce_2S_3 dissolved up to 10 per cent in cub.-SrS).[717]

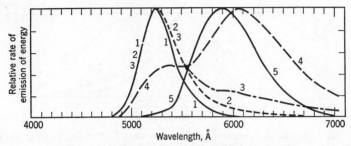

Fig. 60. Cathodoluminescence emission spectra of:

1. rbhdl.-$Zn_2SiO_4 \cdot 0.012MnSiO_3$, 1250°C, 1 hr
2. ↑ $8ZnO \cdot CdO \cdot 5.5SiO_2 \cdot 0.41MnO$, 1075°C, 1 hr
3. $5ZnO \cdot 5CdO \cdot 7.75SiO_2 \cdot 0.41MnO$, 1075°C, 1 hr
4. ↓ $ZnO \cdot 8CdO \cdot 8.55SiO_2 \cdot 0.41MnO$, 1075°C, 1 hr
5. rhomb.-$CdSiO_3 \cdot 0.006MnSiO_3$, 1125°C, 1 hr

Relative Peak Outputs
under 6-kv 1-μa cm^{-2} CR

1. 100
2. 41
3. 20
4. 7
5. 30

A brief review of some of the general preparative aspects of O-dominated phosphors may be advantageous at this point.

1. General Absence of Fluxes. All the O-dominated phosphors described in this section were prepared *without fluxes*, because fluxes generally decrease the luminescence efficiencies of the resultant phosphors. Where the efficiencies are decreased, the common fluxes appear to react with the host-crystal ingredients to form alkali or alkaline-earth silicates (germanates, borates, and the like) which produce vitreous phases or produce undesirable distortions by being incorporated as structurally dissimilar inclusions. In some cases, however, fluxes are definitely beneficial in preparing O-dominated phosphors.

2. Effect of Atmosphere during Crystallization. All the O-dominated phosphors described in this section were crystallized in a silicon carbide element electrical-resistance furnace with a normal air atmosphere containing small proportions of CO and CO_2 formed by reaction of atmospheric oxygen with the SiC elements. An oxidizing (O_2) or neutral (N_2) atmosphere generally produces higher-efficiency products than a strongly reducing atmosphere (H_2) or a vacuum, at least in the cases of phosphors with Mn activator. Some of the phosphors with

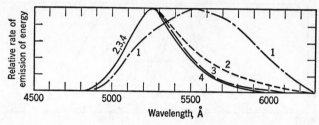

Fig. 61. Cathodoluminescence emission spectra of rbhdl.-$2ZnO \cdot SiO_2$:Mn(0.3) prepared by heating a mixture of $ZnCO_3$, $MnCO_3$ (both precipitated from nitrate solution) and fine-particle SiO_2. (Reflection colors of the cooled products viewed in white light are: 100°C, brown–black; 600°C, dark tan; 700°C, gray; 800–850°C, gray orange–yellow; 1300°C, mottled orange–yellow and white; 1400°C, white.)

1. 700°C, 3 hr
2. 800°C, 2 hr
3. 850°C, 3 hr
4. 1400°C, 1 hr

Peak Outputs under
6-kv 1-μa cm^{-2} CR

1. 6
2. 16
3. 21
4. 79

rare-earth activators may be prepared best by heating in a reducing atmosphere, or else by precrystallization in a neutral or oxidizing atmosphere, followed by heat treatment in a reducing atmosphere to reduce the tripositive rare-earth ions (line spectra) to dipositive rare-earth ions (band spectra).

3. Effect of Temperature, Crystal Structure, and Rate of Cooling. Figures 61 and 62 show the cathodoluminescence emission spectra of some well-known silicate phosphors prepared by the carbonate process, including heating for at least an hour at the indicated temperatures. Below each figure is given a subjective assessment of the reflection colors of the heated products, showing how the brown–black MnO becomes gradually incorporated in the crystallized materials which may be made practically white at sufficiently high temperatures and sufficiently long times of heating. When rbhdl.-Zn_2SiO_4:[Si] is prepared by reaction and crystallization at 1200°C for 90 min, then mixed with 1 weight per cent of Mn as $MnCO_3$ (precipitated from nitrate solution onto the suspended silicate particles), and heated for at least an hour at various temperatures from 700 to 1400°C, there is *no trace of yellow emission*, such as the band shown as curve 1 in Fig. 61. All the heated products, in this case of activator addition by diffusion into the precrystallized host crystal, have

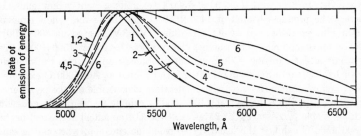

FIG. 62. Cathodoluminescence emission spectra of $8ZnO \cdot BeO \cdot 5SiO_2$:Mn(1.4) prepared by heating a mixture of $ZnCO_3$, $BeCO_3$, $MnCO_3$ (all precipitated from nitrate solution) and fine-particle SiO_2. (Reflection colors of the cooled products viewed in white light are: 100°C, brown–black; 600°C, light gray–brown; 800–850°C, gray; 900°C, tan–gray; 1000°C, dark tan; 1100°C, light yellow–orange; 1200°C, mottled yellow–orange and white; 1300°C, very light gray–white; 33-Z-5A, white.)

1. 900°C (16)	4. 1200°C (79)
2. 1000°C (31)	5. 1300°C (57)
3. 1100°C (55)	6. 33-Z-5A (91)

Relative *visible* cathodoluminescence outputs in parentheses.

the normal green emission band shown as curve 4 in Fig. 61. The reflection colors of the heated and cooled products of this series are: 700–800°C, *light* tan; 900°C, very light tan; 1000°C, cream–white; 1100–1400°C, white (luminescence efficiency data on all these phosphors are presented in Figs. 90 and 92). Figure 62 shows that ZnO reacts more rapidly than BeO with SiO_2, as evidenced by the late appearance of the Be-induced emission band peaked near 6300 Å. Complete reaction and crystallization often requires comminution and reheating to produce uniform high-efficiency products such as the phosphor denoted by 33-Z-5A. When the phosphor host crystal is complex, variations in the crystallization temperature, heating time, and rate of cooling often produce remarkable changes in the properties of the resultant phosphors.

For example: (a) $ZnO \cdot BeO \cdot Al_2O_3$:Mn($N$) heated at 1000–1400°C for an hour exhibits a blue–green emission with no visible persistence, whereas, when this material is heated at 1500–1600°C, the emission color becomes yellower, and a moderate red–orange persistence is observed after cessation of excitation; (b) $2ZnO \cdot ZrO_2 \cdot 2SiO_2 : Mn(N)$ heated at 800–1600°C and cooled in air exhibits green-to-yellow emission, whereas the products heated at 1500–1600°C and quenched by plunging into cold water exhibit red emissions; (c) $ZrO_2 \cdot SiO_2$ (no added activator) exhibits the usual UV + pale-blue emission (Fig. 51) when heated from 1000 to 1600°C; $ZrO_2 \cdot SiO_2 : Mn(N)$ heated at 1000°C exhibits the UV + blue emission but when heated at 1200–1400°C exhibits a pale-green–yellow emission, and when heated at 1600°C exhibits a light-lavender emission; $ZrO_2 \cdot SiO_2 : Mn(5N)$ heated at 1100–1300°C exhibits a light-yellow emission, whereas the products heated at 1400–1600°C have the same yellow emission plus a blue emission; (d) $2CdO \cdot ZrO_2 \cdot 3SiO_2 : Mn(3N)$ heated at 800–1200°C exhibits an orange emission, whereas the products heated at 1300–1600°C exhibit only a blue emission (CdO apparently sublimes rapidly from this material above 1300°C); (e) $4BeO \cdot CdO \cdot 3SiO_2 : Mn(N)$ heated at 800–1000°C exhibits a strong yellow–orange emission with moderately long visible orange persistence, whereas the products heated at 1100–1600°C exhibit feeble purple emissions with no visible persistence; (f) $2MgO \cdot CdO \cdot 2SiO_2$ (no added activator) heated at 800–1500°C exhibits the usual UV + pale-blue emission, whereas the same host crystal with $Mn(N)$ heated at 800°C exhibits a pale-orange emission and becomes increasingly redder on heating at higher temperatures up to 1500°C (light red); and (g) $ThO_2 \cdot Al_2O_3 \cdot 2SiO_2 : Mn(N)$ exhibits a lavender-hued orange emission when heated from 800 to 1000°C, a light-violet–blue emission color when heated from 1200 to 1400°C, and a light-orange emission color when heated at 1600°C, and cooled in air or quenched in water. All these subjective observations were made under excitation by 10-kv cathode rays with the phosphors at room temperature. The designation $Mn(N)$ corresponds to 0.006 mole Mn per mole of cation, for example, $2ZnO \cdot ZrO_2 \cdot 2SiO_2 : Mn(N) = 2ZnO \cdot ZrO_2 \cdot 2SiO_2 \cdot 0.018MnO$, and $2CdO \cdot ZrO_2 \cdot 3SiO_2 : Mn(3N) = 2CdO \cdot ZrO_2 \cdot 3SiO_2 \cdot 0.054MnO$. As additional information on cathodoluminescence emission colors, it may be noted that the emission color of cub.-$MgO \cdot Al_2O_3 : Mn(N)$, 1600°C, is a weak green; $CaO \cdot Al_2O_3 : Mn(N)$, 1600°C, is a stronger yellow–green with a band peaked at 5470 Å, $CdO \cdot Al_2O_3 : Mn(N)$, 1000–1150°C, is a still stronger light orange–yellow; [157] and $CdO \cdot Ga_2O_3 : Mn(N)$, 1500°C, is a dark red. The emission colors of some of these host crystals with Cr, instead of Mn, heated at the same temperatures, are: cub.-$MgO \cdot Al_2O_3 : Cr(N)$ is a weak mottled green and red, $CaO \cdot Al_2O_3 : Cr(N)$ is a very faint blue, and $CdO \cdot Al_2O_3 : Cr(N)$ is a fairly strong light yellow with practically the same two emission bands (peaked at 4960 and 5890 Å) observed for $CdO \cdot Al_2O_3 : Mn(N)$.[157] Also, cub.-$ZnGa_2O_4 : Cr(N)$, 1600°C, exhibits a weak lavender–violet emission. As another example of the effect of crystal structure on luminescence emission, $2ZnO \cdot SiO_2 : Mn(N)$ heated to melting at 1600°C and cooled slowly has the $R\overline{3}$ structure and emits the green band peaked at 5250 Å, whereas a platinum dish of the same material heated to 1600°C and quenched in water contains thinner patches of the yellow-emitting β-form sharply demarcated from thicker patches of the green-emitting $\alpha(R\overline{3})$-form (Figs. 11a and 51).[58,61] An excess of ZnO over the orthoproportion has been found to increase the efficiency of the yield of the β-form, partly because ZnO sublimes rapidly at temperatures above about 1300°C.[58] Fonda has reported that a yellow-emitting form of this phosphor, with largely amorphous structure, can be produced by low-temperature (850°C) reaction in the presence of about 15 per cent of a flux, such as KCl, but there is no clear indication that the high- and low-temperature yellow-emitting forms are identical.[221,223-225] These various results obtained from the same ingredients treated in different ways further emphasize the necessity for *complete* specification of new phosphors and their preparations in order to have interpretable information.

SUMMARY COMMENTS ON EMISSION SPECTRA OF PHOSPHORS. *1. General Remarks, and Action of Temperature.* The emission spectra of phosphors are primarily a function of their compositions and structures. The locations and shapes of the emission lines and *component* emission bands are practically unaffected by the nature or energy of the primary excitant particle, at least at low excitation densities. At high excitation densities, however, the phosphor crystals may become sufficiently heated to produce a relatively small temperature variation of the emission spectra. The different effects of temperature on the emission spectra of typical phosphors were outlined in connection with Figs. 28–30. Because the basic factors in determining phosphor-emission spectra are the natures, interactions, and structural arrangements of the component atoms of the phosphor crystals, a review is presented here to illustrate the salient features of activators and host crystals in their effects on emission spectra.

2. Action of Activator. Activators may be classified in two major groups according to their action on emission spectra.

(a) **Intensifier activators** *evoke or intensify a latent emission line or band* of the host crystal made without added activator; for example, Ag(0.01) intensifies the emission of cub.(or hex.)-(Zn:Cd)S or cub.-Zn(S:Se) and shifts it to shorter wavelengths in these S(Se)-dominated phosphors. Also, Pb(1–10) intensifies the emission of tetr.-$CaWO_4$, tetr.-$CaMoO_4$, rhomb.(or monocl.)-$BaSO_4$, and $Ba_2Si_3O_8$, and Ti(1–10) intensifies the emission of many silicates, such as rbhdl.-Zn_2SiO_4, $ThSiO_4$, and $CaSiO_3$.[34,154,155,158,159,244-250,353] Both Pb and Ti apparently shift the original emission spectrum, usually to *longer* wavelengths in these O-dominated phosphors (except monocl.-$CaSiO_3$:Pb [221,229]). The effect of increasing the Pb content in tetr.-$(Ca:Pb)MoO_4$:[Mo] is to shift the λ_{pk} value from 5340 Å (no Pb) to 6100 Å (25 mole per cent Pb) and then *back* to 5130 Å (no Ca).[155] Here, the Pb plays the dual role of intensifier-type activator and host-crystal substituent.

In most of the silicates, the Si is 4-coordinated by oxygen, whereas Ti tends to be 6-coordinated (as in crystalline TiO_2 and cub.-$CaTiO_3$), and so the TiO_6 group probably *attempts* to form as a distinct entity when Ti is used as an activator in silicates. This notion is strengthened by the experimental observation that several silicate phosphors with Ti activator, for example, rbhdl.-Zn_2SiO_4:Ti(1) and $ThSiO_4$:Ti(1), which have UV + blue emission bands at 300°K (Fig. 31), exhibit the same orange emission color as TiO_2 at 90°K (rbhdl.-Zn_2SiO_4:[Si] and $ThSiO_4$:[Si] exhibit a blue emission at both temperatures). It is possible that Cb activator in tetr.-ZnF_2:Cb intensifies the feeble UV + blue emission of tetr.-ZnF_2.[58] In these cases of activation to produce intensification, it is probable that the primary action of the activator atom is to upset the crystal-field symmetry and the operation of the host-crystal selection rules in the vicinity of the emitting atom or atomic group of the host crystal.

(b) **Originative activators** *produce new emission lines or bands*, usually at the expense of the original emission bands of the host crystal, for example, Cu(0.01) and Mn(1) produce new emission bands in cub.(or hex.)-ZnS, and Sb, Eu, Dy, Cr, and Mn produce new emission lines or bands in phosphors such as cub.-KCl:Sb, cub.-MgS:Sb, cub.-SrS:Eu, $ThSiO_4$:Eu (no change in emission color at 90 and 300°K), $SrSiO_3$:Eu, $Ca_2P_2O_7$:Dy, rbhdl.-Al_2O_3:Cr, rbhdl.-Zn_2SiO_4:Mn, cub.-$ZnAl_2O_4$:Mn, tetr.-ZnF_2:Mn, rhomb.-$CdSiO_3$:Mn, rhomb.-$CdSO_4$:Mn, and monocl.-$MgSiO_3$:Mn. In many of these cases of activation, it is probable that the activator atom is also the chief emitting atom, although it is possible that strongly bound activator atoms may sometimes displace the radiation-transition energy levels of their host-crystal ligands as well as locally altering the general host-crystal selection rules.

Some activators may act primarily as an intensifier in one host crystal and primarily as an originative emitter in another host crystal. For example, an optimum of Bi(0.01) in S-dominated cub.-CaS or cub.-SrS affords a long-persistent blue or green emission at 300°K (short-persistent red emission at 90°K!), whereas an optimum of Bi(1 or more) in O-dominated rhomb.-$CaSO_4$ or rhomb.-$SrSO_4$ (or phosphates of Ca or Sr) affords a yellow, orange, or red emission of very short persistence.[154, 346a] Also, Cu(0.2) apparently intensifies the UV + blue emissions of many silicates (Fig. 31),[346b] whereas Cu(0.01) produces a new emission band in ZnS-type phosphors, excepting cub.-ZnSe (Figs. 32–45), and Cu(> 0.01) intensifies the blue emission band of ZnS:[Zn], despite the fact that Cu(0.01 or less) suppresses the blue emission band (Figs. 45a–c).

Two or more activators in a given host crystal may sometimes be used to produce two or more emission bands by careful choice of the activators and their proportions. Figure 46 shows an example of Ag (intensifier) activator and Cu (originative) activator incorporated in such proportions that their corresponding emission bands appear with approximately equal intensity under the given conditions of excitation. When an added activator produces a new band, the activator proportion may often be adjusted so that both the original host-crystal emission band and the new activator-induced emission band appear (for example, Figs. 42, 45, 52). It is then possible to study simultaneously the effects which various other added ingredients and treatments have on the two emission bands. For example, many oxygen-dominated host crystals, such as the silicates, germanates, borates, aluminates, galliates, phosphates, and sulphates of Group II elements (for example, Mg, Ca, Sr, Ba, Zn, and Cd), when crystallized with about 0.01 weight per cent of Mn activator, may be made to exhibit the characteristic ultraviolet + blue emission band of the complex host-crystal anion (radical) *and* the longer-wave green-to-red emission band produced by the Mn. The UV + blue host-crystal emission apparently originates in the anion radicals (for example, in the SiO_4 tetrahedra in orthosilicates), whereas the Mn-produced emission originates in cation sites (for example, Mn substituted for Zn in zinc-silicate crystals); so variations in the two emission spectra afford a sensitive indication of the relative influences which different ingredients and treatments have on the two different sites in the crystals. By using proper proportions and treatment of an added activator, such as Eu, which affords either line- or band-emission spectra, depending on its degree of oxidation (electropositiveness), it is possible to produce phosphors which emit simultaneously a host-crystal band, an activator-induced band, and activator-induced lines. A particularly interesting phosphor is cub.-ZnS:[Zn]:P, where the presumably induced intensifier-activator [Zn] emission band peaks near 4600 Å and the added originative-activator (P) band peaks near 5600 Å to produce a resultant near-white emission color.[698] This phosphor is especially noteworthy because the originative phosphorus activator is generally classed as a nonmetal, whereas most of the useful added activators are definitely metallic elements.

3. Action of Host Crystal. Insofar as emission spectra alone are concerned, phosphor host crystals apparently perform two major functions: (1) The host crystal may have energy levels such that a suitable radiative transition is *possible* and becomes *probable* when a perturbing intensifier-type activator impurity is incorporated in the crystal, and (2) the host crystal may function primarily as a suspension and energy-transfer medium which surrounds an originative-type activator atom and yet allows radiative transitions to take place chiefly in the field of the activator atom. Intermediate cases are conceivable, of course, but the foregoing simple partition of the functions of host crystals appears to fit qualitatively most of the known phosphor emission spectra. In some cases, such as rbhdl.-$(Zn:Be)_2SiO_4:Mn$, a host-crystal component (Be) may act as an intensifier acting on an originative-activator (Mn) center to selectively enhance one or more of several possible emission lines and bands.

As previously outlined, emission *lines* occur when the radiative transitions take place within shielded incompleted inner shells of virtually independent emitting atoms (compare Figs. 13, 14, and 31), whereas emission *bands* occur when the radiative transitions involve the outermost valence electrons of the emitting atoms which are strongly affected by the proximity, nature, and structural arrangement of the atom's ligands. When either the chief emitting atom (center) or its strongly influential ligands are closely bound to the other atoms of the host crystal, the energy levels of the excited and ground states of valence electrons involved in the radiative transition are usually proliferated into bands, according to the exclusion principle, and the bands are further broadened by thermal agitation and nonuniformities in the sites occupied by the chief emitting atoms (centers). It is inadvisable, particularly in the case of band-emission spectra, to speak of an emission as being due entirely to one or another atom, because the environment of the atom plays a vital part in determining the location and shape of the band (compare the different emissions of the same activators in different host crystals in Table 5 and in Figs. 13, 14, and 31–62). Structural changes affect the energy levels involved in emission according to their relative influences on (1) the configuration and force-field symmetry around the emitting atom or center (including changes in polarization and effective directed bonding of the atoms), and (2) the interatomic spacings and average force field around the emitting atom or center (including increases or decreases in the amount of interaction with the immediate ligands and with the crystal as a whole).

The large electronegative anions in crystals generally dominate the structure of the solid by occupying most of the volume.[39, 41, 47] In many structures, such as spinels, perovskites, and ilmenites, the anions form close-packed structures, leaving the smaller cations to distribute themselves in an excess number of several *ideally* equivalent sites between the close-packed anions. Because of this anion dominance, and the prevalence of oxygen, sulphur, and selenium in useful phosphors, it is convenient to classify the more important host crystals into two groups: (*a*) *the O-dominated group*, and (*b*) *the* S(Se)-*dominated group*. Some fluorides may be classified in the O-dominated group because the fluorine and oxygen atoms are both strongly electronegative [24] and are practically the same size (Table 6), and chlorides and bromides may be tentatively classified in the S(Se)-dominated group for the same reasons (the luminescence characteristics of tetr.-ZnF_2:Mn are similar to those of rbhdl.-Zn_2SiO_4:Mn, rhomb.-$CdSiO_3$:Mn, and rhomb.-$ZnSO_4$:Mn; also, the characteristics of cub.-KCl:Sb are similar to those of cub.-MgS:Sb, and there is a similarity between cub.-KCl:Ag [330] and cub.-ZnS:Ag). Thus far, no useful phosphors have been made from tellurides or iodides, in part because their crystals tend to absorb light strongly, and because they are relatively unstable under heat treatment and intense bombardment by high-energy particles.

A distinctive feature of the O-dominated host crystals is the frequent occurrence of atomic complexes (radicals), such as the structurally *distinct* tetrahedral SiO_4^{4-}, PO_4^{3-}, SO_4^{2-}, and ClO_4^- groups, and the structurally *continuous* arrangements formed by joining together two or more corners of planar BO_3^{3-} groups, tetrahedral SiO_4^{4-} groups, or octahedral WO_6^- groups to form chains or rings which, in the case of chains, are terminated only by the boundaries of the crystal. In contrast to this common occurrence among O-dominated crystals, S(Se)-dominated crystals rarely have MeS_x complexes; in fact, S and Se prefer to act as core atoms in O-dominated complexes, such as SO_4^{2-} and SeO_4^{2-}. (The indicated conventional formal valencies should not be construed as meaning that the groups have the entire given electrostatic charges, because the anion–cation bonds are, as always, partly ionic and partly covalent.) X-ray-diffraction studies show that the interatomic spacing and bonding in complex groups, such as the tetrahedral SiO_4 group in orthosilicates, remains practically constant, irrespective of the nature of the external cation. This is manifested also by the similar UV + blue emission bands of many complex O-dominated crystals made without added activator (Fig. 51).[72] Spectral displacements in the emission spectra of such crystals afford a sensitive indication of structural variations, because readily detectable 10-Å displacements correspond to energy changes of about 0.01 ev, or less than 1 per cent of the energy difference between the excited and ground states involved in the radiative transition.

It is remarkable that (1) the edge (host-crystal) emission bands of phosphors without complex groups, for example, hex.-ZnO:[Zn] and cub.(or hex.)-(Zn:Cd)S:[Zn:Cd], are less than 0.5 ev wide, whereas the longer-wave emission bands of the same phosphors (with or without Ag, Cu, or Mn activators in the case of the sulphides) are usually over 1-ev wide, and (2) the UV + blue host-crystal emission bands of phosphors with complex groups, for example, rbhdl.-Zn_2SiO_4:[Si], ThSiO_4:[Si], tetr.-$CaWO_4$:[W], rbhdl.-Al_2O_3:[Al], and cub.-$ZnAl_2O_4$:[Al], are usually over 2-ev wide, whereas the component longer-wave emission *bands* of these phosphors with activators such as Pb, U, Cr, Sm, Eu, and Mn are usually narrower than the host-crystal emission bands, becoming as narrow as 0.3-ev in the case of cub.-$ZnAl_2O_4$:Mn. The small effect that changes in the core atoms of complex groups have on the cations outside these groups (keeping the space group constant) is indicated by the following examples, where the Mn activator atoms (ions) are substituted for the Zn cations:

	cub.-$ZnAl_2O_4$:Mn →	cub.-$ZnGa_2O_4$:Mn	Difference
λ_{pk} (ev)	5130 Å (2.42)	5060 Å (2.45)	+0.03 ev
Band width in ev	0.27	0.36	+0.09 ev

	rbhdl.-Zn_2SiO_4:Mn →	rbhdl.-Zn_2GeO_4:Mn	
λ_{pk} (ev)	5250 Å (2.36)	5370 Å (2.31)	−0.05 ev
Band width in ev	0.58	0.57	−0.01 ev

Much larger effects are produced by changes in simple anions, such as shown in the following examples, where the activator atoms may occupy either interstitial or substitutional sites:

	hex.-ZnO:[Zn] →	hex.-ZnS:[Zn]	Difference
λ_{pk} (ev)			
Edge emission	3850 Å (3.22)	3400 Å (3.64)	+0.42 ev
Usual emission	5050 Å (2.46)	4420 Å (2.80)	+0.34 ev
Band width in ev			
Edge emission	0.50	0.40	−0.10 ev
Usual emission	1.28	1.13	−0.15 ev

	cub.-ZnS:[Zn] →	cub.-ZnSe:[Zn]	
λ_{pk} (ev)	4660 Å (2.66)	6530 Å (1.90)	−0.76 ev
Band width in ev	1.14	0.70	−0.44 ev

	cub.-ZnS:Ag →	cub.-ZnSe:Ag	
λ_{pk} (ev)	4490 Å (2.76)	5690 Å (2.18)	−0.58 ev
Band width in ev	1.20	0.70	−0.50 ev

The comparison for the S → Se substitution with Cu and Mn activators gives somewhat smaller differences, although still larger than for the Al → Ga and Si → Ge substitutions, and a tabular comparison is omitted because the emission spectra of cub.-ZnSe with [Zn], Cu, and Mn are all practically identical. Changes in the cation components of phosphor host crystals appear to produce relatively small effects on the emission bands associated with substitutionally located activator atoms, such as Mn, and large effects on presumably interstitially located activator atoms, such as Ag, for example:

	hex.-ZnS:Ag →	hex.-CdS:Ag	Difference
λ_{pk} (ev)	4330 Å (2.86)	>7300 Å (<1.7)	>−0.96 ev
Band width in ev	0.96	?	?

	hex.-ZnS:Mn →	hex.-CdS:Mn	
λ_{pk} (ev)	5850 Å (2.12)	7100 Å (1.75)	−0.37 ev
Band width in ev	0.72	0.56	−0.16 ev

	rbhdl.-Zn_2SiO_4:Mn →	rbhdl.-Be_2SiO_4:Mn (extrapolated)	
λ_{pk} (ev)	5250 Å (2.36)	6300 Å (1.97)	−0.39 ev
Band width in ev	0.58	0.69	+0.11 ev

	tetr.-MgF_2:Mn →	tetr.-ZnF_2:Mn	
λ_{pk} (ev)	5900 Å (2.10)	5870 Å (2.11)	+0.01 ev
Band width in ev	0.68	0.68	. . . .

	rhomb.-$ZnSO_4$:Mn →	rhomb.-$CdSO_4$:Mn, or rhomb.-$CdSiO_3$:Mn	
λ_{pk} (ev)	6340 Å (1.96)	5900 Å (2.10)	+0.14 ev
Band width in ev	0.57	0.64	+0.07 ev

Experiments with zinc and cadmium selenate host crystals containing Mn activator have, thus far, been unsuccessful in producing appreciable visible luminescence emission. The remarkable similarity of the emission spectra of tetr.-MgF_2:Mn, tetr.-ZnF_2:Mn, rhomb.-$CdSO_4$:Mn, and rhomb.-$CdSiO_3$:Mn extends also to other luminescence characteristics, such as the long exponential decays, of these phosphors. Co-crystallization of cub.-CaF_2, cub.-SrF_2, cub.-BaF_2, or cub.-CdF_2 with structurally dissimilar tetr.-ZnF_2, containing Mn activator, produces phosphors having lower efficiencies, shorter persistences, and emission bands which are displaced to shorter wavelengths in the case of incorporation of CaF_2, SrF_2, and CdF_2, and are displaced to longer wavelengths in the case of incorporation of BaF_2.[58, 262]

Analysis by Means of Luminescence

Because the luminescences of materials are determined chiefly by their compositions and structures, it is sometimes possible to identify a luminescent material by its distinctive luminescence characteristics. This analytical technique has become of practical consequence in (1) immediate identification of certain luminescent minerals, such as violet-emitting scheelite (tetr.-$CaWO_4$:[W]) and green-emitting manganese-bearing willemite (rbhdl.-Zn_2SiO_4:Mn), which may be excited in the field by a portable source of ultraviolet,[544a] (2) identification of numerous natural and synthetic organic substances and materials, many of which exhibit distinctive changes in their luminescence characteristics when subjected to different treatments (these changes are useful in controlling variations in processing, and in some criminological investigations),[75–84, 144, 318, 533, 563] and (3) identification of certain chemical constituents of a material, for example, by coupling a metallic cation to an organic substance to form a metallo-organic compound having a distinctive luminescence emission (see Figs. 121 and 122). Under favorable conditions, less than one part of copper salt incorporated in a billion parts of otherwise *pure* blue-emitting crystallized zinc sulphide is easily detectable by the green emission (Figs. 45b and 45c) which the copper activator originates both during and (especially) after excitation. As mentioned in the discussion of Fig. 49a, however, it is possible to have even larger proportions of copper incorporated in crystalline zinc sulphide without having the green emission appear. This lack of universal specificity makes analysis by means of luminescence characteristics less useful, in general, than the established analytical methods which depend on chemistry, radioactive tracers, absorption and emission spectroscopy, and absorption or diffraction of x rays, electrons, and neutrons. Luminescence analysis, nonetheless, is increasing in utility because it sometimes provides more positive or more convenient identification than the other methods of analysis.

Luminescence as a Function of Time

There are three obvious time components in an individual one-quantum luminescence process: (1) the energy-absorption or excitation-transition time, (2) the lifetime of the excited state, and (3) the energy-emission or radiative-transition time (see Franck-Condon principle). These three components are usually experimentally inseparable, and so they are treated as a single lifetime (of the excited state). In the case of *nonmetastable* states of isolated atoms, the natural fluorescence lifetime τ_F of the excited state was previously related (eq. 71) to the width $\overline{\Delta E}^*$ of the emission line by

$$\overline{\Delta E}^* \geq \hbar \tau_F^{-1} \tag{86}$$

For the emission of a 10^5-ev x-ray photon, $\overline{\Delta E}^*$ is of the order of 3 ev, and so the fluorescence lifetime τ_F is about 10^{-15} sec, whereas for the emission of a 2-ev optical photon, $\overline{\Delta E}^*$ is of the order of 10^{-7} ev, and the fluorescence lifetime is of the order of 10^{-8} sec. Luminescence emissions which occur at times greater than the natural fluorescence lifetimes, in each case, are called phosphorescence. The natural duration of the excited state is apparently determined by the energy of the emitted photon (which is equal to the energy difference between the excited and ground states), such that the natural duration is inversely related to the energy of the photon.

Phosphorescence denotes a constrained, partially "forbidden," or "unnatural" delay in the radiative return of an excited system to the ground state, for example, the abnormal delays associated with **metastable states** in isolated atoms.[5, 15, 107, 320] Examples of metastable states are found in the mercury atom (Fig. 63) wherein the system may be excited from the $6s^2$, $^1S_0$ ground-state level to the $6s6p$, $^1P_1$ excited-state level, and thence to the $7s$, $^3S_1$ level from which radiative transitions may be made to any of the three lower levels, $6s6p$, $^3P_0$, $6s6p$, $^3P_1$, and $6s6p$, $^3P_2$, of the triplet excited state. The further radiative transition $6s6p$, $^3P_1 \rightarrow 6s6s$, $^1S_0$ is permitted and affords the strong 2537-Å emission line in the mercury spectrum, but radiative transitions from the two metastable states $6s6p$, $^3P_0$ and $6s6p$, $^3P_2$ to the ground-state level $6s^2$, $^1S_0$ are forbidden by the first-order selection rule for the total-angular-momentum quantum number J. An atom in an excited metastable state may remain excited for an indefinitely long time unless the first-order selection rule which prohibits normal dipole radiation can be circumvented by contriving quadripole or octopole radiation,[2] or unless the atom becomes excited to a higher-energy nonmetastable state from which a radiative return to the ground state is permitted. For isolated atoms in a gas, the latter course is usually more probable. For example, an isolated mercury atom in the $6s6p$, $^3P_2$ state may remain there almost indefinitely unless the atom absorbs sufficient energy from a suitable photon or moving material particle (including collisions of the second kind) to raise the system to, say, one of the $6s6d$, $^3D_{1-3}$ nonmetastable triplet states (the radiative transitions $6s6d$, $^3D_2 \rightarrow 6s6p$, $^3P_2$ and $6s6d$, $^3D_3 \rightarrow 6s6p$, $^3P_2$ produce the strong emission lines near 3650 Å in the mercury spectrum). The circumvention of formal selection rules by the action of foreign atoms or other imperfections adjacent the atoms in a phosphor crystal probably operates as a combination of (1) disturbing the periodicity of the structure, thereby circumventing the general rule which requires that the reduced wave-number vector $\mathbf{k}_r$ remain constant during excitation or emission transitions; and (2) altering the relative probabilities of symmetrical and distorted dipole, quadrupole, *etc.*, radiation, thereby circumventing the general first-order rule which requires that the total-angular-momentum quantum number J may not change by more than ± 1 and forbids $J = 0 \rightarrow J = 0$. It is noteworthy that the strong 2537-Å resonance line of the isolated mercury atom occurs between states with *different* multiplicities ($^3P \rightarrow {}^1S$), and so transitions between states with different multiplicities in phosphor centers should be at least as common and prominent.

It is to be expected that the lifetimes of the excited states of atoms combined in a solid should be affected by interaction with neighboring atoms, and the degree of interaction should be greatest for the outermost optical (valence) electrons. Radiative transitions between the inner shielded energy levels of an atom in a solid are little affected; hence, x-ray

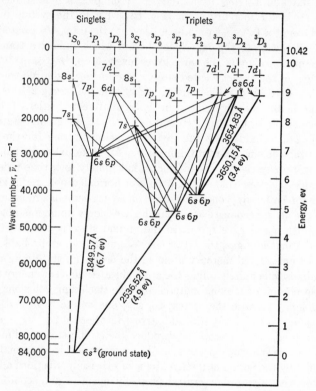

FIG. 63a. Part of the energy-level diagram for an isolated (free) atom of mercury (Hg).

photon emissions are usually fluorescence, and the photon emissions from some fluorophors which emit only line spectra are also fluorescence. It is not necessary, however, to have line-emission spectra in order to obtain fluorescence. For example, some organic molecules, such as fluorescein, eosin, pseudoisocyanin, and chinolin, in dilute solutions (10^{-5} to 10^{-2} mole per liter) exhibit band-emission spectra and are reported to have persistences of less than 5×10^{-9} sec.[75,116c,147] Also, certain solid organic compounds with band emissions (which approximate their emissions in solution [75,145]) have persistences less than 10^{-7}

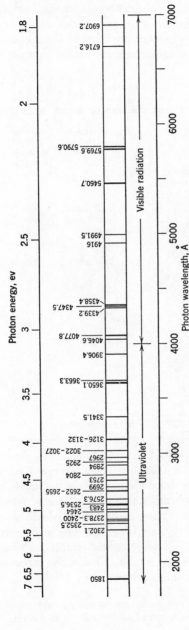

FIG. 63b. Some of the principal lines in the arc emission spectrum of mercury. The lines are distinct at a pressure of 10^{-3} cm (of Hg) but generally broaden and become diffuse bands at 10^4 cm. Also, the mean radiated-power distribution shifts toward longer wavelengths with increasing pressure.

sec. Table 14 gives some data, reported by R. Hofstadter and G. B. Collins (see also H. N. Bose), on several organic crystals which have been found particularly useful in scintillation counters for detecting gamma rays.[292] The compounds are listed in the approximate order of their effective roentgenoluminescence efficiencies when excited by gamma

TABLE 14

LUMINESCENCE CHARACTERISTICS OF SOME ORGANIC SOLIDS

Crystal Structure, Name, and Formula	Structural Formula	Wavelength at Peak of Emission Band(s), Å	Bandwidth at Half Maximum Intensity, Å	Time to Decay to $L_0 \epsilon^{-1}$, sec
Monocl.-stilbene $C_6H_5 \cdot CH{:}CH \cdot C_6H_5$		4200 (weak) 4080 (strong)	360 100	$\approx 10^{-8}$
Monocl.-naphthalene $C_{10}H_8$		3450 ± 50	250	$5.7(\pm 0.5) \times 10^{-8}$
Monocl.-anthracene $C_{14}H_{10}$		4440	60	$1.3(\pm 0.2) \times 10^{-8}$
Monocl.-phenanthrene $C_{14}H_{10}$		4100 4300	100 110	$0.9(\pm 0.2) \times 10^{-8}$

rays and detected by a 1P21 or 1P28 multiplier phototube (H. T. Gittings *et al.*[292]). In these organic crystals, the indicated simple structural units remain distinct, being bonded together by mutually induced fluctuating-dipole attraction energies (van der Waals energies) which are much weaker than the covalent–bond energies within the simple ring-structure units. The distinct simple molecules, then, are the luminescence centers in these organic crystals. (There is considerable doubt about the purity of many of the organic crystals whose luminescences were reported in the early enthusiasm about large-crystal scintillation counters. It is not, therefore, unequivocally established that the reported emission is intrinsic to the host crystal in every case.)

Luminescent organic dye molecules are generally characterized by **conjugated double bonds**,[24,221] where each bond line in a structural formula represents two paired (shared) electrons with antiparallel spins, as in the xanthenes, represented by the fluorescein molecules shown in Fig. 64. The luminescence process in these cases apparently occurs within the closed-ring molecules, presumably by producing changes in the electron spins and distributions associated with the alternating single and double bonds, much as the line-emission luminescence process in atoms with incompleted shells involves changes in electron spins and

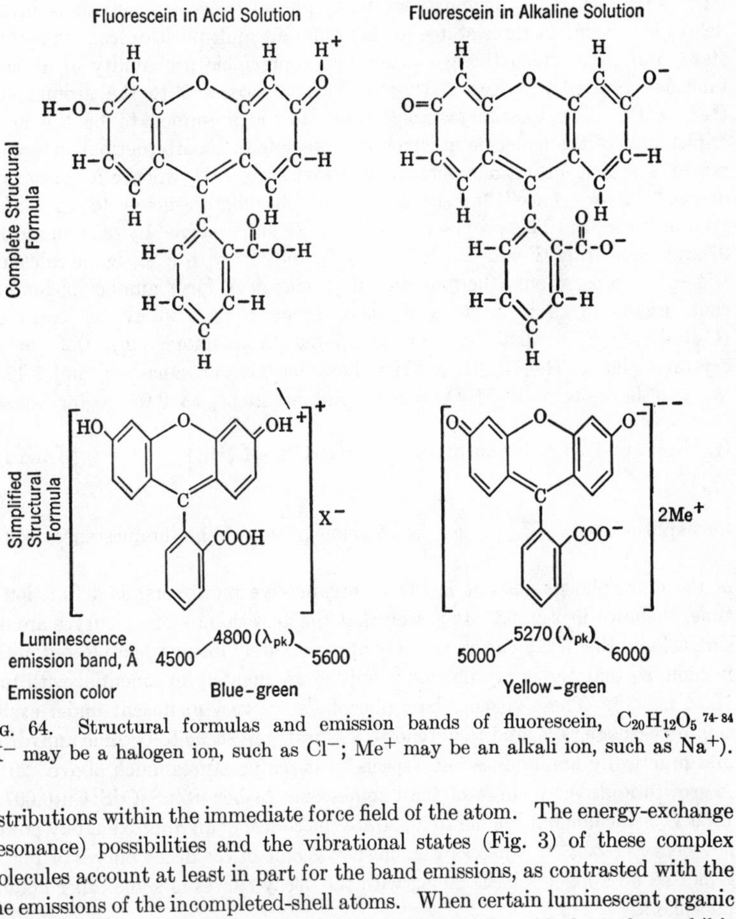

FIG. 64. Structural formulas and emission bands of fluorescein, $C_{20}H_{12}O_5$ [74-84] (X^- may be a halogen ion, such as Cl^-; Me^+ may be an alkali ion, such as Na^+).

distributions within the immediate force field of the atom. The energy-exchange (resonance) possibilities and the vibrational states (Fig. 3) of these complex molecules account at least in part for the band emissions, as contrasted with the line emissions of the incompleted-shell atoms. When certain luminescent organic molecules are incorporated (not just adsorbed) in boric acid glass, they exhibit (1) a very short-persistent short-wavelength emission band which is located near the absorption edge, and (2) a long-persistent emission located at considerably longer wavelengths.[77] (This behavior appears to be analogous to the short-persistent edge emission bands of hex.-ZnO:[Zn], cub.(or hex.)-ZnS:[Zn], and hex.-CdS:[Cd] relative to the longer-persistent longer-wavelength emission bands of these same phosphors). Kasha reports that the lifetimes of the excited states of the long-wave emission bands depend strongly on the molecular structure and the (host-glass) environment, and range from about 10^{-4} to 50 sec.[58,77] It is further reported that acid fluorescein dye in boric acid glass is paramagnetic when excited, and it has been proposed that the short-persistent (fluorescence) emissions involve transitions between states with the same multiplicity ($^1\Gamma^* \rightarrow {}^1\Gamma$), whereas the long-persistent (phosphorescence) emissions involve delays in quasimetastable states having different multiplicities from the ground state, and from which there is a low but appreciable probability of a "spontaneous" radiative transition ("beta phosphorescence") to the ground state ($^3\Gamma^* \rightarrow {}^1\Gamma$). The quasimetastable state, $^3\Gamma^*$, is presumed to be the lowest triplet state of the molecule in which it is possible to absorb thermal energy and raise the system into a nonmetastable singlet state, $^1\Gamma^*$, whence it can make a delayed "fluorescence" transition ("alpha phosphorescence") to the singlet ground state, $^1\Gamma$. This proposed situation may be visualized by referring to Fig. 27 and associating $^1\Gamma$ with E_0, $^1\Gamma^*$ with E^*_1, and $^3\Gamma^*$ with E^*_2. Some calculated $^3\Gamma^* - {}^1\Gamma^*$ separations (thermal activation energies), for a number of luminescent organic dyes in boric acid glass, range from 0.26 ev for eosin dye ($C_{20}H_6Br_4Na_2O_5$ = alkali salt of 2,4,5,7-tetrabromofluorescein), 0.27 ev for crystal violet [$(CH_3)_2NC_6H_4]_3COH$ = hexamethylpararosanaline], and 0.42 ev for acid fluorescein ($C_{20}H_{12}O_5$ = resorcinolphthalein), to 0.97 ev for benzene

(C_6H_6,), 1.25 ev for anthracene (C_6H_4:(CH)$_2$:C_6H_4,), and 1.51

for naphthalene ($C_{10}H_8$,). A semilog plot of the luminescences of two of the more phosphorescent H_3BO_3 + organic-dye phosphors, as a function of time, is shown in Fig. 65. It is seen that the growth and decay curves are dissimilar, and the decay curves are only approximately linear when plotted in this fashion, so that the decay process is only approximately exponential with time ($L \propto L_0\epsilon^{-at}$). These vitreous-base phosphors are very inefficient under excitation by cathode rays, and they tend to lose water in vacuo to become anhydrous and practically nonluminescent, especially at temperatures much above 120°C. A growth and decay curve of the luminescence of hex.-9ZnS·CdS:Cu(0.0073), 1250°C, is included in Fig. 65 to illustrate the course of an approximately power-law decay ($L \propto t^{-n}$), whereas Fig. 66 shows some of the decay curves of Fig. 65 replotted on a log–log basis, along with the decay curves of some other notable phosphors.[58]

Contrary to the relationship (eq. 86) between the lifetime of the excited state and the line width in atomic emission spectra, there is no apparent *general* correlation between the widths of the emission bands of phosphors and their phosphorescence decay times. This is illustrated

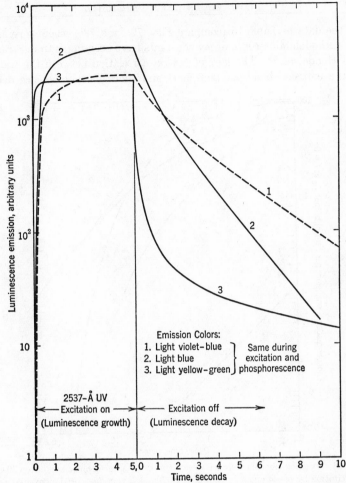

FIG. 65. Semilog growth and decay curves of some phosphors during and after excitation by predominantly 2537-Å UV. Thick layers, measured on excited side.

1. H_3BO_3 glass + 10^{-3} toluic acid *
2. H_3BO_3 glass + 10^{-3} terephthalic acid **
3. hex.-9ZnS·CdS:Cu(0.0073), 1250°C.

Curves 1 and 2 drawn to scale.

Curve 3 not to same scale as 1 and 2.

* p-toluic acid =

** terephthalic acid =

by the data in Table 15 (compare Figs. 79*a* and 79*b*) where several of the cathodoluminescence decay constants have been taken from Strange and Henderson.[366] The lack of *general* correlation between the widths of the emission bands of phosphors and their phosphorescence decay

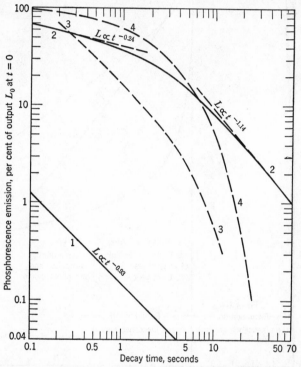

FIG. 66. Log–log phosphorescence decay curves of several phosphors after 30 sec of excitation by predominantly 2537-Å UV. Thick layers, measured on excited side.

1. hex.-ZnS : Ag(0.015), 1250°C
2. hex.-9ZnS·CdS : Cu(0.0073), 1250°C
3. rbhdl.-6ZnO·6SiO$_2$·SnO$_2$·0.036Mn, 1200°C
4. H$_3$BO$_3$ glass + 10^{-3} toluic acid

times is an indication that the excitation and emission processes in phosphors are relatively independent of each other. (It is shown later that there is a fair correlation between the band widths, peak wavelengths, melting points, and hardnesses of ϵ^{-at} decay phosphors with a *given type* of host crystal having the *same* activator and activator proportion).

Before commencing a discussion of the two major decay types, it should be mentioned that, when a phosphor has a single simple emission band, the emission spectrum is the same during excitation and during phosphorescence after cessation of excitation.[102, 355] Phosphors which have complex emission spectra may or may not exhibit the same emission spectrum during and after excitation. For example, the emission spectra

TABLE 15

SOME PHOSPHORS ARRANGED ACCORDING TO DECAY TYPES AND TIME OF DECAY
TO $0.37L_0$, WHERE L_0 IS THE CATHODOLUMINESCENCE EMISSION AT CESSATION
OF EXCITATION

The Lifetimes of the ϵ^{-at} Decays Are Practically Independent of the Conditions of Excitation, Whereas the t^{-n} Decays Are Sensitive to Changes in Temperature and Excitation

No.	Phosphor	Decay Type	Approx. Time to Decay to $L_0\epsilon^{-1}$ sec	Approx. Width of Emission Band, ev	Wavelength of Peak of Emission Band, Å	Melting Point, °K	Moh Hardness of Host Crystal
1	Hex.-ZnO:[Zn] (UV band)	ϵ^{-at}?	10^{-7}	0.50	3850		
2	Hex.-ZnO:[Zn] (Visible band)	ϵ^{-at}?	10^{-6}	1.28	5050		
3	Tet.-CaWO$_4$:[W]	ϵ^{-at}	10^{-5}	2.0	4300		
4	Monocl.-Mg$_2$WO$_5$:[W]	ϵ^{-at}	10^{-5}	1.43	4930		
5	Hex.-ZnS:Mn(1)	ϵ^{-at}	2.5×10^{-4}	0.72	5910		
6	Rbhdl.-Al$_2$O$_3$:Cr(1) (red lines) *	ϵ^{-at}	5×10^{-3}	Lines	6926 & 6941		
7	ZnAl$_4$O$_7$:Cr(1) (red lines) *	ϵ^{-at}	10^{-2} to 10^{-3}	Lines	6926 & 6941		
8	Cub.-ZnAl$_2$O$_4$:Mn(1)	ϵ^{-at}	0.005	0.27	5130	>1900	8
9	Rbhdl.-Zn$_2$SiO$_4$:Mn(5)	ϵ^{-at}	0.002	0.58	5290		
10	Rbhdl.-Zn$_2$SiO$_4$:Mn(2)	ϵ^{-at}	0.01	0.58	5260		
11	Rbhdl.-Zn$_2$SiO$_4$:Mn(1)	ϵ^{-at}	0.015	0.58	5250	1785	5.5
12	Rbhdl.-Zn$_2$SiO$_4$:Mn(0.05–0.5)	ϵ^{-at}	0.019	0.58	5250		
13	2ZnO·B$_2$O$_3$:Mn(1)	ϵ^{-at}	0.018	0.83	5500	1250	<5.5
14	Rhomb.-CdSiO$_3$:Mn(1)	ϵ^{-at}	0.03	0.64	5900	1520	<5.5
15	Rhomb.-CdSO$_4$:Mn(1)	ϵ^{-at}	0.05	0.64	5900	1270	<5.5
16	Tetr.-ZnF$_2$:Mn(1)	ϵ^{-at}	0.1	0.68	5870	1140	≈4
17	Cub.-ZnSe:[Zn]	t^{-n}?	10^{-6}	0.70	6530		
18	Hex.-4ZnS·6CdS:Ag (0.01)	t^{-n}	10^{-5}	0.86	6150		
19	Hex.-7ZnS·3CdS:Ag(0.01)	t^{-n}	10^{-5}	1.15	5220		
20	Hex.-ZnS:Ag (0.01)	t^{-n}	10^{-4}	0.96	4330		
21	Hex.-8ZnS·2CdS:Cu (0.01)	t^{-n}	10^{-5}	0.73	6050		
22	Hex.-ZnS:Cu(0.01)	t^{-n}	10^{-4}	0.85	5190		

* Reportedly exhibits green band, also, with $\tau \approx 5 \times 10^{-4}$.

of Ca$_2$P$_2$O$_7$:Dy and certain rbhdl.-(Zn:Be)$_2$SiO$_4$:Mn phosphors which have two component emission bands are practically identical during and after excitation; whereas, in cub.-ZnS:Ag:Cu, cub.-ZnS:Cu:Mn, and hex.-ZnO:[Zn], each of which may also have two component emission bands, the *longer*-wave band predominates during phosphorescence.[169, 231, 233, 354] In the Ca$_2$P$_2$O$_7$:Dy and rbhdl.-(Zn:Be)$_2$SiO$_4$:Mn phosphors, the excitation and emission spectra are so widely separated that the short-wave emission band does not excite the long-wave emission band of the same phosphor. In the cited phosphors whose long-wave emission bands predominate during phosphorescence, however, the excitation and emission bands overlap, and so the *internal* photon emission from the short-wave emission band is largely reabsorbed in the phosphor crystal and may excite the longer-wave emission band by a **cascade process.**[58, 63] Therefore, the phosphorescence of a phosphor with two emission bands, where the short-wave emission band cascade-excites the long-wave emission band, has a rate which is determined largely by the longer-persisting band. This explains why the persistence of orange-emitting ZnS:Cu:Mn is very long, even though that of ZnS:Mn is short (Table 15); that is, the green Cu emission is internally absorbed, and so the long-persistent t^{-n} decay green Cu band "feeds" the short-persistent ϵ^{-at} decay orange Mn band. In this case, a corresponding effect occurs during luminescence growth; the short-persistent long-wave component band is the first to reach an equilibrium output.[356]

In some cadmium–borate:Mn phosphors having both a green emission and an orange emission band, the orange band predominates during phosphorescence. There are cases also where two different activators in a given host crystal produce two different emission bands which decay practically independently. For example, Studer and Rosenbaum report that the rate constants a for the initial exponential decays of two such phosphors after excitation by 2537-Å UV are: (1) calcium–halophosphate:Sb:Mn, $a_{Sb} = 1500$ sec^{-1}, $a_{Mn} = 75$ sec^{-1}, and (2) calcium–silicate:Pb:Mn, $a_{Pb} = 570$ sec^{-1}, $a_{Mn} = 60$ sec^{-1}. Tetr.-ZnF$_2$:Ti:Mn reportedly exhibits a similar rapid decay of the short-wave (Ti) band as compared with the slow decay ($a_{Mn} \approx 10$ sec^{-1}) of the long-wave (Mn) band.[689] This selective-decay effect, which may occur to a slight extent in rbhdl.-(Zn:Be)$_2$SiO$_4$:Mn phosphors also, does not appear to be a cascade process of the type just described, because there is not the necessary overlap of emission and excitation bands. Instead, the two bands apparently decay almost independently, indicating that different virtually independent activator centers are involved. If energy transfer does take place between centers under these conditions, the agency may be mobile positive holes, excitons, or ψ-overlap energy transfers.

EXPONENTIAL AND POWER-LAW DECAYS. *1. Temperature-independent Exponential Decays.* As an elementary example of this type of decay, assume that Fig. 67a represents an optical absorption E_a and resonance-emission E_e transition of an isolated hydrogen atom (which may be crudely pictured as a radiating dipole whose one end is the positive nucleus and whose other end is the negative electron). If there be N

excited hydrogen atoms at time t, then the luminescence emission (radiance), L, from the N atoms will be given by

$$L = -dN/dt = aN \qquad (87)$$

that is, the number of photons emitted, or atoms de-excited, at time t

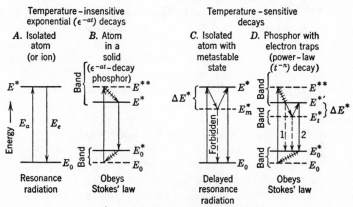

FIG. 67. Simplified energy-level diagrams illustrating electronic transitions involved in the two major decay processes.

is equal to a rate constant a times the number N of excited atoms existing at time t. Integration of eq. 87 yields

$$-\log_\epsilon L = -\log_\epsilon N = at + \text{constant} \qquad (88)$$

where the rate constant a is equal to the reciprocal of the lifetime τ of the excited state,

$$a = \tau^{-1} \qquad (89)$$

and τ is the time taken by the system to decay to $\epsilon^{-1}L_0$ ($= 0.36788L_0$) when L_0 is the luminescence emission at $t = 0$ (for example, the instant of cessation of excitation). On this basis, eq. 88 becomes

$$L = L_0\epsilon^{-at} \qquad (90)$$

that is, a plot of $\log L$ versus t is a straight line for an exponential decay. There are no external constraints on the free atoms assumed in this example, and so τ is about 10^{-8} sec, and eq. 86 is applicable; that is, the lifetime of the excited state *decreases* with increasing energy of the emitted photon E_e, and the line width $\overline{\Delta E^*}$ *increases* with increasing $E_e = E^* - E_0$. Furthermore, the line width is of the order of 10^{-7} ev.

Consider now the optical luminescence of an atom bound in a solid, according to the diagram shown in Fig. 67b. The energy of the photon used for excitation ($E_a = E^{**} - E_0$) is greater than that of the photon emitted ($E_e = E^* - E^*_0$) because some energy is transferred from the atom (center) to the surrounding solid as heat. From Table 15, it may be seen that the listed phosphors have ϵ^{-at} decays with τ ranging from about 10^{-7} to 10^{-1} sec, and with sharp emission lines or with bands whose widths range from 0.27 to 2 ev. An indication of the constraints imposed by the solid, without greatly influencing the *line* widths of spin-change emissions, is given by the 10^5 larger τ of the line-emitting Cr activator in Nos. 6 and 7, as compared with the 10^{-8} sec value of τ for free atoms. Next, specific attention is directed to the *band*-emitting phosphors Nos. 8, 11, 13, 14, 15, and 16 in Table 15, because these materials all have O(or F)-dominated host crystals and all have Mn(≈ 1) as an activator. In this series there is a fairly regular progression of each of the tabulated properties. For example, both τ and the width of the emission band generally *decrease* with increasing energy (decreasing λ_{pk}) of the emitted photons, whereas the melting points and hardnesses of the host crystals generally *increase* with increasing energy of the emitted photon. Incidentally, there is little or no photoconduction associated with the luminescences of these ϵ^{-at} decay phosphors, indicating that there is little or no "internal ionization." [119-122]

With the aid of the simple potential-energy versus configurational-coordinate diagrams of Fig. 68, it is interesting to attempt to deduce a possible relationship between τ and the bonding energy of the host crystal (as measured by heat of formation, melting point, and hardness), and attempt to determine what other related properties might be expected on the basis of these orthodox diagrams (compare Fig. 3). Figure 68a is intended to represent part of the energy-level diagram of a system comprising an Mn activator atom (ion) and its O (or F) ligands. The system is shown as starting in the ground-state level E_0 (see Fig. 67b) and making an excitation transition to the excited-state level E^{**}. Both levels have a dotted schematic indication of the absolute value of the probability distribution function $|\psi|$, whose amplitude represents the probability of finding the system with the configurational coordinates $\bar{x}$ (averaged inter-atomic separations).[3,26] The transition probability **P** between two energy levels, 1 and 2, is proportional to the product of their $|\psi|$ values; that is, using only absolute values,

$$\mathbf{P} \propto \psi_1\psi_2 = \int \psi^* \cdot \psi \, d\tau \tag{91}$$

and a simplified indication of this product is plotted separately below the main figures of Fig. 68. Assuming the situation shown in Fig. 68a, with the system initially at equilibrium in (near) the energy level E_0 ($= a_n$), the excitation transition probability is a maximum on the *compression* side of the vibrational cycle. For present purposes, the transition $E_0 \rightarrow E^{**}$ may be associated with the raising of an electron of the Mn activator atom into a higher energy level (that is, producing a higher energy state of the Mn atom) *without ionization.* There are many more vibrational levels above and below the indicated levels shown for E_0 and E^{**} (Fig. 3), and so there is a certain indeterminacy of finding the system in either level; that is, the system may start from one of several levels lying near E_0 and make transitions to one of several levels lying near E^{**}.

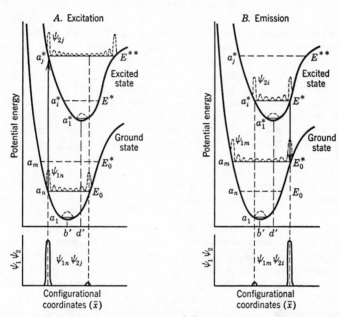

FIG. 68. A diagrammatic representation of the excitation and emission processes in phosphor centers exhibiting only temperature-insensitive ϵ^{-at} decays (see Figs. 3, 16b, and 77).

After the system arrives in, say, level E^{**}, there is a rapid dissipation of some of the excitation energy to the surrounding crystal, and the system comes to an excited-state equilibrium in (near) E^*, the energy difference $E^{**} - E^*$ being spent as heat (phonons). Concomitant with the attainment of an excited-state equilibrium in energy, there is an attainment of a new equilibrium average interatomic spacing $\bar{x}$. The excited-state $\bar{x}$ is shown to be displaced by the amount $\Delta\bar{x} = d' - b'$ in the direction of an expansion of the excited system. [Although it is very probable that there will be a new equilibrium $\bar{x}$ for the excited state, it is not completely certain that the displacement will always be toward larger $\bar{x}$, because at the limit of complete ionization the increase in positive charge of the manganese would result in drawing its negatively charged oxygen (or fluorine) ligands nearer. For present purposes, however, it is assumed that $\bar{x}$

is larger in the excited un-ionized state]. As shown in Fig. 68b, the radiative-transition probability from E^* is greater on the *expansion* side of the vibrational motion of the system, such that the system drops to (near) E^*_0, emitting the energy difference $E^* - E^*_0$ as a luminescence photon. The return to E_0 is then accomplished by further energy loss as heat (compare Fig. 67b).

On the basis of the diagrams of Fig. 68, assume now that (1) the spacing between the ground-state and excited-state energy levels, for example, between a_1 and a^*_1, is determined primarily by the manganese atom (ion), and (2) the equilibrium displacement, $\Delta \bar{x} = d' - b'$, is inversely related to the bonding energy (melting point, hardness) of the host crystal. For the same $MnO_x(MnF_x)$ configuration in host crystals of decreasing bonding energy, then, $\Delta \bar{x}$ should increase and (1) the energy of the emitted photon should decrease as the radiative-transition line from the right end of E^* intersects higher on the ground-state curve, (2) the lifetime τ of the excited state should increase as energy levels with ψ values near ψ_{2i} overlap fewer levels with ψ values near ψ_1, because the decreasing slope of the ground-state curve spreads the higher-lying ψ_1 maxima farther apart as a function of increasing $\bar{x}$, and (3) the minimum energy of excitation should increase as the intersection of the excitation transition from E_0 moves higher on the excited-state curve. These expectations are generally met, with a few obvious exceptions, by the series of phosphors under consideration. Continuing this simplified approach, however, one might expect that the width of the emission band should be greater for smaller $\Delta \bar{x}$ (and correspondingly higher energy of the emitted photon), because a given indeterminacy of the $\bar{x}$ coordinate of the right-side maximum of ψ_{2i} should intersect over a greater potential-energy range on the steeper ground-state curve which obtains at smaller $\Delta \bar{x}$. As shown in Table 15, the width of the emission band generally decreases with increasing bonding energy of the crystal and increasing energy of the emitted photon (decreasing λ_{pk}). This may be interpreted either as a refutation of the proposed mechanism, or as an indication that the width of the observed emission band is determined largely by other factors such as statistical variations in the ground-state levels of the different Mn activator centers distributed throughout the crystal, or a weak interaction of such separated centers to produce a band according to the operation of the exclusion principle. In either of the latter two cases, it is reasonable to expect the observed general decrease in the width of the emission band with increasing bonding energy of the host crystal.

This simple hypothetical correlation between the bonding energy of the host crystal and τ, λ_{pk}, and perhaps the minimum quantum energy required for excitation is offered as a possible practical aid in developing new series of ϵ^{-at} decay phosphors with controllable characteristics. Further refinements, including consideration of crystal structure and degrees of ionic and covalent bonding, may be introduced when and if the hypothesis is found to be valid for the same or other activators in series of other phosphors whose host crystals are dominated by the same or similar anions.

An alternative approach to an explanation of the decrease of τ with increasing bonding energy of the host crystal was offered by Professor E. P. Wigner, of Princeton University, during a discussion of the foregoing material. Wigner suggested that the selection rules for radiative transitions in the Mn activator atoms should be affected more, that is, upset more, when the bonding energy of the host crystal is large. According to this postulate, and the one based on Fig. 68, τ should decrease with increasing crystal anisotropy; that is, τ should be smaller for a given ϵ^{-at} decay activator atom in a triclinic host crystal than in a cubic host crystal when all other factors are held constant. The validity

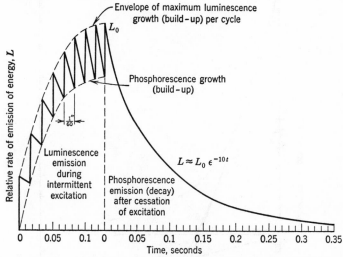

FIG. 69. Growth and decay curves of cathodoluminescence emission of predominantly ϵ^{-at} decay tetr.-ZnF$_2$:Mn(1) excited by nine 1/2000-sec pulses of 6-kv CR (200 μa cm^{-2}) spaced 1/60 sec apart. Measurement on bombarded side of thick layer. (Compare Fig. 19.)

of this prediction cannot be determined from the data in Table 15, because the "other factors" are not held constant on going from cub.-ZnAl$_2$O$_4$:Mn to rhomb.-CdSO$_4$:Mn.

Referring to eq. 87, which is the general equation for simple *temperature-independent* exponential decays, the **decay** of phosphorescence emission is $L = f(t) = dN/dt$, and the **rate of decay** is $dL/dt = d^2N/dt^2$. A **decay curve** is a plot of L versus t during phosphorescence, and the rate of decay at time t is the slope of the tangent of the decay curve at time t. When the decay of a phosphor follows eq. 90, it is found that the growth (during excitation) is also exponential,* and the shape of the growth curve is relatively uninfluenced by changes in the temperature of the luminescing phosphor. Figure 69 shows the cyclic growth (build-up) and short-term decay of luminescence emission from a tetr.-

* $L = acN_0[1 - \epsilon^{-(c+a)t}]/(c+a)$, $c \propto$ excitation density.

ZnF$_2$:Mn phosphor subjected to a short period of intermittent pulses of CR excitation.[58] It is seen that the luminescence output during excitation may increase at a much higher rate than the rate of decrease of the phosphorescence output after the successive excitations. Over the time interval shown, this particular phosphor has very nearly a

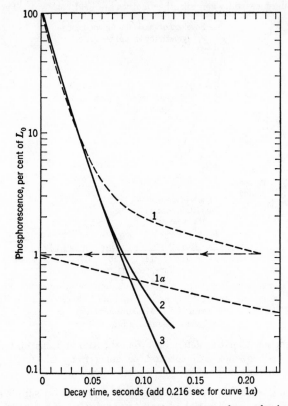

Fig. 70a. Semilog plot of phosphorescence-decay curves of a predominantly exponential-decay rbhdl.-Zn$_2$SiO$_4$:Mn(0.3) phosphor after excitation by (1) UV, and (2 and 3) CR. Measurement on excited side of thick layer.

Excitation:L_0

1. α–Zn$_2$SiO$_4$:Mn(0.3)   Steady 2537 Å:15-ft L
1a. (Continuation of curve 1 after 13/60 sec)
2. α–Zn$_2$SiO$_4$:Mn(0.3)   Steady 6-kv CR:9-ft L
3. α–Zn$_2$SiO$_4$:Mn(0.3)   Steady 6-kv CR:100-ft L

Same phosphor sample used throughout. Thick screen; excitation and measurement both on the same side.

simple exponential decay whose rate constant a is practically independent of moderate changes in the temperature and the degree of excitation of the phosphor. At very long decay times, however, the phosphores-

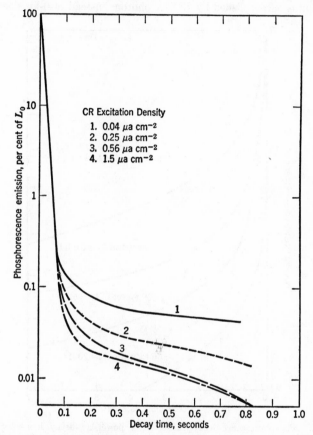

FIG. 70*b*. Decay curves of cathodophosphorescence emission of an aluminized 4-mg cm^{-2} screen of rbhdl.-Zn_2SiO_4:Mn(0.3) excited to equilibrium (at room temperature) by unmodulated 10-kv CR at the indicated current densities. (S. Lasof and R. H. Bube)

cence-decay curve of this phosphor tends to depart from $L \propto \epsilon^{-at}$ and approximate $L \propto t^{-n}$ (so-called **power-law decay**). Power-law decays are quite sensitive to changes in temperature, excitation density, and kind of excitant, as shown in Figs. 70, 71, and 72.[58, 63, 113] Figure 70 shows that the initial exponential (linear) portion of the decay curve of rbhdl.-Zn_2SiO_4:Mn(0.3) is practically unaffected by change in L_0 or in the type of excitant used, whereas the later power-law (nonlinear) portion of the curve appears sooner and is relatively more prominent when L_0 is low, for example, when the excitation density is low, or when the excitation is accomplished by 2537-Å photons instead of 6-kv electrons.

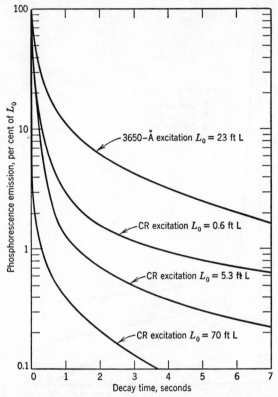

FIG. 71. Phosphorescence-decay curves of a power-law-decay hex.-9ZnS·CdS: Cu(0.0073) phosphor after excitation to equilibrium by steady UV and CR. Measurement on excited side of thick layer.

When the entire decay curve follows a power-law relationship, as in Fig. 71, the rate of decay increases; that is, the tangent to the decay curve becomes steeper, with higher L_0, and with excitation by CR rather than UV. Figure 72a shows that the power-law "tail" on the decay curve of photophosphorescence emission of a phosphor exhibiting an initially exponential decay is absent at very low or very high temperatures, being detectable only at some intermediate temperature.[113] Figure 72b shows that the initial exponential decay of cathodophosphorescence emission of this type of phosphor is also little affected by variation of the operating temperature of the phosphor in the range from 297 to 527°K, whereas the low-intensity power-law "tail" is again strongly influenced by change in temperature. This particular sample

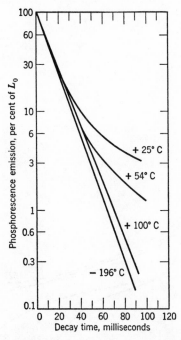

FIG. 72a. Decay curves of a predominantly exponential-decay rbhdl.-Zn_2SiO_4: Mn(0.3) phosphor as a function of temperature during operation. Excitation by 2537-Å UV. The t^{-n} decay "tail" appears at *intermediate* temperatures. (G. R. Fonda)

evidently has one trapping level which empties rapidly at 297°K and another broader and deeper trapping level which does not empty rapidly until the phosphor temperature exceeds 330°K.

As shown in Figs. 72c and 72d, the *apparent* decay constant, $a + a'$, generally exhibits an increase as the operating temperature of the phosphor is increased above the temperature break point T_B. This behavior has been reported also by Kröger, *et al.*, for phosphors such as Mg_2TiO_4:Mn, $(NH_4)_3UO_2F_5$, and $9CaO \cdot MgO \cdot 6Al_2O_3$:Mn excited by 3650-Å UV.[156b] The explanation advanced is that the average lifetime τ of the excited state for *radiative* transitions remains substantially constant with temperature, but an increasing number of excited centers make *nonradiative* transitions to the ground

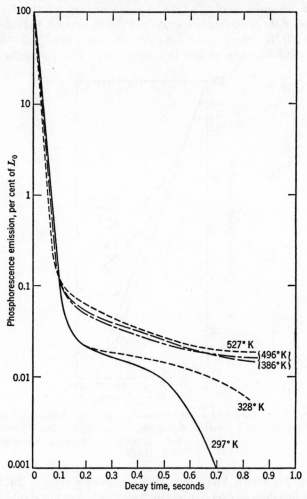

FIG. 72*b*. Decay curves of cathodophosphorescence emission of an aluminized 4-mg cm^{-2} screen of rbhdl.-Zn$_2$SiO$_4$:Mn(0.3) excited by steady 10-kv 0.86-μa cm^{-2} CR at the indicated operating temperatures. (S. Lasof and R. H. Bube)

state as the temperature is increased above T_B. Hence, at temperatures above T_B, eq. 90,

$$L = L_0 \epsilon^{-at}$$

becomes modified to

$$L = L_0 \epsilon^{-(a+a')t} \tag{90a}$$

where a' depends on the temperature according to

$$a' = \nu_a \epsilon^{-E^*/kT}$$

with ν_a being of the order of the vibration frequency of the atoms in the crystal ($\approx 10^{12}$ sec^{-1}) and E^* being the activation energy required to raise the excited system into a state whence it can make a nonradiative transition. For example,

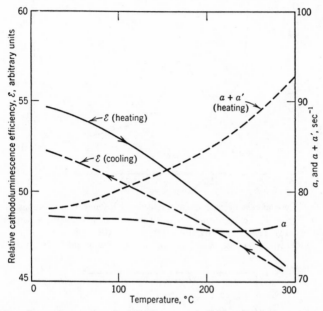

Fig. 72c. Dependence of cathodoluminescence efficiency ε and apparent decay "constant" $(a + a')$, on operating temperature for an aluminized screen of rbhdl.-Zn_2SiO_4:Mn (see Fig. 107) excited by 10-kv 0.2-μa cm^{-2} CR. Observation interval = 1 to 22 msec after cessation of excitation. (S. Lasof and R. H. Bube)

E^* may be visualized as the energy difference between e and f in Fig. 16b, assuming that the excited system initially in state c makes a radiationless transition via c—f—a (there may be a long-wave-infrared radiative transition from E^* to E at f) when raised thermally into the energy level corresponding to f. On this basis, the increase in observed $(a + a')$ with increasing T should be accompanied by a corresponding decrease in luminescence efficiency ε. This is found to be generally true as a rough approximation (Figs. 72c and 72e). In certain cases, however, it has been found that the trend of $(a + a')$ does not mirror the trend of ε, especially when there are sufficient excited trapped electrons to provide a strong power-law tail to the ϵ^{-at} type decay curve. In these cases there are anomalies in the behavior of ε and $(a + a')$ in the temperature region

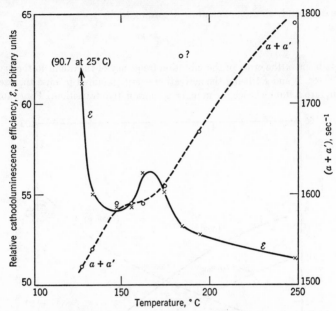

FIG. 72d. Dependence of cathodoluminescence efficiency ε and apparent decay "constant" $(a + a')$ on operating temperature for an aluminized screen of hex.-ZnS:Mn (see Fig. 107) excited by 10-kv 0.2-μa cm^{-2} CR. Observation interval = 1 to 22 msec after cessation of excitation. (S. Lasof and R. H. Bube)

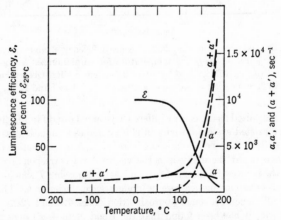

FIG. 72e. Dependence of photoluminescence efficiency ε, apparent decay "constant" $(a + a')$, radiative-transition decay constant a, and nonradiative-transition factor a', on operating temperature for a Mg_2TiO_4:Mn, 1300°C, phosphor excited by 3650-Å UV. Observation interval = 0 to 2 msec after cessation of excitation. (F. A. Kröger et al.)

where the activation energy for the trapped electrons is approximately $30kT$. Such anomalies have been found for hex.-ZnS:Mn (Fig. 72d) and tetr.-ZnF$_2$:Mn under excitation by CR. The tetr.-ZnF$_2$:Mn phosphor (and rbhdl.-Zn$_2$SiO$_4$:Mn excited by either CR or 2537-Å UV) exhibits a marked hysteresis effect in that the curves of ε and $(a + a')$ versus T change as the phosphor is heated and cooled several times. For present purposes, however, we shall refer to ϵ^{-at} decay phosphors as being temperature-independent in the sense that a is independent of T, except that above T_B an additional factor a' must be introduced to account for nonradiative transitions. Below T_B, of course, the observed decay is practically independent of T. In phosphors with only power-law decays, however, the entire growth and decay curves are greatly affected by changes in the operating temperature of the luminescing phosphor both above and below T_B.[110,120]

2. Temperature-dependent Power-law Decays.

There is a plethora of published interpretations of luminescence growth and decay phenomena.[58, 221] These interpretations include (a) attempts to fit all decay curves with one or more simple exponential functions.[115, 126, 357, 358, 366] (b) attempts to explain power-law decays as being (1) basically (internally) bimolecular, $L = -dN/dt = bN^2$ (from which $L = 1/b[(L_0b)^{-\frac{1}{2}} + t]^2$), and obscured by absorption and other phenomena,[166, 342] or (2) combinations of bimolecular and simple hyperbolic functions, where the hyperbolic contribution, $L \propto t^{-1}$, is ascribed to the release of electrons from two-dimensional surface-state traps according to Fermi–Dirac statistics (Fig. 15d, e),[359] and (c) other approaches given in previously cited references and many more dealing with various aspects of phosphorescence.[360-373]

The interpretation of so-called power-law decays which is offered here is based simply on the evidence for electron traps, as shown by photostimulation and glow curves (Figs. 25 and 26), in t^{-n} decay phosphors.[63, 119, 120] A simple example of an electron "trap" is found in the occurrence of a metastable state in an isolated atom in a gas. When an atom is excited into a metastable state, for example, E^*_m in Fig. 67c, it must absorb an amount ΔE^* of activation energy to be raised into the radiative state E^*. If the radiative transition $E^*_m \rightarrow E_0$ were not completely forbidden, this transition would occur as a slow temperature-independent exponential decay according to eq. 90. The transition to E_0 from E^*_m via E^*, however, is strongly temperature-dependent and, *neglecting the lifetime associated with the radiative process*, follows a relation of the type,

$$L = -dN/dt = N\nu_a\epsilon^{-\Delta E^*/kT} \qquad (92)$$

where N is the number of excited atoms in metastable states, ΔE^* is the thermal activation energy required to raise an electron out of the "trap" E^*_m into E^*, and ν_a is the attempt frequency for the transition from E^*_m to E^*. Under these conditions ν_a is the frequency of molecular collisions in the gas ($\nu_a \propto T^{1/2}$). By integration of eq. 92, the phosphorescence output becomes, *in the absence of retrapping*,

$$L = L_0 \epsilon^{-\nu_a t \, \epsilon^{-\Delta E^*/kT}} \tag{93}$$

where L_0 is the output at $t = 0$, and T is the temperature of the gas, in degrees Kelvin.

Equation 93 assumes a single discrete trapping (metastable) level, whereas there may be one or more bands of trapping levels (ΔE^*_t in Fig. 67d) in solids, as demonstrated by the stimulation and glow-curve bands in Figs. 25 and 26. For a solid, therefore, eq. 93 must generally be expanded to

$$L = L_{1_0} \epsilon^{-\nu_{a_1} t \epsilon^{-\Delta E^*_1/kT}} + L_{2_0} \epsilon^{-\nu_{a_2} t \epsilon^{-\Delta E^*_2/kT}} \cdots + L_{N_0} \epsilon^{-\nu_N t \epsilon^{-\Delta E^*_N/kT}} \tag{94}$$

where the subscripts 1, 2, 3 $\cdots$ N denote traps of different depths ΔE^*_1, ΔE^*_2, ΔE^*_3 $\cdots$ ΔE^*_N and attempt frequencies ν_{a_1}, ν_{a_2}, ν_{a_3} $\cdots$ ν_{a_N} which may make different contributions L_{1_0}, L_{2_0}, L_{3_0} $\cdots$ L_{N_0} to the output at time $t = 0$. Randall and Wilkins have shown that, for a hypothetical uniform trap distribution, that is, an equal number N_{E^*} of traps of all depths,[119]

$$L = N_{E^*} kT (1 - \epsilon^{-\nu_a t})/t \tag{95}$$

which reduces to a hyperbolic relationship,

$$L = N_{E^*} kT/t \tag{96}$$

when $\nu_a t \gg 1$. If the trap distribution were exponential with trap depth, that is

$$N_{E^*} = A \epsilon^{-\alpha E^*} dE^* \tag{97}$$

where A is a constant, then,

$$L \approx f(\nu_a, kT) Bt^{-(\alpha kT + 1)} \tag{98}$$

which reduces to eq. 96 when $\alpha = 0$ (uniform trap distribution) and expresses an ideal bimolecular decay ($L \propto t^{-2}$) when $\alpha kT = 1$. An ideal bimolecular decay, according to

$$L = 1/b[(L_0 b)^{-1/2} + t]^2 \tag{98a}$$

would obtain if there were two equally abundant *remote* recombination partners, that is, "free" excited electrons and positive holes, which required no activation energy to become sufficiently mobile to "seek each other out." This mobility requirement is met in bimolecular chemical reactions in fluids, where the "constant" b varies with the temperature- and viscosity-sensitive mobilities of the reacting atoms or ions, but the experimentally observed trapping of (at least) electrons in t^{-n} decay phosphor solids precludes the applicability of eq. 98a until the trapped electrons or positive holes are released by an additional activation energy $E^* = \Delta E^*$, according to eqs. 93 and 94. In practice, then, the temperature-sensitive phosphorescence mechanism comprises an activated release of trapped excited electrons or positive holes, followed (at least in the case of remote trapping) by a bimolecular-type recombination of those electrons and positive holes which have sufficient mobility to reach their opposite recombination partners without being retrapped, and concluded by a monomolecular-type radiative transition. If it is assumed, for simplicity, that there are equal numbers of trapped electrons and positive holes, and that the positive holes remain trapped, then the phosphorescence decay will proceed largely according to eq. 94 if the rate of release of electrons from traps is very slow relative to the rate of radiative recombination of the released "free" electrons with the positive holes. This appears to be generally true, and so thermal release is usually the rate-determining step. On the other hand, the influence of eq. 98a should become larger when the rate of release of electrons from traps is very fast relative to the rate of radiative recombination of the "free" electrons and holes. In general, at the instant of cessation of excitation, there will be certain proportions of (1) "free" electrons with various velocities and at various distances from suitable positive holes, (2) trapped electrons in traps of various depths and at various distances from positive holes, (3) mobile positive holes with various velocities, and (4) trapped positive holes. Phosphorescence decay during the first instants after cessation of excitation arises largely from the radiative recombination of nearby mobile excited electrons and positive holes; then the phosphorescence output comes mainly from the release of electrons from the shallowest traps, and during the later stages of decay the phosphorescence contributions come from successively deeper traps. On this basis, it is understandable that the phosphorescence decay curves of t^{-n} decay phosphors are usually quite complex, even neglecting (a) retrapping, (b) nonuniformities in the degree of excitation within the volume of the phosphor, and (c) optical complexities due to scattering and absorption. Also, the brief initial nonexponential decay of some predominantly ϵ^{-at} decay phosphors (compare Fig. 79) is probably due to the radiative return of "free" excited electrons which happen to be in very shallow traps or the conduction band at the instant of cessation of excitation [referring to Fig. 16c, electrons in E^*_C or $E^*_{I_3}$ (where $E^{**}_{I_3} - E^*_{I_3} \ll kT$) first drop to the $E^*_{I_1}$ level and then make a radiative transition from $E^*_{I_1}$ to E^*_I].

Some phosphors, such as cub.-KCl:Tl, cub.-KCl:Ag,[330] and strontium silicate:Eu,[119, 162, 375] exhibit simple temperature-dependent exponential decays according to eq. 93, but most of the t^{-n} decay phosphors have complex trapping distributions which for practical purposes are generally approximated over the major part of their useful decay times

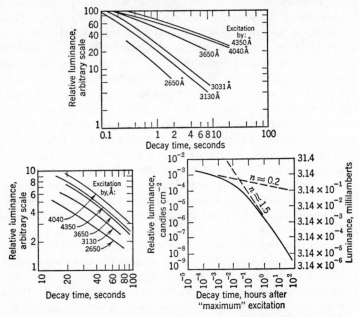

Fig. 73a. Log–log plot of decay curves of hex.-ZnS:ZnO(2):Cu(0.01). Luminescence output at 10^{-5} hr (0.036 sec) $\approx 2.17 \times 10^{-3}$ candle cm^{-2} = 7 mL = 100 lumergs cm^{-2} sec^{-1} $\approx 2.5 \times 10^{13}$ 2.5-ev photons cm^{-2} sec^{-1}. The value of the exponent n, in $L \propto t^{-n}$, is given by the ratio (Δ abscissa)/(Δ ordinate) for a straight line on a log–log plot. (M. Schilling [181])

by an empirical expression of the type,

$$L = L_0[b/(b + t)]^n \qquad (99)$$

where both b and n may vary considerably with large changes in L_0, T, t, and the type and energy of excitant (Figs. 70, 71, and 73).[58, 63] For practical purposes, b is usually smaller than 10^{-3} sec, and n varies from about 0.2 to 2. A more satisfactory empirical equation has been developed by L. S. Nergaard for some of the hex.-ZnS:Cu-type phosphors. The general equation is

$$\log L = b_1(1 - b_2 \log [1 + t/t_0] - b_3[\log (1 + t/t_0)]^2) \qquad (100)$$

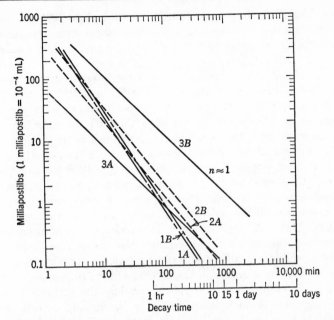

FIG. 73b. Log–log plot of decay curves of:

1. hex.-ZnS:Cu(0.006)   Riedel, Grün N
2. hex.-ZnS:Cu(0.008):Co(0.00007)*   E. Podschus, I.G. Farbenindustrie
3. hex.-ZnS:Cu(0.0008):Co(0.0015)**   E. Podschus, I.G. Farbenindustrie

Excitation by:

A. 5-min exposure to a 100-watt tungsten-filament lamp 50 cm distant (* indicates the optimum phosphor for this excitation).

B. 5-min exposure to a 15-watt mercury-vapor lamp with Schott BG12 filter 20 cm distant (** indicates the optimum phosphor for this excitation or for excitation by daylight). (E. Podschus, per C. H. Love [210a])

Using eq. 100, the decay curve given by Schilling [58, 181] for a "completely excited" hex.-ZnS:Cu phosphor (Fig. 73) is fitted from 10^{-5} to 10^{2} hr by

$$\log_{10} L = 13.4(1 - 0.0334 \log_{10} [1 + t] - 0.008[\log_{10} (1 + t)]^{2}) \quad (101)$$

where L is expressed in photons cm^{-2} sec^{-1}. Integration of eq. 101 gives the **phosphorescence light sum,**

$$L_{total} = \int_{t=0}^{t=\infty} L dt \quad (102)$$

that is, the total number of photons cm^{-2} emitted during the entire decay time, which in this case is about 10^{15} photons cm^{-2}. The penetration "limit" of the UV excitant is less than about 0.01 cm, and so the total number of photons *externally* emitted per cubic centimeter of excited volume is about 10^{17} to 10^{18} photons cm^{-3}. The number of *internally* emitted photons per cubic centimeter may be larger than these calculated values, because some of the internal photons may be reabsorbed and converted into heat. The figure of 10^{18} photons cm^{-3} may be interpreted as the maximum concentration of traps per cubic centimeter in a fully excited hex.-ZnS:Cu(0.01) phosphor, and this trap concentration is the same order of magnitude as the concentration of Cu activator atoms in the phosphor (1 cm^3 of hex.-ZnS:Cu(0.01) contains about 4×10^{-2} mole of ZnS and 4×10^{-6} mole of Cu, or 2×10^{18} Cu atoms cm^{-3}). A similar apparent correlation has been obtained between the total IR-stimulated phosphorescence per unit excited volume and the concentration of Sm activator atoms in a cub.-SrS:SrSO$_4$:LiF:Sm:Ce phosphor.[343] In this case, the **dominant activator** (or **emission activator**) Ce determines the spectral emission characteristic, while the **auxiliary activator** (or **trapping activator**) Sm determines the spectral stimulation characteristic.[171-180] In the hex.-ZnS:Cu phosphor, however, the Cu largely determines the emission *and* the phosphorescence characteristics, probably by forming centers of the type sketched in Fig. 16c (without the highly localized $E^*_{I_1}$ level).

It appears from the two cited cases that the deep traps, with *thermal* trap depths which are very large relative to kT, may be associated with multivalent impurities, for example, electron $+ Cu^{2+} \rightarrow Cu^+$ (trapping state), and electron $+ Sm^{3+} (Sm^{2+}) \rightarrow Sm^{2+} (Sm^+)$ (trapping state). It is possible, also, to trap positive holes, and to have traps associated with crystal imperfections other than those produced by impurities. For example, it is reported that nonphosphorescent natural calcite crystals may be made phosphorescent by mechanical or thermal treatment which introduces observable cracks, and the intensity of phosphorescence is reported to be proportional to the number of macroscopically detectable induced flaws.[371] The thermostimulated light sum emitted from a t^{-n} decay phosphor at a given temperature is generally proportional to (1) the concentration of traps which are sufficiently deep to hold excited electrons for times longer than 10^{-8} sec at the given temperature, (2) the penetration of the excitant, and (3) the duration of excitation below saturation. The phosphorescence light sum is usually decreased by (1) increasing the temperature of the phosphor, that is, increasing the probability of nonradiative transitions, and (2) increasing the energies of individual photons used for excitation (Fig. 73) or stimulation (Fig. 26), that is, increasing the energy deficit which produces heat, or increasing the energy excess over that required to release trapped electrons, thereby increasing quenching relative to stimulation.[372]

During excitation of an initially de-excited phosphor having several trap depths (distinct glow-curve bands or photostimulation bands), the electrons which are captured in the deepest traps are the last to be released and manifest themselves as phosphorescence emission. At the beginning of excitation of such a de-excited phosphor which has deep

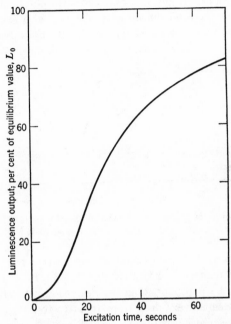

Fig. 74a. Growth curve of luminescence emission during excitation of a thick layer of cub.-SrS:SrSO$_4$:CaF$_2$:Sm:Eu (Std. VI, R. Ward) by blue light at 254°C. (R. E. Shrader)

Excitation by blue light from an incandescent lamp with Wratten filter No. 47.

unfilled traps, there is not the simple exponential growth of luminescence emission exhibited by phosphors with temperature-independent exponential decays, because the deep traps must be filled before the luminescence output can attain equilibrium with the excitation input. A particularly noticeable delay in the growth of luminescence emission is observable during the first excitation of certain freshly prepared deep-trap phosphors, such as CdO:SiO$_2$:Mn [357] or hex.-ZnO:[Zn], which may later exhibit either ϵ^{-at} or t^{-n} decays, depending on the phosphor, the phosphor temperature, and the time after excitation.

Growth-delay effect. This is evident in Fig. 74a which shows a typical S-shaped curve of luminescence growth for a relatively deep-trap t^{-n} decay phosphor (see dotted glow curve in Fig. 74b) under excitation by a low intensity of UV photons.[180] The initial slow growth of externally emitted luminescence photons represents the interval during which the deep traps are being filled. When the deep traps have been filled, the luminescence output increases more rapidly, because the shallow traps release electrons more rapidly under the influence of thermal agitation (thermostimulation). Many hex.-(Zn:Cd)S:Cu phosphors which have become partially oxidized during crystallization also exhibit a visibly detectable growth delay when compared with the unoxidized phosphors. This indicates that the oxidation promotes the formation of moderately

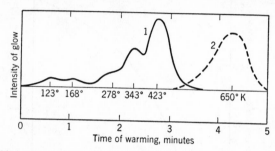

FIG. 74b. Glow curves of (1) cub.-SrS:SrSO$_4$:LiF:Sm:Ce (Std. VII), and (2) cub.-SrS:SrSO$_4$:CaF$_2$:Sm:Eu (Std. VI) after excitation at 77°K. (F. Urbach [221])

deep traps in these phosphors. When the thermal trap depth E^* is only very large relative to kT (that is, $E^* \gg 30kT$) at the operating temperature of the phosphor, then the delayed-growth effect may not be observable because very deep traps (relative to kT!) do not empty sufficiently between successive excitations. When E^* is only very shallow relative to kT (that is, $E^* \ll 30kT$), then a delayed-growth effect may occur, but detection of the effect requires a detector with higher time resolution than is allowed by the 0.1-sec persistence of vision when the human eye is used as the detector. When a phosphor has a wide range and wide density distribution of trap depths, then the growth curves may vary greatly with the temperature of the phosphor, the time between successive excitations, and the type and intensity of excitation. General, or selective, emptying of traps by steady or intermittent photostimulation or photoquenching may also be used to vary the shapes of the growth curves.

At any instant during equilibrium of excitation and emission, that is, when L_0 remains constant with increasing t, practically all the deepest traps will be occupied, whereas most of the shallowest traps will still be unoccupied unless the density of "free" excited electrons produced per unit time is greater than the density of shallow traps divided by the average lifetime of electrons in such shallow traps. As the excitation density is increased, more (shallower) traps become occupied per unit excited volume at any instant during equilibrium. Because the rate of emptying increases as the trap depth decreases, the initial rate of decay of phosphorescence increases as the excitation density is increased. The later rate of decay, at long times after cessation of excitation, however, is little affected by the excitation density as long as equilibrium has been attained; that is, when all the deep traps have been filled.

According to Fig. 22, the excitation density decreases with increasing distance along the path of a beam of excitant particles penetrating a phosphor crystal. This means that the phosphor crystal is nonuniformly excited along the path of the beam, even when equilibrium obtains. Hence, at cessation of excitation, each point along the path of the excitant beam within the crystal has been excited to a different L_0, the highest L_0 being near the point of entry of the beam, and the lowest detectable L_0 being near the penetration "limit." The externally observed phosphorescence output, then, is always the sum of many different internal phosphorescence outputs, each starting from a different L_0 and each representing a different local distribution of excited electrons in traps of various depths. This, of course, is an added complication to eq. 94.

Where there is only a simple temperature-independent ϵ^{-at} decay, the resultant externally observed phosphorescence also follows a simple exponential relation. Where there is a range of trap depths and densities (with varying probability of retrapping, and varying L_0 as a function of distance), however, the resultant decay is necessarily complex, and so the "constant" b and the exponent n in eq. 99 vary with the time of decay as well as with the type and energy of excitant, the intensity and time of excitation, and the temperature of the phosphor (Figs. 70–73, 75, and 76). The different maximum phosphorescence outputs shown in Fig. 73 for the same phosphor excited by photons with different energies may be accounted for by the different depths of penetration and different energy deficits ($h\nu_{absorbed} - h\nu_{emitted}$) of the different low-energy primary photons. Increasing penetration usually increases the phosphorescence output, because more traps become available in the increased excited volume, but absorption losses (especially in thick screens of small scattering crystals) eventually limit this increase. Decreasing the energy deficit usually increases the phosphorescence output, because less heat residue is produced to eject electrons thermally from traps during the excitation time interval. In this respect, it should be noted that there is no direct relationship between the duration of excitation and the duration of detectable phosphorescence of a t^{-n} decay phosphor; indeed, an excited electron produced during an excitation interval of, say, 10^{-12} sec may produce a luminescence photon in about

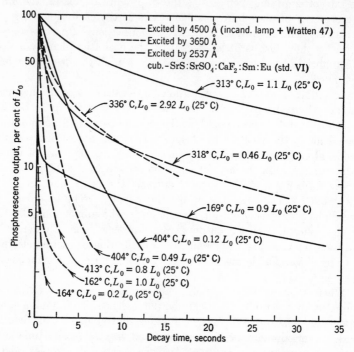

FIG. 75. Decay curves of a deep-trap phosphor, cub.-$SrS:SrSO_4:CaF_2:Sm:Eu$ (Std. VI), after excitation by photons with several different wavelengths when the phosphor is maintained at several different temperatures. At each temperature the phosphor has a different efficiency (see Fig. 76), and so the relative outputs, L_0, at $t = 0$ differ from those obtained at room temperature, as indicated on each curve. Measurements made on the excited side of a thick screen. (R. E. Shrader)

10^{-8} sec, or it may be held in a very deep trap for years, especially at low temperatures.

An example of the influence of the operating temperature of the phosphor and the quantum energy of the primary photons on (1) the intensity of phosphorescence emission, and (2) the shape of a t^{-n} decay curve is given in Figs. 75 and 76 for the case of a phosphor which has a relatively high concentration of chiefly deep traps (Sm centers); [180] compare Fig. 74b. It may be noted that the later stages of the long decays tend to be parallel; these late phosphorescence emissions arise from the return of excited electrons which were detained in deep traps relative to the operating temperature. There is an optimum temperature for maximum phosphorescence L at any particular t, and this optimum temperature is generally different for different t's and for different excitations of a given phosphor. As shown in Table 16, tem-

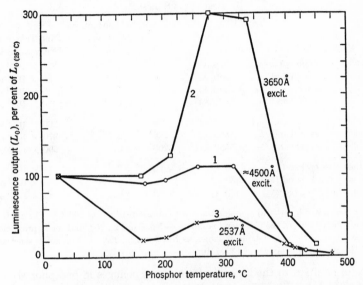

Fig. 76. Luminescence output (excited side) of a deep-trap phosphor, cub.-SrS: $SrSO_4$:CaF_2:Sm:Eu (Std. VI), during excitation by photons with several different wavelengths, when the phosphor is maintained at several different temperatures. (R. E. Shrader)

1. Excited by blue light (incand. lamp plus Wratten No. 47 filter)
2. Excited by 3650-Å UV
3. Excited by 2537-Å UV

perature variations greatly affect phosphorescence emissions associated with traps whose thermal depths E^* are about the same or are somewhat greater than $30kT$ (for example, the yellow emission band in Table 16), whereas the phosphorescence emissions associated with very shallow traps (blue band in Table 16) or very deep traps are little affected by changes in temperature, as long as $kT \gg E^*_{shallow}$, or $kT \ll E^*_{deep}$. It may be seen from Table 16 that higher excitation densities bring more of the faster-emptying shallow traps into play, thereby increasing the *initial* decay rate.[58, 110–114] (Incidentally, there was no indication of any exponential decay, at least after about 30×10^{-6} sec, in any of the measurements made in connection with Table 16.)

TABLE 16

Measurement of the Time Taken to Decay to $L_0/2$, after Excitation to Equilibrium by 10-kv CR (Aluminized Screen) at 25 and 200°C, for the Blue ([Zn]) Emission Band and the Yellow (Cu) Emission Band of hex.-9ZnS·CdS:[Zn]:Cu(0.007), 1250°C

Data from S. Lasof

Phosphorescence Emission Band	T,°C	D-C Current Density, $\mu a\ cm^{-2}$	Time Taken to Decay to $L_0/2$
Yellow emission band ($\lambda_{pk} \approx 5580$ Å)	25	2	230×10^{-6} sec
Yellow emission band ($\lambda_{pk} \approx 5580$ Å)	25	0.5	900×10^{-6} sec
Yellow emission band ($\lambda_{pk} \approx 5580$ Å)	25	0.1	3850×10^{-6} sec
Yellow emission band ($\lambda_{pk} \approx 5580$ Å)	25	0.02	$>7000 \times 10^{-6}$ sec
Yellow emission band ($\lambda_{pk} \approx 5580$ Å)	200	2	200×10^{-6} sec
Yellow emission band ($\lambda_{pk} \approx 5580$ Å)	200	0.5	370×10^{-6} sec
Yellow emission band ($\lambda_{pk} \approx 5580$ Å)	200	0.1	900×10^{-6} sec
Blue emission band ($\lambda_{pk} \approx 4430$ Å)	25	2	15×10^{-6} sec
Blue emission band ($\lambda_{pk} \approx 4430$ Å)	25	0.5	65×10^{-6} sec
Blue emission band ($\lambda_{pk} \approx 4430$ Å)	25	0.2	85×10^{-6} sec

Efficiency of blue-band emission too low to obtain readings at 200°C. The approximate glow curve of the blue band is shown as No. 1 in Fig. 25d, and the approximate glow curve of the yellow band is shown as No. 3 in Fig. 25c. The emission spectrum of this phosphor is shown in Fig. 110.

In addition to the immediate *temporary* changes in phosphorescence wrought by changes in the operating temperatures of trap-type phosphors, there may be *permanent* changes caused by temperature cycling. For example, the intensity of the *initial* room-temperature afterglows of certain ZnS:[Zn]–type phosphors is sometimes decreased by heating the phosphor to 300°C and then cooling again to room temperature. This indicates that annealing allows some activator and trapping centers to assume more nearly the equilibrium configurations that they were prevented from forming by rapid cooling after high-temperature crystallization. In readjusting during annealing, the energy levels of the centers (traps) may move upward, and the shallow traps at the top disappear into the conduction band (see Fig. 16c). Some rapidly cooled phosphors, especially those with low bonding energies and low melting points, may alter appreciably on standing or being excited at room temperature. [The cub.-CsCl:Sb phosphor [65] and the rapid crystallization of (non-luminescent) amorphous selenium during exposure to light at 90°C are examples of such effects.[507a]]

A finite number of excited electrons is required to fill completely all the available traps, of a given depth, in the excited volume of a phosphor, and so equilibrium between excitation and emission is attained in shorter excitation times as the excitation density is increased, and as the density and depth of traps are decreased (also, as the temperature of the phosphor is increased). At very high excitation densities, and particularly with very short-decay shallow-trap phosphors, equilibrium may be attained well within a millisecond. In this respect, it is important to note that the decay of a phosphor from a given value of L_0 may be quite different, depending on the manner in which the L_0 value is attained. If a previously de-excited (for example, heated and cooled) phosphor is excited to a given L_0 with a short-duration burst of high-intensity excitation, such that equilibrium is not attained, the initial decay will be more rapid and the later phosphorescence emission less intense than if the given L_0 were attained as an equilibrium state with lower-intensity excitation of longer duration to maximize the probability of filling the deep traps. Correspondingly, the growth of potential phosphorescence (especially the long-duration phosphorescence) may be much more rapid than the initial growth of luminescence during excitation, particularly at low excitation densities.

3. Pictorial Comparisons of Luminescence Mechanisms. (a) *Mechanism affording temperature-independent* ϵ^{-at} *decays.* Figure 77 is an expanded version of Figs. 67b and 68 with an indication of the similarities and differences among three different graphic representations of the presumed luminescence process in simple ϵ^{-at} decay phosphors. The system described comprises an activator atom, for example, substitutionally located Mn (shown shaded in the center sequence), tetrahedrally coordinated by four ligands, for example, four F, O, or S atoms, within a surrounding host crystal which is not shown. Each of the three sequences of diagrams is a series of simplified schematic "snapshots" of the system at various stages of the luminescence process, and the diagrams are based on the two assumptions that (1) the excited electron does not leave the field of its parent atom (little or no photoconduction accompanying luminescence), and (2) the configurational-coordinate spacing, that is, the average center-to-center interatomic spacings between the emitting atom and its ligands increases when the emitting atom is in the excited state. On this basis:

(i) The upper sequence of energy-level diagrams follows the general pattern established in Figs. 4, 9, and 16c, it being noted that the normally filled ground-state level E_0 is that of the activator center (that is, E_0 in Fig. 77 = E_I in Fig. 16c, and the normally unoccupied higher levels E^*_1 and E^*_2 are also associated with the activator center. The level E^*_3 or the higher discrete level may be an exciton level, and E^*_C is a conduction band of the host crystal. The activator levels should be

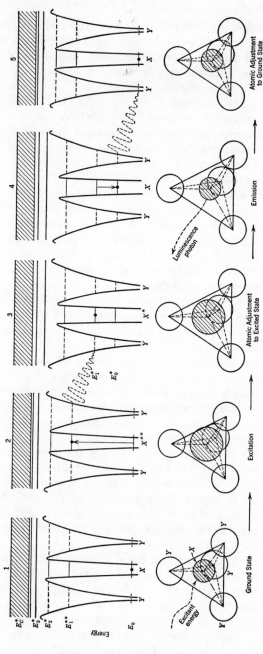

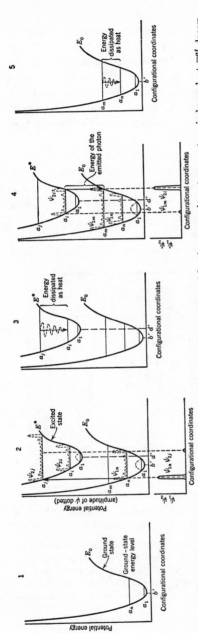

FIG. 77. Chronological schematic sequence of the presumed mechanism of luminescence in a temperature-independent—at decay phosphor center. *Upper row:* conventional energy-level diagrams; *center row:* simplified perspective views of activator atom (ion) surrounded by four tetrahedrally arranged ligands (the atoms are *not* spherical, as shown, and the central activator atom, in particular, may have very knobby outer electron "clouds" whose shapes and sizes change during the luminescence process); *lower row:* hypothetical potential-energy versus configurational-coordinate diagrams corresponding to the two upper rows. (See Figs. 16b and 16c. Figs. 16c are correct for the usual case of X^+ and Y^-. Figures 77 and 78 are drawn for X^- and Y^+.) The shapes of the potential barriers in Fig. 16c.

283

imagined as bands for the crystal as a whole, because there is a deviation in the relatively discrete levels of differently situated individual activator centers in the crystal; also, there is a broadening particularly of the higher levels introduced by some interaction of the activator atoms with each other and with their ligands and a small broadening introduced by local fluctuations in thermal energy. Losses of excitation energy as heat are indicated by dotted curves of damped oscillations to new equilibrium spacings between the atoms.

(ii) The center sequence of hypothetical structural drawings is self-explanatory in indicating the absorption of excitation energy, which may be of primary or secondary origin, followed by localized excitation and then the emission of a luminescence photon of smaller energy than that supplied for excitation (Stokes' law obeyed).

(iii) The lower sequence of potential-energy *versus* configurational-coordinate diagrams was discussed in connection with Fig. 68 where

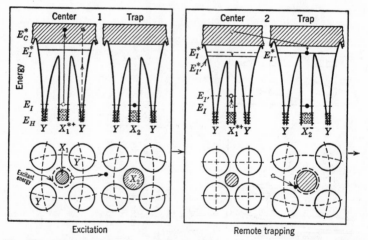

Excitation   Remote trapping

FIG. 78. Chronological schematic sequence of the presumed mechanism of lumines *row:* conventional energy-level diagrams; *lower row:* simplified two-dimension

Legend: Excitation. Absorbed energy raises impurity E_I (*or host-crystal* E_H) in the drawing, remains as a positive hole in the ground-state level E_I (or E_H).

Remote Trapping. The excited system loses some energy as heat while the a remote impurity atom (ion) X_2. [*If the "free" electron were raised from* E_H, *then replenish a* Y).] Both groups of atoms move to new equilibrium positions (heat to $E^*_{I'}$.

Stimulation. Absorbed energy, for example, heat or infrared, raises trapped librium positions (heat liberated).

Emission. The released "free" electron loses energy by inelastic collisions, dropping $E_{I'}$, directly [or, more probably, through intermediate levels (for example, $E^*_{I'}$)], The luminescence-center atoms return to their original equilibrium positions (heat the transition probability was related to $\psi_1\psi_2$, and an attempt was made to relate this product to the bonding energy of the host crystal. If $\psi_1\psi_2$ were large on both sides of the diagram, it would be possible to have two strong excitation or emission bands (cub.-KCl:Tl? Table 5). By drawing the lower sequence of diagrams so that E^* and E_0 approach each other with increasing configurational coordinate, as shown in Fig. 16*b*, it is possible to account for *nonradiative* transitions and for the variation of luminescency efficiency with variation in the operating temperature of the phosphor crystals. This feature is omitted in Figs. 68 and 77 because they are concerned chiefly with *radiative* transitions. In any event, Fig. 77 is included chiefly as a general cross-reference of different commonly used energy-level diagrams and for purposes of visualizing some salient features of the presently assumed mechanisms of temperature-independent ϵ^{-at} decay phosphors.

(*b*) *Mechanism affording temperature-dependent* t^{-n} *decay.* Figure 78 shows sequences of hypothetical simplified schematic "snapshots" of

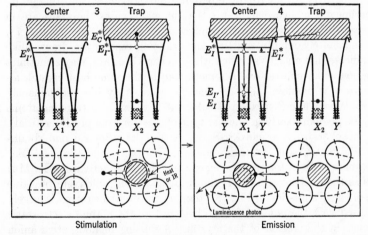

Stimulation   Emission

cence with remote trapping in a temperature-dependent t^{-n} decay phosphor. *Upper* al plan views of activator atom (ion) with its ligands (see caption to Fig. 77).

electron (·) into excited-state conduction band E^*_C. An ionized atom (ion) X^*_1

"free" electron loses energy by inelastic collisions and becomes trapped by or near *the positive hole could wander to* X_1 *and be trapped (by having* X_1 *lose an electron to* liberated). Ground-state level E_I shifts to $E_{I'}$, and excited-state level E^*_I shifts

electron back into conduction band. Trapping-center atoms move to original equi-

to the lower edge of E^*_C. The "free" electron becomes bound to X^*_1 and drops to and the system emits the energy difference $E^*_C - E_{I'}$ or $E^*_{I'} - E_{I'}$ as a photon. liberated), and $E_{I'}$ returns to E_I.

the energy-level diagrams (top row) and atomic sizes and interatomic spacings (bottom row) during one luminescence act in a photoconducting phosphor exhibiting temperature-dependent t^{-n} decay of phosphorescence. For simplicity, only one trapping level is shown (E^*_I- in trap). The *assumedly* interstitial atom X_1 of the emitter center may be excited directly by primary or secondary excitants, as shown, or by having an electron in one of the filled bands of the host crystal (for example, E_H) raised into an excited-state level, whereupon the resultant positive hole in E_H may wander near X_1 and excite it by the indirect process of having an electron drop from the E_I level to fill the positive hole in E_H (thereby trapping the positive hole in E_I). The excited electron may be raised to an exciton state (compare Figs. 4 and 9) or to the conduction band, as shown. In either case, the excited electron may travel away from its parent atom until it is trapped. A trap may be a crystal imperfection, including a surface discontinuity, an anion defect, or (as shown) an impurity atom. Although it is possible that the trap may be many atom-spacings distant from the parent atom or eventual emitting atom, experimental evidence indicates that in many cases the traps are in or near the emitting centers as indicated in Fig. 16c.[145, 238] In Fig. 78, the indefinitely remote emitting and trapping centers are drawn alike to indicate that an activator impurity, such as Cu or Mn, may sometimes function as an emitter *and* a remote trap in the same host crystal, although it is possible that the emitting atoms may be predominantly in one type of site (for example, substitutional), and the trapping atoms may be predominantly in another type of site (for example, interstitial). In phosphors such as cub.-Sr(S:Se):SrSO$_4$:CaF$_2$:Sm:Eu and rbhdl.-(Zn:Be)$_2$SiO$_4$:As(or Sn):Mn, however, one activator (Sm, As, Sn) apparently acts as the trapping agent and the other activator (Eu, Mn) acts as the emitting agent. If the trap is a vacant anion site (imagine one of the Y's missing in Fig. 78), the ψ of the excited electron is largely localized near the anion defect to help restore the normal electrostatic balance in that region of the crystal. A missing mononegative anion may trap one electron, a missing dinegative anion may trap two electrons, and so on. When the excited electron is trapped very near its parent atom, it may be imagined as being in a metastable-state level of the center (Fig. 16c), or it may be thought of as a quasi-exciton where the positive hole remains "frozen" at X_1 in Fig. 78 and the ψ of the excited electron travels around X_1 and its immediate neighbors. In all these cases, it is necessary to supply additional energy to release (stimulate) the trapped electron, as distinguished from the spontaneous radiative transition shown in Fig. 77 for simple ϵ^{-at} decays. Such additional stimulation energy may come from phonons (heat), primary photons (for example, infrared), or direct or secondary action of primary charged material particles. The stimulation may be to the quasi-exciton or conduction levels, and the released electron has again the ability to move away with various probabilities of (1) being re-trapped,[232-239] (2) losing all of its energy by nonradiative processes, and (3) making a radiative recombination with a positive hole. [It is probable that there is a short-duration pause of the returning electron in a normally unoccupied impurity level ($E^*_{I'}$ in Fig. 78–4) before making a spontaneous radiative transition to the ground-state level.[35, 251] This preradiative "capture" is a means of increasing the wave-function overlap, that is, increasing $\psi_1\psi_2$ of the excited electron and the trapped positive hole. Potential-energy versus configurational-coordinate diagrams were not drawn in Fig. 78 because they would be quite complex, and the local processes of readjustments and radiationless energy losses following electronic transitions have been exemplified in Fig. 77.]

(c) *Discussion.* Diagrammatically, the basic difference between phosphors having ϵ^{-at} decays and those having only t^{-n} decays is that the latter phosphors do not have highly localized nonmetastable excited-state energy levels, such as $E^*_{I_1}$ in Fig. 16c and E^*_1 in Fig. 77, to which the centers may be excited directly. The remote trapping shown in Fig. 78 is an extreme case of the t^{-n} decay mechanism which, in general, includes also trapping within the parent center, as shown in Fig. 16c.

Although the two luminescence mechanisms shown in Figs. 16c, 77, and 78 appear to be distinct and experimentally separable by, for example, their dependence on temperature and intensity of excitation, such a separation is difficult when the duration of detectable phosphorescence approaches 10^{-8} sec. As the lifetime of the excited state approaches 10^{-8} sec, the t^{-n} decay type should approach the ϵ^{-at} decay type as a limit; that is, there is an increasing overlap of the ψ of the excited electron with the field of its parent atom, and the trap depth approaches zero. This raises the question whether, in the phosphor centers at least, there may not be local fine structure above the lower edge of the host-crystal conduction band,[281] perhaps in the form of metastable-state levels from which trapped electrons could be released to either *higher or lower* energy levels in which there are weighted probabilities of (1) making a local radiative transition, (2) making a local nonradiative transition, and (3) wandering away from the center through the host-crystal conduction band. On this basis, a heat-actuated release of a trapped electron to lower levels could be effected by high-amplitude temperature fluctuations which upset the strict operation of selection rules and in some cases could increase the probability of radiationless return to the ground state (thermoquenching). In any event, many phosphors, such as rbhdl.-Zn_2SiO_4:Mn (Figs. 70 and 72), rbhdl.-$(Zn:Be)_2(Si:Sn)O_4$:Mn (Fig. 66 [58]), and tetr.-$(Zn:Mg)F_2$:Mn, exhibit strong initial ϵ^{-at} decays followed by weaker long-duration t^{-n} decays *without change in spectral distribution of emission.* This means that both radiative transitions take place from the same energy level ($E^*_{I_1}$ in Fig. 16c, E^*_1 in Fig. 77), from which it may be argued that t^{-n} decay mechanisms involve a preradiative pause on an intermediate energy level, for example, the level shown as $E^*_{I'}$ in Fig. 78 (leaving open the possibility that E^*_I may be overlapped by the conduction band in some cases). On the other hand, some phosphors, such as hex.-9.5ZnS· 0.05CdS:Ag(0.02):Cu(0.005) and hex.-ZnS:[Zn]:Cu(0.001) (Figs. 45 and 46), exhibit two t^{-n} decay emission bands where the blue ([Zn], Ag) band decays at a much faster rate than the green (Cu) band (Fig. 66 and Table 16).

This may be interpreted as indicating either (1) that excited electrons released from traps wander freely through the host crystal, providing the observed electric conduction paralleling phosphorescence,[123] and that these released electrons have a higher probability of recombination with an ionized [Zn] or Ag center than with an ionized Cu center, or (2) that excited electrons are trapped in or near the emitting centers, and the trap depth associated with [Zn] or Ag centers is smaller than the trap depth associated with Cu centers. Alternative 1 allows retrapping, whereas alternative 2, which is similar to Lenard's *"Zentren verschiedener Dauer"* (centers having different lifetimes of the excited state),[145] makes retrapping less probable. Garlick's studies of the dielectric changes of certain ZnS-type phosphors in alternating electric fields tend to support the second alternative, that excited electrons are trapped in or near the eventual emitting centers, although there has been some controversy about the relative importance of retrapping during phosphorescence.[221, 232–239, 655] According to Garlick and Levshin, there is little or no detectable retrapping during normal (thermostimulated) phosphorescence of many ZnS-type phosphors, but W. L. Parker (private communication) has obtained experimental evidence for about 10 per cent probability of retrapping during photostimulation of certain IR-stimulable phosphors, such as Nos. 20–22 in Table 5.[177a]

It is probable that both of the phosphorescence mechanisms are operative to some degree in phosphor crystals, with (1) some long-range motion of excited electrons through the host crystal accounting for any nonpolarizing photoconduction, and (2) short-range motions of excited electrons between parent atoms and nearby traps (probably within the small distorted domains which are called centers) accounting for the large polarizing photoconduction which is observed with many t^{-n} decay phosphors.

In a phosphor system which has energy-level diagrams such as those shown in Fig. 16c, with an appreciable concentration of trapping states suitable for providing phosphorescence delay at room temperature, it is to be expected that excitation by long-wave photons could afford ϵ^{-at} decay (excitation from E_I to $E^{**}{}_{I_1}$), whereas excitation by short-wave photons could afford longer-duration t^{-n} decay (excitation from E_I to $E^{**}{}_{I_2} \cdots E^{*}{}_C$, and subsequent trapping). This is apparently what occurs in cub.-KCl·0.03TlCl (made by grinding together KCl and TlCl) which exhibits (a) a pale-yellow luminescence (band) emission under excitation by 2537-Å UV, with no perceptible phosphorescence emission, and (2) an identical pale-yellow emission under excitation by 1850-Å UV, x rays, or CR, followed by a long t^{-n} decay yellow phosphorescence emission. The 4.9-ev 2537-Å primary photons are apparently unable to raise electrons from E_I to $E^{**}{}_{I_2}$ or $E^{*}{}_C$ whereas primary excitant particles with energies equal to or greater than 5.4-ev 2300-Å photons can raise electrons from E_I to $E^{*}{}_{I_2}$ or higher levels. Similar effects occur in $SrSiO_3$:Eu.[237, 238]

4. *Influence of Phosphor Composition and Structure on Phosphorescence.* It is not yet possible to give a quantitative explanation of the growth and decay characteristics of different phosphors in terms of their chemical compositions and crystal structures. Many of the phosphors which were measured and reported in the past were of doubtful composition or structure, and there is still a dearth of accurate growth and decay curves of well-defined phosphors measured after various excitations at various temperatures and over several orders of magnitude of decay time and intensity. There is a need for thorough studies, making different measurements *on the same sample*, to correlate luminescence efficiency, growth, and decay under different excitants and excitant densities (at different temperatures) with chemical composition, crystal structure, crystal size, photoconduction, glow curves; absorption, excitation, stimulation, quenching, and emission spectra; dielectric and magnetic properties; and photoelectric- and secondary-emission characteristics of the phosphor. Such fundamental experimental studies are needed to resolve the many conflicting interpretations of the fragmentary experimental information which is now available.

In general, chemical composition (including activator), rather than gross crystal structure, is the major factor in determining whether a phosphor exhibits predominantly ϵ^{-at} or t^{-n} decay. For example, cub.-ZnS:Mn, hex.-ZnS:Mn, tetr.-ZnF_2:Mn, and $Ca_2P_2O_7$:Mn all exhibit strong initial ϵ^{-at} decays, whereas cub.-ZnS:[Zn], hex.-ZnS:[Zn], cub.-ZnS:Cu, hex.-ZnS:Cu, cub.-ZnS:Ag, hex.-ZnS:Ag, tetr.-ZnF_2:Cb, and $Ca_2P_2O_7$:Dy all exhibit t^{-n} decays, at least after the first few microseconds after cessation of excitation.[58] Similarly, rbhdl.-Zn_2SiO_4:Mn

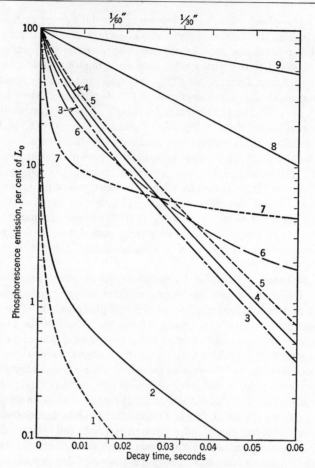

Fig. 79a. Semilog plot of the phosphorescence decays of several phosphors after excitation to equilibrium by 10^{-7}-sec pulses of 10-kv 10^{-2}-amp cm^{-2} CR repeated 30 times a second. Measurement on bombarded side.

1. cub.-ZnS:Ag(0.01)
2. hex.-ZnS:Cu(0.01)
3. rbhdl.-8ZnO·BeO·5SiO$_2$:Mn(1.4)
4. rbhdl.-Zn$_2$SiO$_4$:Mn(0.3)
5. β-Zn$_2$SiO$_4$:Mn(0.3)
6. rhomb.-CaSiO$_3$:Mn(1)
7. rhomb.-Mg$_2$SiO$_4$:Mn(1)
8. rhomb.-CdSiO$_3$:Mn(1)
9. tetr.-ZnF$_2$:Mn(1)

(uniaxial) and the structurally dissimilar β-Zn_2SiO_4:Mn (biaxial [220]) have practically identical ϵ^{-at} decays (Fig. 79a), whereas partial substitution of Sn for Si in rbhdl.-$(Zn:Be)_2SiO_4$:Mn (Fig. 66), or of Mg

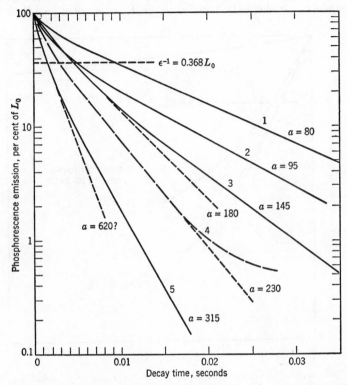

FIG. 79b. Semilog plot of the cathodophosphorescence decays of (1) rbhdl.-$Zn_2SiO_4 \cdot$ 0.012$MnSiO_3$, 1250°C; (2) rbhdl.-$Zn_2GeO_4 \cdot$0.012$MnGeO_3$, 1000°C; (3) cub.-2$ZnO \cdot 3Al_2O_3 \cdot$0.012MnO, 1600°C; (4) cub.-$ZnO \cdot Ga_2O_3 \cdot$0.006$MnO \cdot Ga_2O_3$, 1600°C; (5) cub.-2$ZnO \cdot 3Al_2O_3 \cdot$0.12MnO, 1600°C.

Excitation by 10^{-4}-sec pulses of 6-kv 10^{-6}-amp cm^{-2} CR repeated 60 times a second. Measurement on bombarded side.

for Zn in tetr.-ZnF_2:Mn, produces a very prominent t^{-n} decay "tail" on the initial ϵ^{-at} decay curve.[58,63]

It is shown in Fig. 79b that τ is decreased (a is increased) when the larger and heavier Ge is substituted for Si in rbhdl.-Zn_2SiO_4:Mn, or Ga is substituted for Al in cub.-$ZnAl_2O_4$:Mn. The melting point of rbhdl.-Zn_2SiO_4 is 1785°K and that of rbhdl.-Zn_2GeO_4 is 1763°K [220], and so the change in τ with change in melting point is contrary to the sequence in Table 15, but the changes are much too small to affect seriously the previous correlation of τ with the bonding energy of the host crystal (similar considerations apply to the aluminate and galliate phosphors). It should be noted that the decay curve of hex.-ZnS:Mn,

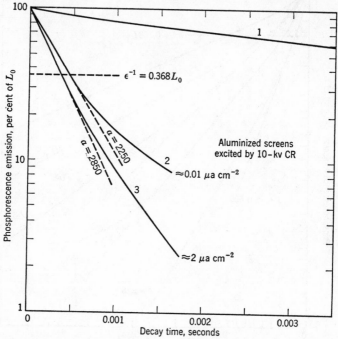

FIG. 79c. Semilog plot of the cathodophosphorescence decays of (1) rbhdl.-Zn_2SiO_4: Mn(0.3), (2 and 3) hex.-ZnS:Mn(1).

in Fig. 79c, is initially exponential and that $a(=\tau^{-1})$ apparently varies with the intensity of excitation. This apparent variation of τ with excitation density in the S-dominated host crystal, as contrasted with the substantially constant τ for Mn activator incorporated in O-dominated phosphors (Figs. 70 and 72), is a further indication of the large difference between these major host-crystal classes, although it is possible that the variation in initial τ may be caused by the different contributions of the later t^{-n} decay "tail" under the two conditions of excitation.

Although the chemical identity of the activator seems to be a major factor in determining the *type* of decay (ϵ^{-at} or t^{-n}), the composition of the host crystal and, to a lesser extent, the structure and size of the host crystal are important in determining the intensity and duration of detectable phosphorescence. The influence of the composition of the host crystal on some temperature-independent ϵ^{-at} decays arising from Mn activator was outlined in connection with Table 15 and Fig. 68. As shown in Fig. 79a, the room-temperature decays of "monocl."-CaO·SiO₂:Mn(1), 1300°C, and "rhomb."-Mg₂SiO₄:Mn(1), 1400°C, are not simple exponentials over the 0.06-sec observation interval. In these cases, it is probable that the observed decay of each phosphor is the resultant of two or more ϵ^{-at} decays plus some latter-stage t^{-n} decay. This is particularly probable in the case of the $CaO \cdot SiO_2$:Mn phosphor whose heterogeneous batches exhibit several stoichiometric combining proportions, that is, $CaO/SiO_2 = 3/1$, $2/1$, $3/2$, and $1/1$, and several different crystal structures as evidenced also by the multiband emission spectrum.[72, 157, 221] The apparent double-exponential decay ($\approx \epsilon^{-1000t} \rightarrow$ $\approx \epsilon^{-10t}$) of rhomb.-Mg₂SiO₄:Mn (also monocl.-MgSiO₃:Mn) may be correlated with the two emission bands (Fig. 58) and two combining proportions, ortho = rhomb.-Mg₂SiO₄ and meta = monocl.-MgSiO₃, which may appear together (especially when the initial ingredients are not completely reacted). Exponential decays with smaller a values than those exhibited by tetr.-ZnF₂:Mn and rhomb.(monocl.)-MgO:SiO₂:Mn may be obtained from $Th(NO_3)_4$ at 90°K and from SrSiO₃:Eu at 90–500°K ($a \approx 0.3$),[203, 375] but the luminescence efficiencies of these phosphors are very low.

Further examples of the effects of composition and structure on phosphorescence are:

(a) *Phosphors with initial ϵ^{-at} decays.* 1. As shown in Table 15, the proportion of Mn activator has little effect on the decay constant, a ($= \tau^{-1}$), up to the optimum proportion of about 1 weight per cent, but higher proportions of Mn greatly increase the decay constant (shorten the detectable phosphorescence). (As shown in Figs. 52 and 53, the *position* of the emission band is similarly unaffected by the Mn proportion up to the optimum, whereafter there is a shift to longer wavelengths with increasing Mn content of the host crystal.) This behavior is observed also in other systems, such as cub.-ZnAl₂O₄:Mn (Fig. 79b), CdO:SiO₂:Mn, and tetr.-ZnF₂:Mn, and the variation of decay with activator content serves as a useful means for adjusting the phosphorescences of phosphors of this type. It is noteworthy that, in these cases, the lifetime τ of the excited state *decreases* as the bonding energy of the resultant crystal decreases with increasing Mn content. No useful explanation of this effect is available at present, apart from the obvious conclusion that interaction between Mn centers increases as the Mn proportion increases, and that the increase in average proximity and interaction of the Mn centers operates to decrease τ.

2. The incorporation of SnO_2 in rbhdl.-$(Zn:Be)_2SiO_4$:Mn greatly enhances the room-temperature t^{-n} decay "tail" on the phosphorescence curves of these phosphors [58], but SnO_2 has little effect on the *room-temperature* phosphorescence of rbhdl.-$(Zn:Be)_2GeO_4$:Mn because the Sn introduces chiefly very deep traps in the latter infrared-stimulable system (see next section).

3. The t^{-n} decay "tail" of rbhdl.-Zn_2SiO_4:Mn is enhanced by substituting up to about 20 mole per cent of Be or Cd for Zn, or Sn for Si, especially when the ratio ZnO–SiO_2 is less than 2.[58, 63]

(b) *Phosphors with t^{-n} decays (at least after about 10^{-6} sec).* 1. The relatively short and weak blue phosphorescence emission of cub.-ZnS:[Zn] is changed into a strong long-duration green phosphorescence by incorporating 0.003 weight per cent of Cu activator in the phosphor, the Cu influencing both the emission spectrum and the trapping (Figs. 25d and 45). Increases in Cu content above about 0.003 weight per cent decrease the amplitude of the Cu-induced trap and decrease the intensity and duration of phosphorescence emission (Fig. 80a).

2. A further large increase in the intensity and duration of phosphorescence emission of ZnS:Cu is obtained by crystallizing the material in the high-temperature hexagonal form, which forms above about 1020°C (Fig. 80b; compare with Fig. 25b).

3. The rate of decay of the ZnS:Cu phosphors shown in Fig. 80 decreases; that is, the duration of detectable phosphorescence increases, as the average crystal size increases for a given crystal structure (compare curves 1 and 2, and 3 and 4 in Fig. 80b).[373]

4. The incorporation of about 2 per cent of oxide or fluoride in hex.-ZnS:Cu further increases the intensity and duration of phosphorescence.[63, 374]

5. The intensity and duration of phosphorescence emission of either cub.- or hex.-ZnS:[Zn], or ZnS:Ag, or ZnS:Cu are decreased by increasing substitution of Cd for Zn (Fig. 81), or Se for S.[58]

6. The incorporation of 0.01 weight per cent of Cu in cub.-ZnS greatly increases the intensity and duration of phosphorescence emission, whereas the same proportion of Cu in cub.-Zn(S:Se) phosphors has little effect on the phosphorescence, and the same proportion of Cu in cub.-Zn(S:Se:Te) greatly *decreases* the moderately long and intense phosphorescence of the host-crystal material crystallized without Cu.

Present phosphors with *useful* temperature-independent ϵ^{-at} decays have decay constants a which range from about 10^7 sec^{-1} (UV emission band of hex.-ZnO:[Zn] [354]) to about 10 sec^{-1} (orange emission band of

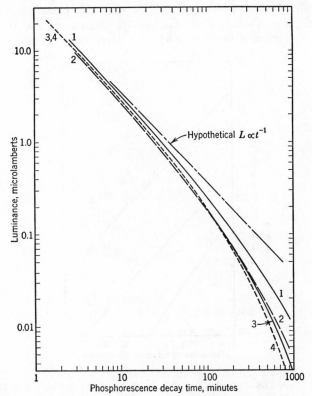

Fig. 80a. Log–log plot of the photophosphorescence decays of some t^{-n} decay hex.-ZnS:Cu phosphors. Measurements on the excited side of thick layers.

1. hex.-ZnS:Cu(0.003):[NaCl(2)], 1240°C
2. hex.-ZnS:Cu(0.005):[NaCl(2)], 1240°C
3. hex.-ZnS:Cu(0.007):[NaCl(2)], 1240°C
4. hex.-ZnS:Cu(0.01):[NaCl(2)], 1240°C

Excitation to equilibrium by 20 min under 13.7 ft-candles of light from incandescent lamps.

Detector = 931-A multiplier phototube + Wratten No. 4 filter.

tetr.-(Zn:Mg)F$_2$:Mn), although some low-intensity temperature-dependent ϵ^{-at} decay phosphorescences have been observed with $a < 0.3$.[203,330,375] That is, there are detectable ϵ^{-at} decay phosphorescences lasting from less than 10^{-7} sec to several seconds, whereas present phosphors with t^{-n} decays have detectable phosphorescences which range from about 10^{-8} sec (alpha-particle excitation of cub.-ZnS:Ag) to indefinitely long periods (years!) for deep-trap phosphors, such as Nos. 20–22 in Table 5, depending on the temperature and trapping-state distribution of the phosphor; the nature, intensity, and duration

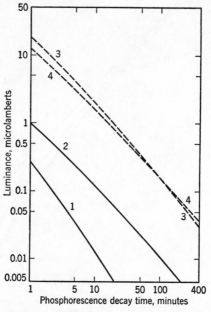

FIG. 80b. Log–log plot of decay curves of selected members of the t^{-n} decay phosphor family ZnS:Cu(0.003) crystallized with 2 per cent NaCl at 730 to 1360°C. Note the marked change in phosphorescence intensity which occurs on making the transition from the $F\bar{4}3m$ (cubic) to the $C6mc$ (hexagonal) structure. Measurements made on excited side of thick screens.

1. cub.-ZnS:Cu(0.003), 730°C
2. cub.-ZnS:Cu(0.003), 918°C
3. hex.-ZnS:Cu(0.003), 1100°C
4. hex.-ZnS:Cu(0.003), 1360°C

Excitation to equilibrium by 5 min under 13.7 ft-candles of light from incandescent lamps.

Detector = 931-A tube + No. 4 Wratten.

Average crystal size: $1(\approx 1\mu) < 2 < 3 < 4(\approx 20\mu)$.

of excitation; and the type of radiation (if any) [377-379] to which the phosphor is exposed during the decay period.

Of the few phosphors known to have temperature-dependent ϵ^{-at} decays, none has had sufficient luminescence efficiency to be of practical importance.[162-166,330,375] Some of these phosphors, however, are of considerable interest as representing simple limiting cases of the general category of temperature-dependent t^{-n} decays. Kato has reported that centimeter-size single crystals

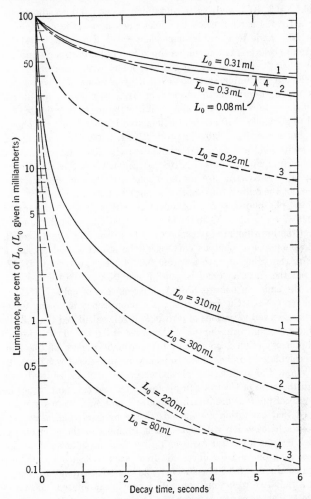

FIG. 81. Semilog plot of the decay curves of several t^{-n} decay phosphors after excitation to equilibrium by 3650-Å ultraviolet. Measurements made on the excited side of thick screens. Note the accelerated *initial* decays from the higher L_0's, and the approximately parallel *later* decays from different L_0's for the same phosphor.

1. hex.-ZnS:Cu(0.005), 1220°C
2. hex.-9ZnS·1CdS:Cu(0.0073), 1250°C
3. hex.-8ZnS·2CdS:Cu(0.004), 1200°C
4. hex.-ZnS:Ag(0.015), 1250°C

of cubic NaCl, KCl, and NaBr containing about 0.001 weight per cent of Cu or Ag activator exhibit near-ultraviolet phosphorescence emissions which are exponential and temperature-dependent.[330,691]   Unexcited crystals of cub.-KCl:Ag are colorless but acquire a broad induced absorption band peaked near 4200 Å when the crystals are excited by photons with wavelengths shorter than about 2000 Å (no luminescence is excited by photons with wavelengths falling within a longer-wavelength absorption band starting at about 2350 Å and rising sharply to a peak at 2100 Å). When the previously excited crystals are heated to above 60°C or are photostimulated by blue light, the phosphor emits a phosphorescence band extending from 2500 to 3200 Å with a peak near 2720 Å. These excitation, emission, and stimulation characteristics are very similar to those shown in Fig. 26 for cub.-Sr(S:Se):SrSO$_4$:CaF$_2$:Sm:Eu, except that the spectra of cub.-KCl:Ag are moved about 2 ev toward shorter wavelengths; for example, the optical trap depth is about 3 ev for this phosphor. The thermostimulated phosphorescence is exponential, with a decay "constant" a varying from 0.0037 sec^{-1} at 62°C to 0.039 sec^{-1} at 90°C. From light-sum measurements of thermostimulated phosphorescence, it was found that over 10^{15} potential phosphorescence photons cm^{-3} could be stored in the crystal. In cub.-NaCl:Ag, the value of a varies from 0.0024 sec^{-1} at 17°C to 0.087 sec^{-1} at 52°C, and the emission band is peaked near 2490 Å. (In cub.-NaCl:Cu, the peak of the emission band is 3580 Å, whereas that of cub.-KCl:Cu is 3960 Å, and the cub.-NaCl:Cu has a long phosphorescence at room temperature.) From the observed variations of a with T, it was calculated that all these phosphors have thermal trap depths of the order of 0.8 ev. It was reported, also, that photoconduction accompanies the phosphorescences of these phosphors, and the calculated number of electrons flowing per second was found to be only about $\frac{1}{1000}$ the number of phosphorescence photons emitted per second, although the ratio of electrons to photons remained constant, regardless of the wavelength of photons used during photostimulation. From these data it was concluded that excitation corresponds to the removal (ionization) of an electron from a halogen ion with subsequent trapping of the electron by an Ag$^+$ ion which became a neutral Ag atom. Blue light, or sufficient heat, ionizes the presumed Ag atom and the liberated "free" excited electron contributes to the observed conduction (at 900 volts cm^{-1}) as it travels through the crystal before making a radiative recombination with a previously ionized halogen ion (positive hole). The disparity between the number of calculated conduction electrons and emitted phosphorescence photons was accounted for by assuming that the "free" electrons travel on the average only $\frac{1}{1000}$ of the distance between the electrodes, that is, just a few microns, before making a radiative transition during capture by a positive hole. On this basis, it was determined that the average distance traveled by the electrons decreased as the excitation density increased.

It should be kept in mind that the duration of detectable phosphorescence emission from a phosphor depends on the sensitivity of the detector as well as on the decay curve and intensity of phosphorescence. The oft-repeated question, "How long does the phosphor persist (glow, phosphoresce)," is quite ambiguous unless the sensitivity of the detector and the conditions of excitation and operation are specified. For example, the average well-dark-adapted (scotopic) human eye can readily detect the feeble phosphorescence emission (about 3×10^{-9} lambert) from a UV-excited hex.-ZnS:Cu phosphor which has been phosphorescing at room temperature in the dark for over 4 days, but the average well-light-adapted (photopic) human eye cannot perceive the much stronger phosphorescence emission (about 10^{-5} lambert) from the same phosphor a few minutes after cessation of excitation.[58] The variable sensitivity of the partially dark-adapted (mesopic) human eye, which is intermediate to the photopic and scotopic eye, provides additional indeterminacy to nonspecific questions about the duration of phosphorescence. With a well-refrigerated multiplier phototube, for example, the 931-A, it is possible to detect the emission of single photons.[376] By this means, the time "limit" of detectable room-temperature phosphorescence from deep-trap phosphors such as hex.-ZnS:Cu, cub.-(Ca:Sr)S:Bi, and rbhdl.-(Zn:Be)$_2$(Si:Sn)O$_4$:Mn may be extended to many years. H. E. Millson reports that some specimens of apparently nonradioactive fluorite minerals afford phosphorescence emission detectable by the *scotopic* eye 6 years after a 1-min excitation by 2537-Å UV, and he has made contact photographs of such minerals *in color* by thousand-hour exposures 4 to 5 years after such a brief excitation.[82a]

Stimulation and Quenching of Phosphorescence

This subject has been treated briefly in the discussion of Table 10 and Fig. 26. Appreciable photostimulation or photoquenching has not been observed with phosphors which have only temperature-independent ϵ^{-at} decays, although it is sometimes possible to make a predominantly ϵ^{-at} decay phosphor photostimulable by altering its composition to provide an appreciable density of traps. This may be done, for example, by incorporating SnO$_2$ in the ϵ^{-at} decay rbhdl.-Zn$_2$GeO$_4$:Mn phosphor. The green-emitting phosphor rbhdl.-3ZnO·2GeO$_2$·0.06SnO$_2$·0.02MnO, 950°C, has little thermostimulated phosphorescence at room temperature, but it has considerable photostimulated phosphorescence under infrared ($\approx 10{,}000$ Å) irradiation.[180] The incorporation of SnO$_2$ in rbhdl.-(Zn:Be)$_2$GeO$_4$:Mn phosphors produces deeper traps than in the case of rbhdl.-(Zn:Be)$_2$SiO$_4$:Mn phosphors. The germanate phosphors with SnO$_2$ must be heated above room temperature to observe the enhanced thermostimulation (t^{-n} decay "tail") which is obtained at room temperature in the case of the silicate phosphors with SnO$_2$. It is important to note that, in both these cases, the emission spectrum is the same during excitation (predominantly ϵ^{-at} decay processes) and during photostimulation or thermostimulation (predominantly t^{-n} decay processes). This indicates that the emitting centers make radiative transitions between the same energy levels whether the returning excited electron has been (1) firmly bound to the parent atom during the entire excited lifetime (ϵ^{-at} decay = unimolecular or first-order process), or (2) removed from the (ionized) parent atom and trapped by near-by or remote atoms (t^{-n} decay = *ideally* a bimolecular or second-order process).

In cub.-CaO:Pr:Sm, where there are two activators, both the Pr and Sm centers produce characteristic emissions during excitation by high-energy particles (for example, CR), whereas only the Pr emission appears during phosphorescence when the Sm centers act as traps which supply stored energy to the emitting Pr centers.[65] Similarly, in alkaline-earth sulphide phosphors containing Eu and Sm, the band emission of the Eu and a weak line emission of Sm appear during excitation, whereas only the Eu band emission appears during phosphorescence when the stored energy in the Sm-center traps is supplied to the Eu centers to produce luminescence emission.[221] These cases emphasize the multiple roles which may be played by an activator (for example, the Sm) in a given host crystal.

In general, for a given phosphorescing t^{-n} decay phosphor, predominant photoquenching is produced by irradiation with photons whose energies are equal to or slightly greater than the energies of the luminescence-emission photons, whereas predominant photostimulation is produced by irradiation with lower-energy photons which are just able to release the trapped electrons (Fig. 26). At intermediate energies of the photons used for irradiation, both quenching and stimulation may be operative. It may be recalled that the phenomena of stimulation and quenching were "explained," in Fig. 27, by a drafting artifice. Another "explanation" might be based on the *Vielfachstösse* hypothesis which suggests that "free" electrons, which are raised over 1 ev above the lower edge of the conduction band, may lose all their free energy by multiple collisions with each other.[102] This is apparently discredited by the observation that quenching radiation *decreases* the number of conduction electrons (curve 2, Fig. 82 [123]) as well as the number of phosphorescence photons. In general terms, it seems easy to picture the release of a trapped electron by the absorption of a just sufficient amount of infrared-photon energy, and so the released excited electron may return to a positively charged ionized atom (positive hole) and allow the bound combination to return to the ground state by an intermediate radiative transition. If the energy of the infrared photon were insufficient to release the electron from the trap, no stimulation would occur, and the trapped electron might remain as a source of potential energy which could be converted into a luminescence photon. If the energy of the infrared photon is increased beyond the point of being just sufficient to release the trapped electron, then at least two things may occur which are detrimental to the radiative process: (1) the return-

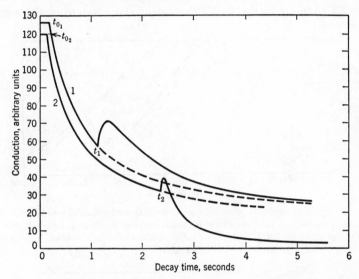

FIG. 82. Decay of electric conduction of two t^{-n} decay sulphide-type phosphors after excitation by 3650-Å ultraviolet. The unoxidized phosphor is stimulated, whereas the oxidized phosphor is largely quenched by the red + infrared irradiation. As shown in Figs. 116 and 117, the luminescence growth and decay parallel the changes in electric conduction of these phosphors. (A. E. Hardy)

1. hex.-9ZnS·1CdS:Cu(0.0073), 1250°C, synthesized to avoid oxidation
2. hex.-9ZnS·1CdS:Cu(0.0073), 1250°C, slightly oxidized during crystallization

t_{0_1} and t_{0_2} = 3650-Å UV turned off.
t_1 and t_2 = red light (>5900 Å) turned on.
Dotted lines show curves which are obtained when red-plus-infrared irradiation is omitted.

ing electron has appreciable kinetic energy which increases interaction and scattering by the vibrating crystal atoms, and makes it difficult to satisfy conservation of momentum upon being captured and making a radiative transition, and (2) the excess energy of the returning electron is converted into heat which disturbs the sensitive emitting centers, and may raise their energies to the radiationless crossover level (f in Fig. 16b). That heat is deleterious is indicated by the fact that the photo-stimulated light sum is usually 5 to 30 times as great as the thermostimulated light sum in the case of efficient IR-stimulable phosphors, such as Nos. 20–22 in Table 5 (see Fig. 88).[343]

PHOTOCONDUCTION AND PHOTOSTIMULATION. Some exemplary information on the spectral-distribution and intensity-versus-time relationships between photoconduction and photostimulation are shown in Figs. 83–86 (compare Fig. 25g). These data were obtained by Professor

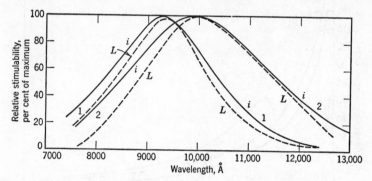

FIG. 83. Spectral correspondence between the relative stimulabilities of the photocurrents i and phosphorescence emissions L of previously excited (1) cub.-Sr(S:Se): SrSO$_4$:CaF$_2$:Sm:Eu (B1), and (2) cub.-SrS:SrSO$_4$:CaF$_2$:Sm:Eu (Std. VI). (W. L. Parker)

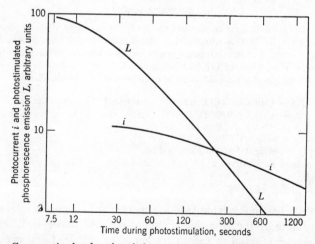

FIG. 84. Comparative log–log plot of photocurrent and photostimulated phosphorescence emission as a function of time during stimulation of a previously excited cub.-Sr(S:Se):SrSO$_4$:CaF$_2$:Sm:Eu (B1) phosphor. The stimulation was effected by radiation extending from 7000 to over 14,000 Å. Figure 85 shows these data replotted with i as a function of L. (W. L. Parker)

W. L. Parker on deep-trap phosphors furnished by Professor R. Ward, both of the Polytechnic Institute of Brooklyn. Figure 83 shows that the spectral distribution of photon-stimulated phosphorescence is practically identical with the spectral distribution of photon-stimulated

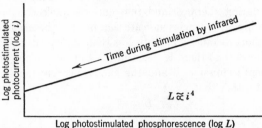

FIG. 85. Log–log plot of the time and intensity relation between the photocurrent and the photostimulated phosphorescence emission of a previously excited cub.-Sr(S:Se):SrSO$_4$:CaF$_2$:Sm:Eu (B1) phosphor being stimulated by infrared radiation (7000 to 14,000 Å). (W. L. Parker)

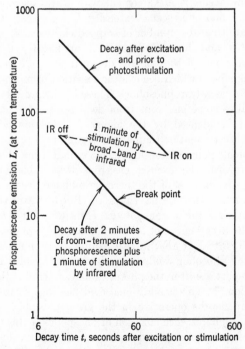

FIG. 86. Log–log plot of phosphorescence emission of cub.-SrS:SrSO$_4$:CaF$_2$:Sm:Eu (Std. VI) before and after stimulation by infrared. (W. L. Parker)

photoconduction for two phosphors which had been previously excited by violet light. This spectral correspondence and the similarity between the growth and decay curves of photostimulated phosphorescence and photoconduction (Fig. 84) are proof that some of the electrons released from traps move through the crystal, but the strong polarization of conduction during thermostimulation and the dielectric behaviors of phosphors at high frequencies indicate that most of the released electrons travel only short distances, perhaps less than 10 interatomic spacings.[238] Figure 85 shows that there is not a linear correspondence between the phosphorescence emission L and the simultaneously measured photocurrent i but rather that

$$L \propto i^n \tag{103}$$

where n is about 4 for the curve of Fig. 85. In other words, the photostimulated phosphorescence intensity decreases with time much faster than the concomitant photocurrent, indicating that at least some of the current-carrying electrons may not contribute to the phosphorescence emission. (This result is in agreement with Garlick's report that the *thermostimulated* phosphorescence intensity L decreases with time at a much faster rate than the number of trapped electrons N. For samples of hex.-ZnS:Cu and hex.-9ZnS·CdS:Cu at 291°K, $L \cong t^{-0.7}$ and $N \cong t^{-1.7}$ from about 4 to 1000 sec decay time.[238]) Figure 86 shows two interesting effects which occur after cessation of incomplete photostimulation of a deep-trap phosphor. These effects are: (1) The photostimulation raises some electrons from deep traps into shallower traps (retrapping), as evidenced by the higher L value immediately after cessation of photostimulation than at the beginning of photostimulation, and (2) the break point of the lower curve, indicating that two main trap depths furnish the excited electrons responsible for the light output during this interval of the thermostimulated decay.

DECAY CURVES AND LIGHT SUMS OF PHOTOSTIMULATED VERSUS THERMOSTIMULATED PHOSPHORESCENCE EMISSION. It was shown in Fig. 75 that the decay curve of *thermostimulated* phosphorescence emission of cub.-SrS:SrSO₄:CaF₂:Sm:Eu exhibits a pronounced break at 169°C when the phosphor was previously excited by blue light. Also, Urbach *et al.* report a shift in the photostimulation band of this phosphor during excitation.[221] This is additional evidence that there are at least two major trap depths operative in the given phosphor. The room-temperature phosphorescence emission of phosphors of this type usually decreases in a fraction of a second to a vanishingly low value. Furthermore, the phosphorescence decay curves are of the t^{-n} type, just as reported by Fonda for phosphors such as hex.-ZnS:Cu:Pb (Table 5).

The decay curves of *photostimulated* phosphorescence emission of phosphors may be either of the ϵ^{-at} or the t^{-n} type. Fonda has reported that the IR-stimulated decays of hex.-ZnS:Cu:Pb, hex.-ZnS:Cu:Mn, and hex.-ZnS:Pb:Mn are all initially exponential, although they may exhibit later t^{-n} decay "tails" when photostimulated shortly after excitation.[169] When the phosphors were allowed to stand in the dark for several hours after excitation, no t^{-n} decay "tails" were observed (these later decays were, however, measured during a shorter time interval, and so it is possible that the t^{-n} decay "tails" would always appear if the phosphorescence emission were measured over a sufficiently long time interval). From these results, and the results of experiments on thermostimulation and photostimulation at 77 and 300°K, Fonda concluded that the thermostimulated t^{-n} decay is due to ejection of electrons from shallow traps with attendant retrapping, whereas the photostimulated ϵ^{-at} decay (at 300°K) is due to ejection of electrons from deep traps without retrapping. The observation was made, also, that photostimulated emission at 77°K is not so intense as at 300°K, and it was concluded that photostimulation by 0.4- to 1.2-ev photons at 77°K ejects electrons selectively from the *shallow* traps, whereas the same photostimulation at 300°K ejects electrons selectively from the *deep* traps.

Examples of decay curves of photostimulated and glow-curve-type thermostimulated emissions from deep-trap phosphors, replotted from data by Ellickson,[343] are shown in Figs. 87 and 88. It may be noted that the decay of curve 1 in Fig. 87 is exponential for the first 2.5 min, although the original paper describes the entire decay as being of the power-law type with $L \propto t^{-2}$. The remaining decays are all of the t^{-n} type, being nonlinear on the semilog plot. Figure 88 shows clearly the larger light sum which is obtained by photostimulation at 300°K, compared with the smaller light sum obtained by glow-curve-type thermostimulation when the temperature is raised above 300°K. (It is possible, in some cases, to have a larger thermostimulated light sum than photostimulated light sum if the latter is measured at a low temperature where the phosphor is *less* efficient than at the higher temperatures attained during glow-curve-type thermostimulation; compare the pronounced efficiency maximum above room temperature in Fig. 76.)

Although the higher heating rate affords the larger thermostimulated light sum in Fig. 88b, a still higher heating rate gave a lower light sum; that is, there is an optimum rate of heating for maximum glow-curve-type thermostimulated light sum. This is of importance in estimating the density of trapped electrons from glow-curve data. Similarly, the phosphor temperature is important in determining the light sum during photostimulation or during conventional phosphorescence (thermostimulation at a fixed temperature), because increasing temperature generally tends to depopulate shallow traps and increase the probability of nonradiative transitions.

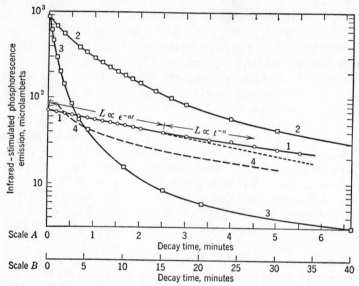

FIG. 87. Semilog plot of the decays of photostimulated phosphorescence emissions of three deep-trap phosphors. (R. T. Ellickson)

	Light sum, microlambert minutes
1. cub.-Sr(S:Se):SrSO$_4$:CaF$_2$:Sm:Eu (B1), Scale A	1250
2. cub.-SrS:SrSO$_4$:LiF:Sm:Ce (Std. VII), Scale A	1200
3. cub.-SrS:SrSO$_4$:LiF:Sm:Ce (Std. VII), Scale B	1200
4. cub.-SrS:SrSO$_4$:CaF$_2$:Sm:Eu (Std. VI), Scale A	

1 and 4. Excited by blue light for 5 min.
2 and 3. Excited by 3650-Å UV for 5 min.
Photostimulation at 300°K by IR from a 10-watt tungsten-filament lamp (with 5-mm Corning 254 filter) placed 46 cm from phosphor.

In the case of the cub.-SrS:SrSO$_4$:CaF$_2$:Sm:Eu and cub.-Sr(S:Se): SrSO$_4$:CaF$_2$:Sm:Eu phosphors, Urbach, Ellickson, and Parker report that the stimulation band is unchanged during the course of thermostimulation or photostimulation by either monochromatic (9000-Å) photons or the usual broad band of 10,000 to 30,000-Å infrared photons.[221,342] This has been interpreted as indicating that only a single optical trap depth is involved, because it is improbable that electrons in traps with a range of depths would be released at rates which are independent of the trap depth and the spectral distribution of the source of photons used for stimulation. The appreciable retrapping in these

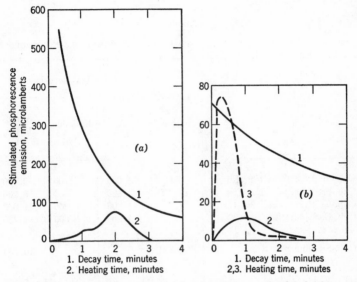

FIG. 88. A comparison of the photostimulated and thermostimulated (glow-curve-type) phosphorescence emissions of two deep-trap phosphors. (R. T. Ellickson)

(a) cub.-SrS:SrSO$_4$:LiF:Sm:Ce
    1. IR-stimulated at 300°K
    2. Thermostimulated by gradually raising the temperature (from 300°K) in the dark
(b) cub.-Sr(S:Se):SrSO$_4$:CaF$_2$:Sm:Eu
    1. IR-stimulated at 300°K
    2 and 3. Thermostimulated (heating rate of 3 greater than that of 2)

phosphors may, however, allow two or more interconnected traps to appear as one.

STIMULATION SPECTRA AND CONSTITUTIONS. The spectral displacement of the photostimulation curves in Fig. 83 is in the same direction as the displacement of the emission spectra of these phosphors (Nos. 20 and 21 in Table 5). In the sequence of SrSe-SrS-CaS-MgS as major host-crystal constituents with Sm and Eu activators, the emission band and the stimulation band are both displaced to longer wavelengths just as the line emission spectra are displaced in Fig. 13. The approximate peak wavelengths of the known excitation, emission, and photostimulation bands of some of the more efficient IR-stimulable phosphors are presented in Table 17. Of the known *efficient* IR-stimulable phosphors, the photostimulation band of hex.-ZnS:Cu:Pb extends to the longest wavelength (approx. 16,000 Å).

TABLE 17

SPECTRAL CHARACTERISTICS OF SOME USEFUL IR-STIMULABLE PHOSPHORS

Phosphor	Common Designation	Peak Wavelengths of		
		Excitation, Å	Emission, Å	Stimulation, Å
1. Cub.-Sr(S:Se):[flux]:Sm:Eu	B1	4,600	5,700	9,300
2. Cub.-SrS:[flux]:Sm:Eu	Std. VI	4,800	6,300	10,200
3. Cub.-Ca(S:Se):[flux]:Sm:Eu		>4,800	>6,300	11,000
4. Cub.-CaS:[flux]:Sm:Eu		>4,800	6,600	11,700
5. Cub.-SrS:[flux]:Sm:Ce	Std. VII	2,900 3,500	4,800 5,400	10,200
6. Hex.-ZnS:Cu:Pb[SO₄]:[NaCl(2)]	101	3,700	4,880	7,500 13,200

Where [flux] is marked, the flux usually contains both O and F, for example, as in $SrSO_4$ and CaF_2 or LiF.[221]

Ward reports that the efficiency of IR-stimulated phosphorescence emission of cub.-SrSe:[flux]:Sm:Eu is increased greatly by substituting SrS (in solid solution) for 4 per cent of the SrSe, and a further increase in efficiency is obtained by substituting CaS for an additional 5 per cent of the SrSe. A similar efficiency increase is obtained with a few per cent of SrO substituted for SrS in cub.-SrS:flux:Sm:Eu, even though x-ray diffraction photographs show no detectable solid solution of SrO in the SrS crystals. The optimum IR-stimulable phosphorescence emission of the phosphor whose host crystal is predominantly SrS is obtained with about 35 parts per million each of Sm and Eu, whereas the optimum proportions when the host crystal is predominantly SrSe is about 180 parts per million of Sm and about 50 parts per million of Eu.[172-176, 221] There is some experimental evidence that the trapping activator, Sm, exists mainly as Sm^{3+}, as indicated by characteristic line emissions reported by Parker, whereas the emission-determining activator, Eu, exists mainly as Eu^{2+}, as indicated by the band emission and by chemical evidence presented by Ward.[177a,221]

PHOTOSTIMULABLE VERSUS PHOTOQUENCHABLE PHOSPHORS. A good photostimulable phosphor has a high density of traps whose thermal depths are very large relative to kT at the operating temperature, that is, depths larger than $30kT$. Such a phosphor may decay rapidly to a negligibly low phosphorescence-emission intensity at room temperature (Fig. 26b) but becomes well thermostimulated when the temperature of the phosphor is raised until kT approaches the principal trap depth (Fig. 75). Conversely, when a phosphor which ordinarily exhibits strong t^{-n} decay (thermostimulated) phosphorescence at room temperature is cooled to lower temperatures, its potential phosphorescence may be "frozen in." Excited electrons "frozen" in shallow traps of a refrigerated phosphor may be photostimulated by lower-energy photons than are required to stimulate an excited phosphor whose traps are all deep relative to room temperature kT. The few tentative identifications of electron traps with certain crystal imperfections, notably multivalent impurities and anion omission defects (F centers [58,66,67]), does not yet provide sufficient information to design phosphors with given trap distributions. A major need is to identify positively and locate the electron traps, in order to depart from the empirical research which has produced the few efficient photostimulable and thermostimulable phosphors now known.

A good photoquenchable phosphor is also a good thermostimulable phosphor, that is, it is a phosphor which has a high density of traps whose thermal depths are somewhat greater than kT at the operating temperature, with the additional feature that irradiation during phosphorescence converts the energy of the trapped electrons into heat or infrared rather than conventional luminescence emission. The quenching feature appears to be largely a function of special impurities which (1) increase the absorption of the phosphor for the photons which produce quenching, and (2) act as poisons (killers) for phosphorescence. An example of 1 is found in the incorporation of oxides or sulphates in hex.-ZnS:Cu, and an example of 2 is found in the incorporation of compounds of Fe, Co, or Ni in the same phosphor.[58,377-379]

Sulphates appear to be particularly beneficial in providing new near-optical vibration modes in crystals of simple binary compounds, such as ZnS and SrS, which are not strong absorbers of infrared. Optical vibration modes, it may be recalled, arise from oppositely disposed motions of different species of particles (atoms or ions) relative to each other, and the number and frequency of vibration modes increases with increasing number of such species.[28,36,300] The optical absorption modes within the SO_4 radical are numerous because there are many exchange (resonance) possibilities in the complex $SO_4{}^{2-}$ ion.[24]

A possible explanation of the potent "killer" actions of iron, cobalt, and nickel salts may be sought in spin-orbit interaction of excited electrons with the unpaired $3d$ electrons of these transition-group elements, such that the excitation energy is dissipated as phonons to the host crystal. For example, an excited electron may become trapped by an Fe^{2+} ion, and the trapped electron may gradually give up energy to the crystal by periodically entering the $3d$ shell of Fe^{2+} as an exchange interaction to balance temporarily an unpaired spin of one of the 6 $3d$ electrons (according to Table 8, Fe^{2+} has $\mu_{eff} \approx 5 = [N(N+2)]^{1/2}$, where N is the number of unpaired electrons). The cobalt ions, Co^{3+} and Co^{2+}, which have 6 and 7 $3d$ electrons, respectively, and have μ_{eff} values near 5, are especially strong "killers" of luminescence in hex.-ZnS:Cu.[377] It is significant that photoquenching of phosphorescence coincides with photoquenching of photoconduction,[121] as in slightly oxidized batches of hex.-ZnS:Cu (Fig. 82), and also that grinding of t^{-n} decay phosphors reduces phosphorescence intensity at a much more rapid rate than the intensity of luminescence emission during excitation. It appears that the imperfections (impurities) which enhance quenching either impede the mobility or radiative return of electrons released from traps, or, more probably, accelerate the return of excited (trapped) electrons to the ground-state level by nonradiative processes, or by processes which produce infrared emission without raising the trapped electrons into the host-crystal conduction band. Again, sufficient experimental information is lacking.

SCOTOPHORS. Curiously enough, the decay of F-center *tenebrescence* (Table 10) of CR-darkened cub.-KCl, under photostimulation by 2.2-ev photons, is greatly *accelerated* by freshly incorporated impurities, such as Th^{4+}, Al^{3+}, Mn^{2+}, and Mg^{2+}, which apparently promote the release of trapped electrons without affecting the spectral distribution of the F-center absorption band.[58, 380] Nottingham found, however, that the accelerating effect of the impurities decreases with time of cyclic darkening and bleaching, until at the end of about 10^6 CR pulses, each providing 1 microcoulomb cm^{-2} at 9 kv and spaced 10 sec apart, the tenebrescences of cub.-KCl scotophor crystals with cocrystallized impurities decay at the same rate as the tenebrescence of pure cub.-KCl. It is necessary, therefore, to continue to supply Th^{4+}, Mg^{2+}, etc. to the screen during operation. Evaporated screens of cub.-KCl are outstanding among presently known scotophors but are difficult to bleach in times less than 1 sec and do not afford such high sensitivities and optical contrasts as many phosphor screens.[58, 63]

The practical superiority of cub.-KCl as a scotophor is due in part to the strong *visible* absorption band induced by CR or x rays, although the superiority may be associated also with the fact that KCl is the only alkali halide with a density less than 2 g cm^{-3}. This low density is related to the abnormally large ionic radii of K$^+$ and, especially, Cl$^-$ in the sequences of the other alkali and halogen ions.[58] The relatively large radius of Cl$^-$ (1.81 Å) evidently facilitates removal of the extra electron which becomes a "free" excited electron when cub.-KCl is excited by CR or x rays. Trapping of the excited electrons, presumably in the sites of missing or displaced Cl$^-$ ions (s–Cl$^-$ + E_a → i–Cl$^\circ$ + electron), gives rise to the characteristic induced F-center absorption band which is about 1 ev wide and is peaked near 5560 Å in the case of cub.-KCl.[58,63,66] The maximum of about 10^{18} F centers per cubic centimeter, which may be induced in cub.-KCl at room temperature, is comparable with the maximum number of phosphoresence photons emitted per cubic centimeter of a fully excited phosphorescent material (compare discussion of eqs. 100–102).[63] When the excited scotophor is irradiated with photons whose wavelengths fall within the induced absorption band, the tenebrescence decays, that is, the darkening caused by absorption of light by the F centers decreases (bleaches). This photostimulated decay of tenebrescence is accelerated by increasing the temperature of the scotophor (thermostimulation of tenebresecence), and the decay is found to approximate the power-law relation,

$$C = C_0 t^{-n} \tag{104}$$

where C and C_0 are the tenebrescences (for example, the relative contrasts, in per cent, of darkened/undarkened areas under illumination) at times t and $t = 0$ (cessation of excitation), respectively, and n decreases from about 1.5 to 0.1 with decreasing temperature or intensity of illumination and with increasing C_0.[58,63] When C_0 is made larger than a few per cent, several additional induced absorption bands associated with ionic displacements may appear. Some of these bands are so difficult to bleach that the darkened material becomes practically a permanent pigment instead of reverting to its pristine colorless condition under appropriate irradiation and moderate heating. Strong heating is then required to effect ionic motion and restore the crystal to its original state. A 5 x 5-cm tenebrescent screen comprising 3 mg cm^{-2} of cub.-KCl evaporated onto a very thin substrate of low heat capacity may be *darkened* by a 10-kv 0.5-ma scanning CR beam and then *bleached* (in about 1 sec) by the same or another CR beam carrying about 10 ma. This is an example of the potent localized heating which can be accomplished by high-intensity CR beams.[58,63,607,699]

Ivey has formulated the following empirical equations which correlate the peak wavelengths λ_{pk} of some of the prominent induced absorption bands with the interatomic spacings x_τ (lattice constants in angstroms) of those alkali halides which crystallize as face-centered cubic structures: [67]

$$\text{U band} \quad \lambda_{pk} = 615 x_\tau^{1.10} \text{ Å} \tag{105a}$$

$$\text{F band} \quad \lambda_{pk} = 703 x_\tau^{1.84} \text{ Å} \tag{105b}$$

$$\text{R}_1 \text{ band} \quad \lambda_{pk} = 816 x_\tau^{1.84} \text{ Å} \tag{105c}$$

$$\text{R}_2 \text{ band} \quad \lambda_{pk} = 884 x_\tau^{1.84} \text{ Å} \tag{105d}$$

$$\text{M band} \quad \lambda_{pk} = 1400 x_\tau^{1.56} \text{ Å} \tag{105e}$$

These difficultly eradicable complex absorptions, which are induced in alkali halide scotophors by high-energy particles, often render the alkali halides unsuitable for study as simple phosphors, because the initially simple crystals become very complex, chemically and optically, during excitation.[157] Furthermore, most alkali halides are hygroscopic and deteriorate in humid atmospheres. Many other crystals, such as tetr.-ZrSiO₄, cub.-CaF₂, and cub.-BaTiO₃, also exhibit visible tenebrescence, but none of these scotophors has been found to produce appreciable phosphorescence emission during either thermostimulated or photostimulated bleaching of tenebrescence. The cited scotophors are similar to photostimulable phosphors to the extent that excitation produces an induced absorption (stimulation) band associated with excited trapped electrons (Fig. 26), but most or all of the trapped electrons released in many scotophors evidently make *nonradiative* transitions to the ground-state level, whereas most of the trapped electrons released in efficient t^{-n} decay phosphors make *radiative* transitions to the ground-state level. These different behaviors are probably associated with the fact that the more efficient bleachable scotophors, for example, cub.-KCl, presumably trap electrons in anion defects which are produced during excitation,[66] whereas the more efficient stimulable phosphors, such as cub.-KCl:Ag, hex.-ZnS:Pb:Cu and cub.-Sr(S:Se):SrSO₄:CaF₂:Sm:Eu, trap electrons in the fields of multivalent cation impurities. The accurate determination of the mechanisms which afford electron trapping and determine the relative probabilities of radiative and nonradiative transitions is a basic problem in the studies of luminescence, tenebrescence, and solarization (difficultly reversible tenebrescence) of phosphors, scotophors, and pigments.

Luminescence Efficiency of Phosphors

GENERAL CONSIDERATIONS. It was previously mentioned that the excitation energy absorbed by a phosphor crystal may be used to produce (1) heat with accompanying thermal radiation, (2) luminescence emission, (3) electron emission, and (4) chemical or structural changes. An efficient phosphor converts a large proportion of absorbed excitation energy into luminescence emission, usually with inappreciable energy losses by electron emission and chemical or structural changes. The chief competition is generally between the undesirable transformation of excitation energy into heat and thermal radiation as contrasted with the desirable transformation of excitation energy into luminescence emission (compare Figs. 17 and 18).

In the broad objective sense, luminescence efficiency ε is defined without regard for the location of the luminescence emission spectrum, that is,

$$\varepsilon = 100 \iint N_e E_e dt dE_e / \iint N_0 E_a dt dE_a \quad \text{(in per cent)} \quad (106)$$

where N_e is the number of luminescence photons of energy E_e emitted per unit time, and N_0 is the number of excitant particles of *absorbed* energy E_a per unit time (remembering that charged material particles may sometimes penetrate a phosphor crystal or screen without giving up all of their energy). In practice, detectors are used whose sensitivities to photons of energy E_e may be written $X_d(E_e)$, and so the luminescence efficiency relative to a given detector becomes

$$\mathcal{E} = 100 \iint N_e E_e \cdot X_d(E_e) \cdot dt dE_e / \iint N_0 E_a dt dE_a \qquad (107)$$

where the sensitivity factor $X_d(E_e)$ has a maximum value of 1 for an objective comparison of the integrated output and input energies. If the average human eye is used as the detector, for example, the only detectable photons have energies in the range 1.7 ev $< E_e <$ 3.1 ev, and the gaussian spectral distribution of sensitivity of the photopic eye to photons of wavelength λ (in angstroms) is, according to Ellickson,[343]

$$X_d(\lambda) = \epsilon^{-2.77 \times 10^{-6} (\lambda - 5550)^2} \qquad (108)$$

whereas the corresponding spectral sensitivity of the scotopic eye is

$$X_d(\lambda) = \epsilon^{-3.75 \times 10^{-6} (\lambda - 5100)^2} \qquad (109)$$

(See Fig. 120.) It is not desirable, however, to restrict the discussion of luminescence efficiency to a detector which is as limited in range and variable in sensitivity as the human eye.[58, 381, 382]

For the purposes of the following discussion of luminescence efficiency, assume a single value of E_a (for example, monochromatic primary radiation) and an averaged value of $\bar{E}_e$ (for example, the photon energy corresponding to the peak of a single simple emission line or band). Then eq. 107 may be simplified to

$$\mathcal{E} = 100 \cdot X_d(\bar{E}_e) \cdot \bar{E}_e \int N_e dt / E_a \int N_0 dt \qquad (110)$$

Equation 110 is subject to many interpretations, for example, (1) $\bar{E}_e \int N_e dt$ may be chosen to be (a) the integrated energy of the luminescence photons emitted *within* the phosphor crystals, or (b) the integrated energy of the emitted luminescence photons which appear *without* the phosphor crystals (sometimes restricted to those photons which escape in a given direction or given solid angle); (2) $E_a \int N_0 dt$ may be chosen to be (a) the integrated excitation energy *absorbed* by the phosphor crystals, or (b) the entire *expended* excitation energy, regardless of the proportion which is absorbed by the phosphor crystals (Figs. 20 and 22); and (3) the integration limits may be chosen such that (a) both $\bar{E}_eN_e$ and E_aN_0 are measured only during the time interval of excitation (neglecting subsequent phosphorescence emission), or (b) $\bar{E}_eN_e$ is measured during and after excitation (including phosphorescence emission), whereas E_aN_0, of course, is measured only during the excitation interval. Most luminescence-efficiency measurements are made by measuring $\bar{E}_eN_e$ (escaping from the phosphor) and E_aN_0 (absorbed by the phosphor) during equilibrium between unmodulated excitation and emission, so that residual phosphorescence is canceled out; or by measuring $\bar{E}_eN_e$ and E_aN_0 (defined as before) averaged over the entire period containing a large number of repetitive excitations after cyclic equilibrium has been established; that is, no further growth of luminescence or phosphorescence is observed (such a condition would obtain after about two seconds of the given repetitive excitations and decays for the phosphor shown in Fig. 69). Results of measurements chiefly of this type are reported as **luminescence efficiency** in this book. When the phosphor has low absorption of its luminescence emission, the luminescence efficiency so defined closely approximates the internal luminescence efficiency. When the phosphor has appreciable absorption of its own luminescence emission, the measurement of internal luminescence efficiency should be made with thin single crystals of the phosphor to minimize absorption losses which are intensified in layers of tiny scattering crystals. Most phosphor screens are several average crystal diameters thick, and so luminescence photons which are emitted within the screen are strongly scattered by multiple reflections from the many crystals. This scattering makes an average internally emitted photon traverse an optical path which may be many times the thickness of the screen. Under these circumstances, and particularly when the phosphor crystals are small and the penetration of the excitant is many crystal diameters, even a very small absorptivity of the phosphor for its own emission occasions a large loss in externally evident luminescence output.

For development or engineering purposes, it is sometimes convenient to speak of **phosphorescence efficiency.** For example, while $E_a\int N_0dt$ should still be measured as the integrated absorbed primary energy, $\bar{E}_e\int N_edt$ (escaping from the phosphor) may be chosen to be the *total* phosphorescence light sum (eq. 102),

$$\bar{E}_e\int N_edt = \int_{t=0}^{t=\infty} Ldt \tag{111}$$

where $t = 0$ is the instant of cessation of excitation, or $\bar{E}_e \int N_e dt$ may be chosen to be only a *portion* of the phosphorescence light sum emitted within a finite (useful) observation interval,

$$\bar{E}_e \int N_e dt = \int_{t_1}^{t_2} L dt \qquad (112)$$

Phosphorescence efficiency is generally low, because phosphorescence output is usually low relative to the (here neglected) luminescence output during prolonged excitation. Relatively high phosphorescence efficiency, according to eq. 111, may be obtained by brief low-intensity excitations of previously de-excited phosphors which have many deep traps (see the discussion of the initial part of the growth curve in Fig. 74), or by brief excitations of ϵ^{-at} decay phosphors such that the duration of excitation Δt_{ex} is very short relative to the lifetime τ of the excited state, that is

$$\Delta t_{ex} \ll \tau = a^{-1} \qquad (113)$$

In practice, the **relative efficiencies** of different phosphors are often determined under identical conditions of temperature and excitation. By this method, the quotient of the luminescence **radiances** (rates of emission of energy as radiation) of two phosphors may be used directly to assess their relative efficiencies under the same given conditions. The correlation is not strictly valid, except in those cases where the different phosphors have practically identical crystal sizes, refractive indices, absorption coefficients (for the given excitant, and for their own emissions), screen (layer) thicknesses, and screen textures. It is common practice to determine the relative efficiencies of phosphors deposited in thick layers, by measuring the luminescence radiance from the excited side of the screens. A very thick screen, relative to the penetration of the excitant, assures complete absorption of the unreflected portion of the excitation energy, and the deeper unexcited part of the screen tends to reflect and scatter the luminescence emission back toward the screen side impinged by the excitant. Appreciable differences in the aforementioned crystal sizes, absorption coefficients, and other properties of different phosphors, however, can lead to erroneous conclusions regarding the intrinsic efficiencies of individual phosphor crystals. For these reasons, and because the accuracy of radiometric and photometric measurements is often poor (particularly at low radiances), data on the intrinsic luminescence efficiencies of phosphors are rarely accurate within better than ± 20 per cent. Also, all the present useful detectors have nonuniform sensitivities, as a function of photon energy or wavelength, and are usually quite selective in responding to only a part of the 1- to 10-ev range of (conventional) luminescence emissions (see Fig. 120). Hence, $X_d(\overline{E}_e)$ is less than one, in practice, and an actual measurement affords an **effective luminescence efficiency** (or effective phosphorescence efficiency) relative to the given detector. Effective luminescence efficiency values are often of direct practical importance, but they may be translated into objective luminescence efficiency only when the spectral distribution of sensitivity of the detector and the spectral distribution of the entire luminescence emission of the phosphor have been accurately determined.

SURVEY OF LUMINESCENCE EFFICIENCY AS A FUNCTION OF THE TYPE AND ENERGY OF THE PRIMARY EXCITANT PARTICLE. Although it is difficult to make an accurate comparison of the luminescence efficiencies of various phosphors excited by different types and energies of primary particles, Fig. 89 gives an approximate comparison of the relative effects of various excitants on phosphors in general.[34] It is assumed that the most efficient phosphors are used for each excitant and excitant energy, and that the intensity of excitation is well below the point of saturation of luminescence observed near room temperature. Variations in efficiency, caused by the discontinuous energy levels of solids, are omitted for simplicity (see Figs. 21, 23, and 143).

Excitation by Photons. As shown in Fig. 89, the luminescence efficiency $\mathcal{E}$ of phosphors excited by photons varies initially as $h\nu_e/h\nu_a$, where $h\nu_e$ is the energy of the emitted luminescence photon (fixed by the phosphor), assuming monochromatic emission for simplicity, and $h\nu_a$ is the energy of the absorbed primary photon, where $h\nu_a \geq h\nu_e$ by Stokes' law. Maximum luminescence efficiency is obtained when $\nu_e = \nu_a$; that is, resonance radiation obtains, and every absorbed photon produces an emitted luminescence photon; in other words, 100 per cent quantum efficiency obtains. Almost 100 per cent quantum efficiency is obtained in, for example, "fluorescent" lamps whose phosphor coatings are excited to visible emission by 2537-Å photons from an electric discharge in low-pressure mercury vapor.[383-390] Assuming $h\nu_e \approx 2$ ev, photoluminescence efficiency decreases as ν_a increases until, theoretically, the efficiency should be 0.2 per cent or less for $h\nu_a$ greater than 10^3 ev. Actually, roentgenoluminescence efficiencies of 0.5 to 20 per cent have been reported for phosphors excited by x-ray photons with energies of the order of 10^5 ev.[75, 102, 391-397] In practice, then, the course of the photoluminescence-efficiency curve is dominated by the ratio ν_e/ν_a, with increasing ν_a, until some point x (Fig. 89) beyond which **secondary excitants** become increasingly important. The secondary excitants in this case are chiefly excited "free" electrons, although excitons may play a part in some cases, and so roentgenoluminescence is a type of cathodoluminescence where the cathode rays are generated within the phosphor crystals by the primary x-ray photons. Roentgenoluminescence differs from photoluminescence in that the number of emitted luminescence photons exceeds the number of absorbed effective primary x-ray photons. Each absorbed x-ray photon produces only *one* "free"

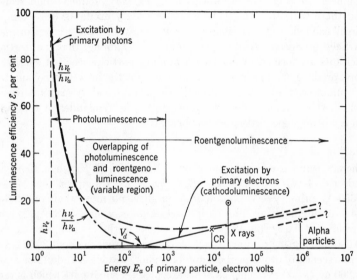

FIG. 89.   Generalized curves of luminescence efficiency of the better phosphors (emitting photons of energy $h\nu_e$) excited by a range of energies of different primary particles. In general, there is considerable structure in the photoluminescence curve (see Fig. 23).

internal excited electron, but this electron gives up its energy bitwise and may produce many internal secondary electrons.[34] The intermediate multiplication of photons in roentgenoluminescence, therefore, is accomplished by the bitwise conversion of secondary-excitant electron energy into luminescence photons. Excitation of luminescence by very high-energy gamma rays leads to roentgenoluminescence after the primary gamma-ray photon has lost sufficient energy by Compton scattering to become, in effect, a primary x-ray photon.

On the low-energy side of the point x in Fig. 89, that is, when $\nu_e \approx \nu_a$, an absorbed primary photon produces *one* luminescence photon directly, or through the medium of one secondary-excitant energy "bit." On the high-energy side of the point x, that is, when $\nu_e \ll \nu_a$, an absorbed primary photon of energy E_a may produce *one or more* secondary-excitant energy "bits" with different energies E^*; that is,

$$E_a = E^*{}_1 + E^*{}_2 + E^*{}_3 + \cdots + E^*{}_N \qquad (114)$$

where $N = 1$ for $\nu_e \approx \nu_a$, and $N > 1$ for $\nu_e \ll \nu_a$. Secondary-excitant energy "bits" with $E^* < h\nu_e$ are incapable of producing the given luminescence emission, those with $E^* \geq h\nu_e$ *may* produce luminescence, and those with $E^* \gg h\nu_e$ may produce **tertiary excitants** (for example, "free" excited electrons, photons, excitons), just as the original primary particle produced secondary excitants. Assuming that no tertiary excitants are formed, N_0 absorbed primary particles, each of energy E_a may produce N_e luminescence photons directly or indirectly, and/or $N_{e'}$ luminescence photons indirectly through secondary-excitant energy "bits" which numerically exceed N_0, with the remainder of the input energy being expended largely as heat; that is

$$N_0 E_a = (N_e + N_{e'})h\nu_e + E_{\text{heat}} \qquad (115)$$

where, assuming efficient phosphors, $(N_e + N_{e'}) \leq N_0$ when $E_a \approx h\nu_e$, and $(N_e + N_{e'}) > N_0$ when $E_a \gg h\nu_e$. As examples, one 3.4-ev (3650-Å) primary UV photon may produce no more than one 2.8-ev (4430-Å) luminescence photon in a crystal of, say, hex.-ZnS:Ag, leaving a residue of 0.6 ev (≈ 18 per cent) as heat in the crystal, whereas one 90,000-ev (1.38-Å) primary x-ray photon may produce about 3000 2.8-ev luminescence photons in the same phosphor, leaving a residue of about 81,600 ev (≈ 90 per cent) as heat, except for a small part which is radiated as x-ray fluorescence [391-397] or dissipated in producing photoelectrons or chemical changes.

There are very few reliable determinations of luminescence efficiency for primary-photon energies greater than about 5 ev. The inception of the point x in Fig. 89 and the course of the curve beyond x have not been determined quantitatively for even one phosphor. Different phosphors may have quite different curves of luminescence efficiency as a function of the energy of the primary photon. For example, tetr.-$CaWO_4$:Pb is well excited by all primary photons with energies from about 4 ev to over 10^6 ev, whereas hex.-$ZnS \cdot CdS$:Ag is well excited by ≈ 4-ev photons and 10^5-ev x-ray photons but is scarcely excited in the intermediate region of primary photons with energies of the order of 10^2 ev. This is a fertile field for phosphor research, if the difficulties which were discussed at the beginning of the section on excitation spectra can be overcome.

Excitation by Charged Material Particles. It may be seen from Fig. 89 that the curve of luminescence efficiency ε, versus E_a for phosphors excited by charged material particles varies in a strikingly different manner from that obtained with excitation by photons. As indicated in Fig. 89, *cathodoluminescence* efficiency starts at a vanishingly low value for $E_a \approx h\nu_e$ and then rises very slowly to a point V_d (the extrapolated "dead voltage" [72,145,324]) beyond which the maximum observed ε increases rapidly up to about 7 per cent for E_a of the order of 10^4 ev. Considerably higher efficiencies may be possible at still higher primary voltages, but accurate results are lacking. The low ε for $E_a < V_d$ is attributed to (1) the complete absorption of the low-energy primary electrons in the relatively distorted, imperfect, and inefficient surface layers of the phosphor crystals (see the penetration curves in Fig. 23), and (2) the 2 to 10-ev energy loss per ejected secondary electron (an average of at least one secondary electron per primary electron must be ejected from the phosphor screen, or the screen will become negatively charged and electrostatically repel further primary electrons). That the surface layers of crystals differ from their volumes is indicated by electron-diffraction measurements on crystals of zinc oxide and several alkali halides whose surface layers, and perhaps the volumes, apparently become less dense (lattice spacings increase) as the size of the crystal decreases.[398] This is a natural consequence of the long-range nature, or additivity, of the interatomic attraction energies in crystals; that is, the surface atoms on large crystals are bound more firmly (packed more densely) than they are on smaller crystals.

At sufficiently high energies of the primary electrons, $E_a > V_d$, the more regular and efficient interiors of the phosphor crystals are penetrated, and, at these high primary-particle energies, secondary excitants are produced as expressed by eq. 114. At equilibrium, and again neglecting tertiary excitants, N_0 primary electrons, each of absorbed energy E_a, may produce N_e luminescence photons directly, produce $N_{e'}$ luminescence photons indirectly through secondary excitants, eject N_0 secondary electrons to maintain electrical equilibrium, and dissipate a major portion of the input energy as heat; that is,

$$N_0 E_a = (N_e + N_{e'})h\nu_e + N_0 \bar{E}_s + E_{\text{heat}} \tag{116}$$

where $\bar{E}_s$ is the average energy of the ejected secondary electrons, which is of the order of 5 ev. The electron-emission term $N_0 \bar{E}_s$ was not included in eq. 115 because an isolated crystal which emitted photoelectrons would soon become charged sufficiently positive to prevent any more electrons from escaping. Some t^{-n} decay phosphors emit photoelectrons during excitation (not during phosphorescence), but the photoelectric effect from cub.-CaS:[flux]:Bi, for example, is reportedly somewhat less than that obtained from a freshly polished surface of magnesium metal (that is, over 1000 primary photons per emitted photoelectron), and the insulator quality of phosphor crystals greatly limits their steady-state photoelectric-emission currents even when they are in electric contact with a metallic cathode.[16,145] In eq. 116, the term $N_0 \bar{E}_s$ constitutes an important energy loss only at low E_a when $E_a \approx E_s \approx h\nu_e$.

Returning to our previous example of hex.-ZnS:Ag, for primary electrons with $E_a = 5$ ev, very few of the absorbed electrons will produce the given 2.8-ev luminescence photons, and each primary electron which does produce a luminescence photon will produce but one. Most of the absorbed energy is expended in (1) ejecting secondary electrons, and (2) increasing thermal agitation in the crystallographically imperfect surface layers of the crystal. Transfer of excitation energy, by secondary excitants moving from the random points of primary absorption in the host crystal to the relatively few luminescence centers, is subject to great attenuation (energy degradation) in the distorted surface layers. For primary electrons with $E_a = 10,000$ ev, each primary electron may produce, on the average, as many as 300 2.8-ev luminescence photons. This leaves a residue of about 9,160 ev (≈ 92 per cent) of which less than 1 per cent is expended in ejecting secondary electrons and producing soft x rays, and the remaining 91+ per cent is consumed chiefly in heating the crystal. Of the many measurements which have been made of cathodoluminescence efficiency, with E_a ranging from 5 to over 10^5 ev, none has given a reliable value higher than 25 per cent, although higher efficiencies are theoretically possible.[755]

IONOLUMINESCENCE. The luminescence efficiencies *reported* for excitation by alpha particles ($E_a \approx 10^6$ ev) range up to 100 per cent,[102,305,399] but a critical evaluation of the experimental data indicates that the actual efficiencies are probably less than about 20 per cent.[34,400] At 10 per cent luminescence efficiency, one 10^6-ev alpha particle may produce 35,000 2.8-ev luminescence photons in a 30-micron crystal of hex.-ZnS:Ag, leaving a residue of 9×10^5 ev (≈ 90 per cent) to be expended in producing x rays, electron emission, chemical changes, and chiefly heat. Ionoluminescence efficiency should vary with the energy E_a of the primary ion in much the same manner as that shown for cathodoluminescence efficiency, except that (1) the minimum E_a (for example, ev$_d$) for appreciable ionoluminescence from phosphors should be considerably larger than that for cathodoluminescence from the same phosphors, since the larger ions penetrate less than electrons with the same primary energy (see Fig. 23), and (2) the energy losses (per unit distance) of primary ions differ from those of electrons penetrating solids, because ions have larger masses and may have larger charges (note the effect of changing m_p and z in eq. 78). At high energies of primary ions, $E_a \gg ev_d$, their relatively small penetrations are an advantage for efficient excitation of present fine-crystal phosphor screens. The small penetration minimizes optical scattering and resultant absorption losses of luminescence photons emitted near the penetration "limit" of the primary particle. When the penetration "limit" of a primary particle is much greater than the average crystal diameter, improved luminescence efficiency may be obtained by (1) using oblique (for example, grazing) incidence rather than normal incidence of the primary particle on the phosphor screen, or (2) passing the primary particle repeatedly through the edge-supported screen until the energy of the particle is completely absorbed.[401]. When it is possible to obtain and use phosphors in single crystals whose thicknesses exceed the penetration "limit" of the primary particle, the intensified absorption loss due to scattering is, of course, eliminated.

THE NATURE OF THE SECONDARY EXCITANTS. A further point of interest in Fig. 89, is the substantially similar maximum $\mathcal{E}$ values (5 to 20 per cent) of efficient crystalline phosphors excited by high-energy photons, electrons, or ions. The similarity leads to the conclusion that a common secondary excitant is active in all these cases. In the case of a primary x-ray photon, the primary particle gives up all of its energy in one absorption act to produce a high-energy internal (photo-) electron which then excites cathodoluminescence. A high-energy primary electron, or internal photoelectron, or primary ion apparently gives up its free energy in random-sized "bits" (averaging about 10 to 30 ev) chiefly to the larger population of host-crystal atoms. Energy transport from the abundant host-crystal atoms to the sparse luminescence centers may be accomplished by several conceivable secondary or tertiary excitants, such as "free" internal electrons, photons, excitons, and resonance transfers. Although it is not known with certainty which of these possible energy-transfer agents is most influential in the luminescence mechanisms of phosphors in general, most of the evidence favors the "free" internal electrons. Particularly convincing evidence in favor of energy transfer by electrons rather than by excitons is given by Haynes' experiments on the print-out effect in crystals of silver chloride excited by microsecond pulses of light and subjected to microsecond applications of a high electric field.[221, 667] The mobility of the resultant darkening, at about 8 cm sec^{-1} per unit gradient of 1 volt cm^{-1}, in a direction opposite to that of the applied electric field shows that the energy transport is by "free" electrons rather than by neutral excitons. Further support for the importance of electrons as energy-transfer agents is obtained from photographs by Coolidge and Moore of the luminescence scintillations produced by bombardment of calcite crystals with 900-kv electrons. These photographs show dotted lines which the writer has interpreted as evidence that secondary internal electrons act as the energy transfer agents.[34] More quantitative measurements of concomitant luminescence, dielectric changes, magnetic changes, and electric conductions (including Hall effect and thermoelectric effect) should be made with different phosphor crystals excited by high-energy photons and charged material particles to determine whether the charged "free" electrons (or positive holes) are generally the primary energy-transfer agents, or whether neutral photons, excitons, or ψ–overlap transfers may sometimes be important secondary excitants.

EFFECT OF CRYSTAL STRUCTURE ON EFFICIENCY. It may be emphasized at this point that phosphor crystals must be highly crystalline to exhibit high luminescence efficiency under excitation by charged material particles or high-energy photons. Some vitreous or distorted luminescent materials, such as uranium glass or cub.-$SrS:SrSO_4:CaF_2:Sm:Eu$, exhibit high luminescence efficiency under excitation by low-energy near-UV photons ($E_a \approx h\nu_e$) which may be absorbed preferentially by the luminescence centers, but none exhibit high luminescence efficiency when excited by high-energy particles ($E_a \gg h\nu_e$) which give up energy rather indiscriminately to all the atoms in the material. In the latter case of high-energy primary particles, a high degree of crystallinity is necessary to allow efficient energy transfer between the random points of primary absorption and the luminescence-active centers. Certain vitreous luminescent materials, such as boric-acid, borate, silicate, germanate, or phosphate glasses with or without activators, such as organic dyes, U, Eu, *et al.*,[65, 74-84, 144, 402-411] may be well excited by low-energy UV photons because most of the primary photons are absorbed directly by the luminescent centers, thereby eliminating the energy-transfer process. Some vitreous materials containing embedded phosphor crystals of the same refractive index as the "host glass" may appear to be entirely vitreous, but actually exhibit the properties of both crystalline and glassy luminescent materials with respect to excitation by low- and high-energy primary particles.

PHOSPHORESCENCE EFFICIENCY. *Phosphors with Temperature-independent ϵ^{-at} Decays.* The phosphorescence efficiencies of phosphors which decay according to eq. 90 are directly related to their luminescence efficiencies, because the decay constant a changes little with temperature or with the type or intensity of excitation. Hence, to obtain maximum L at a given time t after cessation of excitation, one seeks to decrease a and increase L_0. The decay constant a is a function of the phosphor composition and structure, and the initial radiance L_0 is a function of $\mathcal{E}$ and the type, intensity, and conditions of excitation.

Phosphors with t^{-n} Decays. The phosphorescence efficiencies of phosphors which decay according to eqs. 93–100 (with added complications due to different L_0's at different depths of excitation and to retrapping) may vary independently of the luminescence efficiencies, because the trapping and emission processes are substantially independent of each other in the phosphor crystals. In an effort to obtain maximum L at a given time t after cessation of excitation, one might increase L_0, for example, but it has been found that increasing L_0 increases the rate of decay, especially when the excitant is changed from low-energy to high-energy particles (Figs. 70, 71, 73, 75, and 81). Although the problem is complex, it is essentially a matter of maximizing the number of traps of thermal depth large relative to kT, then exciting the phosphor to fill all the available traps, and finally seeking to thermostimulate or photostimulate the phosphor so that a maximum number of traps empty (without quenching) at the given time t.

In some cases where a well-crystallized homogeneous phosphor, such as hex.-ZnS:Cu (principal thermal trap depth ≈ 0.5 ev [221]), has traps which are moderately deep, long intense phosphorescence is observed after excitation by UV, shorter phosphorescence is observed after excitation by CR (Fig. 71), and the phosphorescence after excitation by alpha particles is almost undetectable after about 10^{-6} sec. It is reasonable to assume that the larger heat residue per primary particle and the higher power input per unit excited volume largely account for the decreased trapping and increased quenching on going from UV to CR to alpha particles as primary excitants. In the cases of some heterogeneous deep-trap phosphors, such as cub.-CaS:Na$_2$SO$_4$:CaF$_2$:Bi and cub.-Sr(S:Se):SrSO$_4$:CaF$_2$:Sm:Eu, phosphorescence efficiencies higher than 50 per cent may be obtained after very brief excitations by low-energy near-UV photons. The deep traps of the latter phosphor, at least, are readily filled during excitation by high-energy beta particles and alpha particles, although the actual efficiencies are probably quite low.[221, 343] Here, the traps are apparently deep enough so that at room temperature and at low excitation intensities the localized heat residues from excitation by high-energy particles eject excited electrons chiefly from the medium-deep traps which make the predominant contribution during phosphorescence at room temperature. When the temperature of the entire phosphor crystal is raised, then more excited electrons are ejected from deeper traps because the localized heat residue after excitation adds to the average temperature of the crystal.

EFFECT OF PHOSPHOR COMPOSITION AND STRUCTURE ON LUMINES-
CENCE EFFICIENCY DURING EXCITATION. The following comments on
the effects of phosphor composition and crystal structure on luminescence efficiency $\mathcal{E}$ apply for phosphors *excited at room temperature*
(compare Fig. 18) *and at excitation densities well below saturation.* Con-

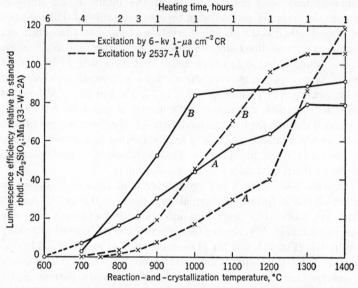

FIG. 90. Variation in the relative luminescence efficiency of several rbhdl.-Zn_2SiO_4:
Mn(0.3) phosphors prepared from (*A*) previously unheated ingredients, and (*B*)
rbhdl.-Zn_2SiO_4 (previously crystallized at 1200°C, 90 min) mixed with $MnCO_3$
[made by carbonate precipitation from $Mn(NO_3)_2$]. The same reference-standard
phosphor was used for all the measurements. (Reflection colors of the *B*-series phosphors under white light are: 100°C, brown–black; 700–800°C, very light tan; 900°C,
lighter tan; 1000°C, cream–white; 1100–1400°C, white.)

siderable data on this subject have already been given in Table 5 and
in many figures in this chapter, particularly for the (Zn:Cd)(S:Se)
system of phosphors (Figs. 32–49).[221] It should be emphasized, again,
that specific values of $\mathcal{E}$ cited for phosphors identified by conventional
shorthand formulas, for example, hex.-ZnS:Ag and rbhdl.-Zn_2SiO_4:Mn,
may be accurate for the *particular sample* which was measured under
the given conditions, but are not to be construed as fixed values which
obtain for all samples designated by the simplified formulas. This will
become more apparent in the following discussion which uses as examples
the zinc-silicate:Mn type of phosphors whose genesis and properties
were described in part in connection with Figs. 50–56 and 61–62.

Silicate Phosphors. Figures 90 and 92 show curves of relative cathodoluminescence and photoluminescence efficiencies versus crystallization temperature for the same samples of phosphors whose spectra are shown in Figs. 61 and 62, with the addition of a B series of rbhdl.-Zn_2SiO_4:Mn phosphors prepared by heating a mixture of $MnCO_3$ and precrystallized rbhdl.-$2ZnO \cdot 1.012SiO_2$, 1200°C, at the different tem-

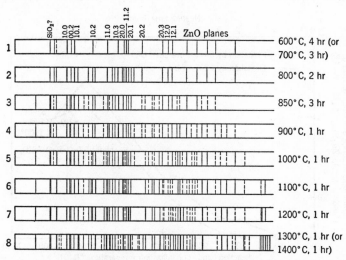

Fig. 91. X-ray diffraction patterns of the phosphor samples in the *A* series in Fig. 90, showing the disappearance of the hex.-ZnO pattern and appearance of the rbhdl.-$Zn_2SiO_4 \cdot 0.012MnSiO_3$ pattern with increasing temperature of reaction and crystallization. Copper-$K\alpha$ radiation, nickel filter (dotted lines indicate very weak lines on the original negatives); compare Fig. 61.

peratures and times indicated. The measurements were made on the excited sides of thick layers. All the phosphors in the B series had the same emission band as curve 4 in Fig. 61, that is, there was no visible evidence of any yellower emissions under either CR or 2537-Å UV, and this was substantiated by spectroradiometric measurements. Figure 90 shows that appreciable cathodoluminescence efficiency CR ε appears at about 700°C, whereas appreciable photoluminescence efficiency UV ε does not appear until about 800°C, regardless whether the phosphors are (1) synthesized from the previously unheated basic ingredients, or (2) made by diffusing MnO into the previously crystallized rbhdl.-Zn_2SiO_4 host crystal. It is noticeable that the ε values for the B series increase at a faster rate than for the A series (except for the crossover of low CR ε values near 730°C), and the CR ε reaches its maximum value much sooner than the UV ε in both series.

Comparison of the x-ray diffraction patterns of the A-series phosphors in Fig. 91 with their ε values plotted in Fig. 90 shows that complete reaction and crystallization is not attained until the given ingredients have been heated to over 1300°C for longer than an hour. The x-ray diffraction pattern of the 1000°C B-series phosphor was apparently

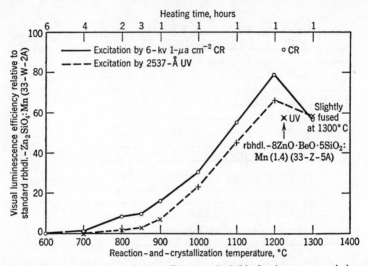

FIG. 92. Variation in the relative efficiency of visible luminescence emission of rbhdl.-$Zn_8BeSi_5O_{19}$:Mn(1.4) prepared by heating a mixture of $ZnCO_3$, $BeCO_3$, $MnCO_3$ (precipitated from nitrate solution) and finely divided SiO_2 (compare Fig. 62).

intermediate between those of the 1100°C and 1200°C A-series phosphors. The effect of Mn is to expand slightly the structure of rbhdl.-Zn_2SiO_4, because Mn is larger than Zn. In Fig. 92, which shows similar data for rbhdl.-$Zn_8BeSi_5O_{19}$:Mn prepared only from previously unheated basic ingredients, the CR ε and UV ε curves run substantially parallel with the same prior appearance of appreciable CR ε as in Fig. 90. These figures clearly indicate the widely different ε values which a phosphor symbolized by, for example, rbhdl.-Zn_2SiO_4:Mn(1) may have when prepared from different ingredients at different temperatures. Further variations may be produced by varying the chemical natures, particle sizes, previous heat treatments, and degrees of admixture of the ingredients.

Figures 93–95, which are drawn with the same CR ε scale as Figs. 52–56, show the decrease in relative cathodoluminescence efficiency which occurs on going from the "parent" phosphor, rbhdl.-Zn_2SiO_4:Mn, to rbhdl.-Zn_2GeO_4:Mn, to β-Zn_2SiO_4:Mn, and then to rbhdl.-$(Zn:Be)_2SiO_4$:Mn (note the low ordinate values in Fig. 55). Figures 55, 95, and 96 show some of the pronounced variations in CR ε and emission spectra which may be produced by varying (1) the acti-

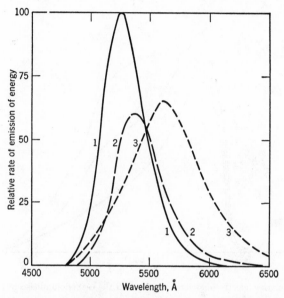

FIG. 93. Relative spectral-distribution curves of the cathodoluminescence emissions (bombarded side) of the low (rbhdl.)- and high (β)-temperature forms of Zn_2SiO_4:Mn and of rbhdl.-Zn_2GeO_4:Mn.

1. rbhdl.-Zn_2SiO_4:Mn(0.3), 1250°C, 1 hr
2. rbhdl.-Zn_2GeO_4:Mn(0.3), 1100°C, 1 hr
3. β-Zn_2SiO_4:Mn(0.3), 1600°C, $\frac{1}{4}$ hr, quenched

vator proportion, (2) the crystallization temperature, (3) the crystallization time, and (4) the host-crystal proportions of rbhdl.-ZnO:BeO:SiO_2:Mn phosphors. The proportion of Mn activator in these phosphors is based on a "normal" proportion N of 0.006 mole Mn per mole of cation(s) (for example, Zn + Be). The "normal" proportion corresponds to about 0.3 weight per cent of Mn in rbhdl.-$Zn_2SiO_4 \cdot 0.012$Mn-(SiO_3), or about 0.5 weight per cent of Mn relative to Zn.[61,72,93] It is generally found that increasing the chemical complexity and structural heterogeneity of phosphors decreases their relative peak outputs (ε values) under excitation by charged material particles, or by high-energy primary particles in general. This is shown by the rapid decrease in the ordinate values of the phosphors in the sequence rbhdl.-$2ZnO \cdot SiO_2$:Mn (= 100 in Fig. 93), rbhdl.-$2ZnO \cdot BeO \cdot 3SiO_2$:Mn (= 7.5 in Fig. 55), and $5ZnO \cdot 3BeO \cdot ZrO_2 \cdot TiO_2 \cdot 6SiO_2$:Mn ($\approx$ 6 in Fig. 56).

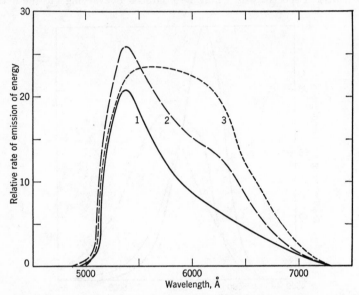

FIG. 94. Relative spectral-distribution curves of the cathodoluminescence emissions (bombarded side) of three phosphors formed by the reaction and crystallization of $9ZnO \cdot BeO \cdot 6SiO_2 \cdot 0.48MnO$ at different temperatures for 1 hr each.

1. rbhdl.-$9ZnO \cdot BeO \cdot 6SiO_2 \cdot 0.48MnO$, 1100°C, 1 hr
2. rbhdl.-$9ZnO \cdot BeO \cdot 6SiO_2 \cdot 0.48MnO$, 1200°C, 1 hr
3. rbhdl.-$9ZnO \cdot BeO \cdot 6SiO_2 \cdot 0.48MnO$, 1250°C, 1 hr

DEFICIENCY STRUCTURES. An *apparent* exception to the general decrease in CR ε with increasing complexity of the phosphor is found in the case of rbhdl.-ZnO:SiO_2:Mn phosphors prepared with an excess of silica over orthoproportions, that is, where the ratio of ZnO/SiO_2 is less than 2/1. It has been determined that (1) the optimum proportion of Mn for maximum CR ε from these phosphors increases as the ratio ZnO/SiO_2 is decreased below 2/1 (Fig. 97), and (2) the crystal structure of the resultant phosphors remains the same ($R\bar{3}$) even when the ZnO/SiO_2 ratio is 1/1, that is, when there is a 100 per cent excess of SiO_2 over the stoichiometric-orthoproportion compound (Figs. 11a and 54).[72] Evidently, the orthosilicate ($R\bar{3}$) structure continues to form even when a large excess of SiO_2 is present. Therefore, the orthosilicate structure with excess SiO_2 may be more appropriately described as being deficient in ZnO; that is, the crystal has a **deficiency structure** wherein

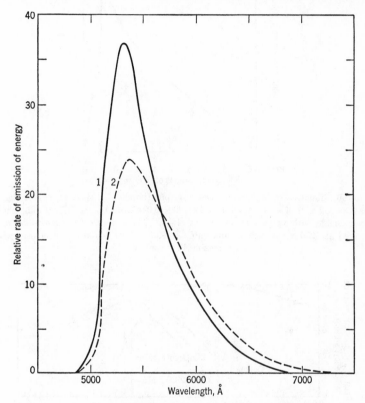

Fig. 95. Relative spectral-distribution curves of the cathodoluminescence emissions (bombarded side) of two phosphors formed by the reaction and crystallization of $9ZnO \cdot BeO \cdot 6SiO_2 \cdot 0.24MnO$ at 1200°C for 1 hr and for 6 hr.

1. rbhdl.-$9ZnO \cdot BeO \cdot 6SiO_2 \cdot 0.24MnO$, 1200°C, 1 hr
2. rbhdl.-$9ZnO \cdot BeO \cdot 6SiO_2 \cdot 0.24MnO$, 1200°C, 6 hr

more MnO units may profitably occupy the vacant ZnO sites.[72] The rbhdl.-$(Zn:Be)_2SiO_4:Mn$ and rhomb.-$CdSiO_3:Mn$ phosphors are further examples of luminescent solids which are efficient when made with deficiency structures (excess SiO_2) and Mn proportions up to about 20 times the "normal" proportion of 0.006 mole Mn per mole of cation(s). These deficiency-structure phosphors, including rbhdl.-$ZnO \cdot SiO_2:[Si]$, often have pronounced t^{-n} decay "tails" on their predominantly ϵ^{-at}

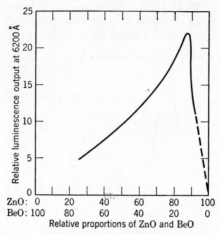

FIG. 96. Relative cathodoluminescence output (bombarded side) at 6200 Å (band pass = 230 Å [93]) of some rbhdl.-$(Zn:Be)_2SiO_4 \cdot 0.044MnSiO_3$, 1225 °C, 90 min, phosphors made with various ratios of ZnO/BeO. The output of the emission band peaked at 5250 Å has a maximum value of 100 at ZnO = 100 and decreases with increasing BeO content.

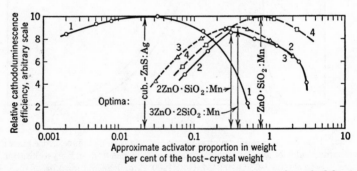

FIG. 97. Semilog plot of the influence of activator proportion on the cathodoluminescence efficiencies of (1) cub.-ZnS:Ag, 940 °C, 2 hr (see Fig. 44 for emission bands), (2–4) rbhdl.-$ZnO:SiO_2:Mn$, 1200 °C, 1 hr (emission band remains unchanged for these phosphors made with different ratios of ZnO/SiO_2). Measurements on the bombarded side of very thick coatings at 6 kv and about 2 μa cm^{-2} (unmodulated stationary CR beam). Curves 2, 3, and 4 are drawn to the same scale, and the maximum energy efficiency of 1 is about 4 times that of 2 under the given conditions.

1. cub.-ZnS:Ag[NO₃]:[NaCl(2)], 940 °C
2. rbhdl.-2ZnO·SiO₂:MnSiO₃, 1200 °C
3. rbhdl.-3ZnO·2SiO₂:MnO, 1200 °C
4. rbhdl.-ZnO·SiO₂:MnO, 1200 °C

decay phosphorescences. This indicates that, in addition to local trapping (Fig. 16c), some excited electrons may become trapped in vacant oxygen-ion sites, much as electrons are trapped in F centers in the alkali halide scotophors. (There is some evidence that many of the shallower traps in ZnS-type phosphors are also anion vacancies in the host crystal.)

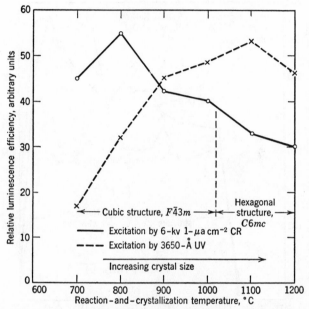

FIG. 98. Variation in the relative luminescence efficiency of ZnS:Mn(2.5), [NaCl(2)] heated to the indicated temperatures and held for about 30 min before cooling. The Mn was added as $MnCl_2$.

With respect to terminology, the vacant ZnO sites in a deficiency-structure rbhdl.-$ZnO \cdot SiO_2$:Mn phosphor are *abnormal* omission defects which should not be confused with the *normal* alternately vacant Zn (and S) sites in cub.-ZnS (Fig. 12). A Mn atom which occupies an *abnormally vacant* Zn site in ZnO-deficient rbhdl.-$ZnO \cdot SiO_2$:Mn forms an **s** center, whereas a Mn atom which occupies a *normally vacant* Zn site in cub.-ZnS (or hex.-ZnS) forms an **i** center (an Mn atom which locates in a *normally occupied* Zn site in cub.-ZnS, or hex.-ZnS, or rbhdl.-$ZnO \cdot SiO_2$:Mn forms an **s** center. It may be recalled that the optimum concentration of phosphor activators in structurally compatible **s** centers is usually about 100 times the optimum concentration of phosphor activators in structurally incompatible **i** centers (Fig. 97). Assuming a random distribution of the activator atoms in the host crystals, calculations presented in Appendix 2 show that these optimum proportions of activators are generally so low that over 95 per cent of the activator atoms are structurally isolated from each other in the sense that they do not occupy adjacent available sites (for example, the Zn sites may be

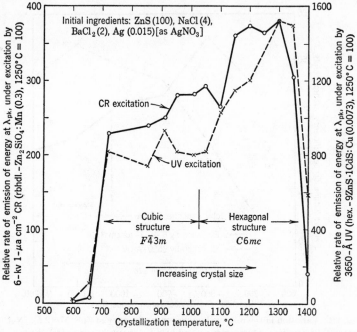

Fig. 99. The effect of increasing temperature of crystallization on the cathodoluminescence and photoluminescence efficiencies of ZnS:Ag(0.015), [NaCl(4)], [BaCl$_2$(2)]. Note the increase in efficiency on going from the $F\bar{4}3m$ (cubic) to the $C6mc$ (hexagonal) structure (compare with Fig. 49).

considered as being available sites for Mn atoms in rbhdl.-Zn$_2$SiO$_4$:Mn). In the case of ZnS-type phosphors with an optimum of about 0.01 per cent of Ag or Cu activator, about 99.9 per cent of the activator atoms are nonadjacent, and so one is led to conclude that these presumably interstitial activator atoms influence each other appreciably when separated by several atomic spacings (note, in Figs. 45b–c, the large changes wrought by increasing the Cu proportion above 0.01 weight per cent).

SULPHIDE PHOSPHORS. Figures 98 and 99 show curves of relative CR ε and UV ε values for some S-dominated phosphors, comprising cubic and hexagonal ZnS host crystals with Mn and Ag activators, as a function of crystallization temperature. Figure 98 shows that maximum CR ε is obtained with the low-temperature cubic form of ZnS:Mn, whereas maximum UV ε is obtained with the high-temperature hexagonal form. An added feature of this phosphor system is that optimum UV ε is obtained with a halide flux, as shown in Fig. 98, whereas optimum CR ε is obtained with an oxygen-containing flux, such as a nitrate or sulphate. Figure 99 shows some relative-efficiency data for the same ZnS:Ag phosphors whose emission-band shifts were presented in Fig. 49. Comparison of the two figures shows that the hexagonal-structure phosphors have higher CR ε and UV ε and shorter-wavelength emission bands, than the cubic-structure phosphors. In the case of hex.-ZnS· CdS:Ag(0.01), whose crystal structure remains hexagonal from 700 to over 1250°C, the peak-output values remain practically unchanged for a series of identically compounded phosphors crystallized in the given temperature range. Similarly, when the structure of cub.-ZnS:Ag is converted to hex.-(Zn:Cd)S:Ag by increasing substitution of Cd for Zn in the host crystal (Fig. 33), the peak-output values decrease, then rise, and then decrease steadily,[312] whereas the peak-output values decrease steadily when the structure remains hexagonal throughout (Fig. 36). It should be noted that the foregoing data on luminescence efficiency are for phosphors which have not been subjected to deleterious chemical or mechanical (grinding) treatment after crystallization. Such treatment may greatly alter the efficiencies of phosphors (see Figs. 114 and 115).[58, 72]

EFFECTS OF DETRIMENTAL IMPURITIES ON EFFICIENCY. As might be expected, the sensitivities of phosphors to injurious impurities (**"poisons,"** or **"killers"**) vary inversely with their optimum activator proportions; that is, an s-center phosphor can usually tolerate about 100 times as much of a given detrimental impurity as an i-center phosphor. This is exemplified in Fig. 100 [58, 416] which shows that a cub.-ZnS: Ag(0.003) phosphor, with about 6×10^{17} [i-center] Ag atoms cm^{-3} is approximately 100 times as sensitive to traces of cocrystallized Ni (added as a compound) as a rbhdl.-ZnO·SiO$_2$:Mn(2) phosphor, with about 8×10^{20} s-center Mn atoms cm^{-3}. From the data in Fig. 100 it may be calculated that 4×10^{18} Ni atoms cm^{-3} (10^{-2} per cent Ni) almost completely suppress the luminescence emission due to the 6×10^{17} [i-center] Ag atoms cm^{-3} in cub.-ZnS:Ag(0.003), whereas the same proportion (4×10^{18} Ni atoms cm^{-3}) of Ni decreases the luminescence efficiency of 8×10^{20} s-center Mn atoms in rbhdl.-ZnO·SiO$_2$: Mn(2) by only 25 per cent. The detrimental effects of poisons are usually more pronounced at low excitation densities, particularly with t^{-n} decay phosphors,[221] although Kröger reports for an ϵ^{-at} decay rbhdl.-Zn_2SiO_4: Mn phosphor that T_{50} (the temperature at which the luminescence efficiency has fallen to 50 per cent of the maximum) is displaced from about 370°C for the material made without iron (Fe) impurity to 270°C for 10^{-3} mole per cent Fe and to 130°C for 10^{-2} mole per cent Fe.[154]

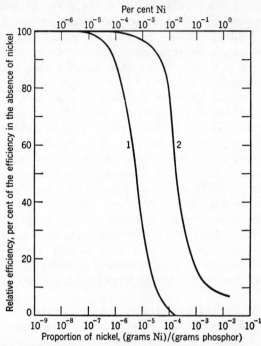

FIG. 100. Poisoning effect of small proportions of cocrystallized nickel compounds incorporated in two representative phosphors. (D. R. Hale and J. W. Marden)

1. cub.-ZnS:[Zn], or cub.-ZnS:Ag(0.003) excited by CR (Hale)
2. rbhdl.-ZnO·SiO₂:Mn(2) excited by 2537-Å UV (Marden and Meister)

Klasens reports that 2×10^{-4} weight per cent of Ni decreases the temperature break point T_B of ZnS:Ag(0.01) by about 50°C, and 10^{-3} weight per cent of Ni decreases the T_B of the same phosphor by about 100°C.[233]   (It may be noted that Nail, Pearlman, and Urbach report that the T_B of a ZnS:Cu(0.01) phosphor, without Ni impurity, is decreased by decreasing the intensity of 3650-Å UV excitation.[221])   When a phosphor such as cub.-ZnS:Ag is operated at temperatures near T_B, therefore, the effect of poisoning impurities is particularly pronounced (Figs. 100 and 103).

The decreased efficiency occasioned by Ni impurity is more pronounced for phosphorescence than for luminescence during excitation.

It has been found that as little as 2×10^{-5} per cent of cocrystallized Ni (8×10^{15} Ni atoms cm^{-3}) in hex.-ZnS:Cu(0.003) markedly decreases the intensity of long-duration phosphorescence (at room temperature) without greatly decreasing the luminescence efficiency during excitation.[370, 417] As shown in Fig. 25a, the Ni impurity either (1) decreases the number of *deep* traps, or, more probably (2), provides a selective mechanism whereby electrons in *deep* traps can make radiationless transitions to the ground state. In either case, only the relatively shallow traps are left, and these traps afford rapidly decaying phosphorescence at room temperature. As little as 2×10^{-5} per cent of Ni, therefore, strongly affects electrons in deep traps without greatly affecting electrons in shallow traps, and without affecting the emitting atoms (the shape and location of the emission band are practically unaltered by this small amount of Ni). This indicates that the deep traps are farther from the emitting atoms and are more susceptible to perturbation than the shallow traps. Hence, the low thermostimulated phosphorescence light sums of deep-trap phosphors, compared with the higher photostimulated light sums of these phosphors, appears to be the result of an increasing probability of radiationless transitions of excited electrons from deep traps as these traps are increasingly perturbed by thermal agitation. The effect of small traces of poisoning impurities is usually much larger on long-duration deep-trap phosphorescence emission than on ϵ^{-at} decays or short-duration shallow-trap phosphorescences. [Garlick reports that traces of Co selectively *decrease* the number of potentially useful shallow traps in some ZnS:Cu phosphors,[221] whereas Klasens believes that Co *increases* the number of shallow traps,[233] and E. Podschuss deliberately incorporates traces of Co in hex.-ZnS:Cu phosphors to facilitate production of the hexagonal form [272] and intensify phosphorescence emission at room temperature (Fig. 80b).] Also, the poisoning effect is usually found to be much less pronounced, particularly on ϵ^{-at} decays and very short t^{-n} decays, when the phosphors are excited to high radiances (high L_0) by high-energy particles.

In general, paramagnetic ions, that is, ions which have unfilled inner shells and which strongly absorb visible radiation, are more potent poisons than nonparamagnetic ions. Studies by Marden and Meister on the poisoning effects of small proportions of compounds of several elements incorporated in some O-dominated phosphors, such as rbhdl.-$ZnO \cdot SiO_2$:Mn, $2CdO \cdot SiO_2$:Mn, $CdO \cdot B_2O_3$:Mn, tetr.-$CaWO_4$:[W], and $MgWO_4$:[W] excited by 2537-Å UV, show that paramagnetic Ni, Fe, Cr, and Cu cause a detectable reduction in photoluminescence efficiency when present in proportions as low as 0.001 weight per cent, whereas nonparamagnetic Ag, Al, Pb, Sn, and Zr generally do not cause appreciable reduction in efficiency in proportions less than about 0.1 weight per cent.[416] It is noteworthy that 0.01 weight per cent of Cu impurity, which may be either dia- or paramagnetic (Table 8), acts as (1) an *intensifier-type activator* which enhances the UV + blue emission bands of some silicates (Fig. 31); (2) a *new-band-type activator* which suppresses the original emission band and produces a new emission band in some ZnS-type phosphors (Fig. 45) when the activator proportion is Cu(≤ 0.01), and *intensifies* the original emission band at higher proportions of Cu [even when Mn(1) is incorporated in the phosphor] (Figs. 45b–c); (3) a *trapping agent* in the latter phosphors (Fig. 25); and (4) a *poison* which reduces the photoluminescence efficiency of the previously cited O-dominated phosphors by as much as 50 per cent.[416] This example illustrates the large and vastly different effects which small traces of a given impurity may produce in different host crystals. The example indicates, also, the probability that one species of activator atoms (ions) in a given host crystal, such as Cu^{n+} in hex.-ZnS:Cu, may function as *activators* (promoters of radiative transitions), *traps*, *and poisons* (promoters of radiationless transitions). Whether these different functions are performed chiefly by differently situated Cu atoms in, for example, hex.-ZnS:Cu, or whether a given Cu atom can perform all of the three functions in time sequence, has not been determined.

Another example of multiple functioning of an activator element is found in the incorporation of several weight per cent of TiO_2 in rbhdl.-$(Zn:Be)_2SiO_4$:Mn where the Ti atoms act as (1) *activators* which intensify the UV + blue emission band of the silicate made with or without Mn, (2) *trapping agents* which greatly increase the room-temperature phosphorescence emission associated with the Mn activator, and (3) *poisons* which decrease the luminescence efficiency of the Mn emission band. These data emphasize the versatility and importance of small proportions of impurities in the luminescence of solids and re-emphasize the importance of extraordinary care in purifying, compounding, and crystallizing phosphors.

The mechanism of poisoning, as well as the decrease in efficiency beyond the optimum proportion of activator (Fig. 97), have an analogue in the concentration quenching of organic luminescent materials in solutions and host glasses.[74–84, 340, 367] A clear understanding of the mechanism of efficiency reduction by the proximity of another activator impurity or a foreign poison-type impurity has not yet been obtained. It is possible to speak in general terms of energy degradation (1) by exchange (resonance) between luminescence-active and luminescence-inactive atoms or centers whose excited-state wave functions overlap in a solid, or (2) by radiationless transitions down a "ladder" of closely spaced levels introduced into the forbidden region between the excited and ground-state levels by an impurity atom, or (3) by means of impurity-influenced intersections of excited-state and ground-state curves in diagrams such as the one shown in Fig. 16*b*, but a quantitative solution to the general problem of loss of luminescence efficiency as caused by traces of impurities in solids is still lacking. In the case of some tungstate and molybdate phosphors, Kröger reports that the detrimental action of Cr impurity is primarily to absorb the internally emitted radiation.[154] In tetr.-$CaWO_4$:[W] the absorption is selective, as indicated by the fact that the externally observed emission band shifts its peak from about 4200 to 4700 Å as the Cr content is increased from 0 to 0.1 mole per cent, without displacing the long-wave part of the emission band. The apparent shift in the peak of the emission band of rbhdl.-Zn_2SiO_4:Mn with increasing Mn shown in Fig. 53 may be caused by selective optical absorption by luminescence-inert Mn centers.

To summarize the roles of impurities and imperfections which function as poisons in phosphors, it appears that a poison may reduce luminescence efficiency by:

1. Decreasing the solubility of the activator compound in the host crystal during formation of the phosphor.

2. Absorbing primary excitation energy directly and converting it into heat.

3. Dissipating (as heat) excitation energy which has been absorbed by the host crystal and is intercepted by the poison center while the excitation energy is *en route* to or from activator centers or traps.

4. Interfering with activator centers or traps, such as to increase their probability of making nonradiative transitions, for example, by lowering the crossover point *f* in Fig. 16*b*.

5. Taking excitation energy from nearby activator centers, by ψ-overlap energy transfer, and converting the potential luminescence energy into heat (especially when the poison center has *f* below *c* in Fig. 16*b*).

6. Optically absorbing luminescence radiation which is seeking to escape from within a phosphor crystal or screen.

For any given phosphor, several of these mechanisms may operate to give the observed poisoning effect.

EFFECT OF ACTIVATOR(s) ON EFFICIENCY. *Single Activators.* In general, maximum luminescence efficiency is obtained with an optimum proportion of *one* added activator. Additional added activator impurities tend to act as poisons by (1) usurping sites which could be occupied profitably by the initial activator, or (2) functioning as poisons according to the other mechanisms proposed in the previous section. When only a single simple emission band is enhanced or produced by the added activator, as in the case of rbhdl.-Zn_2SiO_4:Ti or rbhdl.-Zn_2SiO_4:Mn, the optimum proportion of activator may be readily determined by making a series of phosphors with different initial activator proportions and plotting their relative peak outputs as shown in Fig. 97. When the [activator] is not deliberately added, but is produced by chemical action or by structural or atomic dislocations during crystallization, as in hex.-ZnO:[Zn] and tetr.-$CaWO_4$:[W], it is very difficult to estimate and control the optimum proportion of presumed [activator]. When a single added activator produces two (or more) emission bands, the relative efficiencies of the two emission bands may change with variations in the activator proportions and degree of oxidation or reduction during crystallization. As examples:

1. The *line* emission spectrum of $ThSiO_4$:Eu (Fig. 31) is favored by crystallization in a strongly oxidizing atmosphere ($\rightarrow Eu^{3+}$), whereas the emission *band* peaked near 6200 Å is favored by crystallization in a reducing atmosphere ($\rightarrow Eu^{2+}$).

2. In the series $SrSiO_3$:Eu(0.02), $SrSiO_3$:Eu(0.3), and $SrSiO_3$:Eu(1), all crystallized at 1250°C in a normal SiC-element furnace atmosphere, the luminescence efficiency (during excitation) increases with increasing Eu content under excitation by 2537-Å UV (*red emission*) and decreases with increasing Eu content under excitation by 6-kv CR (*green–yellow emission*). None of these phosphors has an afterglow longer than the persistence of vision, whereas the same phosphors heated in hydrogen at lower temperatures have low-intensity afterglows which are detectable for several seconds.[203, 375]

3. The *short-wave* emission bands of hex.-ZnS:Pb, cub.-CaO:Pb, $CaO:SiO_2$:Mn, $MgO:SiO_2$:Mn (Fig. 58), and cub.-RbI:Tl are favored by *low* activator proportions (< 0.2 weight per cent), whereas the *long-wave* emission bands of these two-band phosphors are favored by *high* activator proportions (> 1 weight per cent).[221] In these cases, it is possible that different atoms of a given activator may have different effective valence states and different coordination numbers in a given host crystal, so that different states of the same species of activator atom may sometimes act as poisons for each other.

Two (or More) Activators. It is sometimes possible to obtain increased luminescence efficiency (during excitation) and increased phosphorescence efficiency by incorporating two (or more) added activators in a given host crystal.[58, 63, 226–229, 411–415] Two effects may be distinguished: (1) when the different activators do not appreciably interfere or cascade with each other, and (2) when the activators cascade, or when one activator "sensitizes" another; that is, one activator serves to convert a given primary excitation energy into secondary nonphoton excitation energy which excites luminescence whose spectral character is determined ("dominated") by another activator. Case 1 is practically nonexistent, except perhaps for very low (inefficient) activator proportions, because the luminescence efficiency of one activator is almost invariably decreased by the addition of another noncascading and non-sensitizing activator (compare Figs. 42, 45, and 52.)[63,72] This seems to be true, regardless of whether the different activators occupy similar sites (for example, all substitutional sites) or different sites (for example, one activator in s sites, and another activator in i sites) in the host crystal. Despite the lowering of luminescence efficiency associated with one species of activator by the incorporation of another species of activator in the same host crystal, phosphors with multiple activators are sometimes advantageous. An outstanding example is hex.-$3Ca_3(PO_4)_2 \cdot Ca(F:Cl)_2:Sb:Mn$ whose complementary blue (Sb) and yellow (Mn) emission bands afford a resultant white emission in "fluorescent" lamps.[389] Here, each of the two activators, Sb and Mn, decreases the luminescence efficiency of the other activator, but the total efficiency remains sufficiently high for practical use.

Case 2 has been amply demonstrated for (a) *phosphorescence;* for example, the thermostimulated phosphorescence (at 100°C) of glassy $1.27ZnO \cdot B_2O_3:Mn(0.2)$ is enhanced by the incorporation of $Ce(0.05)$;[411] and the enhanced photostimulated phosphorescence of phosphors Nos. 20–22 in Table 5 is obtained by the incorporation of a small proportion of Sm (the Ce and Sm in these examples produce trapping centers in the given phosphors), and (b) *luminescence during excitation;* for example, (i) the UV-excited orange luminescence emission (several lines near 6000 Å) of cub.-$CaS:Sm(0.2)$ is greatly intensified *at room temperature* by the incorporation of $Bi(0.05)$, and (ii) a similar intensification is obtained at 100°C (*but not at room temperature*) by the incorporation of $Pb(0.15)$ in the same cub.-$CaS:Sm(0.2)$.[412] The enhancement produced by a second activator is apparent both during and after excitation of long-phosphorescing cub.-$CaS:Sm:Bi$, whereas the enhancement has been determined only during excitation of fast-decay phosphors, such as cub.-$SrS:Ag(0.1):Pr(0.6)$.[412]

Two distinct luminescence processes may be distinguished when phosphors with two or more *cooperating* activators operate efficiently during excitation. These are:

1. *Cascade luminescence.* In some phosphors with two or more activators, the short-wave emission band determined (dominated) by one activator A may overlap the excitation band of another activator B which determines an emission band located at longer wavelengths.

Under these circumstances, an internal cascade process may occur, whereby the luminescence emission associated with A is absorbed in the phosphor crystal and excites luminescence emission associated with B, as in the hex.-ZnS:Ag:Cu phosphor, where the violet Ag emission band cascade-excites the green Cu emission band,[58, 63] or in the hex.-ZnO:[Zn] phosphor, where the UV edge emission band cascade-excites the visible emission band (Fig. 50).[354]

2. *Sensitized luminescence.* Some phosphors with a single activator, and which are practically unexcited by primary photons of a given energy, may become well excited by the given primary photons when a second activator is added. As examples, (a) the orange emission of cub.-CaS:Sm is excited by UV, but not by blue light, whereas the *same* orange emission is well excited in cub.-CaS:Bi:Sm by blue light;[412] and (b) the orange emission of rbhdl.-CaCO$_3$:Mn is excited by CR, but not by UV, whereas the same orange emission is well excited in rbhdl.-CaCO$_3$:Pb:Mn, or rbhdl.-CaCO$_3$:Tl:Mn, or rbhdl.-CaCO$_3$:Ce:Mn by UV.[227-229a] Schulman *et al.* have reported that a mechanical mixture of, for example, rbhdl.-CaCO$_3$:Mn and rbhdl.-CaCO$_3$:Pb is not excited by UV, whereas the rbhdl.-CaCO$_3$:Pb:Mn phosphor is well excited by UV.[227] This result eliminates the possibility of a cascade mechanism, and makes it appear that the Pb centers perform the dual function of (a) introducing new absorption bands in the UV region (Fig. 101), and (b) transforming primary UV photons into nonphoton secondary excitants which can excite the Mn activator centers to luminescence. Figure 102 shows some relative excitation spectra of several of these sensitized phosphors under excitation by a hydrogen-discharge lamp. The figure shows the virtual absence of excitation by UV when only Mn is present in the host crystal and shows the strong sensitization which is produced when Pb, Tl, or Ce are incorporated as second activators.

A similar sensitization is obtained in the case of monocl.-CaSiO$_3$:Pb: Mn where the Pb produces a strong absorption of 2537-Å photons and a weak emission band extending from about 2800 to 4200 Å, with a peak near 3500 Å, while the strong Mn doublet emission band extends from about 5500 to over 7000 Å.[221, 229, 229a] As another example, Fonda reports that tetr.-ZnF$_2$:Mn, which is well excited by 2200-Å photons but not by 2537-Å photons, becomes much more efficient under excitation by 2537-Å photons when Pb, W, Ce, or Ti is added as a second activator.[223] The Ti and W reportedly increase the photoluminescence efficiency under 2537-Å photons 4- to 6-fold without altering the orange emission band or the long ϵ^{-at} decay associated with the Mn activator in this phosphor.

From the evidence available, it appears that the sensitizer impurity centers absorb the primary photon energy directly and become excited but have a low probability of making radiative transitions or dissipating

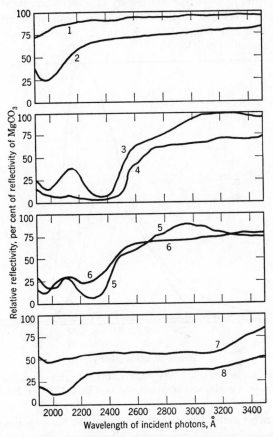

FIG. 101. Reflection spectra of (1) rbhdl.-CaCO₃ (nonluminescent), (2) rbhdl.-CaCO₃:Mn, (3) rbhdl.-CaCO₃:Pb, (4) rbhdl.-CaCO₃:Pb:Mn, (5) rbhdl.-CaCO₃:Tl, (6) rbhdl.-CaCO₃:Tl:Mn, (7) rbhdl.-CaCO₃:Ce, and (8) rbhdl.-CaCO₃:Ce:Mn. (J. H. Schulman *et al.*)

the excitation energy as heat. The transfer of energy from the excited sensitizer centers to the activator emitter centers apparently takes place by exchange. This requires an overlap of the ψ's of the two types of centers when they happen to be sufficiently excited and near each other (see Appendix 2).[229, 229a]

In cascade or sensitized luminescence the over-all luminescence efficiency, under given conditions, is the *product* of (*a*) the efficiency of the primary activator (for example, Pb in rbhdl.-CaCO$_3$:Pb:Mn) in converting the primary excitant energy into secondary excitant energy, and (*b*) the efficiency of the secondary ("dominant") activator (for example, Mn in rbhdl.-CaCO$_3$:Pb:Mn) in converting the secondary

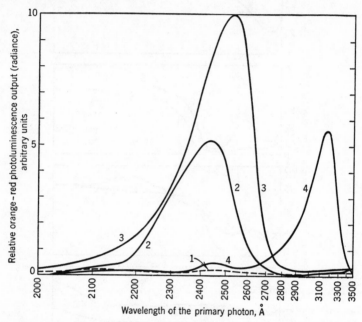

Fig. 102. Excitation spectra of (1) rbhdl.-CaCO$_3$:Mn, (2) rbhdl.-CaCO$_3$:Tl:Mn, (3) rbhdl.-CaCO$_3$:Pb:Mn, and (4) rbhdl.-CaCO$_3$:Ce:Mn. (J. H. Schulman *et al.*)

excitant energy into luminescence emission. Each activator has a different dependence of luminescence efficiency on temperature, and so the optimum operating temperature for maximum luminescence efficiency under a given type and intensity of excitation is a compromise which affords a maximum *product* of the two efficiencies.

EFFECT OF OPERATING TEMPERATURE ON EFFICIENCY. It may be re-emphasized that, unless otherwise indicated, general information on luminescence efficiency pertains to measurements made with the phosphor at or near room temperature. If the temperature of a luminescing phosphor be increased or decreased, with respect to room temperature, luminescence efficiency may be increased or decreased according to the composition of the phosphor (Figs. 18, 28a, 50b, 76, 103, 104, 105) and, in the case of t^{-n} decay phosphors, according to the intensity of the excitation (Fig. 106).[58, 63, 119, 221, 233, 310, 311, 384]

Most of these phosphors exhibit small but detectable *permanent* changes in ε on temperature cycling up to about 300°C; the small changes being caused by atomic readjustments to more nearly approach equilibrium configurations at the operating temperature. The measured efficiencies sometimes increase and sometimes decrease, depending on the nature and past history of the phosphor, but the ε changes in the better phosphors rarely exceed 15 per cent unless the operating temperature and excitation density are very high.

For the purposes of discussion, phosphors are divided into two major groups whose members are represented in Figs. 103 and 104.

S-*Dominated Phosphors Exhibiting* t^{-n} *Decays.* Figure 103 shows plots of ε vs. T curves for a number of S-dominated phosphors excited under identical conditions by UV from a 100-watt CH-4 lamp (with Corning 5840 and 5970 filters) placed about 30 cm distant from a very thick layer of the phosphor placed within an air oven with a Vycor lid.[58] An outstanding characteristic of these ε vs. T curves is the *maximum* near 280°K for several of the phosphors with Ag activator, and the *maximum* near 480°K for several of the phosphors with Cu activator. Changes in the crystal structure and cationic composition of the host crystal are seen to have a relatively small effect on the temperature break point T_B, compared with the large effect obtained by incorporating a few thousandths of a per cent of Ag or Cu. As shown by curves 5 and 6, however, partial substitution of anionic Se for S greatly affects the ε vs. T curve, shifting T_B to lower temperatures and greatly reducing the temperature interval between the T_B's of phosphors with Ag and Cu activators.[170] Efficient phosphors made by choosing a given host-crystal proportion in the general system (Zn:Cd)(S:Se) made with or without Ag, Au, or Cu activator are found to have the peak wavelengths of their emission bands λ_{pk} and the temperatures T_{50}, at which their luminescence efficiencies have decreased to 50 per cent of maximum arranged in approximately the same order,[170] for example:

Phosphor	λ_{pk}	T_{50}
Cub.-ZnS:Ag(0.003), [NaCl(2)], 950°C	4620 Å	350°K
Cub.-ZnS:[Zn], [NaCl(2)], 940°C	4700 Å	350°K
Cub.-ZnS:Au(0.002), [NaCl(2)], 940°C	4730 Å	440°K
Cub.-ZnS:Cu(0.003), [NaCl(2)], 660°C	5280 Å	550°K

Variations in the flux proportion used to prepare cub.(or hex.)-ZnS:[Zn] at 940°C and 1200°C have small but measurable effects on the ε vs. T curve when the flux proportion is changed from 0.02 to 20 weight per cent of NaCl.[58] As previously mentioned, Klasens reports that the T_{50} value of ZnS:Ag(0.01), under UV excitation, is shifted about 80°C to lower temperatures by the incorporation of Ni(0.001).[221, 233] Also, the

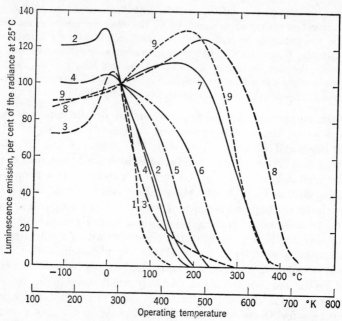

FIG. 103. Variation of photoluminescence efficiency (during excitation) of several S-dominated phosphors, all excited by the same intensity of predominantly 3650-Å UV (from a CH-4 lamp with filters), as a function of the temperature of the phosphor during luminescence.

1. cub.-ZnS:Ni(10^{-6})	[NaCl(2)],	730°C
2. cub.-ZnS:Ag(0.003)	[NaCl(2)],	950°C
3. hex.-ZnS:Ag(0.015)	[NaCl(2)],	1240°C
4. hex.-ZnS(48)CdS(52):Ag(0.01)	[NaCl(2)],	940°C
5. cub.-ZnS(60)ZnSe(40):Ag(0.005)	[NaCl(2)],	730°C
6. cub.-ZnS(60)ZnSe(40):Cu(0.005)	[NaCl(2)],	730°C
7. cub.-ZnS:Cu(0.003)	[NaCl(2)],	660°C
8. hex.-ZnS:Cu(0.003)	[NaCl(2)],	1240°C
9. hex.-ZnS(88)CdS(12):Cu(0.008)	[NaCl(2)],	1220°C

T_B and T_{50} values of S-dominated phosphors and of phosphors in general are usually shifted to higher temperatures when the measurements are made under CR excitation.[58, 63, 170]

Schoen has reported that the photoluminescence efficiencies of several (Zn:Cd)S-type phosphors, with and without Cu activator, are

2 to 3 times *higher* at 83°K than at 293°K when the phosphors are excited by 3025-Å photons, which are absorbed by the host-crystal atoms, whereas the efficiencies are 0.65 to 0.8 times *lower* at 83°K than at

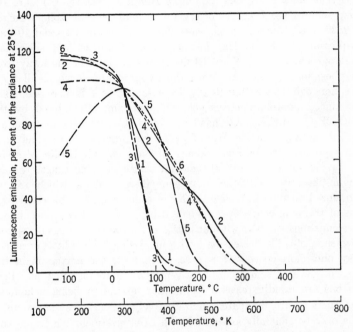

Fig. 104. Variation of photoluminescence efficiency (during excitation) of several O-dominated phosphors as a function of the temperature of the luminescing phosphor.

		Excited by
1.	hex.-ZnO:[Zn], 1000°C in CO (visible band)	3650-Å UV
2.	rbhdl.-Zn_2SiO_4:Mn(0.3), 1250°C	2537-Å UV
3.	rbhdl.-Zn_2GeO_4:Mn(0.3), 1050°C	2537-Å UV
4.	rbhdl.-$9ZnO \cdot BeO \cdot 6SiO_2$:Mn(1.5), 1150°C	2537-Å UV
5.	$3CdO \cdot 2B_2O_3$:Mn(1), 850°C	2537-Å UV
6.	tetr.-$CaWO_4$:[W], 950°C	2537-Å UV

293°K when the phosphors are excited by 4358-Å photons which are absorbed by the (presumably) i-center activator atoms [58, 307] (compare Fig. 18c). From these data it appears that, in general, normal (substitutional) host-crystal atoms have an increasing relative probability of dissipating excitation energy by nonradiative processes as the temperature of the crystal is increased, whereas interstitial atoms have an *initially* decreasing relative probability of degrading excitation energy into heat as the temperature of the crystal is increased (these probabilities are relative to the probability of effecting or aiding radiative transitions).

O-*Dominated Phosphors Exhibiting Initial* ϵ^{-at} *Decays.* Figure 104 shows some $\mathcal{E}$ vs. T curves of several O-dominated phosphors excited by 2537-Å UV (except in the case of curve 1) from a special 10-watt low-pressure (≈ 8-μ of Hg) mercury-vapor lamp with a silica-glass envelope (and Corning No. 9863 filter) placed about 30 cm away from the phosphor as before.[58] With the exception of curve 5, all these phosphors exhibit a rather steady decrease of $\mathcal{E}$ as T is increased. The pronounced photoluminescence-efficiency maximum exhibited by the $3CdO \cdot 2B_2O_3$:Mn phosphor has been reported by Thorington to obtain also for cadmium-silicate:Mn(0.5), and less pronounced maxima have been reported for phosphors identified only as zinc-borate:Mn(0.5), zinc–beryllium silicate:Mn(3), calcium tungstate, zinc tungstate, and calcium-phosphate:Tl(0.3).[221] In general, phosphors which exhibit pronounced maxima in their $\mathcal{E}$ vs. T curves have appreciable concentrations of trapping centers, as evidenced by long t^{-n} decay phosphorescence emissions (with or without initial ϵ^{-at} decays), whereas phosphors which show steadily decreasing $\mathcal{E}$ with increasing T have little or no t^{-n} decay phosphorescence. It is possible that (1) the maximum in the $\mathcal{E}$ vs. T curve of some phosphors represents a temperature at which electrons are rapidly released from traps instead of being held until their energies are dissipated nonradiatively, or (2) the maximum $\mathcal{E}$ value may be generally associated with i centers (including traps) and the steadily decreasing $\mathcal{E}$ with increasing T may be generally associated with s centers. In the latter case, the strong coupling of the s center to the host crystal may render this type of center particularly susceptible to thermal agitation, whereas the weakly coupled i centers may experience a decreasing perturbation with increasing T (for example, by lattice expansion) until the point T_B is reached where the vibrational excursions of the ligand atoms of the i-center atom become large enough and rapid enough to make excitation-energy loss as phonons more probable than by photons during the lifetime of the excited state of the emitting atom.[418] The tendency for the emission bands of s-center phosphors to shift toward shorter wavelengths (Figs. 28c and 30), and for the emission bands of i-center phosphors to shift toward longer wavelengths (Fig. 29) lends credence to the notion that the effective volume occupied by an s-center atom decreases, whereas the effective volume occupied by an i-center atom increases as the temperature of the phosphor crystal is increased.

Figure 105 shows the results of some measurements made by Kröger which indicate that the ε vs. T curves of many tungstate phosphors are strongly influenced by the temperature at which the material is crystallized.[155] It may be seen from the figure that the T_B and T_{50} values of the precipitated and merely dried tungstates occur at very low temperatures and that increasing the crystallization temperature increases T_B

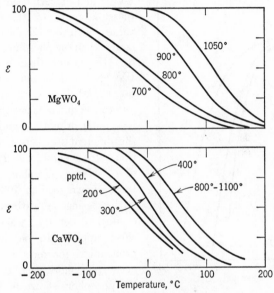

Fig. 105. Photoluminescence efficiencies ε of two tungstate phosphors as a function of operating temperature (abscissa) and crystallization temperature (noted in the drawings, in degrees centigrade). Excitation by 2537-Å UV. (F. A. Kröger)

and T_{50}. These variations are not due to changes in spectral characteristics, because the emission band of the magnesium tungstate phosphor is practically the same when the material is crystallized at 700 and 1100°C, and the emission band of the cooled calcium tungstate phosphor is shifted only slightly toward shorter wavelengths on going from 200 to 1100°C. Hence, the increase in T_B is attributed to an increase in the degree of crystallinity at higher crystallization temperatures; that is, the more regular structures made at high temperatures offer fewer opportunities in the form of distortions, strains, and other imperfections, for degeneration of primary excitation energy into heat instead of luminescence.

A shift of T_B to higher temperatures with increasing degree of excitation occurs for photoconducting t^{-n} decay phosphors, such as hex.-

ZnS(95)CdS(5):Cu and the phosphor shown in Fig. 106, but the shift does not occur for nonphotoconducting ϵ^{-at} decay tetr.-CaWO$_4$:U according to Garlick.[221] Kröger, however, reports that the T_B values of some tetr.-(Ca:Pb)MoO$_4$:[Mo] phosphors are considerably larger when the phosphors are excited by 3650-Å UV instead of by 2537-Å UV,[155] and it has been found in these laboratories that the T_B value of rbhdl.-Zn$_2$SiO$_4$:Mn is larger when the phosphor is excited by CR rather than by 2537-Å UV.[58, 63, 221] It is not known at present whether these effects are due to the different energies of the individual primary particles or to the different intensities of excitation. In general, the T_B values of t^{-n} decay phosphors, such as those shown in Fig. 103, exhibit an increasing T_B with increasing excitation density, whereas this effect appears to be much less pronounced for most of the ϵ^{-at} decay phosphors, such as many of the phosphors shown in Fig. 104. The shift of T_B to higher temperatures in the case of t^{-n} decay phosphors occurs because excited electrons in shallow traps (at high excitation densities) have shorter lifetimes and, hence, less time for their excitation energy to be converted into heat instead of luminescence emission.

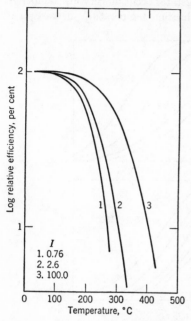

FIG. 106. Curves showing the decrease in photoluminescence efficiency of ZnS:Cu(0.01) as a function of the temperature of the luminescing phosphor, and as a function of the intensity I of the 3650-Å UV used for excitation. (N. R. Nail, D. Pearlman, and F. Urbach)[221]

During intense excitation, in particular, the efficiencies of both ϵ^{-at} and t^{-n} decay phosphors are decreased in proportion to the average lifetime of the excited state(s). This is so because centers that are already excited cannot utilize additional excitation energy expended by the primary excitant particles in the crystal (the probability of super-excitation is usually very low, because excited nontrapped electrons are so loosely bound that it is difficult to satisfy conservation of momentum on further excitation [34]). The efficiency decrease is particularly noticeable when the phosphor has a large proportion of very deep traps which hold excited electrons for long times and thereby greatly reduce the density of excitable centers.

There is a great need for quantitative information on the mechanisms and relative probabilities of radiative and nonradiative transitions at each step of the different luminescence processes, for example, during (1) absorption of primary energy, (2) electronic excitation, (3) local and long-range crystal rearrangements after excitation (or emission), (4)

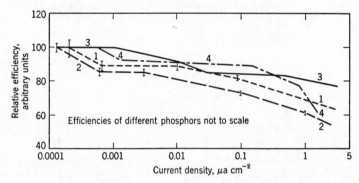

FIG. 107. Relative cathodoluminescence efficiencies of aluminized screens of (1) hex.-ZnO:[Zn] (visible band), (2) hex.-ZnS·0.018Mn(S), (3) rbhdl.-Zn$_2$SiO$_4$· 0.012MnSiO$_3$, and (4) tetr.-ZnF$_2$·0.005Mn(F$_2$). Excitation by various intensities of 10-kv (curve 4 = 15 kv) stationary unmodulated CR. (S. Lasof)

energy transfer, (5) temporary storage of excitation energy, and (6) return of the system to the ground-state level.[221, 419] It may be reasonable to focus attention on the luminescence (radiative) processes, *per se,* when the luminescence efficiency is very high (for example, under UV excitation), but considerably more attention should be given to understanding and controlling the nonradiative processes when the luminescence efficiency is low (for example, under CR excitation).

EFFECT OF EXCITATION DENSITY AND DURATION ON EFFICIENCY. **Excitation density** is here defined as the number of excitant free-energy "bits" (each capable of producing a luminescence photon) per unit volume per unit time. The effects of excitation density and duration on luminescence efficiency ε are various, depending on the nature, crystal size, and temperature of the phosphor, and on the nature and energy of the primary particles used for excitation. As shown in Fig. 107, many phosphors exhibit constant or slowly *decreasing* ε with increasing excitation density, whereas Figs. 108 and 109 show that some types of phosphors exhibit *increasing* ε with increasing excitation density until saturation effects become predominant.[221] (Figure 110 shows, in particular, the two-band emission of the phosphor whose characteristics are shown in Fig. 109).[421]

The maximum cathodoluminescence efficiency values of the specific aluminized (≈ 1000-Å thick Al) screens listed in Figs. 107–110, measured at room temperature under the given conditions, are given in Table 18.

TABLE 18

EFFICIENCY OF VISIBLE CATHODOLUMINESCENCE EMISSION, AND OTHER CHARACTERISTICS OF THE PHOSPHOR SCREENS USED IN FIGURES 107–109

Fig.	Curve	Phosphor	Approx. Average Particle Size, microns	Screen Density, mg cm^{-2}	Accelerating Voltage, kv	Maximum Measured Efficiency $\mathcal{E}$, candles watt^{-1}	Current Density at the Maximum Efficiency, μa cm^{-2}
107	1	Hex.-ZnO:[Zn]	3	4	10	1.8	2.5×10^{-4}
107	2	Hex.-ZnS:Mn	8	4	10	7.0	1.3×10^{-4}
107	3	Rbhdl.-Zn$_2$SiO$_4$:Mn *	3	4	10	7.6	10^{-4}
107	3	Rbhdl.-Zn$_2$SiO$_4$:Mn *	3	4	15	12.7	3.0×10^{-4}
107	4	Tetr.-ZnF$_2$:Mn	3	4	15	0.5	6.0×10^{-4}
108	1	Hex.-1.33ZnS·CdS:Ag	8	4	10	5.0	3.3
108	1	Hex.-1.33ZnS·CdS:Ag	8	4	15	12.3	0.22
109	1	Hex.-9ZnS·CdS:[Zn]:Cu	15	4	10	6.1	10^{-3}
109	1	Hex.-9ZnS·CdS:[Zn]:Cu	15	4	15	12.6	6.0×10^{-4}

* For a current density of 0.002 μa cm^{-2}, the $\mathcal{E}$ values at 10, 15, and 20 kv are in the ratio of about 1:2:2.5.

With respect to the initially increasing $\mathcal{E}$ shown in Fig. 108, Nail, Pearlman, and Urbach have reported that a ZnS:Ag(0.04) phosphor exhibits increasing $\mathcal{E}$ with increasing intensity of UV at low excitation densities, whereas the same material with Ni(0.00003) requires about 10 times as much UV intensity before commencing to show increasing $\mathcal{E}$, and the material with Ni(0.0003) requires about 1000 times as much UV intensity before commencing to show increasing $\mathcal{E}$ with increasing excitation density.[221] To summarize the effect of traces of Ni on this type of t^{-n} decay phosphor, the Ni (1) reduces phosphorescence, (2) decreases T_B and T_{50} (thereby decreasing $\mathcal{E}$ measured above T_B), and (3) necessitates higher excitation densities to obtain increasing $\mathcal{E}$ as a function of excitation density.

Predominantly ϵ^{-at} decay s-center phosphors, some of which are shown in Fig. 107, generally exhibit monotonically decreasing CR $\mathcal{E}$ with increasing excitation density. The decrease in $\mathcal{E}$ may be attributed to the increased thermal perturbation of the strongly coupled substitutional emitting atoms caused by increased residual excitation energy at the higher excitation densities. It may be noted that a decrease in the lifetime τ of the excited state of a potential emitting atom would tend to increase ε by decreasing the time interval during which excitation energy may be degraded into heat. For a simple ϵ^{-at} decay phosphor, however, τ is practically independent of excitation density (Fig. 70).[34, 65, 112–120] In these cases, τ remains substantially constant while the amplitude and frequency of thermally induced collisions between an excited atom and its neighbors increase with increasing excitation den-

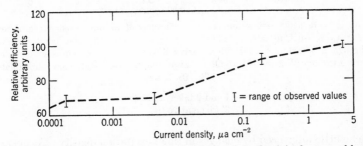

FIG. 108. Relative cathodoluminescence efficiency of an aluminized screen of hex.-1.33ZnS·CdS:Ag(0.01), 950°C, excited at various current densities by a 10-kv beam of stationary unmodulated CR. (S. Lasof)

sity. Therefore, the relative probability of energy loss as phonons increases and ε decreases with increasing excitation density.

In the case of (presumably) i-center phosphors with t^{-n} decays, the observed changes in ε with increasing excitation density appear to be the *net* result of (1) a decrease in the average $\bar{\tau}$, by bringing a larger proportion of the shallower traps into action; (2) a possible increase in the effective "room" around the interstitial emitting atom; and (3) an increase in thermal perturbation of the interstitial emitting atom. Effects 1 and 2 tend to increase ε, whereas effect 3 tends to decrease ε. When measurements are made on t^{-n} decay phosphors at temperatures well below T_B (for example, hex.-ZnS:Cu or hex.-(Zn:Cd)S:Cu at room temperature), the photoluminescence and cathodoluminescence efficiencies are observed to remain fairly constant or *decrease* with increasing excitation density below saturation.[420–422] When these t^{-n} decay phosphors are measured at temperatures above T_B (for example, 270°C), however, their ε's *increase* with increasing excitation density below saturation [422] (compare curves 2 and 2′ in Fig. 109). It appears from these data that the beneficial effect of decreasing $\bar{\tau}$ (and increasing the lattice spacing), as the excitation density is increased, outweighs the detrimental effect of increasing thermal perturbation when the phosphor is operated above T_B. At operating temperatures well below T_B, the relative probabilities of radiative and radiationless transitions appear

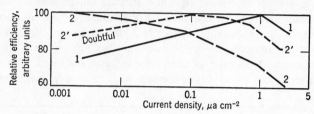

FIG. 109. Relative cathodoluminescence efficiency of an aluminized screen of hex.-9ZnS·CdS:[Zn]:Cu(0.0073), 1250°C. Curve 1 = blue emission band observed through a Wratten No. 47 filter, curves 2 and 2' = yellow emission band observed through Wratten No. 15 filter. Excitation by 10-kv stationary unmodulated CR. (S. Lasof)

1 and 2 taken at 25°C
2' taken at 200°C

to be little influenced by changes in (low) excitation density. At operating temperatures above T_B, where the probability of radiationless transitions is strongly dependent on T, however, a decrease in $\bar{\tau}$ increases ε

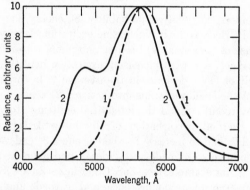

FIG. 110. Cathodoluminescence emission spectra of aluminized screens of (1) hex.-1.33ZnS·CdS:Ag(0.01), 950°C and (2) hex.-9ZnS·CdS:[Zn]:Cu(0.0073), 1250°C (plus some change during processing of the experimental tube). Excitation by 10-kv unmodulated CR at about 1 μa cm^{-2}. (S. Lasof)

by decreasing the time during which excitation energy may be degraded into heat. It should be noted that $\bar{\tau}$, at a given excitation density, is shorter at or above T_B than at a temperature less than T_B (see Table 16), and so luminescence emission during excitation at or above T_B involves chiefly very shallow traps in the case of t^{-n} decay phosphors.

Complete saturation of the luminescence mechanism occurs when the excitation density exceeds the maximum number of excitant energy "bits" which a unit volume of the phosphor can transform into luminescence photons per unit time *under the given conditions*. The maximum number of excitation and emission acts which a given atom can perform per unit time is inversely proportional to τ. Because the excitation density is greatest near the point of entry of the primary particles and decreases to practically zero at the penetration "limit," complete saturation is reached gradually as the intensity of the primary excitant beam is increased (it being remembered that intensity is the number of particles passing unit area in unit time). Hence, the layers nearest the primary-energy-input side of the crystal may be completely saturated while the inner (penetrated) layers of the crystal are far from saturated. Practically complete saturation may be attained at relatively low excitation intensities when the penetration x_l of the primary particles is much less than 0.1 micron, whereas complete saturation may be practically unattainable when x_l is much more than 10 microns. That is, for $x_l \gg 10$ microns, present sources of excitation energy cannot provide sufficient beam power to saturate completely the penetrated phosphor layers near the penetration "limit" x_l.

During inefficient luminescence, when the phosphor crystals and the penetration of the primary particles are both small (less than 0.1 micron), the tiny (insulator) crystals may be heated quickly to a temperature above T_B by a high-intensity primary beam, thus occasioning a decrease in ε. The rate of temperature rise, for a given primary beam *power*, may be decreased by increasing x_l, or by increasing the sizes of the phosphor crystals so the deeper-lying unexcited portions of the larger crystals may act as "heat sinks." [58] If, during prolonged excitation, the temperature of the phosphor crystal is raised far above T_B, the crystal may undergo some recrystallization with accompanying changes in ε (Figs. 90–94, 98, and 99). At high operating temperatures, also, the luminescence efficiencies of phosphors may be more rapidly and permanently changed by the chemical actions of certain primary particles, or adsorbed impurities, or the surrounding atmosphere. The resistance of a phosphor to chemical changes which may be wrought by, for example, captured primary electrons (reducing agents) or alpha particles (oxidizing agents) is generally proportional to the bonding energy of the host crystal; that is, the more stable phosphors usually have high melting points, high heats of formation, and high hardnesses. As shown by the examples in Figs. 18, 103, 104, and 105, most of the efficient phosphors have T_B's below about 500°K, but there are a few phosphors which have appreciable efficiencies at higher temperatures. Among the phosphors which are relatively stable and have useful efficiencies at temperatures as high as 800°K are rhomb.-$BaSO_4$:Pb and rbhdl.-Al_2O_3:Cr (The room-temperature quantum efficiency of rbhdl.-Al_2O_3:Cr under 3650-Å UV is about 35 per cent, and the relative cathodoluminescence outputs of rbhdl.-Al_2O_3:Cr at 6840 Å and rhomb.-$BaSO_4$:Pb at 4100 Å are about 50 per cent and 5 per cent, respectively, of the peak output of rbhdl.-Zn_2SiO_4:Mn at 5250 Å. These measurements were made with the recording spectroradiometer [93] having a slit width of 1 mm and using 6-kv 1-μa cm^{-2} CR).

The upper (saturation) limit of luminescence output *during excitation* of a phosphor has different values depending on whether unmodulated or pulsed excitation is used. Theoretically, the two saturation outputs would be the same at 100 per cent luminescence efficiency. In this *theoretical* case, assuming that there are no absorption losses, the maximum luminescence output L_m for 100 per cent ε and unlimited intensity of excitation would be given by

$$L_m = \delta x_l (h\bar{\nu}_e)\bar{\tau}^{-1} \text{ erg cm}^{-2} \text{ sec}^{-1} \qquad (117)$$

where δ is the maximum number of *operating* emitting atoms (centers) per cubic centimeter in the phosphor crystal, x_l is the penetration "limit" of the excitant, $h\bar{\nu}_e$ is the average energy of the emitted luminescence photons, and $\bar{\tau}$ is the average lifetime of the excited state.[72] Assuming $\delta \approx 10^{21}$ emitting centers cm^{-3}, $x_l \approx 2.5 \times 10^{-4}$ cm ($\approx$ 20,000-v CR, eq. 77), $h\nu = 3.54 \times 10^{-12}$ erg (that is, assuming monochromatic emission, for simplicity, with $\lambda = 5560$ Å for maximum photopic response), $\bar{\tau} \approx 10^{-2}$ sec (for example, β-Zn_2SiO_4:Mn), and sufficiently high excitation intensity, eq. 117 gives

$$L_m \approx 10^8 \text{ ergs cm}^{-2} \text{ sec}^{-1} = 10 \text{ watts cm}^{-2}$$

$$= 6850 \text{ lumens cm}^{-2} = 6.85 \times 10^6 \text{ mL} \qquad (118)$$

If the afore-mentioned hypothetical phosphor were operating at, say, 1 per cent effective *visual* luminescence efficiency, the L_m value would be reduced to about 7×10^4 mL, which is somewhat above the maximum *averaged* luminance $\bar{L}_m$, obtained from present aluminized [63] projection-CRT screens operated at 20 to 30 kv under television scanning conditions (that is, an interlaced scanning raster where the intense scanning CR beam cyclically excites each elemental area of the phosphor screen for about 10^{-7} sec 30 times a second [72, 423]). The maximum luminance *during excitation* is considerably greater than the average luminance, be-

cause each elemental area of the phosphor screen is excited for only
30×10^{-7} sec during each second; therefore, each screen element is
emitting a decaying output of relatively low-intensity phosphorescence
during most of the averaged observation (measurement) interval. The
ratio, $L_m/L_{0.033 \text{ sec}}$, of maximum luminance during excitation to the
luminance at 0.033 sec after excitation (just prior to re-excitation in
television), is about 1.7 for tetr.-ZnF_2:Mn ($L \approx L_0\epsilon^{-10t}$, Figs. 69 and
79), about 17 for rbhdl.-Zn_2SiO_4:Mn ($L \approx L_0\epsilon^{-100t}$), and may be well
over 10^4 for hex.-ZnO:[Zn] (and for many t^{-n} decay phosphors, such as
cub.(or hex.)-(Zn:Cd)(S:Se), with or without Ag or Cu, *at high excitation
densities*). Hence, under television conditions, the ratio $L_m/\bar{L}_m$ may be
about 1.3 for tetr.-ZnF_2:Mn, 13 for rbhdl.-Zn_2SiO_4:Mn, and over 10^4
for many fast-decay phosphors. The instantaneous L_m values of efficient fast-decay phosphors, such as hex.-ZnO:[Zn] and cub.(or hex.)-
(Zn:Cd)(S:Se):Ag, may exceed 10^7 mL in television projection CRT.

It must be emphasized that phosphors cannot endure prolonged
steady excitation (for example, a stationary unmodulated high-intensity
CR beam) at the excitation densities required to produce over 10^7 mL.
Such instantaneous excitation intensities, in the case of a 25-kv projection CRT beam, may exceed 25,000 watts cm^{-2}, which gives excitation
densities in excess of 10^7 watts cm^{-3} in most phosphors.[58] Practically
any material would vaporize under such high power input, if prolonged.
It is seen, then, that very high instantaneous L_m values ($> 10^7$ mL)
may be obtained for very short excitation times, whereas, (1) much lower
L_m values ($\approx 10^4$ mL) obtain under continued unmodulated excitation,
and (2) correspondingly low $\bar{L}_m$ values ($\approx 10^4$ mL) obtain under high-
intensity cyclic excitations of the television type just discussed. Higher
L_m (and $\bar{L}_m$) values may be obtained by (1) increasing δ and x_l, (2) decreasing $\bar{\tau}$, (3) using a more efficient excitant (UV would be most useful
if more intense sources were available), and (4) cooling the phosphor
crystals to maintain high efficiency.[424] In current practice, L_m does not
increase linearly with increasing x_l, because greater penetration increases
scattering and absorption losses of light emitted deeper in the fine-
crystal phosphor coating. This difficulty would be largely obviated by
the development and use of thick single-crystal screens with optically
rough surfaces on the side to be observed.[324]

If phosphors absorbed *none* of their own luminescence emission, then
the luminescence output from the roughened end of an otherwise
polished long thin phosphor crystal, excited along its entire length,
would be directly proportional to the length of the crystal, that is,
infinite length would yield infinite luminescence output.[219] In such a
hypothetical phosphor-rod luminescence source, total internal reflection

would allow each area perpendicular to the axis to contribute the same photon intensity to the end area, regardless of distance from the end (in the *complete* absence of absorption). Experiments on a 1.7-cm long thin crystal of rbhdl.-Zn_2SiO_4:Mn have given end luminances about 100 times as high as the side luminances under 40-kv CR excitation.[219, 220] A similar large increase is obtained under excitation by 2537-Å UV,

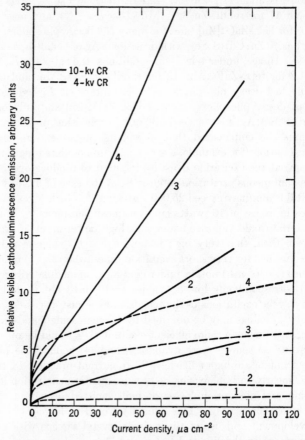

FIG. 111. Cathodoluminescence outputs of some O- and S-dominated phosphors as a function of applied voltage and current density of the unmodulated stationary CR beam. Measurement on the bombarded side of thick screens.

1. cub.-ZnS:Ag(0.015), 950°C
2. rbhdl.-Zn_2SiO_4:Mn(0.3), 1250°C
3. hex.-ZnS:Cu(0.01), 1200°C
4. hex.-ZnS(47)CdS(53):Ag(0.01), 950°C

which also penetrates well into the phosphor, whereas the increase is much smaller under excitation by 6-kv CR whose penetration is only about 0.2 micron according to eq. 77.

Some examples of the increased maximum luminescence outputs which can be obtained by increasing the penetration of the primary particles into the phosphor crystals are shown in Fig. 111 for several thick phosphor screens excited by unmodulated 4-kv CR ($x_l \approx 10^{-5}$ cm = 0.1 micron) and 10-kv CR ($x_l \approx 0.6$ micron). The average particle sizes of these phosphors range from about 2 microns for rbhdl.-Zn_2SiO_4:Mn to about 10 microns for the other phosphor crystals.[58] It may be seen that all the phosphor screens are practically saturated at 10 μa cm^{-2} with 4-kv primary electrons, whereas saturation is not evident even at 70 μa cm^{-2} with the deeper-penetrating 10-kv primary electrons. Assuming a uniform distribution of the primary excitation energy throughout the penetrated volume, instead of the actual exponential distribution shown in Fig. 22, the approximate power dissipated per unit of penetrated volume is 4000 watts cm^{-3} at 4 kv and 10 μa cm^{-2}, and is only 1700 watts cm^{-3} at 10 kv and 10 μa cm^{-2}. The higher power input per unit *area* in the second case becomes a lower power input per unit *volume* because the penetration increases as the square of the voltage accelerating the primary electrons (eq. 77). At sufficiently high current densities in either case the luminescence outputs of all the phosphor screens will decrease because the residual heat after excitation raises the temperature of the phosphor enough to lower the efficiency at saturation.[221]

It is not possible to make effective use of very high excitation densities when the excitation is unmodulated and prolonged, because the phosphor screen soon becomes heated to a temperature well above T_{50}, but it is possible to use very high excitation densities when the excitation comprises intermittent pulses with sufficient rest intervals to allow the phosphor crystals to dissipate most of their excess heat by conduction and radiation. Figure 112 shows some results obtained with aluminized screens [100] of two t^{-n} decay phosphors (Nos. 1 and 2) and an ϵ^{-at} decay phosphor (No. 3) excited by intense CR pulses lasting about 10^{-7} sec and repeated every $\frac{1}{30}$ sec. Under these conditions, there is a slow warming of the phosphor screen, but the rate of warming is not rapid enough to preclude making photometric measurements of luminescence output during the first few seconds of excitation. An outstanding feature of these data is the large difference between the saturation characteristics of the t^{-n} decay phosphors with about 0.01 per cent activator and the ϵ^{-at} decay phosphor with about 1 per cent activator, especially when the phosphors are compared under focused and defocused conditions. The luminescence outputs change instantly on changing the focus of the primary CR beam, thereby eliminating the possibility that the effect is due simply to a change in temperature.[34] The results,

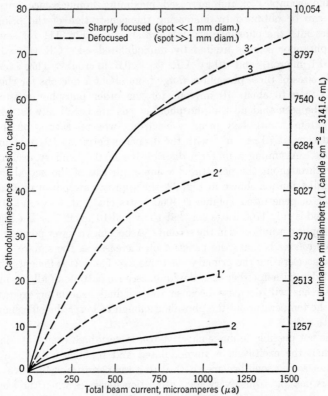

Fig. 112. Cathodoluminescence outputs of aluminized screens of some O- and S-dominated phosphors as a function of excitation density (data obtained with the cooperation of D. W. Epstein and P. J. Messineo).

1. cub.-ZnS:Ag(0.0025), 850°C, 7 mg cm^{-2}
2. hex.-ZnS:Ag(0.015), 1250°C, 8 mg cm^{-2}
3. rbhdl.-Zn$_2$SiO$_4$:Mn(0.3), 1250°C, 5 mg cm^{-2}

All aluminized screens.
Measurements at 21 kv with 5 × 5-cm scanned pattern (525 lines).

then, approximate the attainment of maximum luminescence output L_m, according to eq. 117. Under the focused condition, with a beam current of 500 μa, the excitation density exceeds 10^7 watts cm^{-3}, and it is found that only phosphors with a high proportion of *operating* activator centers exhibit little difference between their outputs under focused and defocused CR beams. In practice, this means that phosphors with high activator proportions, for example, rbhdl.-$(Zn:Be)_2SiO_4:Mn(1–10)$, tetr.-$ZnF_2:Mn(1–10)$, $3CdO \cdot 2B_2O_3:Mn(1)$, rbhdl.-$(Zn:Be)_2SiO_4:Ti(1)$, and monocl.-$CaMg(SiO_3)_2:Ti(1)$, may give much larger luminescence outputs at such high excitation densities than phosphors with low activator proportions, for example, hex.-$ZnS:Cu(0.01)$ and hex.-$(Zn:Cd)S:Ag(0.01)$. This is true even though the latter phosphors may exhibit higher luminescence efficiencies at low excitation densities (Fig. 111), and may exhibit increasing instead of decreasing efficiency with increasing (low) excitation density (compare Figs. 107 and 108). It is important to note that, according to eq. 117, the proportion of *operating* emitting atoms per unit volume must be large to obtain a large luminescence output. In some cases, such as hex.-$ZnS:Mn(2)$ and $Ca_3(PO_4)_2:Bi(4)$, the proportion of activator may be optimum and of the order of magnitude of that in, for example, rbhdl.-$Zn_2SiO_4:Mn(1)$, yet the former phosphors exhibit current saturation characteristics more nearly like those of the t^{-n} decay phosphors, Nos. 1 and 2, shown in Fig. 112. This indicates that only a small proportion of the Mn and Bi activator atoms in hex.-$ZnS:Mn(2)$ and $Ca_3(PO_4)_2:Bi(4)$ are operating efficiently (some of the remainder may act as poisons). Figure 113 contains the data of Fig. 112 replotted to show the variation of cathodoluminescence efficiency as a function of increasing excitation density. The efficiency data on the rbhdl.-$Zn_2SiO_4:Mn$ screen are low because the measurements were made on a cathode-ray tube which had been operated so long that the glass envelope had become darkened by the action of soft x rays produced during electron bombardment. A fresh screen and tube afford an efficiency of over 9 candles watt^{-1} ($\approx$ 4-per cent energy efficiency) at low current densities at the given voltage. At *very low* current densities, it is possible to obtain 15 candles watt^{-1} from aluminized screens of rbhdl.-$Zn_2SiO_4:Mn$ operated at about 20 kv. (Despite these consistently low measured values of effective cathodoluminescence efficiency, estimates of the maximum intrinsic efficiencies of phosphors excited by high-energy particles range up to about 80 per cent.[292, 615])

The foregoing data on saturation and maximum luminescence outputs were restricted to excitation by primary electrons because present sources of primary photons are not sufficiently intense to saturate the luminescences of efficient phosphors. In this respect, phosphors excited by intense beams of charged material particles may become important photon sources with radiances larger than other photon sources, especially if long single-crystal phosphor rods be developed to obtain intensification by cumulation of light through total internal reflection.[219]

This procedure offers considerable potential advantage over present gas-discharge or heat-actuated photon sources, in that the phosphor source may be readily pulsed or modulated at megacycle frequencies by

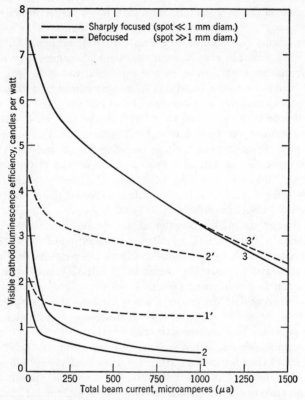

FIG. 113.   The data of Fig. 112 replotted to show the variation in cathodolumines-cence efficiency as a function of excitation density.

1. cub.-ZnS:Ag(0.0025), 850°C, 7 mg cm^{-2}
2. hex.-ZnS:Ag(0.015), 1250°C, 8 mg cm^{-2}
3. rbhdl.-Zn$_2$SiO$_4$:Mn(0.3), 1250°C, 5 mg cm^{-2}

All aluminized.
Measurements at 21 kv with 5 × 5-cm scanned pattern (525 lines).

grid- or deflection-controlled modulation of the beam(s) of charged material particles.

The final item to be considered in this section is the selective effect which may occur, with increasing excitation density, when a phosphor has more than one (simple component) emission band. Where there are two pronounced emission bands, as in certain hex.-ZnS:[Zn]:Cu (Figs. 45a–c and 48), cub.(or hex.)-ZnS:Ag:Cu, cub.(or hex.)-ZnS:[Zn]:Mn, and hex.-ZnO:[Zn] (Fig. 50) phosphors, the long-wave emission band usually predominates at low excitation densities, and the short-wave emission band predominates at high excitation densities.[72, 354] In the cited cases, with the exception of hex.-ZnO:[Zn] (Fig. 50b), the luminescence efficiency of the short-wave band decreases faster than that of the long-wave band as the temperature of the phosphor is increased (at constant excitation density). That is, the short-wave band generally, but not always, has a lower T_B than the long-wave band. Kröger has reported narrow short-wave edge-emission bands, similar to the narrow UV edge-emission band of hex.-ZnO:[Zn], for cub.-ZnS:[Zn] and hex.-CdS:[Cd] phosphors excited at 93°K by UV.[154] The low-temperature short-wave edge-emission band of cub.-ZnS:[Zn] is reported to extend from about 3300 to 3600 Å ($\lambda_{pk} \approx 3400$ Å), and that of hex.-CdS:[Cd] to extend from about 5000 to 5500 Å ($\lambda_{pk} \approx 5170$ Å), with indications of some fine structure in the bands. In these cases, as in the case of hex.-ZnO:[Zn], the narrow short-wave emission bands are located just on the edge of the main host-crystal absorption bands. This fact, coupled with the exceptionally short excited lifetime ($\approx 10^{-7}$ sec) of the short-wave edge emission band, indicate that the luminescence emission may be the result of a highly localized excitation followed by a radiative transition involving imperfection-perturbed excited normally situated atoms of the host crystal (rather than abnormal activator atoms). Referring to Fig. 16c, such a radiative transition might occur at A_2 or $A_{2'}$, under certain conditions, from an excited-state level near E^*_C to the topmost normally filled band E_H. If the radiative transition is from E^*_C, then in terms of the crystal model in Fig. 12, the radiative transition may be the return of an excited electron from a Zn^+ ion to a ligand S^- ion (assuming Zn^{2+} and S^{2-} is the ground-state condition). In any event, this is not a true resonance process, because work is done on the crystal during excitation and emission. For this reason, these short-wave emission bands must be excited by still shorter-wave (higher-energy) photons or material particles.

Manifestation of the foregoing short-wave emissions *outside* the phosphor crystals is hindered by nonradiative absorption processes *in* the crystals, especially when the refractive index is high and the emission band overlaps the strong host-crystal absorption. When there is a considerable overlap, as in hex.-ZnO:[Zn] and hex.-ZnS:[Zn], the short-wave (UV) emission is observed outside the crystals only when (1) the internal emission density is made high enough to exceed the saturation point of the nonradiative absorption process (for example, in the case of hex.-ZnS:Ag or hex.-ZnO:[Zn][354]); or (2) the temperature of the phosphor is made low enough to decrease the probability of energy degradation into heat (for example, in the case of hex.-ZnS:[Zn] [154]). The possibility exists also of shifting the host-crystal absorption to shorter wavelengths by chemical or structural changes. This is done, for example, by burning zinc vapor in oxygen to form a hex.-ZnO phosphor which emits the narrow UV band *alone* at relatively low excitation densities and at room temperature.[58, 354]

CHAPTER 6

GENERAL PROPERTIES OF PHOSPHORS
(RÉSUMÉ OF USEFUL PHOSPHORS)

Brief Summary of Phosphors

By definition, **phosphors** are materials which (1) absorb quanta of energy usually larger than about 1 ev from primary excitant particles, such as photons or charged material particles, and (2) convert a portion of the absorbed energy into detectable *luminescence* photon emission which (a) exceeds the thermal radiation from the material at the given temperature (Figs. 17 and 18), and (b) persists for a time interval longer than the natural lifetime τ_F of the excited state of isolated nonmetastable atoms considered as dipole radiators (eqs. 69–71). *Fluorescence* is luminescence whose excited-state duration is equal or less than τ_F, where τ_F is about 10^{-15} sec for nuclear gamma-ray emission, about 10^{-14} sec for hard x-ray emission, and about 10^{-8} sec for the near-optical emissions which comprise *conventional luminescence* (see Appendix 4). Fluorescence is a limiting case of *phosphorescence* which is defined as delayed luminescence emission whose excited-state duration exceeds τ_F (compare Table 10).[678]

The foregoing definition of phosphors includes some organic crystals,[78, 292] some hydrated semiorganic materials, such as boric acid containing traces of certain organic substances (Figs. 65 and 66),[58, 74–83, 375] and some glassy inorganic materials,[402–411] but most of the commercially useful phosphors are artificial anhydrous crystalline inorganic materials which are synthesized by heating well-purified and accurately compounded ingredients at temperatures above 800°C (Table 5). Useful phosphors are generally tiny crystals, with diameters of the order of 10^{-4} cm, which comprise one or more high-melting cocrystallized pure inorganic compounds (the **host crystal**) containing a small proportion of cocrystallized compounds of certain multivalent cationic impurities called **activators**. Activators may be introduced into the host crystal by (1) partial decomposition or atomic dislocations in the host crystal during high-temperature crystallization, or (2) deliberate addition (or unwitting inclusion!) of small amounts of compounds of foreign multi-valent cationic elements during crystallization or recrystallization. Activator atoms may lodge in substitutional sites in place of regular host-crystal atoms, thereby forming s-centers; or, activator atoms may lodge in interstitial sites between regular host-crystal atoms, thereby forming i centers. There appears to be a greater tendency for the formation of relatively high optimum proportions of s centers in *oxygen* *(fluorine)-dominated* phosphors, such as rbhdl.-Zn_2SiO_4:Mn(1), rbhdl.-Al_2O_3:Cr(1), and tetr.-ZnF_2:Mn(1), as contrasted with the tendency to form much lower optimum proportions of (presumed) i centers in *sulphur(selenium)-dominated* phosphors, such as hex.-ZnS:Ag(0.01), hex.-(Zn:Cd)S:Cu(0.01) and cub.-ZnSe:Ag(0.01), *where the numbers in parentheses represent weight per cent of activator element based on the weight of the host crystal.*

An activator element may promote luminescence emission by (1) *intensifying* a latent emission band of the host crystal (Fig. 31), (2) *originating* a new line or band emission spectrum which is often characteristic of the activator element (Figs. 13, 14, 31, and 45), or (3) *sensitizing* a luminescence emission attributable to another activator (Figs. 101 and 102). Furthermore, activator elements may introduce electron or positive-hole *trapping states* (Fig. 25d), and may act predominantly as *poisons* (1) when present in excess in a phosphor in which they function as normal activators in small proportions (Fig. 97), and (2) when incorporated in small proportions in certain phosphors in which they are incompatible.[416]

The chief function of phosphor formulas is to indicate the general crystal structure and essential ingredients of the material (compare Chapter 3, page 75). In the conventional though nonstandardized symbolic formulas of phosphors, the proportions of the host crystals and activators may be denoted on a complete or partial mole basis or weight basis; for example, the following formulas are equivalent: rbhdl.-$2ZnO \cdot SiO_2 \cdot 0.012MnSiO_3$ = rbhdl.-$2ZnO \cdot SiO_2$:Mn(0.3) = rbhdl.-Zn_2SiO_4:Mn(0.3) = rbhdl.-$ZnO(162.76)SiO_2(60.06)$:Mn(0.3), where the first (mole proportion) formula is readily interpretable, and the subsequent formulas imply 0.3 weight per cent of Mn activator (incorporated as $MnSiO_3$) based on the total weight of the host crystal $2ZnO \cdot SiO_2$. Where the activator proportion is very small, as in hex.-ZnS:Cu(0.003) and cub.-MgS:Sb(0.01), the activator is generally indicated on a weight basis, but when the activator proportion is large, as in rbhdl.-$2ZnO \cdot SiO_2 \cdot 0.02TiO_2$ or rhomb.-$BaSO_4 \cdot 0.3PbSO_4$, the activator is often indicated on the same mole basis as the host crystal. It is sometimes convenient to assign different formula designations to different classes of phosphors being studied simultaneously in a given laboratory; for example, all-sulphide phosphors may be designated entirely on a weight per cent basis, and all-silicate phosphors may be designated entirely on a mole proportion basis. Also, intensifier-type activators may be included on a mole basis in the host-crystal formula while new-band-type activators may be separated from the host-crystal formula by the conventional colon and indicated on a weight basis; for example, in rbhdl.-9ZnO · BeO · 5SiO$_2$ · 0.2TiO$_2$: Mn(0.5) the Ti enhances the UV + blue emission band of the silicate, and the Mn and Be collaborate in producing a new orange–red emission band (Figs. 31, 51, and 56). When the Mn and Be contents are reduced, the Ti-intensified UV + blue silicate emission may be made to predominate, and, when the Ti content is reduced, the orange–red emission band associated with the Mn and Be may be made to predominate.[34,61,429,430]

Most of the useful phosphors have band spectra, rather than line spectra, of absorption (excitation) and emission. When line spectra are observed in solids they are generally attributed to electronic transitions of electrons redistributing in incompleted inner shells, such that the transitions involve changes in the resultant spins and total angular momentum quantum numbers. When band spectra are observed, they are attributed to electronic transitions involving the outermost valence electrons whose resultant principal, angular-momentum, and spin quantum numbers may change during the transitions.

The lower limit of energy for a primary particle capable of **exciting** (conventional) luminescence is about 2 ev, whereas the upper energy limit, if any, lies above 10^7 ev (Fig. 89).[34,63] It is possible to pre-excite some phosphors which have deep electron traps and then **stimulate** or **quench** their latent phosphorescence emissions by phonons (heat) or by particles having energies *lower* than the energies of the emitted luminescence photons (Figs. 25, 26, and 83–86). In general, the excitation and emission spectra, luminescence efficiency, phosphorescence, stimulability, and quenchability of phosphors are determined *jointly* by (1) the *chemical composition* of (a) the host crystal and (b) the activator(s), (2) the *crystal structure and lattice spacing* of (a) the host crystal and (b) the activator and its ligands (the luminescence center), and (3) the *physical conditions of operation*, especially the temperature of the luminescing phosphor.

If a plot were to be made of the number N of efficient phosphors available for operation at different temperatures T, the curve would probably approximate an expression of the type,

$$N = N_0 \epsilon^{-\alpha T} \tag{119}$$

where N_0 may be of the order of 10^6 and α is probably of the order of magnitude of 10^{-2} °K^{-1}, because very few phosphors are known to function efficiently at temperatures above 800°K. (This hypothetical equation applies only to conventional luminescence, because gamma-ray and x-ray emission processes are practically uninfluenced by variation of the temperature of the emitting material below 1000°K.) In practice, this means that the probability of obtaining a suitable phosphor for a particular use generally increases rapidly as the operating temperature of the phosphor is decreased.

The time dependence of phosphor luminescence, that is, the growth of radiance L during excitation and the decay during phosphorescence, may be classified according to two simplified major decay types as:

1. Temperature-independent *exponential decays* proceeding according to

$$L = L_0 \epsilon^{-at} \quad (a = \text{constant}) \tag{120}$$

where the luminescence centers are apparently excited without ionization, and there is a spontaneous radiative return to the ground state (eqs. 90 and 90a).

2. Temperature-dependent *power-law decays* proceeding according to

$$L \propto t^{-n} \quad (n = \text{variable}) \tag{121}$$

where "internal ionization" (electric conduction) accompanies luminescence and electron trapping apparently occurs intermediate to excitation and emission. Many phosphors, including those with traps, emit considerable luminescence radiation during the first 10^{-8} sec of excitation, especially when high-energy excitant particles are used.[292, 293, 680] For example, the **rise time** of growth of luminescence emission during excitation is less than 10^{-8} sec for cub.-ZnS:Ag excited by CR and for hex.-ZnS:Ag excited by alpha particles. In the absence of adequate information about the exact natures of traps and the probability of retrapping, attempts have been made to account for the observed complex power-law decays by postulating series or combinations of exponential decays (eq. 94) or by postulating a basic bimolecular recombination mechanism according to

$$L = 1/b[(L_0 b)^{-\frac{1}{2}} + t]^2 \tag{122}$$

which may be partially obscured by absorption and scattering phenomena (see discussion of eq. 94).

In temperature-independent ϵ^{-at} decays, the constant a for the radiative process has known values ranging from about 1 to over 10^7 sec^{-1} and is practically unaffected by moderate changes in temperature or in the type, duration, or degree of excitation. This type of decay is exhibited mainly by O(F)-dominated phosphors, although the later stages of these decays usually change over to the t^{-n} type. In temperature-dependent t^{-n} decays, the exponent n varies greatly with temperature and with the type, duration, and degree of excitation. This type of decay is displayed mainly by S(Se)-dominated phosphors, with the exception of those containing Mn activator. Present data indicate that there may be over 10^{17} traps cm^{-3}, with optical trap depths ranging up to about 3 ev, in some t^{-n} decay phosphors. Such phosphors, when fully excited, may store over 10^{17} potential luminescence photons per cubic centimeter of excited volume, and the duration of detectable luminescence emission after an excitation pulse as short as 10^{-9} sec may range from about 10^{-7} sec to many years. As the depths of the traps and the distances of the traps from the parent atom are decreased, the t^{-n} decay type approaches the ϵ^{-at} decay type as the limit for persistences approaching fluorescence in duration.

With respect to luminescence efficiency $\mathcal{E}$, the primary excitation energy may be converted into (1) luminescence emission, (2) heat, (3) electron emission, or (4) chemical or structural changes in the phosphor crystals. Degradation of primary energy into heat is the chief competitor of luminescence, and it is found that the $\mathcal{E}$ values of phosphors generally decrease with increasing temperature of the luminescing phosphor, except when intermediate trapping plays an important role in the luminescence. In the latter case, $\mathcal{E}$ has a maximum value at an optimum temperature which increases with increasing trap depth and is sometimes near room temperature (Figs. 103 and 104). Most phosphors are operated at room temperature, and it is found that $\mathcal{E}$ values in excess of 60 per cent can be obtained from some materials excited by photons with energies only slightly greater than the energy of the emitted photons, but it has been difficult to obtain $\mathcal{E}$ values even as high as 20 per cent when phosphors are excited by high-energy particles, that is, photons or charged material particles with energies in excess of about 10 ev (Fig. 89).

The maximum luminescence output (radiance) from phosphors *during* excitation depends on the depth of penetration of the primary excitant particles, the excitation density, the intrinsic efficiency of the phosphor under the given conditions, and the losses due to absorption of photons attempting to escape from the phosphor crystals (eq. 117). In practice, it has been possible to obtain continuous average area luminances up to about 30,000 mL from screens of rbhdl.-Zn_2SiO_4:Mn excited by intense scanning beams of 100-kv cathode rays, although the screens tend to lose efficiency rather rapidly at such high excitation intensities. This amounts to an average emission intensity per unit surface of about 10^{17} photons cm^{-2} sec^{-1}, or an average emission density per unit of *excited* volume of about 10^{20} photons cm^{-3} sec^{-1}. Under pulsed excitation, with long rests between brief (10^{-7}-sec) excitation pulses, it is possible to obtain instantaneous (spot) luminances in excess of 10^7 mL, with intense CR beams supplying over 25,000 watts cm^{-2} at 50 to 100 kv. It is found that highest sustained (area) luminances are obtained from O-dominated ϵ^{-at} decay phosphors with high activator proportions, although it may be possible to obtain highest instantaneous (spot) luminances from short-persistent S-dominated t^{-n} decay phosphors.

The maximum phosphorescence output from phosphors *after excitation* depends strongly on the time which elapses between cessation of excitation and observation or measurement of phosphorescence emission. If the waiting interval is very long, for example, minutes or hours, then only the t^{-n} decay phosphors are serviceable, and these phosphors can store up to about 10^{18} potential phosphorescence photons per cubic centimeter penetrated by the primary particles (or secondary excitants). The intensity of phosphorescence emission from such a phosphor at any instant of time is a complicated function of the excitation, the temperature of the phosphor, and the time (Figs. 71, 73–76, and 81). If the phosphorescence output is measured after a very short time interval, for example, a few microseconds, then either ϵ^{-at} decay or t^{-n} decay phosphors may be serviceable. In the ϵ^{-at} decay phosphors, the phosphorescence output at a given time after cessation of excitation is directly proportional to L_0 and inversely proportional to a, whereas the phosphorescence output of a t^{-n} decay phosphor at a given time after cessation of excitation increases nonlinearly with L_0 because the initial decay becomes more rapid as L_0 increases.

Correlation of Host-Crystal and Phosphor Properties

Some of the more useful and interesting host crystals, and several of their derivative phosphors, are presented in Table 19. Table 19 lists the phosphors in three groups: (1) The phosphors in the top group exhibit predominantly ϵ^{-at} decays whose durations increase on going downward in the table; (2) the phosphors in the middle group exhibit initial ϵ^{-at} decays followed by strong t^{-n} decays; and (3) the phosphors in the bottom group exhibit only t^{-n} decays (at least after a few microseconds) whose durations increase on going downward in the table. A general correlation between some of the luminescence and chemicophysical characteristics of useful phosphors may be obtained by referring to the exemplary data in Table 19 (also Table 5) and to standard tables and handbooks of chemistry and physics. Reading from left to right, in Table 19, these correlations include:

I. Host-crystal Properties. *1. Crystal Structures.* The most efficient phosphors, under general excitation by photons and material particles, have structures of high symmetry. This is shown by the preponderance of cubic, hexagonal, rhombohedral, and tetragonal structures listed in Tables 5 and 19 (and throughout this book).

2. Chemical Compositions. Group II elements (Be, Mg, **Zn,** Ca, Sr, and Ba) predominate as cation constituents, whereas group VI elements (O, S, and Se) predominate as anion constituents, with B, Al (Ga), Si (Ge), P, S (Se), and W (Mo) occupying key positions in the formation of oxygen-containing radicals, and F (Cl) finding occasional use as an anion constituent.

3. Melting Points. The melting points of the listed host crystals range from 870°C (ZnF_2) to 2050°C (Al_2O_3). Where mixed host-crystal compositions are used, the melting points of the mixtures are usually lower than those of the single compounds; for example, the melting-point curve of the system $Zn_2SiO_4:Be_2SiO_4$ decreases from over 1510°C, at the ends, to a eutectic point of about 1170°C near $4Zn_2SiO_4 \cdot 6Be_2SiO_4$ (determined in these laboratories).

4. Reflection Colors. All the listed host crystals, except CdS, are practically colorless; that is, they have little or no absorption in the visible region of the spectrum. It is, of course, axiomatic that a host crystal should not be strongly absorbing in the spectral region where efficient luminescence emission is to be produced. Hence, phosphors which are to luminesce efficiently in the ultraviolet or infrared regions of the spectrum should have host crystals with low absorption coefficients for radiations in those invisible regions.

5. Optical Anisotropies and Refractive Indices. Except for the optically isotropic cubic crystals, all the host crystals exhibit some bi- or trirefringence; that is, they have different densities of optical dispersion electrons in different crystallographic directions.

As a crude picture of refraction, it may be imagined that each atom having a certain oscillator frequency along a given crystal axis absorbs and reradiates photons of the corresponding frequency transmitted through the crystal in the given direction. This absorption and reradiation process would require a finite though small time interval, and so the light travels on the average more slowly in a direction of high density of such oscillator atoms (direction of high refractive index) than in a direction of low density of such oscillator atoms (direction of low refractive index). Some of the highly ionic host crystals, such as cub.-NaCl and cub.-KCl, have low refractive indices, whereas the more homopolar crystals, such as cub.-ZnS and hex.-CdS have high refractive indices. In cub.-ZnS, the refractive index for visible light increases rapidly as the temperature of the crystal increases, and some degree of correlation has been found between the optical and general physical properties of several cubic crystals.[283,428]

TABLE 19

A Tabular Summary of Some of the Host-crystal and Luminescence Properties of Several Representative Phosphors Arranged according to Type and Duration of Persistence

Compare Table 5

	Host-crystal Properties							Phosphor Properties			Excited by				Emission Color	ϵ^{-at} Decay. a in sec^{-1}	References; F = Fig.; T = Table
No.	Structure and Formula	Melting Point °C	Reflection Color	Refr. Index (* = Aniso-tropic)	Density, g cm^{-3}	[Hardness] (Moh Scale). Solubility, g 100^{-1} cm^{-3} water	Activator (wgt. per cent)	Flux (wgt. per cent)	Photolysis, or Dec. in Moist Air	Reflection Color	CR	Photons, Å 1	2537	3650			
1	Hex.-ZnO	>1800	Colorless	2.02 *	5.6	[4.5] 10^{-4}	[Zn]	....	....	Colorless (white)	x	x	x	x	1. UV 2. Blue-green	1. 10^7 2. 10^6	58, 102, 425; T5; F50, 104
2	Rhomb.-BaSO$_4$	tr. 1149-monocl. 1580	Colorless	1.64 **	4.5	[3.5] 10^{-4}	Pb(1–10)	....	....	Colorless (white)	x	x			UV + violet	10^6	246, 369; T5
3	Tetr.-CaWO$_4$	1400	Colorless	1.92 *	6.06	[5] 0.2	[W], Pb(1)	....	....	Colorless (white)	x	x	x		UV + violet	10^5	118, 353, 358, 366, 369; T5
4	Monocl.-Mg$_2$WO$_5$	1460	Colorless	*	5.6	Insol.	[W]	....	....	Colorless (white)	x	x	x		Pale blue	10^5	118, 187–190; T5; F30

No.	Substance	M.P.	Color	n	Density	Solubility	Activator (2)	[NaCl(2)] (optional)	Fluor. color					Emission	Value	References
5	Cub.⎫Hex.⎭-ZnS	tr. 1020-hex. 1850	Colorless	2.37 *	4.1	[4] 10⁻⁴		....	Pale orange	x	x	x		Orange	630	366; T5; F41, 42, 79c
6	Rbhdl.-Al₂O₃	2050	Colorless	1.77 *	4	[9] 10⁻⁴	Cr(0.5)	....	Red	x	x	x	x	Red	240	366, 194-196; T5
7	Cub.-ZnAl₂O₄	>1600	Colorless	1.78 *	4.58	[8] Insol.	Mn(0.1-2)	....	Colorless		x	x	x	Green	140	366; T5; F58, 79b
8	Rbhdl.-Zn₂SiO₄	1510	Colorless	1.71 *	4.2	[5.5] Insol.	⎱Mn(0.1-10)	....	⎱Colorless		x	x	x	⎱Green to orange-red	70 to 120	366; T5; F79b, 28, 30, 51, 52-56, 62, 90-96
9	Rbhdl.-Be₂SiO₄	>1600	Colorless	1.66 *	3.0	[8] Insol.	⎰	....	⎰			x	x	⎰		366; T5
10	Tricl.-Zn₃B₄O₉	980	Colorless	*	4.22	Sl. Sol.	Mn(≈1)	....	Colorless	x	x	x	x	Green-yellow	70	366; T5
11	Rhomb.-CdSiO₃	1240	Colorless	*	4.93	Sl. Sol.	Mn(≈1)	....	Colorless	x	x	x	x	Orange	40	114; T5; F79a
12	Monocl.-MgSiO₃	1560	Colorless	*	3.28	[6] Insol.	Mn(≈1)	....	Colorless	x	x	x	x	Red + infrared	10³ → 10	58, 63, 600, 601; F58, 79a
13	Rhomb.-CdSO₄	1000	Colorless	*	4.69	140	Mn(≈1)	Dec.	Colorless	x	x	x	x	Yellow-orange		T5
14	Tetr.-ZnF₂	870	Colorless	*	4.84	Sl. Sol.	Mn(0.1-10)	(+)	Colorless			x	x	Orange	10	58, 63; F69, 79a; T5

TABLE 19 (Continued)

A Tabular Summary of Some of the Host-crystal and Luminescence Properties of Several Representative Phosphors Arranged According to Type and Duration of Persistence

Compare Table 5

Initially ϵ^{-at} Decay Phosphors Having Long Pronounced, t^{-n} Decay "Tails"

‡ Indicates Decay Varies with Temperature, and with Type and Intensity of Excitation

No.	Structure and Formula	Host-crystal Properties					Phosphor Properties										
		Melting Point °C	Reflection Color	Refr. Index (* = Anisotropic)	Density, g cm⁻³	[Hardness] (Moh Scale). Solubility, g 100⁻¹ cm⁻³ water	Activator (wgt. per cent)	Flux (wgt. per cent)	Photolysis or Dec. in Moist Air	Reflection Color	CR	1	2537	3650	Emission Color	$\epsilon^{-at} \to t^{-n}$ Decay	References; F = Fig.; T = Table
15	Rbhdl.-$(Zn:Be)_2SiO_4$			See Nos. 8 and 9.			$Mn(1) + As(0.05)$	....	....	Colorless	x		x	x	Green to orange	Long ‡	426, 427
16	Rbhdl.-$(Zn:Be)_2SiO_4$			See Nos. 8 and 9.			$Mn(1) + Sn(5)$	....	....	Colorless	x		x	x	Green to orange	Long ‡	58, 665; F66
17	Tetr.-ZnF_2			See No. 14.		[5]	$Mn(1)$										
	Tetr.-MgF_2	1400	Colorless	1.38 *	3	0.01		....	(+)	Colorless	x		x	x	Orange	Very long ‡	58

Power-Law (t^{-n}) Decay Phosphors

No.	Material	M.p.	Color	n	Density	Solubility	Activator	Base	(+)	Color				Red + infrared (emission)	t^{-n} Decay	References
18	Hex.-CdS	1750	Orange	2.52 *	4.82	10^{-4}	[Cd], Ag(0.01), or Cu(0.01)	[NaCl(2)]	+	Orange-red	x	x	x	Orange	Very short ‡	F32-37
19	Cub.-ZnSe	>1400	Colorless	2.89	5.42	Insol.	[Zn], Ag(0.01), or Cu(0.01)	[NaCl(2)]	+	White or yel.(Cu)	x	x	x	Orange	‡	T5; F38-43
20	Cub.-ZnS	tr. 1020 →	Colorless	2.368	4.1	[4] 10^{-3}	[Zn], or Ag(0.01)	[NaCl(2)]	3+	Colorless	x	x	x	Blue	‡	110–121; T5; F32-44
21	Hex.-ZnS	1850	Colorless	2.356 * / 2.378	4.09	10^{-3}	[Zn], or Ag(0.01)	[NaCl(2)]	2+	Colorless	x	x	x	Blue–violet	‡	366; T16; F81
22	Cub.-ZnS	See No. 20.					Cu(0.01)	[NaCl(2)]	+	Green	x	x	x	Green	‡	F34
23	Hex.-ZnS	See No. 21.					Cu(0.01)	[NaCl(2)]	+	Green	x	x	x	Green	‡	T5; F37, 73, 80, 81
24	Tetr.-ZnF₂	870	Colorless	*	4.84	Sl. Sol.	Cb(0.3)	....	....	Colorless	x	x	x	Blue	‡	58
25	Ca₂P₂O₇	1230	Colorless	*	3.09	Insol.	Dy(0.05)	....	....	Colorless	x		[x]	White	‡	58; T5
26	Cub.-SrS	>1200	Colorless		3.7	Dec.	Bi(0.01)	CaF₂(2) Na₂SO₄(6)	Dec.	Colorless	Wk.	Wk.	x	Blue-green	‡	145,119,181–186
27	Cub.-CaS	>1200	Colorless	2.137	2.18	Dec.	Bi(0.01)	CaF₂(2) Na₂SO₄(6)	Dec.	Colorless	Wk.	Wk.	x	Violet	‡	145, 119, 181–186; T5
28	Cub.-SrS	>1200	Colorless		3.7	Dec.	Eu(0.03) + Sm(0.03)	CaF₂(6) SrSO₄(6)	Dec.	Yellow	Wk.	Wk.	x	Orange	Very long ‡	T5; F74, 75

373

The listed host crystals have refractive indices ranging from 1.38 (MgF_2) to 2.89 (ZnSe), with one index for cubic crystals, two for uniaxial crystals (hexagonal, rhombohedral, and tetragonal), and three for biaxial crystals (rhombic, monoclinic, and triclinic) [see Table 4]. The large host-crystal anions are major factors in determining the refractive index of the crystal, and it may be noted that the refractive index decreases in the anion order: Se^{2-} - S^{2-} - O^{2-} - F^-.

6. *Densities.* The densities of the host crystals listed in Table 19 range from 3 (Be_2SiO_4, MgF_2) to 6.06 ($CaWO_4$), the densities usually being dominated by the concentration and atomic weight of the heaviest atom. Density values are important in calculating the energy losses of material particles in phosphor crystals (eqs. 77 and 78).

7. *Solubilities and Hardnesses.* By cross-comparing the melting points and solubilities (in water) of the host crystals, it may be observed that the more insoluble host crystals have high melting points, whereas the more soluble host crystals have relatively low melting points. As shown in the periodic chart at the end of this book, hardness also increases with increasing melting point; hence, the higher-melting crystals are the harder ones. High melting point, high heat of formation, low solubility, and high hardness are indications of strong chemical and structural bonding of the host-crystal ingredients and betoken high chemical and physical stability (high resistance to decomposition and disruption).

II. PHOSPHOR PROPERTIES. *1. Crystal Structures and Chemical Compositions.* Table 19 shows that gross crystal structure alone does not determine whether a phosphor has ϵ^{-at} or t^{-n} decay. It is apparent, however, that there is a preponderance of oxygen(fluorine)-containing host crystals in the ϵ^{-at} decay group, and a preponderance of sulphur (selenium)-containing host crystals in the t^{-n} decay group. In this respect, it is interesting to note that the ionic radii of O^{2-} and F^- are practically the same (1.40 and 1.36 Å, respectively) whereas the ionic radii of S^{2-} (1.84 Å) and Se^{2-} (1.98 Å) are considerably larger. Since anions are generally much larger than cations, the host-crystal anions are major factors in determining the structures and lattice-spacings of phosphor crystals.

2. Activators. All the listed pure host crystals may be made to luminesce, after high-temperature crystallization, without added activators. The less-stable host crystals, in particular, undergo a slight partial decomposition during high-temperature crystallization (and reduction, in the case of hex.-ZnO). The more volatile decomposition products, that is, usually the anion elements, may selectively vaporize, leaving a *presumed* slight excess of one or more of the cation elements (for example, [Zn], [Cd], [W], [Si], [Al]). The luminescence emission bands of the host-crystal phosphors, made without added activators, are generally at shorter wavelengths than those which are produced by the added activators. Both the host-crystal and the activator emission bands may be made to appear selectively, or simultaneously, in certain cases by (a) using low concentrations of added activators (Figs. 45 and 52), (b) using high excitation densities (Fig. 50), (c) using very low operating temperatures,[154] or (d) adding an activator which intensifies a latent host-crystal emission (for example, Ti enhances the violet–blue emission of "pure" silicates, even in the presence of Mn,[34, 61, 154, 157–159, 429, 430] and Pb apparently enhances the violet–blue emission of certain "pure" sulphates,[246] tungstates,[155, 353] and silicates.[229, 247, 248])

All the *added* activators in useful phosphors are multivalent. The activator Mn is outstanding in giving ϵ^{-at} decays, with Sb, Cr, Tl, and Eu finding lesser use. The list of added activators affording t^{-n} decays is quite large, including Ag, Cu, Bi, Cb, Tl, and many rare-earth elements. Where both ϵ^{-at} and t^{-n} decays are desired, it is possible to produce pronounced t^{-n} decay "tails" on initially ϵ^{-at} decay phosphorescences by providing suitable traps in the phosphor. This may be done, for example, by (a) adding trapping "activators" (for example, As or Sn incorporated in certain silicates, such as Nos. 15 and 16 in Table 19), (b) producing anion omission defects (for example, making silicates with deficiency structures), or (c) introducing a smaller highly polarizing cation into the lattice to provide a polarization (distortion)-type trap (for example, substituting Be in part for Zn in rbhdl.-Zn_2SiO_4:Mn, or substituting Mg in part for Zn in tetr.-ZnF_2:Mn.

It should be noted that the *added* activators of ϵ^{-at} decay phosphors are added in proportions generally *exceeding* 0.3 weight per cent, whereas the *added* activators of t^{-n} decay phosphors are added in proportions *below* 0.3 weight per cent. It should be noted, also, that (a) in the case of ϵ^{-at} decay phosphors, the ionic radii of the activator and at least one of the host-crystal cations are usually within 15 per cent of each other, thereby allowing intersubstitution, whereas (b) in the case of t^{-n} decay phosphors, the ionic radii of the activator and the host-crystal cations may be quite different, since these activators presumably occupy interstitial sites where the available room is determined by the particular crystal structure and lattice spacing (dominated by the anion constituents). *Note: Consult Table 6 and the periodic chart at the end of this book for information on ionic radii.*

3. Fluxes. Table 19 shows that efficient ϵ^{-at} decay phosphors are generally made *without* fluxes, whereas t^{-n} decay phosphors are generally made *with* fluxes (the fluxes usually are not incorporated in ZnS-type phosphors, but they are incorporated in alkaline-earth–sulphide-type phosphors). Fluxes are apparently necessary to promote low-temperature crystallization of the relatively unstable S(Se)-dominated phosphors which would swiftly decompose if heated near their melting points.[678] Most O(F)-dominated phosphors may be heated above their melting points without undue decomposition, although selective sublimation may occur when the host crystal is formed from two or more simpler ingredients (for example, ZnO selectively sublimes from zinc-silicate batches at temperatures above 1200°C [58]).

4. *Decomposition or Photolysis in Moisture.* Phosphors whose host crystals have solubilities much greater than 10^{-2} g 100^{-1} cm^{-3} of water are generally hygroscopic or even deliquescent in moist air. The oxygen-dominated ϵ^{-at} decay phosphors are usually quite stable in ordinary atmospheres, whereas the less stable sulphur-dominated t^{-n} decay phosphors tend to decompose in air-borne moisture, evolving obnoxious telltale hydrogen sulphide (this is especially true of the alkaline-earth sulphides and selenides). Predominantly ionic crystals, such as many of those dominated by halogen ions, are very susceptible to photodecomposition. The photodecomposition apparently proceeds by photon-produced transfer of electrons from the halogen anions to the metal cations, for example, $Me^{2+}X^{2-}$ becomes (locally) $Me^{+}X^{-}$, and the X^{-} may be left excited in the case of weakly polar crystals (for example, ZnI_2, $CdCl_2$), or it may be left unexcited in the case of strongly polar crystals (for example, MgX_2, BaX_2).[431]

Some of the sulphur-dominated phosphors exhibit pronounced **photolysis**; that is, their chemical decomposition is promoted and accelerated by the simultaneous action of shortwave UV ($\lambda < 3000$ Å) and atmospheric moisture.[432–438] If a hex.-ZnS:[Zn] or hex.-ZnS:Ag phosphor is completely dry while it is exposed to 2537-Å UV, or if it is moist but is not exposed to the UV, there is practically no discoloration of the crystals, but, if the phosphor is exposed to 2537-Å UV and humid air simultaneously, the crystals darken in a few minutes or seconds, depending on the intensity of the UV. The rate of darkening increases with increasing intensity of the short-wave UV and is accelerated by the presence of certain halide salts (for example, residual NaCl flux!) in the hydrosphere around the moist crystals. The cubic form of ZnS is reported to be less susceptible than the hexagonal form is to photolysis, and it is further reported that the darkening of hex.-ZnS:Cu phosphors in sunlight is greatly decreased by coating the crystals with potassium silicate and suspending them in an oil which selectively absorbs radiation below about 3200 Å.[435] Photolytic darkening is attributed to electrolytic production of free zinc (and free sulphur) at the moist surfaces of

ZnS crystals.[437] The electrolytic potentials necessary for the electrolysis are believed to arise as the result of a two-quantum photoexcitation process. This process presumably produces "free" internal electrons and positive holes which wander to the moist crystal surfaces to provide separated oppositely charged Zn^+ and S^- ions. These ions attract and neutralize each other to form free Zn and S on the crystal surfaces. The darkening is spontaneously reversible if the darkened specimens are left in the dark in humid air, but darkened and immediately dried crystals do not decolorize when kept dry in the dark. The decolorization is accelerated if the darkened crystals are warmed and moistened with an aqueous solution of sodium chloride.

Small traces of certain multivalent elements (for example, Co or Cu incorporated in ZnS [170, 435, 436]) inhibit photolysis, presumably by disposing of the energy of internal excited "free" electrons and positive holes by radiative or nonradiative transitions before the charges can wander to the crystal surface (note, also, that ZnS:Cu has a higher T_B than ZnS:[Zn] and ZnS:Ag, Fig. 103). In this respect, it has been reported that small proportions of Ce greatly reduce the solarization of some glasses.[407] Of interest, also, is the fact that cub.-MgS:Sb(0.01) does not darken when exposed to intense 2537 Å UV under normal atmospheric conditions, despite the high solubility and low dissociation temperature of the host crystal. In general, however, the greatest chemical and physical stability is obtained from very hard, high-melting, and insoluble phosphor crystals which are as near exact chemical-combining proportions as possible, and which have been heated to as high a temperature as possible without vitrifying or unduly decomposing.

5. *Reflection Colors.* By comparing the reflection colors of the host crystals and their phosphor counterparts, it may be seen that the small proportions of activators usually leave the host-crystal color unchanged. Some activators, notably Cr and Cu, strongly color their host crystals, whereas Mn produces a noticeable color in only ZnS and ZnSe among the listed host crystals. In general, activators produce a noticeable change in host-crystal color only when the long-wave absorption edge of the host crystal is in or near the visible spectrum (this occurs with crystals of high refractive index).[170] In such cases, the small absorption "tail," which the activator produces on the long-wave side of the host-crystal absorption band, produces a visible change in color. Most of the useful ϵ^{-at} decay phosphors have about 1000-Å separation between their absorption and emission bands, whereas most of the t^{-n} decay phosphors have practically overlapping absorption and emission bands.

6. *Effective Excitants and Luminescence Efficiencies.* Charged material particles, represented by CR in Table 19, excite all the listed phos-

phors to some degree, although some of the more complex, heterogeneous, and distorted phosphors, such as Nos. 26–28, are weakly excited. A "bridge" between excitation by fast-moving charged material particles and by high-energy photons is obtained by recalling that these high-energy particles excite luminescence through intermediate secondary excitants (for example, "free" electrons). A phosphor which is efficiently excited by CR and alpha particles is generally, but not always, well excited by gamma rays and x rays. Host-crystal elements of high atomic number promote absorption of high-energy photons. A phosphor which is efficiently excited by long-wave UV, however, may be quite inefficient under x rays and charged material particles, especially when the phosphor is vitreous or is chemically complex and structurally irregular (compare patterns 8 to 10 in Fig. 11a). It may be noted that there is a preponderance of useful t^{-n} decay phosphors which are well excited by 3650-Å UV, whereas a lesser proportion of ϵ^{-at} decay phosphors is efficiently excited by 3650-Å UV (note, however, phosphors Nos. 1, 26, 36, 37, 50–54, 56, and 60 in Table 5). In general, the excitation and emission spectra of t^{-n} decay phosphors are practically contiguous, as in the case of cub.(or hex.)-(Zn:Cd)(S:Se) with Ag or Cu activator, and cub.-KCl with Ag or Cu activator,[330] whereas the excitation and emission spectra of temperature-independent ϵ^{-at} decay phosphors are widely separated, as in the case of cub.-KCl:Mn,[330] rbhdl.-Zn_2SiO_4:Mn, and tetr.-$CaWO_4$:[W] (compare Fig. 24).

7. *Emission Colors.* More quantitative descriptions of the emission colors listed in Table 19 may be found by using the reference column and the cited literature references. There is little general distinction between the room-temperature emission spectra of ϵ^{-at} decay and t^{-n} decay phosphors, although they often have different characteristics as a function of the temperature of the luminescing phosphor (Figs. 28–30 and 50). A large number of emission colors, covering the visible and near-visible spectrum, is already available in both groups of phosphors. Rhomb.-$BaSO_4$:Pb emits predominantly in the UV, with just the long-wave end of its emission band in the violet and blue parts of the visible spectrum. An important UV-emitting phosphor, γ-$Ca_3(PO_4)_2$:Ce(0.5), listed as No. 47 in Table 5, has an ϵ^{-at} decay with $a \approx 10^4$ sec^{-1}, which would place it between Nos. 4 and 5 in Table 19. An efficient unlisted orange-emitting cub.-MgS:Sb(0.01) phosphor, listed as No. 60 in Table 5, has an initial decay as rapid as that of the visible band of hex.-ZnO:[Zn] and has a T_B about 150°C higher than that of the hex.-ZnO: [Zn] phosphor.[439, 440] The cub.-MgS:Sb phosphor (and possibly hex.-ZnO:[Zn]) may be examples of the limiting case of shallow-trap t^{-n} decay phosphors which should approximate fast ϵ^{-at} decays.

8. Phosphorescence Decays. Among the efficient phosphors, the shortest detectable phosphorescences are obtained from phosphors such as hex.-ZnO:[Zn], cub.-MgS:Sb, hex.-CdS:[Cd], and cub.(or hex.)-Zn(S:Se):[Zn] (or Ag), which decay to half their peak emissions (that is, to $L_0/2$) at cessation of excitation in times of the order of 10^{-6} sec, *particularly under high excitation densities produced by primary charged material particles.* The phosphorescence decays of almost all the ϵ^{-at} decay phosphors are relatively unaffected by moderate changes in phosphor temperature, or by changes in the type, duration, or intensity of excitation, whereas the decays of the t^{-n} decay phosphors are accelerated (that is, the persistences are shortened) by (a) increasing the phosphor operating temperature, (b) increasing the excitation density, and (c) using primary charged material particles or high-energy photons (for example, x rays) instead of low-energy photons (for example, long-wave UV) as primary excitants. The longest useful temperature-independent ϵ^{-at} decay phosphorescence, at present, is obtained from tetr.-ZnF$_2$:Mn whose exponential-decay component becomes undetectable after about 1 sec of afterglow. A strontium-silicate:Eu phosphor [203] has been found to have a temperature-independent exponential-decay component lasting several seconds ($\tau = 2.3$ sec [375]), but the intensity of this phosphorescence emission is so low that it is barely discernible to the well-dark-adapted (scotopic) human eye. The longest useful room-temperature t^{-n} decay phosphorescences, at present, are those of the alkaline-earth–sulphide-type phosphors, such as Nos. 26 and 27 in Table 19. The room-temperature phosphorescence emissions of these phosphors are discernible by the *scotopic* human eye for over a week, although the luminances have then decreased to such low values that details and colors are unrecognizable.[58, 181–186] The phosphorescence emission of a hex.-ZnS:Cu phosphor has been found to be detectable by the *scotopic* eye for over 4 days, and phosphors Nos. 15–17 in Table 19 also afford phosphorescence emissions which are detectable for several days. Some of the infrared-stimulable phosphors with deep traps, for example, cub.-Sr(S:Se):[flux]:Sm(Tb):Eu and rbhdl.-(Zn:Be)$_2$(Ge:Sn)O$_4$:Mn, may be made to phosphoresce visibly for several hours at temperatures well above room temperature (Fig. 75). By cooling these deep-trap t^{-n} decay phosphors below 100°K, during and after excitation, it is possible to store their potential phosphorescence emissions indefinitely.

In Table 15 it was shown that there is an apparent correlation between the bonding energy of the host crystal and the lifetime τ of the excited state of some temperature-independent ϵ^{-at} decay phosphors with Mn activator. It was found that τ decreases with increasing bonding energy of the host crystal, and it was mentioned that the effect may be caused by an increasing perturbation of the activator atom (ion) by the closer ligands in crystals with higher bonding energy. In this respect, assuming simple ions, it is interesting to note that partial substitution of the highly polarizing Be^{2+} (charge/radius ratio = 2/0.31 = 6.5) for Zn^{2+} (charge/radius ratio = 2/0.74 = 2.7) in rbhdl.-Zn_2SiO_4:Mn (a) produces a new longer-wavelength emission band with τ about 50 per cent smaller than that of the phosphor without beryllium, (b) increases the relative intensity of the temperature-dependent t^{-n} decay "tail" on the predominant temperature-independent ϵ^{-at} decay phosphorescence curve, and (c) increases the resistance of the phosphor to loss of luminescence efficiency during operation.[58,119,221,366,462] Apparently, (a) the interaction of an Mn^{2+} and its four surrounding O^{2-} ligands is decreased by substituting Be^{2+} for Zn^{2+} as a ligand adjacent (outside) the tetrahedral array of O^{2-} around the Mn^{2+}, (b) the local distortion introduced by a strongly polarizing undersized Be^{2+} produces an electron trap, perhaps by increasing the probability of capture of an excited "free" electron by an unexcited Mn^{2+}, and (c) the introduction of the higher-melting rbhdl.-Be_2SiO_4 into the rbhdl.-Zn_2SiO_4 host crystal increases the stability of the system.

III. OTHER PHOSPHOR PROPERTIES, which are not listed in Table 19, are:

9. Crystal Sizes and Shapes (compare Fig. 10.) The high bonding energies of most of the useful host crystals make it very difficult to grow large crystals of phosphors. Growth from aqueous solution is made difficult by the low solubilities (high covalent bonding energies) of the host crystals and their corresponding tendency to nucleate prolifically near the saturation point. Growth from a melt is similarly difficult, because the density of crystal nuclei is high in the liquid phase, especially when the host crystal has a high melting point (high bonding energy). During most syntheses of phosphor crystals, by the solid-state reaction of finely divided solid ingredients, the crystal sizes and shapes are dictated largely by the sizes and shapes of the ingredient particles, their degree of contact and admixture, their rates of reaction (in the presence or absence of fluxes), and the growth of larger host crystals at the expense of smaller crystals during the heating process (reaction *and* crystallization). The intrinsic crystal habit, corresponding to the particular space group of a given phosphor, often has little influence in determining the shapes of phosphor crystals (compare Figs. 10 and 11). This is particularly true in rbhdl.-Zn_2SiO_4:Mn grown from finely divided ingredients, whereas hex.-(Zn:Cd)S phosphors, grown with fluxes which are afterwards washed off,[444] often exhibit very regular shapes of hexagonal symmetry.[58,63,678] Most phosphors have a wide variety of irregular crystal sizes and shapes in each batch, and many phosphors are optically anisotropic and selectively absorb their own luminescence emission. The general problem of scattering and absorption of photons in conventional randomly oriented fine-crystal phosphor screens is, therefore, very complex.

Despite the general difficulties of growing large phosphor crystals, centimeter-size single crystals have been grown of efficient cubic alkali halide:Tl(or Ag, or Cu, or Mn) phosphors,[51, 162, 293, 330] monocl.-$CdWO_4$: [W], tetr.-$CaWO_4$:[W], rbhdl.-Al_2O_3:Cr,[677] rbhdl.-Zn_2SiO_4:Mn,[219, 220] hex.-CdS:[Cd], monocl.-stilbene, rhomb.-chrysene, and monocl.-naphthalene.[291, 292] Although these represent only a few of the known phosphor systems, the great interest in obtaining large phosphor crystals for detecting and counting high-energy particles should accelerate expansion of the list.

10. *Resistance to Loss of Luminescence Efficiency during Comminution, Application, and Operation.* A very few phosphors may be synthesized directly in screen form; for example, hex.-ZnO:[Zn] may be condensed from a smoke made by burning zinc vapor in a slightly reducing oxygen-containing atmosphere,[215] and tetr.-ZnF_2:Mn may be formed by evaporating zinc and manganese fluorides simultaneously onto the surface to be coated,[58, 262] but these processes are difficult to control and have not afforded products with luminescence efficiencies so high as those which are regularly obtained by the conventional preparative procedures outlined in Chapter 3. The general method of condensation on a cold substrate is disadvantageous in that the crystals are not so regular and free of distortion as those produced at high temperatures. Heating the condensed materials sometimes improves their luminescence efficiencies, but the materials are very susceptible to decomposition and contamination when thus heated in the form of thin coatings. It was found that evaporated and condensed screens of tetr.-ZnF_2:Mn were at best about 10 per cent as efficient as conventional tetr.-ZnF_2:Mn screens, and the evaporated screens lost efficiency very rapidly during operation under CR excitation.

Most phosphor batches are prepared by solid-state reactions of finely divided ingredients at high temperatures, and so the individual crystals in the cooled batch are usually aggregated into irregularly shaped particles, lumps, or cakes. In order to apply these phosphors as thin screens (coatings) of uniform texture, it is often necessary to crush the cakes and deaggregate the particles and lumps by grinding (for example, by ball milling).[58, 324] General methods for applying deaggregated phosphor batches include: (a) *air settling* or *dry spraying* from a cloud suspension produced by sieving or "atomizing," (b) *wet spraying* a suspension of crystals in an easily volatilized liquid (for example, anhydrous acetone containing a fraction of a per cent of pure acetic acid to inhibit reaggregation), (c) *electrophoresis*, (d) *flowing on* a viscous suspension [for example, a suspension of phosphor crystals in a mixture of butyl acetate (100), nitrocellulose (pyroxylin) (1), and synthetic camphor (4)] whereby a smooth film of the viscous suspension remains on the surface to be coated (the organic matter may be burned away by heating in air at about 550°C), and (e) *liquid settling* (with or without centrifugation), for example, allowing the phosphor particles to settle out of a suspension in amyl acetate or pure distilled water (usually containing a few per cent of dissolved electrolytes or binders, such as ammonium carbonate, lithium germanate, colloidal silica, or potassium silicate [58, 100b, 324, 441–445]). The surface to be coated should be thoroughly clean, and contamination from dust or from impurities in the suspension media must be scrupulously avoided to obtain screens of practically uniform texture and high luminescence efficiency.[58] It should be noted that there are always statistical variations in texture over the surface of a given screen. For example, in a screen averaging 4 particles in thickness, there are some thin spots having only 1 particle (or none) and other thick spots having particles heaped up 7 (or more) high. Additional textural variables are the distribution and range of crystal sizes and their degree of aggregation as particles in a particular phosphor (Figs. 7 and 10).

Each different phosphor requires a different technique of deaggregation and application for best results. This part of phosphor technology is still very much of an art, and the reader is referred to other sources for detailed descriptions of some of the procedures which have been devised for commercial phosphors.[58, 324] Best results are generally obtained by crystallizing the phosphors in the form of loosely aggregated crystals whose average size is already optimum for the particular primary excitant and excitant penetration and for the required resolution (that is, the degree of detail to be observed on the phosphor screen). In the case of insoluble host crystals grown with the aid of soluble fluxes, for example, hex.-ZnS:Ag(0.015), [NaCl(4), BaCl$_2$(2)], 1250°C, the phosphors should be carefully elutriated to remove all traces of flux, because the fluxes generally accelerate photolysis and afford halogen ions which are injurious to oxide-coated thermionic cathodes in cathode-ray tubes. The cited phosphor crystallizes as a granular cake which crumbles into separate crystals, averaging about 15 microns in size, when the cake is washed in water containing S^{2-} ions to inhibit decomposition.[444] The washing dissolves out the flux coatings which cement the crystals together, and leaves the material free flowing. When the crystals are very small, for example, less than about 1 micron in size, it is more difficult to obtain a free-flowing product because the smaller crystals have a greater tendency to cling to each other by the action of surface forces and electrostatic attraction; also, it is more difficult to remove completely all the residual flux from the less porous small-crystal batches which have a larger ratio of crystal surface to crystal volume.

In many cases, it is necessary to ballmill the phosphor batch in order to deaggregate the crystals. This should be done as *gently* as possible

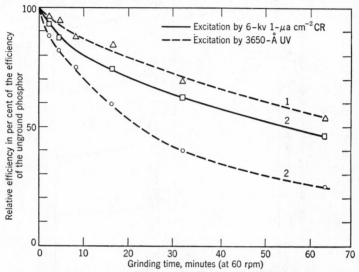

Fig. 114. Luminescence efficiencies of two S-dominated phosphors as a function of time of gentle milling in a distilled-water suspension in a 6-cm inside-diameter 20-cm-long Pyrex ball mill with 35 Pyrex balls (prolate spheroids) 1 cm × 3 cm, weighing approximately 3 g each. The mill was rotated at about 1 revolution per second.

1. cub.-ZnS:Ag(0.003), 850°C, $\bar{x}_c \approx 3$ microns
2. hex.-ZnS:Ag(0.015), 1250°C, $\bar{x}_c \approx 15$ microns

to minimize the reduction in luminescence efficiency which occurs when strains or distortions are introduced into the phosphor crystals by mechanical action. Figures 114 and 115 show some examples of the reduction in efficiency which occurs when phosphors are ballmilled. Figure 114 shows that the large-crystal ($\approx$ 15 microns) hex-ZnS:Ag suffers more than the smaller crystal ($\approx$ 4 microns) cub.-ZnS:Ag during a very gentle short-duration milling in triple-distilled water. For equal weights of phosphor in the two cases, there are more crystals to be reached by the sliding balls in the cub.-ZnS:Ag test, and the smaller crystals are more easily swept out with liquid displaced from under the sliding balls and, hence, the smaller crystals escape grinding more than the larger crystals.

Examination of these S-dominated t^{-n} decay phosphors during and

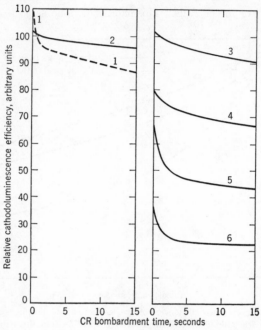

FIG. 115. Effect of vigorous ball milling and intense CR bombardment on the cathodoluminescence efficiencies of some silicate phosphors. Phosphors ball-milled in a rapidly rotated 4-liter porcelain mill with 2-cm flint balls. Excitation by a stationary unmodulated 6.4-kv CR beam delivering about 0.03 watt mm^{-2}. Measurement on the bombarded side of thick screens.

	Grinding Time, hours
1. rbhdl.-$ZnO \cdot SiO_2$:Mn	0
2. rbhdl.-$2ZnO \cdot SiO_2$:Mn, Lot A	0
3. rbhdl.-$2ZnO \cdot SiO_2$:Mn, Lot B	0
4. rbhdl.-$2ZnO \cdot SiO_2$:Mn, Lot B	16
5. rbhdl.-$2ZnO \cdot SiO_2$:Mn, Lot B	24
6. rbhdl.-$2ZnO \cdot SiO_2$:Mn, Lot B	64

after excitation under 3650-Å UV disclosed, as reported by other observers, that the photophosphorescence efficiencies decreased at a much faster rate than the photoluminescence efficiencies as the grinding time was increased.

Reference samples of these same phosphors were placed in triple-distilled water for 64 min, without grinding, and found to have unaltered luminescence efficiencies. Also, a sample of rbhdl.-Zn_2SiO_4:Mn was subjected to the same treatment as that given the phosphors in Fig. 114, and the luminescence efficiency of the silicate phosphor was found to be unaltered. This silicate phosphor generally requires much more mechanical action for effective deaggregation. Figure 115 shows the effects of long and vigorous grinding with heavy tumbling balls in an acetone suspension.[72] The left side of the figure shows that unground non-stoichiometric-proportion rbhdl.-$ZnO \cdot SiO_2$:Mn may have higher initial efficiency than the stoichiometric-proportion rbhdl.-Zn_2SiO_4:Mn, but the former loses its efficiency faster during operation. The right side of the figure shows the results of grinding another lot of rbhdl.-Zn_2SiO_4: Mn for various lengths of time. It is apparent that the grinding lowers the initial cathodoluminescence efficiency and tends to increase the rate of efficiency loss during intense excitation by cathode rays. The measurements in Figs. 114 and 115 were made on the excited sides of very thick patches of the phosphors coated on Pyrex plates.[93]

The major causes of the observed luminescence-efficiency decreases during grinding are probably:

(a) *Distortion and strain introduced into the phosphor crystals.* Microscope studies of hex.-ZnO:[Zn] crystals show that when these crystals are subjected to mechanical pressure, the crystals may become nonluminescent throughout their entire volumes without being decreased in size. The distortions and strains introduced into phosphor crystals by mechanical action tend to reduce luminescence efficiency much as imperfect crystallization (Figs. 90, 91, and 105), or crystallization in low-symmetry classes, reduces luminescence efficiency. The general effect is to increase the number of dislocations and imperfections which increase the probability of nonradiative dissipation of the primary excitation energy. (In phosphors, such as cub.-KCl:Sb, formed by grinding together an activator and host-crystal compound, the *beneficial* mechanical action apparently (a) promotes diffusion of the activator into the host crystal and (b) helps the activator atom to rearrange its ligands to form a suitable center. For a given phosphor, therefore, grinding may produce both adverse and beneficial effects, with the former generally predominating.)

(b) *Increased absorption of luminescence photons.* This is evidenced by the fact that some crystals, such as hex.-ZnO:[Zn], which have absorption edges near 4000 Å tend to become yellowish on grinding. It is not necessary for the crystals to become colored in order to become nonluminescent, however, because hex.-ZnO:[Zn] crystallized with a trace of Cu can be white and nonluminescent, whereas hex.-ZnO:[Zn] crystallized with traces of Mn or Sb can be yellow or black, respectively, and yet have weak luminescences.

Even in the absence of increased intrinsic absorption, a reduction in crystal size by comminution will increase optical scattering and increase the average path length of luminescence photons attempting to escape from a fine-crystal phosphor screen. It is this increase in optical absorption loss, due to increased scattering, which accounts in large measure for the more intense phosphorescence emission of *unground* large-crystal fractions of hex.-ZnS:Cu which are separated from and compared with the unground small-crystal fractions.[435] Here, the increased scattering occasioned by the smaller particles decreases the penetration of the primary excitant photons and increases the absorption of the phosphor for its own phosphorescence emission. The data in Table 20, taken from E. Podschus (PB 74886 [435]), are illustrative.

TABLE 20

A Comparison of the Phosphorescence Intensities (1 min after Cessation of Excitation) of Various Sieve Fractions of an Unground hex.-ZnS:Cu Phosphor Excited by 3 min of Exposure under a 100-watt Incandescent Lamp, 60 cm Distant

E. Podschus

Phosphorescent Material	Weight, Per cent of Total	Relative Phosphorescence Intensity, Arbitrary Units		
		As Dry Powder	In Poly-styrol	In Nitro-cellulose
1. Original unsieved hex.-ZnS:Cu	100	52	58.5	45
2. Portion retained by 4900 meshes cm^{-2}	5	55	62	46.5
3. Portion retained by 6400 meshes cm^{-2}	10	54.5	61	45.5
4. Portion retained by 10,000 meshes cm^{-2}	20	52.5	59	45.5
5. Portion passed by 4900 meshes cm^{-2}	95	52	58.5	45
6. Portion passed by 6400 meshes cm^{-2}	90	51.5	58	43
7. Portion passed by 10,000 meshes cm^{-2}	80	44	49.5	37

After a phosphor batch has been deaggregated with as little disruption of the individual crystals as possible, the crystals must be deposited according to the technique which best suits the average particle size and chemical reactivity of the particular phosphor.[58, 324, 441-445] Very small particles, below 0.1 micron, are usually wet-sprayed, flowed on, or liquid-settled (with centrifugation); medium-sized particles, between 0.1 and 5 microns, are usually air-settled, wet-sprayed, electrophoretically deposited, flowed on, or liquid-settled; and large particles, above 5 microns, are usually dry-sprayed, flowed on, or liquid-settled. After liquid settling, the liquid may be removed by decantation, siphoning, or evaporation. The aqueous liquid-settling technique has been used extensively in preparing cathode-ray-tube screens, and the organic-suspension flow-on technique has been used extensively in making phosphor coatings in "fluorescent" lamps. The liquid-settling technique is advantageous in making screens on flat or only slightly curved substrates, because the weight of phosphor per unit area can be accurately predetermined, whereas the flow-on and spray techniques are advantageous in coating curve-shaped objects, such as cylinders and spheres.

During settling, the larger particles deposit at a faster rate than the smaller particles because phosphor crystals obey approximately Stokes' law for the rate of sedimentation v_s of spherical particles of density σ_p in fluids of density σ_f (in grams per cubic centimeter) [58, 324]

$$v_s = r_p^2(\sigma_p - \sigma_f)g/4.5\eta \text{ cm sec}^{-1} \tag{123}$$

where r_p is the radius of the (spherical) particle (in centimeters), g is the gravitational constant (981 cm sec^{-2}), and η is the viscosity of the fluid (in poises = dyne seconds per square centimeter = grams per centimeter per second). By using centrifugation, the effective g value in eq. 123 may be increased several thousand times, and centrifugation is especially effective in accelerating sedimentation of particles with $r_p \ll 10^{-4}$ cm, that is, when the distance traveled as a result of Brownian motion equals or exceeds the distance traveled owing to gravitational attraction. A consequence of the different rates of settling of large and small particles is that the resultant screens have predominantly large particles next to the substrate and predominantly small particles in the top layer.[72] During spraying, the smaller particles tend to be siphoned first from the supply vessel, and different particle-size distributions are usually found in the center and periphery of the spray cone. When a mixture of two or more phosphors with different average particle sizes, shapes, densities, or particle-size distributions (compare Fig. 7) is to be deposited, selective deposition effects tend to produce screens whose compositions vary throughout the depth of the screen and, especially in the case of spraying, over the area of the screen. The production of uniform-texture screens from mixtures of fine-particle phosphors requires careful study and control of particle size, shape, density, and degree of aggregation as well as critical control of the fluid dynamics, reaggregation during deposition, and geometrical and mechanical features of the coating process.

When the phosphor crystals are water-soluble, *anhydrous* organic media and completely dehydrated vessels must be used during application. When the phosphor particles are very small, binders are generally not needed, but, when the particles are very large, for example, greater than 10 microns, a binder such as potassium silicate or magnesium sulphate is necessary to affix the particles permanently to the substrate.[58, 442, 445] For testing purposes, it is recommended that, whenever possible, phosphors be deposited on horizontal substrates, so that no binders need be used (binders can falsify efficiency and secondary-emission measurements), and it is recommended also that mercury-vapor pumps with liquid-air traps be used instead of oil-diffusion pumps for evacuation of test vessels. Oil vapors are readily adsorbed on phosphor crystals, and the oil films often decompose during heating or electron bombardment to leave a dark carbonaceous residue (some oils even react with certain phosphors under these conditions).

After the phosphor screen has been prepared and incorporated in an operating device where the screen is excited by suitable primary particles, the luminescence efficiency of the phosphor may increase ("bright-burn" under CR) or decrease ("dark-burn" under CR) during operation.[58] An efficiency increase may be occasioned by volatilization of an adsorbed layer acquired during application, or by recrystallization of the surface layers of the phosphor during excitation. In general, however, luminescence efficiency decreases during operation, especially when the phosphor becomes increasingly coated with an absorbent film of dust or other material, or when the screen material is bombarded by intense beams of high-energy particles which may heat the phosphor to sufficiently high temperatures to cause atomic dislocations and oxidize or reduce the phosphor (chemically speaking, a captured extraneous electron may act as a reducing agent, and a positron or a positive ion may act as an oxidizing agent [34]).[58, 102, 240–242, 446–469] Quantitative analytical studies of the specific causes of changed luminescence efficiency are lacking in most cases. Loss of luminescence efficiency may be occasioned by (a) an increase in the degree of imperfection of the phosphor crystals, thereby increasing the probability of nonradiative transitions; (b) a decrease in the proportion of luminescence-active centers; (c) increasing the absorption of the crystals, or its surface films, for the luminescence emission; and, in the case of CR excitation, (d) reducing the secondary-emission ratio of the screen.[72, 124–134] When a phosphor has an appreciable "dead-voltage layer" under excitation by charged material particles (Fig. 89), it is sometimes possible to remove part of the layer by chemical action and thereby increase the luminescence efficiency of the phosphor.[470]

In general, the harder higher-melting water-insoluble phosphors are the most resistant to loss of luminescence efficiency during comminution, application, and operation. Some of the simpler host crystals, such as the cubic, univalent, ionic alkali halides, may appear more advantageous for studies of the fundamental phenomena in phosphors, but these low-melting water-soluble crystals readily dissolve in adsorbed films of atmospheric moisture and generally dissociate and discolor (tenebresce) when excited by high-energy primary particles.[63, 66, 67, 157] The stronger partially covalent bonding of the more complex host crystals comprising multivalent elements makes their phosphors more satisfactory because their mechanisms are almost purely electronic, being without the atomic (ionic) displacements which alter the chemical and structural constitutions of weakly bonded ionic alkali-halide crystals during excitation by high-energy particles. When pulses of very high-intensity excitation are used, it is often an advantage to use phosphor crystals which are large relative to the penetration "limit" of the primary particles, because the unpenetrated portions of the crystals act as "heat sinks" which minimize the temperature rise of the crystals during the excitation pulse. In cases where photolytically susceptible or water-soluble phosphors, particularly sulphides, are to be excited by photons and exposed to actinic radiations, for example, the short-wave UV in sunlight, it is sometimes possible to provide protection for the sensitive crystals by embedding them in organic plastics or lacquers or, preferably, in inorganic enamels.[147, 158, 471–481] As previously mentioned, it has been found beneficial to coat hex.-ZnS:Cu crystals with alkaline potassium silicate and surround the crystals with a film of an oil which selectively absorbs UV below about 3200 Å. If it is desired to use an "extender," that is, a filler material to conserve the photoluminescent phosphor while preserving the hiding power of the coating, it is reportedly advantageous to mix the phosphorescent material with nonabsorbing materials of low refractive index, for example, rhomb.-$BaSO_4$. Twenty per cent of admixed $BaSO_4$ reportedly reduces the phosphorescence intensity of hex.-ZnS:Cu only 10 per cent, and 40 per cent of admixed $BaSO_4$ reduces the phosphorescence intensity of the material only 20 per cent.[435] Such inert admixtures are, of course, undesirable when phosphors are excited by high-energy primary particles.

The usually high melting points of inorganic luminescent materials stand in sharp contrast with the usually low melting points of organic luminescent materials. Most of the useful inorganic luminescent materials have melting points above 800°C, whereas aliphatic (open-chain) organic materials seldom have melting points above 150°C, and aromatic (closed-ring) organic materials seldom have melting points above 550°C.[482] Materials which melt at temperatures below about 600°C are difficult to incorporate in, for example, electronic vacuum tubes which must be heated to about 500°C to remove adsorbed gases from tube parts during evacuation.

11. *Electrical Characteristics.* Almost all phosphor crystals are insulators at or below room temperature, with specific resistivities usually
greater than 10^8 ohm cm, dielectric constants greater than 8, and very
low magnetic susceptibilities.[65, 145, 237, 279, 483-492]   At elevated temperatures,
some phosphor crystals become electronic semiconductors with resistivities between 10^0 and 10^8 ohm cm. This is particularly true of hex.-
ZnO which may be made with a room-temperature resistivity extending
from 0.2 to 10^8 ohm cm, depending on the treatment of the crystals.[215b, 490]
A plot of the logarithm of the resistivity versus the reciprocal of the
temperature of the crystals affords a straight line, such that resistivity
decreases (conductivity increases) with increasing temperature, although
Hall-effect measurements show that the mobility of the electrons in the
hex.-ZnO crystals is little affected by temperature.[490]   Accurate measurements of the electrical characteristics of most fine-crystal phosphors are
made difficult by uncertainty in the ratio of surface to volume effects.
In the determination of photoconduction, for example, Voyatzakis claims
that the high photoconduction currents obtained with certain phosphorescent sulphide and fluoride phosphors disappear when the materials
are completely dried, for example, *in vacuo*,[487] whereas Randall and
Wilkins, Hardy, Parker, and others report large photoconduction currents for similar phosphors when dried or *in vacuo*.[119, 123]   In general,
phosphors with low-melting host crystals have low resistivities, for
example, tetr.-ZnF_2:Mn has a high conduction in the dark at room
temperature, and the resistivities of phosphors increase as the hardness
and melting points of their host crystals increase.

Garlick and Gibson report that the dielectric constants of hex.-ZnS:
Cu phosphors, *in vacuo*, increase from about 8 to 14 under moderate
intensities of UV excitation, and report a corresponding increase in
dielectric loss; for example, there was a change from an equivalent ac
conductance of less than 10^{-8} ohm^{-1} cm^{-1} for the unexcited phosphor
to 10^{-5} ohm^{-1} cm^{-1} for the excited phosphor.[238]   The dielectric effects
are apparently due chiefly to *trapped* electrons, rather than "free" electrons en route to or from traps. Calculations from the experimental
data show that the relaxation times of the trapped electrons in phosphors
such as ZnS:[Zn], ZnS:Ag, and ZnS:Cu are of the order of 10^{-8} sec,
with very little dependence on temperature over the range from 90 to
460°K. During excitation below saturation the changes in dielectric
constant $\Delta\epsilon_D$ and dielectric loss $\Delta\gamma_D$ were found to follow the expression,

$$\Delta\epsilon_D (\text{or } \Delta\gamma_D) = A \log (I' + 1) \qquad (124)$$

where A is a constant and I' is proportional to the excitation intensity
for excitation by photons. The dielectric changes were little dependent

on the wavelength of the photons used for excitation in the range from 3500 to 4100 Å but decreased greatly on going to 2500 Å, because the short-wave primary photons did not penetrate very far into the phosphor crystals. Also, the decrease in the dielectric constant ϵ_D and dielectric loss during thermostimulated phosphorescence decay at room temperature proceeds at a much slower rate than the decrease in intensity of phosphorescence emission L (compare Figs. 84 and 85). With hex.-9ZnS·CdS:Cu, for example, $\epsilon_D \gtrsim t^{-1.3}$, whereas $L \gtrsim t^{-0.7}$ over the decay interval from about 4 to 1000 sec.

With respect to magnetic properties, there is a paucity of data on phosphors. It is reported that cub.-CaS:Bi becomes slightly more diamagnetic in the excited state.[279] This is interpreted as evidence that an excited electron in this photoconducting t^{-n} decay phosphor does not leave its parent atom completely, but merely moves in a much larger orbit than in the unexcited state; that is, the ψ of an excited electron may extend over a much greater domain in the crystal but remains anchored to a positive hole (compare Fig. 16c). If the system were originally nonparamagnetic, that is, all the spins were paired in the unexcited state, and excitation were to result in ionization, then the system should become slightly paramagnetic (or less diamagnetic) because there would be an unpaired spin left behind by the excited electron. The excited and freed electron may increase [700] or decrease the paramagnetism of the center where it is trapped, according to whether the excited electron is bound without pairing its spin or pairs its spin with that of an electron bound to one of the ligand atoms in the trap. Experiments reported by Rupp indicate that (1) cub.-ZnS, 1000°C (no flux) is more paramagnetic than the initial precipitated and dried ZnS, (2) grinding decreases the paramagnetism of phosphors, (3) host crystals with the incorporated activators Bi, Cu, and Ag are generally made more diamagnetic, whereas those with incorporated Mn, Sm, and Ni are more paramagnetic, (4) excitation makes cub.-ZnS:Cu and cub.-ZnS:Mn more paramagnetic, (5) increasing operating temperature decreases the paramagnetism of both excited and unexcited phosphors, and (6) alkaline-earth sulphide-type phosphors with Sm activator become increasingly paramagnetic as the principal host-crystal anion is changed from O to S to Se.[145] An increase in paramagnetism on excitation is reported also for tetr.-ZnF_2:Mn,[263] and for fluorescein incorporated in boric-acid glass.[77, 367]

Various researchers have reported that strong electric or magnetic fields, when applied suddenly to phosphorescing crystals, cause a sudden increase in phosphorescence emission.[102, 145, 146, 485, 486, 491, 492] The hex.-ZnS:Mn and hex.-Zn(O:S):Cu phosphors, in particular, are reported to show such stimulated phosphorescence when subjected to applied electric fields of the order of 15,000 volts cm^{-1}, or to applied magnetic fields of the order of 30,000 oersteds, but experiments in these laboratories have failed to disclose appreciably large stimulation effects with samples of these materials and of several other phosphors subjected to even higher field strengths. What effect there may be is undoubtedly weak, at least at room temperature, for it seems unlikely that electrons in deep traps could be raised out by the application of such external fields, especially magnetic fields which deflect moving electrons but do not increase their energies greatly even when the magnitude of the field is being altered. An external electric field F_e applied to a crystal is internally diminished by polarization; that is, F_e is divided by ϵ_D to give an effective internal electric field $F_i = F_e \epsilon_D^{-1}$. With respect to an electron, of charge e, the effective internal field may lower the activation energy, that is, the height of the potential barrier between two atoms a distance x apart, by an amount of the order of $eF_i x$. For most crystals, F_e may not exceed about 10^6 volt cm^{-1} without producing dielectric breakdown, and so the maximum internal potential gradient is about 10^5 volt cm^{-1}. For $x \approx 2 \times 10^{-8}$ cm, $eF_i x$ is at most about 2×10^{-3} ev which is insignificant relative to kT at room temperature. It is concluded, therefore, that electric fields below breakdown do not produce appreciable direct electronic excitation in insulators. It is possible, however, that previously excited phosphorescence may be stimulated by an electric field applied to a phosphor crystal. The applied field may change the activation energy for release of a trapped electron by altering (a) the shapes and distributions of the electron clouds of atoms (atomic polarization), (b) the spacings and orientations of the atoms (crystal polarization), and (c) the crystal structure (see ferroelectric effect [64b] and electrets [692]). In addition to facilitating release of trapped electrons, electrically produced polarizations may sometimes measurably alter the luminescence characteristics of solids, but the effects should generally be small for insulator crystals of high bonding energy.

Destriau reported in 1942–48 that certain micro-crystalline phosphors, particularly zinc sulphide hosts with copper activator, luminesce when embedded in thin organic films in condenser cells driven by 100 volts alternating potential.[491] In the 1950's this *electroluminescence* became an outstanding newcomer in luminescence. Later work established the presence of threads of relatively conducting copper sulphide in the insulating zinc sulphide host, and Fischer, Garnell, and Dillson showed that electroluminescence occurred alternately around the ends of the conducting threads in the applied alternating field.[754,756] Thus, it is field emission and injection from sharp points in an inhomogeneous material that gives excitation energies in excess of 2 ev, not pre-breakdown field effects in a homogeneous medium. A superior electroluminescent phosphor is hex.-ZnS:Cu:[I], with a high proportion of copper activator, and sophisticated incorporation of the iodine coactivator. Cells with this phosphor have given over 1000 millilamberts, upon application of several hundred volts at several thousand cycles per second, but efficiencies have been only about one per cent.

Destriau has reported, also, that a crystal of hex.-ZnO:[Zn] exhibits anisotropic release of trapped electrons on application of a 15,000-volt cm^{-1} electric field when the crystal is immersed in oil.[491,492] According to this report, (a) sudden application of the field in a given direction across a previously excited crystal of hex.-ZnO:[Zn] causes a sharp burst of increased phosphorescence emission, (b) reversal of the polarity of the electric field (or rotation of the crystal through 180° and reapplication of the field) causes no further burst, (c) rotation of the crystal by 90° followed by reapplication of the field causes a new burst, and (d) reversal of the field (or the crystal), as before, causes no further burst. These reported results, if confirmed, would indicate that electrons released from traps tend to be released in preferred directions or to travel along "tracks," presumably determined by the major crystal axes, in these anisotropic crystals. Experiments in these laboratories have shown that some batches of hex.-ZnO:[Zn] apparently have deep traps, despite the microsecond initial decay of this phosphor. This is readily demonstrated by cleaving a freshly crystallized phosphor cake under weak UV and observing in some cases a slow growth of luminescence emission (compare Fig. 74). When the deep traps have become filled, the slow-growth effect disappears and does not occur on subsequent re-excitations even after long waiting intervals in the dark. Attempts have been made to obtain glow curves of these slow-growth hex.-ZnO:[Zn] phosphors by heating previously excited samples in the dark to temperatures several hundred degrees above room temperature, but none of the samples emitted sufficient thermostimulated phosphorescence to be detected by either the photopic eye or a detector using a blue-sensitive multiplier phototube (No. 931-A).

With respect to reported bursts of phosphorescence emission produced by the sudden application of an electric field, A. E. Hardy has reported that a weak flash may be produced by one sudden reversal of the polarity of the electrodes of a photoconduction cell containing hex.-9ZnS·CdS:Cu(0.0073), 1250°C.[123] Further polarity reversals produced no further visible effect. In this experiment, about 70 microns thickness of the phosphor crystals was held firmly in contact with the electrodes maintained at 400 volts. As shown in Figs. 116 and 117, the concurrent growth and decay of photoluminescence emission and electric conduction of hex.-9ZnS·CdS:Cu follow approximately the same intensity–time relationships (compare Fig. 81), and Fig. 118 shows that red-plus-infrared ($\lambda > 5900$ Å) irradiation of this phosphor during simultaneous excitation by UV decreases (quenches) the photoconduction (compare Figs. 26 and 82). The striking parallelism of luminescence and conduction which occurs in the case of some t^{-n} decay phosphors has led to the plausible notion that the same "free" electrons are responsible for both effects, that is, that in both cases electrons are freed from their

parent atoms and travel considerable distances through the crystal
before being trapped or making a radiative or nonradiative recombination with an ionized center.   It is by no means certain, however, that
the same electrons or mechanisms are involved in both photoconduction
and photoluminescence, even though the parallelism shown in Figs. 116
and 117 invites such a unified view of the two phenomena.   For example,
Hardy has reported that the infrared-excited photoconduction of hex.-
CdS:Cu decays to its half-value in about 20 sec, whereas the half-life
of phosphorescence emission of this phosphor after excitation by UV or

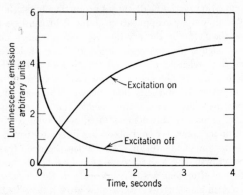

Fig. 116.   Growth and decay of luminescence emission during and after excitation
of hex.-9ZnS·CdS:Cu(0.0073), 1250°C by 3650-Å UV.   (A. E. Hardy)

visible light is only a few milliseconds (compare Fig. 81).   Furthermore,
it is reported that among the ZnS-type phosphors those with the shortest
persistences have the highest photoconductions,[121] and excellent photoconductions may be obtained from practically nonluminescent crystals
of hex.-CdS.[292]   These results indicate that, although the conduction
accompanying the luminescences of some t^{-n} decay phosphors may be a
necessary part of the luminescence mechanism, in other cases the conduction may be largely independent of the luminescence, and so the
magnitude and duration of conduction is not a clear indication of the
electron motions during luminescence.

12. *Polarization of Radiation, and Changes in Optical Properties during
Excitation.*   A simple picture of radiation polarization, apart from that
produced by optically anisotropic solids, may be obtained by imagining
that a free atom at the first instant of excitation has been transformed
into an electric dipole whose dipole axis coincides with the orientation
of the electric vector of the primary photon which was annihilated in
the excitation act.   If the excited (dipole) atom rotates before emitting,

the emitted luminescence photon will not be oriented in the same direction as the primary photon, but, if the emission process is completed before appreciable rotation occurs, the polarization direction of the emitted radiation will be determined by that of the primary radiation used for excitation. When the excited atoms are closely associated with other atoms, as in liquids and solids, thermal agitation tends constantly to disorient the excited atoms, and so polarization phenomena are best

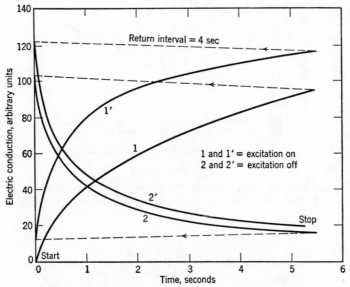

FIG. 117. Growth and decay of electric conduction during and after excitation of hex.-9ZnS·CdS:Cu(0.0073), 1250°C by 3650-Å UV. (A. E. Hardy)

observed at very low temperatures and with very rigid luminescent materials. In luminescing liquids, the degree of polarization is found to increase from vanishingly low values, for low-viscosity liquids, up to almost 50 per cent for high-viscosity liquids containing luminescent organic dyes.[145, 152, 340] In solids, the easy directions of polarization generally correspond to the directions of the major crystal axes, and polarized luminescence emission may be obtained by exciting certain *cooled short-persistent* ϵ^{-at} *decay phosphor crystals* by primary radiation polarized in the directions of some of the major crystal axes. The general diagram of Fig. 119 may be used to help visualize some of the observed results. In the case of hydrated crystals of magnesium platinocyanide, where the axes should be pictured as orthogonal, (*a*) primary photons polarized in the *Z* direction reportedly produce yellow luminescence emission with the emitted photons likewise polarized in the Z direction when observed along the Y direction, whereas (b) primary photons polarized in the Y direction produce red luminescence

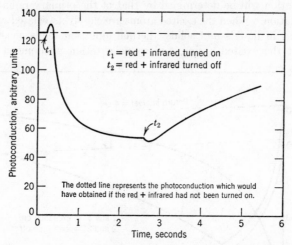

t_1 = red + infrared turned on
t_2 = red + infrared turned off

The dotted line represents the photoconduction which would have obtained if the red + infrared had not been turned on.

Fig. 118. Effect of red and infrared irradiation on the photoconduction of hex.-9ZnS·CdS:Cu(0.0073), 1250°C during sustained excitation by 3650-Å UV. (A. E. Hardy)

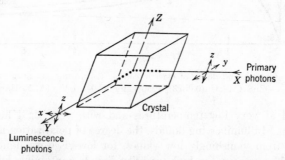

Fig. 119. General diagram representing the excitation of a phosphor crystal by polarized primary radiation and the emission of polarized luminescence radiation. X, Y, Z = major crystal axes.[107,152]

emission with the emitted photons polarized in the X direction when observed along the Y direction.[152] This is an unusually close association of the excitation and emission processes. The platinocyanides reportedly luminesce only in the crystalline state, and their spectral characteristics appear to be distinctive properties of the particular molecular complex (including the water of crystallization) in much the same way that the emission bands of many luminescent silicates are distinctive characteristics of the tetrahedral arrangement of oxygen atoms around silicon in the SiO_4 group (Fig. 51). Other luminescent materials which have been observed to provide polarized luminescence emission lines when excited at very low temperatures are several uranyl salts and rbhdl.-Al_2O_3:Cr.[152, 340, 493–497] The low-temperature line emission of rbhdl.-Al_2O_3: Cr, which is polarized parallel to the major crystal axes, exhibits an isolated instance of Zeeman-effect splitting of the lines in a magnetic field. This is a further indication that the Cr activator atoms (ions) are substitutionally located in the host crystal rather than located in irregular omission defects or interstitial sites. No Zeeman effect has been observed for the sharp-line emissions of rare-earth activators in their phosphor host crystals.[152] Individual crystals of fast exponential-decay s-center phosphors, operated at low temperatures, may be expected to exhibit polarized luminescence emission, but, the longer the lifetime of the excited state, the greater the probability of disorientation by thermal agitation, emission by multipoles higher than a dipole, or exchange (resonance) perturbations involving neighboring atoms. Long-persistent phosphors with i-centers and t^{-n} decays should exhibit practically no polarization of luminescence other than that produced by double refraction and by emission at large angles from the surface normals of the crystals (light emitted from uranium glass at 80° is reported to be 35 per cent polarized [145]). Also, conventional phosphor screens, comprising myriads of randomly oriented crystals, should exhibit negligible polarization effects. As a research possibility, some phosphors may exhibit enhanced emission of preferentially polarized radiation when their crystals are grown or operated in strong electric or magnetic fields,[145] especially if the field directions be made parallel to the major crystal axes which afford easy polarization.

In general, the optical reflectivities and absorptivities of phosphors are little changed on going from the unexcited to the excited state.[498] This is so because the relative proportion of excited atoms or "free" excited electrons during even intense excitations is usually much less than about 10^{-5}. Insofar as refractive index and dielectric constant are related, the refractive index of an excited phosphor may increase on excitation (when the dielectric constant may be doubled [238]). The major optical absorption effect which has been susceptible to investigation is the possible appearance of an induced absorption band, of longer wavelength than the emission band, in phosphors with high trap densities (compare Fig. 26a).[330]

13. Heat Capacities. The heat capacities of phosphors (materials!), that is, the amounts of heat required to raise one gram of material one degree centigrade, have not been measured, but they probably approximate the specific heats of their host-crystal substances. The specific heats (in cal g^{-1} $°C^{-1}$) of host-crystal substances are generally quite low in the range of phosphor operating temperatures, for example, rbhdl.-Al_2O_3 (18°C) = 0.181, hex.-ZnO (16°—99°C) = 0.125, rhomb.-$ZnSO_4$ (22°—100°C) = 0.174, $CaSiO_3$ (25°C) = 0.183, $Ca(PO_3)_2$ (15°—98°C) = 0.199, cub.-ZnS (0°—100°C) = 0.115, and hex.-CdS (26°C) = 0.091 (Landolt–Börnstein). Phosphors with low heat capacities tend to be raised to higher temperatures by a given amount of heat dissipated per unit volume (residue of the luminescence process) than phosphors with high heat capacities.

CHAPTER 7

USES OF PHOSPHORS

General Capabilities

This concluding chapter outlines some of the chief capabilities and uses of phosphors. Emphasis is placed on the broad aspects of requirements and limitations, with a few specific phosphors cited as examples of the best materials known *at present*. The quantitative data obtain for certain *present* phosphors using *present* techniques of preparation, application, excitation, and detection. These data are constantly being superseded as improvements are made in existing phosphors and in the techniques of their application and use and also, as new phosphors, techniques, and uses are devised by over 200 phosphor researchers in domestic and foreign laboratories. It must be kept in mind that most phosphors specified according to simplified notations, such as hex.-ZnS:Ag or rbhdl.-$(Zn:Be)_2SiO_4:Mn$, may differ greatly in crystal size, luminescence efficiency, phosphorescence, or other important characteristics when prepared by different researchers or different manufacturers. For the phosphors cited herein, it is assumed that the greatest care has been taken to use highly purified ingredients which are accurately compounded, thoroughly mixed, and well reacted and crystallized at high temperatures, without acquiring detrimental impurities.

Phosphors, which were unimportant alchemical novelties during the 17th and 18th centuries, found their first important uses in the 19th century as visible indicators of certain invisible energetic particles, such as ultraviolet photons (**UV**), cathode rays (**CR**), x-ray photons, and alpha particles. In recent years of the 20th century, the *direct* conversion of the energies of these invisible particles into light, at operating temperatures near room temperature, has become the major commercial function of phosphors. Meanwhile, phosphors are finding increasing scientific use in detecting these particles and others, including infrared and gamma-ray photons, fast-moving ions, and even neutrons, by converting their energies into radiations which the human eye may detect directly, or indirectly through other photosensitive devices, such as photographic films, tenebrescent materials, phototubes (usually coupled with oscilloscopes or meters), or other phosphors.[58, 63, 75, 102, 145–147, 154, 221, 237, 499–515]

Until the early 1930's, only very small quantities of phosphors were produced, largely for use in x-ray fluoroscope screens, self-luminescent watch dials and instrument dials (where the phosphors are admixed with radioactive salts of radium or mesothorium), and theatrical "black magic" with long-wave ultraviolet. In all these uses, the phosphor screens operated at such low luminances that they had to be observed in the dark with the partially dark-adapted (mesopic) eye. The low luminances were attributable in part to the low intensities of available sources of excitation and in part to the low luminescence efficiencies of the available phosphors. At that time, phosphors were considered by some authorities to be too inefficient to offer promise for commercial use in illumination,[516, 517] even though the efficiencies of ordinary incandescent lamps were less than 2 per cent, being about 10 lumens per watt, that is, less than a candle per watt. At that time, also, intensive research was undertaken to devise phosphors which could transform about one third of a watt of available cathode-ray-tube **(CRT)** electron-beam power into an image of sufficient luminance to make possible practical electronic television.[72] As a result of this research, many phosphors were devised with cathodoluminescence efficiencies exceeding 5 candles per watt, with some attaining over 10 candles per watt when all the emitted light was measured from both sides of the screen.

Under ideal conditions, (1) assuming a *point source* emitting monochromatic light at 5560 Å, 1 lumen per watt = 0.146 per cent efficiency, and 1 candle per watt = 1.86 per cent efficiency on an energy basis, and (2) assuming an *extended area source* emitting a cosine-law distribution (eq. 130) of monochromatic light at 5560 Å, 1 lumen per watt is still 0.146 per cent efficiency, but 1 candle per watt is then only 0.465 per cent efficiency on an energy basis (compare Appendix 3). The case 2 conversion figures apply to most fine-crystal phosphor screens, and it may be noted that the maximum efficiency of an idealized white-light source *emitting equal energy at all visible wavelengths* is about 200 lumens per watt, which for an extended-area source is about 64 candles per watt.[518] Higher efficiencies are obtainable by concentrating the emission into two optimum complementary monochromatic lines.[312]

Among the new and improved phosphors which were devised during phosphor research for television CRT, one of the most efficient, stable, and versatile was the *family* of phosphors symbolized by rbhdl.-$(Zn:Be)_2SiO_4:Mn$.[191] This phosphor family affords a broad range of emission colors, with energy efficiencies as high as 5 per cent under low-intensity high-voltage CR and quantum efficiencies as high as 95 per cent under 2537-Å UV. Certain yellow- or orange-emitting members of the rbhdl.-$(Zn:Be)_2SiO_4$:Mn phosphor family have been produced in quantities exceeding all other phosphors; the principal uses being (1) admixed with blue-emitting cathodoluminescent materials, such as cub.-ZnS:Ag or monocl.-alkaline-earth-silicates:Ti [154, 158, 159, 679] or rbhdl.-$(Zn:Be)_2SiO_4$:Ti,[61, 429, 430] to form efficient white-emitting CRT screens for electronic television, (2) admixed with blue-emitting photoluminescent materials, such as monocl.-Mg_2WO_5:[W] [187-190, 518] or barium silicate:Pb,[247] to form efficient white-emitting coatings in the low-pressure mercury-vapor electric-discharge lamps which are now known as "fluorescent" lamps, and (3) used alone to provide colored luminescent sign tubing for display purposes.[89, 158, 383-388, 519]

During World War II, allied radar CRT [58, 63, 520-522] and phosphorescent tapes and markers [471-481] used over 100,000 kg of phosphors, with lesser amounts being used in important special devices. One German company, the I. G. Farbenindustrie, manufactured over 135 metric tons of phosphorescent hex.-ZnS:Cu-type phosphor in the third quarter of 1944 (see PB74892, frame 8760, Office of the Publication Board). At present, the chief domestic and foreign production of phosphors is for "fluorescent" lamps (over 200,000 kg of phosphors a year), with television and radar CRT requiring over 20,000 kg a year. Production of phosphors for miscellaneous uses, such as luminescent plastics and tapes, x-ray fluoroscope screens, self-luminescent ("radium") dials, and sundry electronic devices, is not accurately known, but probably exceeds 20,000 kg a year. The retail prices of phosphors depend chiefly on their quality and the quantity in production, with most of the commercially available phosphors presently costing from $2 to $120 a kilogram.

Phosphors will undoubtedly continue to be vital parts of an increasing number of useful devices, especially devices operated at temperatures below about 300°C. This statement is based on the unique capabilities of phosphors, for example, (1) their ability to effect an *efficient direct conversion* of various primary excitant particles, with energies ranging from about 1 to over 10^7 ev, into photons having energies of the order of 3 ev [34, 63] (with the added advantage of being able to exercise considerable *control over the energy distribution* of the emitted photons), and (2) their ability to *store potential radiation energy* for times ranging from less than 10^{-7} to over 10^7 sec. As shown in Fig. 120, the human eye can detect some of the (conventional) low-energy luminescence photons directly and can detect the remainder with the aid of auxiliary photosensitive materials, such as photoelectric emitters or photoconductors used with oscilloscopes or other meters, photographic emulsions, or bleachable dyes (the sensitivities of the latter detectors are not shown in Fig. 120).[518, 523-526]

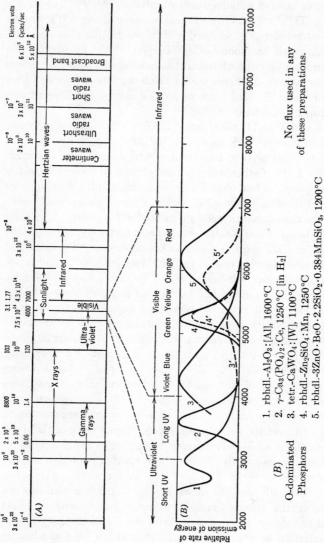

FIG. 120. The electromagnetic radiation spectrum *A*, spectral-distribution curves of the emissions of several representative phosphors *B* and *C*, and the spectral-sensitivity curves of some common detectors *D* (updated information on photocells and phototubes obtainable from RCA Electronic Components, Harrison, New Jersey).

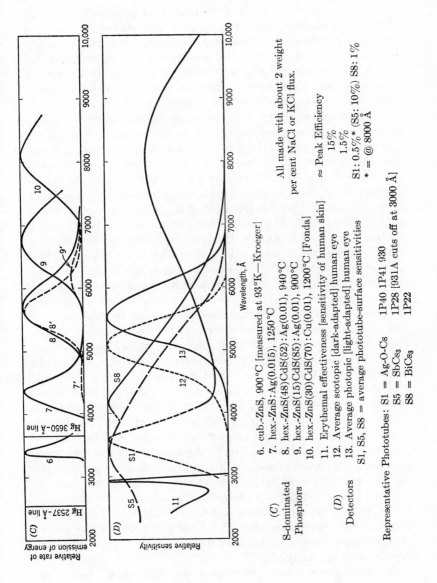

(C) S-dominated Phosphors

6. cub.-ZnS, 900°C [measured at 93°K—Kroeger]
7. hex.-ZnS: Ag(0.015), 1250°C
8. hex.-ZnS(48)CdS(52): Ag(0.01), 940°C
9. hex.-ZnS(15)CdS(85): Ag(0.01), 900°C
10. hex.-ZnS(30)CdS(70): Cu(0.01), 1200°C [Fonda]

All made with about 2 weight per cent NaCl or KCl flux.

(D) Detectors

11. Erythemal effectiveness [sensitivity of human skin]
12. Average scotopic [dark-adapted] human eye
13. Average photopic [light-adapted] human eye
S1, S5, S8 = average phototube-surface sensitivities

≈ Peak Efficiency
    15%
    1.5%
S1: 0.5%* (S5: 10%) S8: 1%
* = @ 8000 Å

Representative Phototubes: S1 = Ag-O-Cs   1P40 1P41 930
                          S5 = SbCs₃   1P28 [931A cuts off at 3000 Å]
                          S8 = BiCs₃   1P22

General Questionnaire for Prospective Phosphor Users

Although tens of thousands of phosphors have been synthesized, few reliable data on their characteristics have been published. For best results in a proposed new device using phosphors, it is usually necessary to custom-make a new or modified phosphor to meet most nearly the various requirements in the order of their importance. In seeking the most satisfactory phosphor for a proposed use, *it is essential to specify completely* **all** *the limitations and requirements,* such as:

1. CHARACTERISTICS OF THE DETECTOR (for example, human eye, phototube, photographic emulsion). If the human eye (and brain), for example, is to be used as the detector of luminescence emission from a phosphor, it is essential to specify, *inter alia:* (*a*) the size and shape of the object or area (screen) to be made luminescent; (*b*) the optical efficiency of the device (including the distance from the phosphor screen to the detector, and the characteristics of any intervening optical media and components); (*c*) the photosensitivity of the detector (as a function of photon energy), for example, the quantum efficiency, which is determined by the degree of dark- or light-adaptation in the case of the eye; (*d*) aesthetic requirements (for example, subjective desires regarding color and luminance); (*e*) resolution (definition) requirements, that is, the maximum tolerable graininess and nonuniformity of texture; (*f*) tolerance of flicker or flacker if cyclic excitation is to be used; [58, 63] and (*g*) the memory (information-storage) requirements, that is, how much longer or shorter the effective persistence of the phosphor should be relative to the persistence of the detector (the persistence of human vision is about 0.1 sec, and the retentivity of a multiplier phototube is somewhat less than 10^{-7} sec, whereas different phosphors have detectable phosphorescences ranging from about 10^{-7} sec to many months).

2. CHARACTERISTICS OF EXCITATION AND OPERATION. It is essential to specify (*a*) the energy and type of excitant particle(s), for example, 1-Å x rays, 2537-Å UV, 6.5-kv CR; (*b*) the intensity of excitation (number of excitant particles passing unit area in unit time); (*c*) whether the luminescent area (screen) is to be scanned by a fine excitant beam or flooded by the excitant; (*d*) whether the luminescent screen is to be viewed from the excited or unexcited side; (*f*) the temperature of the phosphor during operation; and (*g*) the ambient radiation and atmosphere during operation, for example, sunlight, vacuum, mercury vapor, dry air, moist air, dusty air.

3. CHARACTERISTICS OF APPLICATION AND PROCESSING. It is essential to specify (*a*) the chemical, physical, and geometric nature of the substrate on which the phosphor is to be affixed; (*b*) the radiations and

contaminants to which the phosphor may be exposed during processing, for example, sunlight, dust, cesium or barium vapor, and other chemically active vapors from sources in the device; and (c) the temperature to which the phosphor is to be heated or exposed (in this respect, a temperature of about 450°C is necessary to outgas most vacuum tubes [100]).

Uses and Recommended Phosphors

After the pertinent requirements and limitations have been specified, the selection or devisal of a suitable phosphor (and a technique for its application) usually involves research, development, and some engineering compromises. At present, in the absence of a satisfactory compilation of phosphors and phosphor properties, it is usually necessary to enlist the aid of a phosphor expert *with broad experience* in order to assure obtention of the best available phosphor for a particular purpose. Some of the considerations involved in selecting or devising suitable phosphors (and corresponding suitable application techniques) for specific uses are outlined in the following summary of major phosphor uses classified according to detectors, with subclassifications according to excitants and specific devices.

Human Eye as Detector

The *average* human eye (and brain) is one of the most efficient and adaptable photosensitive devices known, but it should be kept in mind that different *individual* eyes (and brains), or the same eye (and brain) *under different conditions*, may differ considerably in color response, visual acuity, and flicker sensitivity, and in *subjective* estimates of intensity.[58, 63, 72, 381–382, 527] The maximum effective quantum efficiency of the color-sensitive *photopic* eye, in the illumination range from 1 to 10^4 millilamberts (mL), is about 1.5 per cent, that is, one 5560-Å photon in about 70 entering the pupil of the photopic eye is used in producing visual sensations.[382] The maximum quantum efficiency of the color-insensitive *scotopic* eye, in the illumination range from 10^{-6} to 10^{-4} mL, is about 15 per cent, that is, one 5100-Å photon in about 7 entering the pupil is used in producing visual sensations. The quantum efficiency and color response of the *mesopic* eye, in the illumination range from 10^{-4} to 1 mL, varies with the state of light or dark adaptation. The scotopic eye cannot detect such small percentage luminance differences as the photopic eye and is less sensitive to flicker than the photopic eye, except for aberrate vision, whereas the photopic eye is a much better detector of color differences and has considerably higher visual acuity than the scotopic eye. The time required for the average eye to become *completely* dark-adapted is about 10 hours in the dark, whereas the light-adapted state may be regained in a few minutes depending on the intensity of illumination.[58] These data do not exhaust the peculiarities of visual processes, but they serve to indicate the wide variations in visual sensations which may be produced by the same modulated or unmodulated luminescence emission viewed by different individuals, or viewed by the same individual at different times or under different conditions. The differences between individuals are largely ignored in the following discussion where reference is made to the *average* human eye (and brain).

The spectral sensitivities of the *average* photopic and scotopic eyes are shown in Fig. 120. The effect of variation of spectral distribution of emission on *effective* (visual) luminescence efficiency is exemplified in curves 3–5 and 7–9, where the dotted curves are the product of the corresponding solid (emission) curves and the sensitivity curve of the average photopic eye (curve 13). The ratio of the relative areas of the dotted and (corresponding) solid curves is a measure of the resultant **spectral efficiency** of the particular phosphor with the given detector.[58] Blue-emitting phosphors (curves 3 and 7) and red-emitting phosphors (curves 5 and 9) are often *called* inefficient, when the real cause of low practical efficiency is the insensitivity of the eye in these outlying regions of the visual spectrum. Green, yellow, and orange emissions (curves 4 and 8) are more efficiently detected by the *average* photopic eye, and blue-green emissions are more efficiently detected by the *average* scotopic eye (see curve 12).

By careful choice of optical filters and phototubes, or other objective photosensitive devices, it is often possible to approximate the spectral sensitivity of the scotopic or photopic eye. For example, Plymale reports that a 4 to 5-mm thickness of Corning filter glass No. 3421 used with a 1P21 multiplier phototube (S4 = S5 photosurface) approximates the average scotopic eye, and 4.1-mm of Corning No. 3307 plus 1-mm of Corning No. 4308 used with a 1P22 multiplier phototube (S8 photosurface) approximates the average photopic eye.[528] Quantitative measurements of light intensity or luminance are best carried out with such *objective* photometers, in order to eliminate the subjective psychological and physiological variables which influence human vision. It must be kept in mind, however, that the data obtained with objective photometers are not always directly interpretable in terms of actual visual response. This is so because there are many intangible factors of human vision which are not assessed by objective photometers. It is known, for example, that the intangible mood or state of health of an individual affects his response to a color, shape, or luminance of an object and, *vice versa*, different colors and illuminations affect the mood (for example, alertness and mettle) of the individual. Hence, while objective photometers are valuable for research and development of phosphors and devices catering to the eye, the ultimate assessment of the phosphor or device should be obtained from the subjective responses of many observers under the proposed operating conditions (actual or simulated). In this respect, it may be noted that efficient phosphor engineers must be experienced in the following complex arts and sciences: (1) *objective* chemical synthesis and physical testing of phosphors, (2) *objective* phosphor application and engineering operation in vacuum tubes and other devices, and (3) *subjective* seeing, including the vagaries of human vision and aesthetic desires in various circumstances.

EXCITATION BY PHOTONS. *1. Photoluminescence Excited by Long-wave Ultraviolet (UV)*. It is remarkable that the most efficient photoluminescences, which are excited by long-wave UV (Fig. 89), have found little practical application. This is due largely to the relatively low efficiency and high cost of primary sources of long-wave UV.[90, 529] The commonly used 100- to 400-watt H4-type mercury-vapor electric-discharge lamps, which are operated at pressures of about 200 cm of mercury (that is, 2.5 atmospheres), emit approximately 10 per cent of their total radiation as the 3650-Å line (Fig. 63).[90] This UV emission line efficiently excites many useful phosphors which have been listed in Tables 5 and 19. The development of a high-efficiency low-cost source of, for example, 3650-Å UV would be a considerable forward step in illumination, because the efficiency of conversion of primary UV energy into light (for example, 5560-Å photons) would be about 65 per cent, instead of the approximately 45 per cent in present "fluorescent" lamps with primary 2537-Å UV. Sunlight and, to a lesser extent, incandescent-lamp light contain small proportions of long-wave UV, but these are inconvenient as well as inefficient sources of UV.[58, 63, 518]

Despite the inefficiency and high cost of commercially available sources of *primary* long-wave UV (for example, argon lamps, and medium- and high-pressure mercury-vapor lamps), there is a growing use of these sources to excite certain phosphors for theatrical effects, dynamic advertising displays, interior decoration, localized illumination during wartime blackouts (this, it may be hoped, is a waning use!), and temporary storage of information on phosphor screens or tapes. A typical theatrical effect is obtained by positioning a person in front of a large screen coated with hex.-ZnS:Cu which is excited, except for the person's shadow, by an EH-4 lamp (3650-Å UV). After a few seconds of excitation, the source may be turned off and the person walk away while his shadow remains on the phosphorescing screen. A strong photoflash lamp may be used in place of the EH-4 lamp.[518, 530] Many tons of phosphorescent tapes and markers, employing chiefly cub.-(Ca:Sr)S:Bi phosphors which are excitable by blue light, were used during the past war, especially in the interiors of ships and power plants where they provided guidance in the event of failure of the normal lighting system. Dynamic theatrical and advertising displays may be produced by intermittent excitation of screens comprising mixtures of long- and short-persistent phosphors with contrasting emission colors. For example, a screen consisting of a properly proportioned mixture of cub.-ZnS:Ag (short-persistent blue) and hex.-9ZnS·CdS:Cu (long-persistent yellow) appears blue under 3650-Å excitation and yellow in phosphorescence, whereas the opposite color change is obtained from a screen consisting of a mixture of hex.-ZnS:Ag (medium-long blue) and hex.-5.5ZnS·4.5CdS:Ag (short yellow). Dynamic effects may be obtained also by using mixtures or separate areas of different phosphors which are excited, stimulated, or quenched to different extents by different superposed or alternating radiations (for example, 2537-Å UV, 3650-Å UV, and infrared). Phosphors excited *in the dark* by well-filtered UV, or observed during phosphorescence, may give contrasts of light to dark exceeding 1000:1, whereas the reflection contrasts obtainable from light shining on ordinary pigments and dyes do not exceed about 25:1. Hence, phosphors are unique in providing vivid displays, such as murals and three-dimensional effects, especially for decor in *dimly lit* interiors.

Screens of certain efficient photoluminescent phosphors, such as hex.-5.5ZnS·4.5CdS:Ag, hex.-ZnS:Cu, and rbhdl.-Zn_2GeO_4:Mn, may be made to emit several hundred millilamberts when placed a few centimeters from a strong source of 3650-Å, such as a 100-watt EH-4 or CH-4 lamp, but the high lamp cost in practical installations where large areas are to be made luminescent usually limits the luminance to less than 20 mL. Although shorter-wavelength 2537-Å UV may be produced more efficiently, the intensity of available 2537-Å sources is even lower than that of the available 3650-Å sources. Furthermore, photons with wavelengths shorter than 3000 Å are injurious to the eye, and are strongly absorbed in ordinary air. Long-wave UV, from 3400 to 4000 Å, has been found to be generally harmless, even in fairly strong doses.[531] The chief disadvantage of moderately large flux densities of long-wave UV entering the eye is the weak blue luminescence which is excited in the vitreous humor of the eye itself. This internal luminescence in the eye makes objects appear as though seen through a light-blue mist, and visual acuity may be seriously decreased if precautions are not taken to minimize the UV reflected or transmitted to the eye by a phosphor screen and its surround.

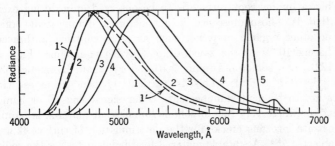

FIG. 121. Photoluminescence emission spectra of various hydrated salts of 8-hydroxy-quinoline: (1) = Ca salt, 1′ = Ca salt after a 30-sec exposure to an unfiltered CH-4 mercury arc lamp, (2) = Mg salt, (3) = Zn salt, (4) = Al salt, and (5) = ZnO: MgO:rare-earth phosphor. Excitation by 3650-Å UV, measurement with the recording spectroradiometer [93] using 0.4-mm slits. (Materials furnished by C. E. Barnett of the New Jersey Zinc Co.)

Relative Peak
Outputs
1. 340
2. 350
3. 240
4. 60
5. 20

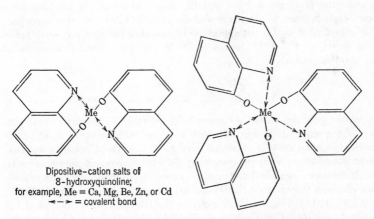

Dipositive-cation salts of
8-hydroxyquinoline;
for example, Me = Ca, Mg, Be, Zn, or Cd
◄—► = covalent bond

Tripositive-cation salt of
8-hydroxyquinoline;
for example, Me = Al
◄—► = covalent bond

FIG. 122. Structural formulas of several luminescent metallo-organic salts of 8-hydroxyquinoline.

When a luminescent surface is to be observed only during excitation, that is, phosphorescence is not required, luminescent organic dyes or other organic material, such as green-emitting anthracene: naphthacene(10^{-3}) or the zinc salt of 8-hydroxyquinoline (Figs. 121 and 122), are sometimes more satisfactory than inorganic phosphors.[74-83, 184, 185, 340, 481, 532-535] This is particularly true in cases where fabrics are to be made luminescent, as for signal flags [473] or laundry marks,[536] or where very fine or soluble luminescent material is required to penetrate into fine cracks for the determination of surface flaws in solid objects.[537, 538] A suitable material for the latter task is the sodium salt of 2-naphthol 3-6 disulphonic acid (R salt) dissolved in methyl alcohol or water.

The structural formula of R salt is $\underset{\text{NaO}_3\text{S}}{\overset{\text{OH}}{\bigodot\bigodot}}$ which has the usual characteristic of being a closed-ring structure with conjugated double bonds (compare Fig. 64), just as occur in the hydrated salts of 8-hydroxyquinoline whose emission spectra are shown in Fig. 121 and formulas are shown in Fig. 122. It may be noted that the Al, Mg, and Zn salts of 8-hydroxyquinoline are stable under intense excitation by a filtered EH-4 lamp, whereas the Ca salt loses efficiency and darkens rapidly while its emission band shifts to longer wavelengths. Much of the stability of these organic compounds has been attributed to exchange (resonance) forces associated with the covalent bonds between the N atoms and the metallic cation, although the predominantly ionic bonds between the O atoms and the metallic cation are also important. The instability of the Ca salt appears to be related to the relatively small polarizing power of the Ca^{2+} ion, as indicated in the following sequence of charge/radius ratios:

$$Al^{3+} \qquad Mg^{2+} \qquad Zn^{2+} \qquad Ca^{2+}$$
$$3/0.5 = 6 \quad 2/0.65 = 3.1 \quad 2/0.74 = 2.7 \quad 2/0.99 = 2$$

At present, only the Zn salt is commercially available. The relative peak outputs of the materials in Fig. 121 were measured with reference to the same standard material used for obtaining the data in Figs. 32–46, remembering that the photoluminescence efficiencies of many of the listed inorganic phosphors can be greatly increased by skillful preparation focused on improving the *photoluminescence* efficiency (for example, the UV-excited output of hex.-9ZnS·CdS:Cu(0.01), [NaCl(2)], 1200°C (Fig. 37) is 39, whereas the UV-excited peak output of the standard hex.-9ZnS·CdS:Cu(0.0073), [NaCl(1), BaCl$_2$(0.5), NH$_4$Cl(1)], 1260°C is 100).

The ZnO:MgO:rare-earth (undisclosed) phosphor shown as No. 5 in Fig. 121 emits strong sharp *lines* in the red region of the spectrum. These lines, which are about 10 Å wide at half their peak intensity, appear as the triangular-shaped emission curves obtained with the recording spectroradiometer.[93] At liquid-air temperature, this phosphor emits a strong yellow *band,* and there is a slight decrease in the width of the red emission line. At temperatures near 150°K the red line emission predominates under weak 3650-Å UV, whereas the yellow band emission predominates under strong 3650-Å UV (in this temperature region, also, the red line predominates during a long t^{-n} decay phosphorescence emission after weak or strong UV excitation). Heat treatment of this phosphor in a reducing atmosphere yields a product which emits the yellow band at room temperature, and this material may be converted back into the original red line emitting material by heating in an atmosphere of oxygen.

The better luminescent organic materials are sometimes less expensive than inorganic phosphors and may afford finer resolution by having practically molecular grain size in solutions, but they do not provide so broad a range of emission colors and are not so efficient, versatile, or durable as the better inorganic phosphors. Some of the most efficient 3650-Å-excited inorganic phosphors, with indications of their persistences and emission colors at room temperature, are:

   A. Persistences long relative to the persistence of vision (that is, $\gg 0.1$ sec): cub.-CaS:Bi,[flux] (very long violet), hex.-ZnS:Ag (medium long violet–blue), cub.-(Ca:Sr)S:Bi,[flux] (very long blue–green), hex.-ZnS:Cu (very long green), hex.-9ZnS·CdS:Cu (long yellow), hex.-8ZnS·2CdS:Cu (medium long orange), compare Figs. 65, 66, 71, 73, 80, and 81; and hex.-ZnS:Cu:Mn (long orange). The blue-emitting phosphors are best suited for observation by the dark-adapted eye (Fig. 120).[181-186] Note the absence of long-persistent *red*-emitting phosphors, and the absence of O-dominated phosphors.

   B. Persistences short relative to the persistence of vision: (1) S-*dominated phosphors* (subject to photolysis): cub.-ZnS:Ag:Ni (blue), cub.-9ZnS·ZnSe:Ag:Ni (blue—compare Fig. 39), hex.-7ZnS·3CdS:Ag (green), cub.-MgS:Sb (green–yellow), hex.-5.5ZnS·4.5CdS:Ag (yellow), hex.-4ZnS·6CdS:Ag (orange), hex.-2ZnS·8CdS:Ag (red) (compare Figs. 33 and 36). (2) O-*dominated phosphors:* (Ca:Sr)SiO₃:Eu (violet–blue), hex.-ZnO:[Zn] (pale blue–green), MgWO₄:U (green), rbhdl.-Zn₂GeO₄:Mn (green–yellow), CdWO₄:U (red–orange), ZnO:MgO:rare-earth (red—Fig. 121), and 4MgO·GeO₂:Mn (red).[58]

   Mixtures of phosphors may be used to obtain *additive* colors, for example, blue + yellow = white, blue + red = purple, and yellow + red = orange, except when one phosphor strongly absorbs the shorter-wave luminescence emission of another in the mixture. When the absorbed luminescence emission from one phosphor excites the absorbing phosphor, a **cascade luminescence** is obtained; for example, the yellow emission of a hex.-5.5ZnS·4.5CdS:Ag phosphor is intensified at the expense of the violet–blue part of the blue emission band of an admixed hex.-ZnS:Ag phosphor. If desirable, this selective absorption may be minimized by depositing the phosphors in separate layers, with the least-absorbing layer facing the detector.[312, 539] Cascade luminescence is sometimes obtainable in one phosphor; for example, the blue emission band (due to Ag) excites the green emission band (due to Cu) of hex.-ZnS: Ag:Cu.[58, 63, 167] Figure 123 shows that the minimum phosphor thickness for maximum blue–green luminescence emission during excitation is about half that required for maximum long-term green phosphorescence emission from a hex.-ZnS:Ag:Cu phosphor with crystals averaging about 15 microns in size.[58] This indicates that most of the emission *during excitation* comes from the layers near the surface, that is, the phosphor layers which are excited to relatively high L_0 values (eqs. 94, 98a, and 99) and, therefore, decay more rapidly than the deeper-lying layers which are excited to lower L_0 values (compare Figs. 22, 71, and 81). Most of the penetrated deeper-lying layers will have their deep traps filled and will contribute almost as much to the long-term phosphorescence emission as the shallower layers, except for the absorption of the material for its own radiation. In the case of the phosphor in Fig. 123, the blue emission (from Ag centers) penetrates deeper than the primary 3650-Å UV, and this blue emission excites underlying layers to green phosphorescence emission (from Cu centers). Also, the predominantly green phosphorescence emission is less strongly absorbed than the blue emission or the primary UV. As shown in Table 20, it is possible to increase the photophosphorescence efficiencies of phosphors by embedding selected crystals in nonabsorbing media of high refractive index (for example, polystyrol).

When the emission from a photoluminescent phosphor screen is to be detected from the excited side, it is important to have a highly reflecting substrate to minimize the amount of phosphor required to coat a given area. Excitation and observation of photoluminescence on the same side of a phosphor screen is generally not so efficient, on an over-all engineering basis, as surrounding the primary excitation source with an optimum thickness of phosphor to utilize all the primary radiation, of which a certain percentage is always reflected from the phosphor (compare discussion of Fig. 20). The relative reflectivities of primary photons by different phosphors are strongly influenced by the compositions and particle sizes of the phosphors, and so phosphors and phosphor screens must usually be custom-made to obtain maximum photoluminescence efficiency under excitation by primary photons with a given energy range.

Among the numerous useful applications of phosphors is a very satisfactory photoluminescent blackboard chalk which has been devised by

Weagle.[540] The chalk is made by mixing and molding 20 parts of hemi-hydrated calcium sulphate with 80 parts of phosphor and a small amount of polyvinyl alcohol. By selecting or mixing phosphors from the fore-going lists, many chalks excitable by 3650-Å UV have been prepared. These luminescent chalks allow lecturers to present information on

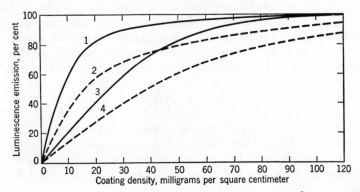

FIG. 123. Photoluminescence emission during excitation by 3650-Å UV (curves 1 and 2), and photophosphorescence emission after excitation (curves 3 and 4), as a function of coating density of hex.-ZnS:Ag(0.02):Cu(0.01), [NaCl(2)], 1250°C. Measurement on the excited side.

	Luminescence	Undercoat	Per Cent Reflectance of Undercoat
1	During excitation	White	90
2	During excitation	Black	4
3 *	Phosphorescence	White	90
4 *	Phosphorescence	Black	4

Excitation: 100-watt EH-4 lamp + Corning No. 587 filter placed 10 cm from phosphor.

Measurement: 931A multiplier phototube + Wratten No. 4 filter.

* Data on average of measurements at intervals from 5 to 960 sec after excitation.

blackboards, in vividly contrasting luminescence colors and persistences, while the room may be kept dark to show lantern slides, or to preserve the dark adaptation of an audience during special demonstrations. A luminescent crayon, which has been devised by Weil, comprises 50 to 100 parts of Japan wax, 25 to 75 parts of ozokerite, 50 to 300 parts of paraffin, and 25 to 250 parts of petrolatum jelly mixed with a small proportion of luminescent material.[541] Phosphors may be molded in various shapes in (preferably nonacid) thermoplastic or thermosetting materials,[480] for example, by mixing 10 to 25 parts of small phosphor crystals with 100 parts of finely divided polystyrene or methyl methacrylate and heating the mixture for several minutes at 140–160°C under an applied pressure of 200 to 700 kg cm^{-2} (3000 to 10,000 lb in^{-2}). Plastics provide a small amount of protection against photolysis, but they are not completely impervious to moisture, and so sensitive phosphors darken in sunlight even when well covered with clear plastic. The incorporation of an ingredient which selectively absorbs photons with wavelengths below about 3200 Å would greatly increase the amount of protection given by plastics. Superior protection against weathering and photolysis may be obtained by embedding ZnS-type phosphors in transparent inorganic enamels.[474-479] An enamel which has been found suitable is prepared by (*a*) melting a mixture of 53 g of Li_2CO_3, 15 g of $CaCO_3$, 16.5 g of CaF_2, 88 g of ZnO, 9 g of Al_2O_3, 73 g of SiO_2, and 300 g of H_3BO_3 at about 1200°C; (*b*) grinding the cooled melt, mixing the ground enamel with about 20 to 50 per cent of phosphor, dusting or sifting the mixture onto an iron plate previously coated with a conventional white ground coat of enamel, and heating at 850°C for 1.5 to 2 min.

Health precautions. When phosphors are used for novelties, especially toys, it is advisable to use only the nontoxic phosphors available. Ingested hex.-ZnS:Cu is generally harmless because it decomposes very slowly, even in dilute acids. Cadmium-containing hex.-(Zn:Cd)S:Cu phosphors, however, are usually somewhat less stable and tend to liberate toxic cadmium salts on decomposition. The alkaline-earth sulphides and selenides, such as cub.-(Ca:Sr)S:Bi, decompose rather quickly on ingestion or exposure to moisture, and are usually recognizable in plastics and tapes by the telltale odor of toxic hydrogen sulphide. Again, it may be emphasized that care should be exercised in working with phosphors and phosphor ingredients whose toxic effects have not been determined.[135-142]

Positive or negative images, or other information records, may be made by reflecting or transmitting exciting, quenching, or stimulating radiations off or through photographic prints or other records onto suitable phosphor coatings, or by "drawing" the information with a modulated beam of the radiation. The phosphor coatings may be in the form of papers, dials, tapes, drums, or other objects incorporating (1) conventional long-persistent phosphors, for example, hex.-ZnS:Cu, cub.-(Ca:Sr)S:Bi, or rbhdl.-$(Zn:Be)_2(Si:Sn)O_4$:Mn (requires 2537-Å UV); (2) readily quenchable long-persistent phosphors, for example, hex.-ZnS:Cu:Fe (Table 5); or (3) stimulable phosphors, for example, Nos. 20–22 in Table 5, or $(Zn:Be)_2(Ge:Sn)O_4$:Mn. The conventional long-persistent phosphors provide a continuously observable, though decaying, record lasting several hours or days, and the useful duration may be varied by changing phosphors, by varying the excitation, or by varying the temperature of the phosphor during or after excitation. The quenchable long-persistent phosphors provide similar continuous decaying records, though of somewhat shorter duration, with the advantage that the recorded information can be erased quickly by radiating the phosphor with photons having wavelengths adjacent the longwave side of the phosphor's emission band (for example, orange, red, or infrared radiations). The stimulable phosphors provide a practically invisible record, lasting many months, which may be brought out periodically by irradiation with long-wave photons (for example, infrared), and may be erased (quenched) by photons with wavelengths lying between the emission and stimulation bands (for example, orange light—compare Fig. 26a).

2. Photoluminescence Excited by Short-wave Ultraviolet ("Fluorescent" Lamps). Photons of ultraviolet with wavelengths less than 3000 Å are poorly transmitted by air, and are injurious to the human eye and skin. For these reasons, short-wave-UV photons are usually generated within partially evacuated electric-discharge tubes whose glass envelopes absorb the injurious short-wave-UV radiations, except when the bactericidal action of these radiations is to be used outside the lamp under controlled conditions.[518] The high efficiency of generation of the 2537-Å emission line of mercury (Fig. 63) in low-pressure electric discharges, coupled with the nearly 100 per cent quantum efficiency of several O-dominated phosphors under excitation by 2537-Å photons, has led to the chief commercial use of phosphors in the modern "fluorescent" lamp.[72, 75, 89, 90, 102, 221, 383-389, 458-463, 518, 542-547] These lamps will probably always be called "fluorescent" even though their luminescent materials are *phosphorescent* (compare Tables 10 and 19). The efficiency of light production of "fluorescent" lamps is about three times that of common incandescent-filament lamps, and "fluorescent" lamps have about three times the operational life of incandescent-filament lamps.

Conventional "fluorescent" lamps are tubular in shape, with electron-emitting (Ba:Sr)O-coated cathodes at each end, and have a 10-micron-thick coating of phosphor crystals covering the inside wall of the tube (about 10^{10} crystals in a 40-watt lamp). The lamp phosphors usually average about 1 to 5 microns in particle size and are applied in coatings containing about 1.7 mg of phosphor per square centimeter of inside surface.[460] [In "fluorescent" lamps, unlike cathode-ray tubes, the energy (penetrating power) of the primary excitant particle is constant, and the intensity of primary 2537-Å radiation is substantially constant when averaged over about a second of operation.] A common method of coating the insides of lamps with phosphors is to force a medium-viscosity suspension of the phosphor particles in nitrocellulose + organic-solvent up into the upright tubes, allow the suspension to drain, dry the thin coating which adheres to the wall, and then invert the tube and repeat the coating operation. The coated lamps are baked and evacuated, then about 0.022 mg of liquid mercury per cubic centimeter of inner volume is admitted through an inside-etched tube, and a rare gas, such as neon, argon, or (preferably) krypton, is admitted to a pressure of 3 to 18 mm of mercury according to the diameter of the lamp.[518] Then the cathodes are activated by heating, and the lamp is sealed off, based, and tested.

For operation, the conventional lamps are started by preheating the oxide-coated cathodes and applying a transient potential surge of about 1000 volts across the terminals. This high potential accelerates thermionic electrons emitted from a cathode coating into the volume of the lamp where some of the rare-gas atoms are excited and ionized and serve to bring the liquid mercury into the vapor phase by collisions of the second kind. Mercury atoms in the vapor phase are struck by accelerated electrons and emit mostly 2537-Å photons (Fig. 63). When the lamp is in operation, usually at an applied potential of 30 to 120 volts, depending on the length of the lamp, the auxiliary cathode heating is discontinued because ion bombardment serves to produce electron emission from the electrodes (each electrode is a cathode during half the alternating cycle and an anode during the other half of the cycle). Some lamps are made to start without preheating, by using a starting-potential surge of about 10,000 volts to initiate electron emission from the cold cathode, and the lamps (especially long sign tubes) may be operated at voltages above 1000 volts.

During normal operation, the surface temperatures of "fluorescent" lamps are about 27°C, and the pressure of mercury vapor is near the optimum of about 7 microns for converting about 60 per cent of the input energy in the discharge into 2537-Å radiation,[548, 549] that is, about 85 per cent of the radiated energy appears as the 2537-Å line.[90] The 2537-Å radiation is absorbed by the phosphor coating which converts the primary energy into photons of longer wavelength. For a conventional 40-watt "fluorescent" lamp as a whole, about 21 per cent of the input energy appears as light, of which 2 per cent is produced by the visible emission lines of mercury, and the remaining 79 per cent is converted into heat which is dissipated by radiation, convection, and conduction.[518]

Most of the commercial "fluorescent" lamps are made with a coating of colorless oxygen-dominated phosphors whose emission, when supplemented by a small contribution from the visible mercury emission lines (Fig. 63), appears white. The majority of white-emitting lamps have

been coated with a mixture of pale-blue-emitting monocl.-Mg_2WO_5:
[W] [187] and orange–yellow-emitting rbhdl.-$(Zn:Be)_2SiO_4$:Mn,[191] for
example, rbhdl.-$9ZnO \cdot BeO \cdot 5.1SiO_2$:Mn(3) or rbhdl.-$3ZnO \cdot BeO \cdot 3SiO_2$:
Mn(1–4).[208] Maximum stability and maximum response to Schumann-
region radiation from mercury is obtained when the silicate phosphor
is very close to orthoproportion.[558] By changing the composition and
proportion of the silicate phosphor, whose emission color is conveniently
adjustable (Figs. 51, 55, 62, and 94–96), "fluorescent" lamps are produced with several color temperatures near the Planckian locus in the
ICI diagram (compare Fig. 128), for example, a 6500°K "daylight"
white, 4500 and 3500°K general-purpose whites, and a "soft" white
which lies on the purple side of 4000°K and has ICI coordinates
$x = 0.369$ and $y = 0.3335$.[518, 550-553] The ICI coordinates of the visible
radiations from phosphor-coated "fluorescent" lamps are, of course,
different from the ICI coordinates of the radiations from the phosphors
alone (for example, under CR excitation), because there is added the
blue–green radiation of the excited mercury vapor. Recently, a barium
silicate:Pb phosphor has been proposed as a possible blue-emitting
component, and a calcium silicate:Pb:Mn phosphor has been reported
as a possible orange-emitting component for "fluorescent" lamps; also,
a single near-white-emitting calcium-halophosphate:Sb:Mn phosphor,
such as hex.-$3Ca_3(PO_4)_2 \cdot CaF_2$:Sb:Mn, has gained most in commercial
use because it is apparently non-toxic.[139, 229, 247, 250, 389, 554]

The foregoing phosphors and several other O-dominated phosphors
are useful alone or with special optical filters (usually incorporated in
the lamp glass) to provide special emissions from "fluorescent" lamps,
for example, (a) the 360BL and RP12 lamps are coated with the γ-
$Ca_3(PO_4)_2$:Ce phosphor (Table 5) and emit long-wave UV which excites
the phosphors described in the preceding section; (b) lamps coated with
monocl.-Mg_2WO_5:[W] or barium silicate:Pb, alone, emit pale-blue light;
(c) lamps coated with rbhdl.-Zn_2SiO_4:Mn emit green light with efficiencies of about 75 lumens per watt; (d) lamps coated with different
rbhdl.-$(Zn:Be)_2SiO_4$:Mn phosphors may be made to emit pink, yellow,
orange, gold, or red light (some of these colors require tinted tubes to
absorb certain of the visible mercury lines, or to absorb selectively some
of the phosphor emission); (e) lamps coated with monocl.-$CaSiO_3$:Pb:
Mn, rhomb.-$CdSiO_3$:Mn, $Cd_2B_2O_5$:Mn, hex.-cadmium-chlorophos-
phate:Mn [addition of V, Cb, Ta, or Sn(< 1) affords increased persist-
ence],[554] or tetr.-$CaWO_4$:Sm [555] emit the orange–red bands of these
phosphors (compare Table 5) along with the usual bluish mercury
spectrum which may be selectively filtered to obtain purer colors in the
orange–red region; and (f) lamps coated with certain UV-emitting

phosphors, such as $Ca_3(PO_4)_2$:Tl ($\lambda_{pk} = 3300$ Å), are useful as suntan lamps.[229b, 248]

The normal operational life of a "fluorescent" lamp is determined chiefly by (a) the rate of decrease of electron emission from the electrodes, and (b) the rate of decrease of photoluminescence efficiency of the phosphor coating. The first factor, that of electron emission from the electrodes, is influenced greatly by the number of times the lamp is turned on and off during use, because repeated heating and cooling of

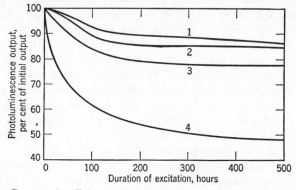

Fig. 124. Decrease in efficiency of some phosphors *in vacuo* when exposed to the radiation from a low-pressure mercury-vapor discharge lamp with a fused-silica envelope. (1) rbhdl.-$(Zn:Be)_2SiO_4$:Mn, (2) rbhdl.-Zn_2SiO_4:Mn, (3) monocl.-Mg_2WO_5:[W], and (4) γ-$Ca_3(PO_4)_2$:Ce. (G. Meister and R. Nagy)

the oxide-coated cathodes reduce their emissivity. Lamps which are operated for 3 hr between starts have a rated life of 2500 hr (to decrease to about 75 per cent of their initial luminance), whereas lamps which are operated for 12 hr between starts have a rated life of 6000 hr.[518] The second factor, that of decreasing efficiency of the phosphor during operation, has been studied by several investigators who are not in agreement concerning the causes and mechanism of loss of photoluminescence efficiency in lamps.[221, 458-463] Figures 124 and 125 show some data, reported by Meister and Nagy, on the efficiency decrease of several lamp phosphors exposed to mercury radiation when the phosphors are (a) *in vacuo*, and (b) in air.[221] It may be seen from the figures that the low-melting phosphate phosphor (No. 4) is the least stable in both cases, and the rbhdl.-$(Zn:Be)_2SiO_4$:Mn phosphor is more stable than the rbhdl.-Zn_2SiO_4:Mn phosphor in both cases. Also, the silicate phosphors are more stable than the tungstates *in vacuo*, whereas the tungstates are the most stable when excited in air. The surface layers of the silicate phosphor crystals darken when the materials are excited *in vacuo*, but the darkened layers become lighter in color, and the photoluminescence efficiency of the darkened material increases when air is admitted, especially when the material is heated to over 500°C in the presence of oxygen. When the various phosphors in Figs. 124 and 125 were exposed to mercury radiation passed through glass which transmitted little radiation below 2000 Å, it was found that the sequence in Fig. 124 obtained; that is, the silicates were the most stable, and all of the phosphors had over 85 per cent of their initial efficiencies at the end of 300

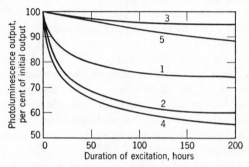

FIG. 125. Decrease in efficiency of phosphors *in air* when exposed to the radiation from a low-pressure mercury-vapor discharge lamp with a fused-silica envelope. Same phosphors as in Fig. 124, plus (5) tetr.-$CaWO_4$:[W]. (G. Meister and R. Nagy)

hr.[221] These data lead to the conclusion that radiations below 2000 Å are largely responsible for the loss in efficiency, presumably by causing a photochemical reduction of the activators Mn, Ce, or [W] (although there may be adsorption phenomena involved in actual "fluorescent" lamps [221, 459-461]). It is reported also that traces of alkali from the glass substrate (tube wall) accelerate the efficiency decrease of phosphors during lamp operation.[463, 556] There have been reports that certain luminescence-inert materials, such as alkaline-earth oxides, carbonates, and phosphates, when admixed with the phosphor particles in "fluorescent"-lamp coatings reduce the loss of efficiency during operation.[58] Lowry, for example, reports that phosphor efficiency decreases faster in lime-glass than in lead-glass envelopes, and reports that lamp phosphors which have been coated with about 0.3 weight per cent of Sb_2O_3 show less rapid loss of efficiency during operation.[462a] The protective film of Sb_2O_3 apparently decreases (1) transmission of photochemically active 1849-Å UV, and (2) chemical interaction of the phosphor particles with (*a*) excited gas atoms, and (*b*) alkali which diffuses out of the glass envelope.

The surface luminances of commercial "fluorescent" lamps range from about 300 mL (0.1 candle cm^{-2}) for red-emitting lamps to about 4700 mL (1.5 candle cm^{-2}) for green-emitting lamps, with white-emitting lamps having about 2500 mL (0.8 candle cm^{-2}).[546] At these luminances, flicker during 60-cycle sec^{-1} a-c operation (50 cycles sec^{-1} in some countries) would be very objectionable if the lamps were truly fluorescent. Instead, the phosphorescence of the rbhdl.-$(Zn:Be)_2SiO_4:Mn$ phosphor, in particular, greatly reduces the flicker in white-emitting lamps by providing light during the relatively dark interval when the potential changes direction and the current goes through zero. The stroboscopic effect from a blue-emitting lamp coated with monocl.-$Mg_2WO_5:[W]$ alone is about 90 (that is, about 90 per cent deviation from the mean light output), whereas the stroboscopic effect from a gold- or pink-emitting lamp coated with rbhdl.-$(Zn:Be)_2SiO_4:Mn$ alone is about 20.[518] By using two lamps with lead-lag ballasts, or three lamps operated on different phases of a three-phase power supply, the stroboscopic effect can be reduced to one-half or one-third that obtained with a single lamp. By using converters to transform 60-cycle sec^{-1} power into 300- to 600-cycle sec^{-1} power, flicker may be completely eliminated, and lamp efficiency is reportedly increased about 25 per cent.[557]

With the quantum efficiencies of phosphor coatings in present "fluorescent" lamps generally exceeding 90 per cent, there is little incentive to obtain further increases in the photoluminescence efficiencies of lamp phosphors.[389] There is considerable interest, however, in (a) decreasing the loss of efficiency of the phosphors during operational life; (b) devising phosphors with longer persistences, and preferably with ϵ^{-at} decays, to reduce flicker;[72] and (c) decreasing the phosphor costs and the costs of applying the phosphor coatings in the lamps. In the absence of cathodoluminescence efficiencies greater than 10 per cent, for the direct production of light from phosphors, further major improvements in lighting efficiency should be sought in the development of more efficient means for converting electric power into UV, preferably long-wave UV. Ultraviolet photons with wavelengths shorter than the 2537-Å line of mercury, for example, the 1850-Å mercury line, and Schumann-region radiations from rare gases, are known to excite many O-dominated phosphors, such as tetr.-$CaWO_4:[W]$ and rbhdl.-$Zn_2SiO_4:Mn$, but these short-wave UV radiations have not been produced with sufficient intensity or efficiency to be of commercial use.[462, 558] In any event, it seems more promising to seek efficient sources of primary long-wave UV, because the energy deficit (eq. 83), on converting UV into visible light by means of phosphors, decreases as the wavelength of the UV increases. The blue-emitting hex.-$ZnO:[Zn]$, and the yellow- to red-emitting germanate and silicogermanate phosphors, such as rbhdl.-Zn_2GeO_4:Mn and rbhdl.-$(Zn:Be)_2(Si:Ge)O_4$:Mn,[191,192] are particularly attractive as stable phosphors which are well excited by long-wave UV and have little or no selective absorption of visible light.

Some of the early researchers on luminescent solids concluded that luminescence emission was the result of a chemical process whereby, for example, one activator center was chemically destroyed for each photon emitted.[469] That this is not true is shown by a simple computation for a "fluorescent" lamp coated with rbhdl.-Zn_2SiO_4:Mn. The actual coating thickness of phosphor is about 10 microns, of which about half the coating volume is void, and so the compacted thickness of rbhdl.-Zn_2SiO_4:Mn is about 5 microns, or 5×10^{-4} cm. The number of Zn atoms (two per simple molecule) in a square centimeter of the coating is calculated by using a density of 3.9 g cm^{-3}, the molecular weight of 222.82 g per mole, and Avogadro's number $N_A = 6.023 \times 10^{23}$ simple molecules per mole. These calculations show that there are

$$2 \times (3.9/222.82) \times 6.023 \times 10^{23} \times 5 \times 10^{-4} \approx 10^{19} \text{ Zn atoms}$$

per square centimeter of coating, or about 10^{17} Mn activator atoms per square centimeter of coating (assuming about 0.04 mole of $MnSiO_3$ per mole of Zn_2SiO_4). During operation, such a coating emits about 4700 mL, and, because 1 mL $\approx 4 \times 10^{12}$ photons cm^{-2} sec^{-1} ($\overline{\lambda} \approx 5250$ Å), this corresponds to an emission density of

$$4700 \times 4 \times 10^{12} \approx 2 \times 10^{16} \text{ photons } cm^{-2} sec^{-1}$$

If the hypothesis of chemical destruction of an Mn activator center for each emitted photon were correct, this would mean that the lamp could operate only about

$$10^{17}/(2 \times 10^{16}) = 5 \text{ sec}$$

before becoming completely nonluminescent insofar as the phosphor coating is concerned. Actually, of course, "fluorescent" lamps have been operated for tens of thousands of hours before failing, and even then the phosphor coatings still had a large percentage of their initial efficiencies. Similar calculations and conclusions may be obtained for the case of excitation by primary electrons, as in the cathode-ray tubes discussed later. Hence, there is no doubt that the luminescence of phosphors is primarily a *physical* action, with chemical changes during operation being largely incidental.

3. Roentgenoluminescence (Luminescence Excited by High-energy Photons). This type of photon-excited luminescence is classified separately from photoluminescence excited by low-energy photons, because high-energy primary photons may each excite more than one luminescence photon through the intermediate action of high-velocity "free" photoelectrons produced within the phosphor crystals by absorbed x rays or gamma rays. For example, a primary 10^5-ev x-ray photon may produce over 3000 emitted roentgenoluminescence photons from a 10 per cent efficient hex.-4ZnS·6CdS:Ag phosphor screen. As indicated in Fig. 89, there is a considerable overlap of photoluminescence and roentgenoluminescence in the "hard" ultraviolet ("soft" x-ray) spectral region where a certain proportion of primary photons of a given energy may each produce only *one* emitted photoluminescence photon, and a certain proportion of the same primary photons may each produce *more than one* roentgenoluminescence photon from a phosphor. There has been little investigation of phosphors excited by photons with energies in the overlap region between photoluminescence and roentgenoluminescence, although this region may prove to be very fruitful for obtaining fundamental information about the mechanisms of phosphors, as well as for providing new technical uses for phosphors.

The chief uses of roentgenoluminescence have been in (*a*) so-called x-ray fluoroscopy, where x rays are converted directly into a visible image by a phosphor screen, and (*b*) intensification of x-ray-produced photographic images, where phosphor intensifier screens are placed adjacent a photographic emulsion and the combination is exposed to x rays. Here, the x-ray-excited luminescence emission from the phosphor screen adds to the relatively feeble photographic darkening produced by the small proportion of x rays absorbed in the emulsion.[102, 369, 391-397, 417] Phosphor screens may be used, also, to convert high-energy photons into radiation which actuates other photosensitive devices, such as multiplier phototubes [292, 504, 505] or image orthicons,[559] whose outputs may be made visible through conventional meters, permanent-recording devices, oscilloscopes, or television CRT. During medical diagnoses, it is essential to keep the x-ray dosage of the patient as low as possible. The average x-ray intensity used in clinical practice is about 5 milliroentgens sec^{-1}; and so the luminances of x-ray fluoroscope screens are usually less than 0.1 mL. Where it is essential to have brighter or larger images, *without delay*, it is possible to obtain intensification and magnification by using an electronic-television-type pickup, amplification, and projection system, or by using electronic image intensifiers.[612a] The reproduced images may then be made with luminances of the order of 50 mL, that is, bright enough to be viewed in a room with normal artificial illumination.

If a phosphor screen of adequate thickness, for example, 2 to 20 mg cm^{-2}, for efficient photoluminescence be excited by primary photons of increasing energy, then, on going from low-energy primary photons to high-energy primary photons, there is a general decrease in reflection loss and an increase in transmission loss of the primary photons (compare Figs. 20–23). It is advantageous to have heavy elements incorporated in roentgenoluminescent phosphors, because the heavy elements increase the absorptivity of the materials for high-energy photons.[2] The absorptivity of matter for x rays generally increases with increasing atomic number (nuclear charge) Z of the constituent elements, because (a) the density of electrons increases with increasing Z; that is, the volumes of atoms do not increase in proportion to the increase in Z, thereby increasing N_σ in eq. 74; and (b) the increased nuclear charge increases the bonding energy of the extranuclear electrons, thereby providing more and higher-energy characteristic frequencies ν_c in eq. 74, to approximate the frequency of the primary high-energy photon. Merely admixed heavy elements are usually of less utility than when incorporated in the phosphor, and reflecting light-metal (for example, aluminum [100, 560]) coatings on the unobserved sides of phosphor screens are generally of minor advantage because the screens are made quite thick to absorb most of the primary x rays.

Short persistence is usually required in roentgenoluminescent screens, in order to minimize blurring of moving image objects during observation. With the hex.-(Zn:Cd)S:Ag phosphors, commonly used for x-ray fluoroscope screens, high cadmium content decreases the persistence, and the incorporation of a few parts per million of nickel (as a compound) provides a further reduction in persistence.[417] A good general-purpose x-ray fluoroscope screen for visual observation may be made by coating about 0.1 g cm^{-2} of green-emitting hex.-7ZnS·3CdS:Ag [209] or orange-emitting hex.-4ZnS·6CdS:Ag (Fig. 36) on leaded glass (to protect the observer from the primary radiation). R. H. Peckham has reported that tests of 8 large-crystal ($\bar{x}_c \approx 40$ microns) hex.-(Zn:Cd)S: Ag(0.006) phosphors showed the composition hex.-ZnS(55)CdS(45):Ag to be superior to those with lower or higher Zn/Cd ratios.[696] According to this report, a 0.03-cm-thick screen of the 40-micron hex.-ZnS(55)-CdS(45):Ag(0.006) phosphor gave 2 to 21 μL under excitation by clinical x rays in the intensity range from 0.25 to 100 milliroentgens sec^{-1}.

Phosphors and thicknesses for intensifier screens, on the other hand, involve engineering compromises, depending on the energy of the primary photon, the spectral sensitivity of the photographic film, and the **resolution** (degree of definition) required in the image. The major conflict is usually between the need for thick screens to absorb the highly penetrating primary radiation (Fig. 23) and the need for thin screens to obtain high resolution. With blue-sensitive photographic emulsions, hex.-ZnS:Ag is a useful intensifier phosphor for low-voltage x rays ($<$ 100 kv), whereas tetr.-CaWO$_4$:[W], tetr.-CaWO$_4$:Pb, rhomb.-BaSO$_4$:Pb, or cub.-BaFCl are useful for high-voltage x rays ($>$ 100 kv), because they contain large proportions of elements with high atomic number.[102, 246, 353, 561] The optimum coating densities are about 40 mg cm^{-2} for intensifier coatings on the side of the photographic plate facing the primary x-ray source, and 80 to 120 mg cm^{-2} for intensifier coatings on the opposite side.[209] By using translucent sheets of short-persistent rubidium-halide or cesium-halide phosphors, prepared by compressing the fine-particle luminescent materials at about 1000 kg cm^{-2} (15,000 lb in.$^{-2}$) and 150°C, it is claimed that improved resolution may be obtained, although these low-melting water-soluble ionic crystals often deteriorate during application and operation.[562] The microsecond decays of some phosphors, such as cub.-NaI:Tl,[293] hex.-ZnO:[Zn],[503] and rhomb.-BaSO$_4$:Pb,[369] are advantageous in pulse-type x-ray stroboscopic techniques and in counting individual primary gamma-ray photons.

The optimum thickness (coating density) of phosphor screens depends chiefly on (a) whether the excitation source and the detector (observer) are on the same or opposite sides of the screen, (b) the penetration of the primary particles into the phosphor (Fig. 23), (c) the absorption of the phosphor for its own radiation, and (d) the crystal size of the phosphor (small crystals increase the optical pathlength of escaping luminescence photons and thereby increase absorption losses). A more detailed discussion of these factors is given in connection with Fig. 129. When the excitation source and the detector are on the *same* side of the phosphor screen, as in Fig. 123, there is a minimum thickness for maximum luminescence efficiency, but no single optimum thickness, because increased thickness generally does not decrease the luminescence output. When the excitation source and the detector are on *opposite* sides of the phosphor screen, as in "fluorescent" lamps, there is a well-defined optimum thickness which must be determined for each phosphor and excitant, because screens which are much thicker than the penetration of the primary particles lose efficiency by absorbing their own radiation in the unexcited layers between the excited portion of the screen and the detector, and the unexcited layers scatter radiation back toward the excitation sources.

The optimum coating densities of photoluminescent screens generally range from 2 to 20 mg cm^{-2}, being of the order of four crystal diameters in thickness, because low-energy primary photons penetrate only a few microns into most phosphor crystals. The optimum thicknesses of roentgenoluminescent screens, however, become many crystal diameters thick as the energy (penetration) of the primary photons increases. Figures 126 and 127 show some data reported by Kallmann on screens of a commercial (Auergesellschaft) hex.-ZnS:Ag phosphor, with crystals averaging about 10 microns in size, and centimeter-size crystals of

10^{-7}-sec persistence UV-emitting monocl.-naphthalene, excited on one side by primary photons with various energies ranging from soft x rays to hard gamma rays, and detected on the opposite side by a UV + blue-sensitive multiplier phototube.[292] (See page 250 for data on the luminescence emission from naphthalene.) Kallmann reports that the fine-crystal hex.-ZnS:Ag phosphor screen affords the higher

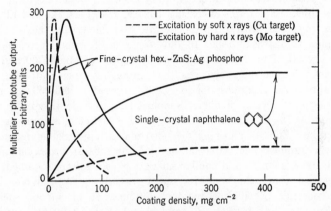

FIG. 126. Relative microsecond roentgenoluminescence outputs of two luminescent materials measured with a UV + blue-sensitive multiplier phototube on the side of the screen opposite the primary x rays. (H. Kallmann)

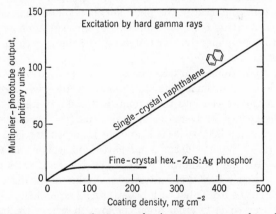

FIG. 127. Relative microsecond roentgenoluminescence outputs of two luminescent materials measured with a UV + blue-sensitive multiplier phototube on the side of the screen opposite the primary gamma rays. Rbhdl.-Zn_2SiO_4:Mn and tetr.-$CaWO_4$:[W] give curves similar to that shown for the hex.-ZnS:Ag phosphor. (H. Kallmann)

roentgenoluminescence output as long as the optimum coating density is less than about 40 mg cm^{-2}, but when the penetration of the primary (gamma-ray) photons greatly exceeds 40 mg cm^{-2}, then the large single crystals of naphthalene afford higher luminescence outputs *within a microsecond detection interval* because they do not scatter their luminescence emission and, hence, absorb less of their own radiation and reflect (scatter) less in a direction away from the detector. Expressing screen thickness in terms of coating density, as in Figs. 126 and 127, the soft x rays and hard x rays penetrate the fine-crystal hex.-ZnS:Ag phosphor to depths of about 15 and 35 mg cm^{-2}, respectively, and the large crystal of monocl.-naphthalene to depths of about 200 and 400 mg cm^{-2}, respectively; whereas the gamma rays penetrate the fine-crystal hex.-ZnS:Ag phosphor to a depth of greater than 250 mg cm^{-2}, and the large crystal of monocl.-naphthalene to a depth greater than 500 mg cm^{-2}. Figure 127 shows that the particular sample of fine-crystal hex.-ZnS:Ag phosphor did not transmit appreciable amounts of its own luminescence radiation beyond thicknesses corresponding to about 100 mg cm^{-2}, and similar saturation values were obtained with fine-crystal rbhdl.-Zn$_2$SiO$_4$:Mn and tetr.-CaWO$_4$:[W]. When large single crystals of efficient roentgenoluminescent phosphors, such as hex.-ZnS:Ag, tetr.-CaWO$_4$:Pb, and rhomb.-BaSO$_4$:Pb, are available (preferably roughened on the detector side), they should provide higher gamma-ray-excited outputs than large naphthalene or stilbene crystals.[292, 677] When the detection interval is extended over most of the phosphorescence interval of the phosphor, it is found that even a fine-crystal hex.-ZnS:Ag phosphor affords the greatest luminescence efficiency of the phosphors thus far tested under gamma rays (or beta rays or alpha particles).

With respect to medical uses of phosphors, it is possible that chemically stable UV-emitting roentgenoluminescent materials may be found useful when introduced into the human body and excited by x rays to provide therapeutic or bactericidal radiations (the bactericidal region extends from about 2000 to 3000 Å, and the therapeutic region lies above about 3200 Å). In particular, if a fine-particle phosphor such as rhomb.-BaSO$_4$:Pb (toxicity unknown!) were to be ingested, as is commonly done with ordinary BaSO$_4$ in radiographic studies of the digestive tract, the body might be irradiated with penetrating x rays when the phosphor is located near some internal organ or growth whose treatment with therapeutic or bactericidal radiations would improve the health of the subject. Such a suggestion was made by Newcomer, who hoped to find a suitable nontoxic organic roentgenoluminescent material for the purpose.[563, 564] Recent studies with gamma-ray-excited stilbene and monocl.-naphthalene,[292] indicate that some organic compounds may be made to emit considerable UV[75] under excitation by high-energy photons, but the intrinsic efficiencies per unit-volume of the organic luminescent materials are considerably lower than those of the inorganic phosphors, because the latter contain elements with much higher atomic number. The possibilities for using roentgenoluminescent phosphors in the body are indicated by the beneficial effect reportedly obtained by UV irradiation of patients' blood (accomplished outside the body) in cases of acute pyogenic infections.[565] The use of phosphors in medicine deserves more research to explore their possibilities in converting (a) high-energy primary photons (for example, x rays) into lower-energy therapeutic radiation by conventional excitation of suitable phosphors, and (b) low-energy photons (for example, infrared) into higher-energy therapeutic radiation by photostimulation of phosphors which have been excited prior to introduction into the body.

EXCITATION BY PRIMARY ELECTRONS (CR). *1. Cathodoluminescence Excited by Low-voltage Electrons ("Magic-eye" Tubes and Oscilloscope CRT).*

Phosphors which are to be excited by primary electrons must generally be contained in well-evacuated vessels ("tubes") because electrons with energies less than 10^4 ev are absorbed in a centimeter or less of ordinary air. It is usual, therefore, to generate thermionic electrons from hot cathodes coated with about 5 mg cm^{-2} of cub.-BaO:SrO in cathode-ray tubes (CRT) evacuated to pressures lower than 10^{-4} mm of mercury, in order to have mean free paths of at least a meter for electrons accelerated by potential differences ranging from about 100 to over 100,000 volts.[58, 100, 423]

Contrary to the results generally obtained with photon excitation, luminescence efficiency under CR excitation decreases as the energy of the primary particle decreases (Fig. 89). At primary-electron voltages near or below the extrapolated "dead voltage" of a phosphor (usually several hundred volts), cathodoluminescence is inefficient because (a) most of the primary energy is dissipated in the distorted surface layers of the phosphor crystals, and (b) the secondary-emission ratio of the phosphor screen may be below unity, in which case the screen tends to become negatively charged so that primary electrons are repelled.[72] A further point is that it is difficult to obtain high CR densities at low voltages, because of electron repulsion due to space charge along the CR beam.[58, 101]

It is possible to obtain detectable cathodoluminescence from certain phosphors, such as hex.-ZnO:[Zn], at primary-electron voltages as low as 5 volts. For practical use, however, voltages below 100 volts are seldom used. An example of low-voltage operation is found in the 6U5/6G5 tube, widely used as a tuning indicator in radios, which operates at 100 to 300 volts.[524] This tube has a metal anode coated with a mixture of rbhdl.-Zn_2SiO_4:Mn and particles of a conducting material (for example, graphite).[566] The conducting particles serve to keep the exposed surface of the coating at the applied potential of 100 to 300 volts. At 100 volts, and current densities of about 25 μa cm^{-2}, luminances of the order of 1 mL may be obtained from rbhdl.-Zn_2SiO_4:Mn. This value corresponds to about 0.4 lumen per watt, or about 0.1 candle per watt.

A practical advantage of low-voltage operation, apart from the relative ease of generating low voltages, is the ease with which low-voltage electrons may be deflected. For the deflection angles used in conventional CRT (up to about 30° from the tube axis), (a) the observed deflection *produced electrostatically* is inversely proportional to the accelerating voltage ($\Delta x \propto V^{-1}$), and (b) the observed deflection *produced magnetically* is inversely proportional to the square root of the accelerating voltage ($\Delta x \propto V^{-\frac{1}{2}}$).[58, 101] When CRT are to be used as simple indicators of slowly fluctuating phenomena, and cathodoluminescence efficiency and luminance are relatively unimportant, it is often advantageous to use low accelerating voltages to obtain maximum deflection sensitivity.

When rapidly fluctuating phenomena are to be depicted, oscilloscope CRT are particularly useful as indicators. A **CR oscilloscope** is generally an evacuated cone-shaped glass bulb with an electron gun mounted in a neck facing the phosphor-coated base inside the cone and with electrodes provided for accelerating the CR beam to the phosphor screen by voltages ranging from about 500 to 20,000 volts.[58, 524] The unmodulated CR beam is deflected by signals applied to electrostatic deflecting plates or (rarely) magnetic deflecting coils. Almost all CR oscilloscopes have electrostatic-deflection plates, because the plates may be used to deflect the electron beam at all frequencies. In general, magnetic-deflection coils are suitable for only a narrow band of frequencies.

CR oscilloscopes may be used to give observable indications of rapid transient phenomena or of fluctuations recurring at rates up to several hundred megacycles a second. When rapid randomly fluctuating phenomena are to be indicated, it is desirable to have short-persistent oscilloscope screens which do not obscure new fluctuations by vestigial traces of old fluctuations. Suitable short-persistent oscilloscope phosphors are, in approximate order of increasing persistence: hex.-ZnO:[Zn] **(P15)**, cub.-MgS:Sb, tetr.-$CaWO_4$:[W] **(P5)**, cub.-ZnS:Ag **(P11)**, rbhdl.-$Zn_8BeSi_5O_{19}$:Mn **(P3)**, and rbhdl.-Zn_2SiO_4:Mn **(P1)**, where the

RMA No.	Representative Phosphor or Scotophor (P10)	Component Cathodolumine Emission Bands, Å
		4000 5000 6000
P1	Rbhdl.-Zn_2SiO_4:Mn(0.3–1)	
P2	Hex.-ZnS:Ag(0.02):Cu(0.01)	
P3	Rbhdl.-$Zn_8BeSi_5O_{19}$:Mn(1.4)	
P4	Hex.-ZnS:Ag(0.015) mixed with hex.-1.3ZnS·CdS:Ag(0.01) (a)	
P4	Hex.-ZnS:Ag(0.015) or P11 mixed with P3 (b)	
P4	Monocl.-$CaMg(SiO_3)_2$:Ti(1) + P3 (c)	
P5	Tetr.-$CaWO_4$:[W]	
P6	Hex.-ZnS:Ag(0.015) + hex.-7ZnS·3CdS:Ag(0.01) + hex.-3ZnS·7CdS:Ag(0.01) (a)	
P6	Silicate:Ti(1) + P1 + P3*[Mn(2.5)] (b)	
P7	Hex.-ZnS:Ag(0.015) layer on top of hex.-9ZnS·CdS:Cu(0.0073)	
P8	Canadian P7 (obsolete) (British M screen)	
P9	Canadian $Ca_2P_2O_7$:Dy (obsolete) (British H screen)	
(P10)	Evaporated KCl (*scotophor*)	Absorp.
P11	Cub.-ZnS:Ag(0.003–0.01)	
P12	Tetr.-$(Zn:Mg)F_2$:Mn(1)	
P13	Monocl.-$MgSiO_3$:Mn(1)	
P14	Hex.-ZnS:Ag(0.015) layer on top of hex.-4.5ZnS·CdS:Cu(0.008)	
P15	Hex.-ZnO:[Zn]	
	Cub.-MgS:Sb(0.01)	

* Updated information obtainable by orde
Jan. 1, 1966, from Electronic Industries Assoc., 2

TABLE 21

.VE BEEN CODED BY THE RADIO MANUFACTURER'S ASSOCIATION *

Color of Emission		Approx. Visible Persistence	Current Density for Saturation	Approx. Average Crystal Size, microns	Chief Uses
ng Excit.	After Excit.				
en	Green	Short, ϵ^{-80t}	High	3	Oscilloscopes, radar
en–blue	Green	Long, t^{-n}	Low	15	Oscilloscopes
en–yellow	Green–yellow	Short, ϵ^{-80t}	High	3	See P4 and P6
ite	White	Short, t^{-n}	Low	8	Black & white television
ite	Green–yellow	Short, $t^{-n} + \epsilon^{-80t}$	P11 low, P3 high	3	Black & white television
ite	White	Short, ϵ^{-80t}	High	3	Black & white television
et–blue	Violet–blue	Very short, $\epsilon^{-10^5 t}$	Med.	1	Oscilloscopes
ite	White	Short, t^{-n}	Low	8	Color television
ite	White	Short, ϵ^{-80t}	High	3	Color television
e–white	Green–yellow	Long, t^{-n}	Low	14	Radar, oscilloscopes
e–white	Green–yellow	Long, t^{-n}	Low	11	Radar, oscilloscopes
ite	White	Long, t^{-n}	Low	1	Radar, oscilloscopes
Magenta absorption		Long, t^{-n}	Low	1	Radar, oscilloscopes
ue	Blue	Short, t^{-n}	Low	3–8	Oscilloscopes
ange	Orange	Med., ϵ^{-10t}	High	3	Radar, teleran
d	Red	Med., $\epsilon^{-10^3 t} \longrightarrow \epsilon^{-10t}$	Med.	3	
rple–white	Yellow–orange	Med. long, t^{-n}	Low	14	Radar
ue–green	Blue–green	Very short, $\epsilon^{-10^6 t}$	Med.	3	Oscilloscopes
een–yellow	Green–yellow	Very short, $\epsilon^{-10^6 t}$	Med.	2	Oscilloscopes

ptical Characteristics of CRT Screens," JEDEC Publication 16A,
e Street, Washington, D. C.

P numbers are the Radio Manufacturers' Association code numbers for CRT screens (Tables 21 and 22, and Fig. 128).[58, 63] The rise time of the P11 screen is less than 10^{-8} sec.[680] When isolated fluctuations or rapid *recurrent* fluctuations are to be indicated, it is often advantageous to have considerable visible persistence of the oscilloscope pattern, in order to study the pattern for a while after the trace has been made. Suitable long-persistent oscilloscope phosphors are: tetr.-$(Zn:Mg)F_2:Mn$ **(P12)**, hex.-$ZnS:Ag:Cu$ **(P2)**, a layer of hex.-$ZnS:Ag$ deposited on a layer of hex.-$8ZnS \cdot 2CdS:Cu$ **(P14)**, and a layer of hex.-$ZnS:Ag$ deposited on a layer of hex.-$9ZnS \cdot CdS:Cu$ **(P7)**.[58, 63] The P7 screen, when well excited, provides a pattern which remains visible for about a minute in a well-darkened room. Longer persistences, providing observable traces lasting for days or months, may be obtained from tenebrescent screens, such as coatings of evaporated and condensed cub.-KCl **(P10)**.[58, 63] These traces are eradicated best by intermittently heating the screen.

The maximum useful **writing speed,** that is, the maximum allowable rate of travel of the CR beam across the CRT screen is proportional to the cathodoluminescence efficiency of the phosphor, the CR-beam current, and V_S^n, where V_S is the screen (accelerating) voltage. In experimental CRT where the luminescence emission from the *bombarded* side of very thick screens is being detected, n is found to vary from about 1.5 (P1 screen) to 2.5 (P2 screen) over the screen-voltage range from 0.5 to 10 kv,[126] but, in conventional CRT whose outputs are detected on the *unbombarded* side of the screen, n is very nearly unity as long as the screen thickness is made optimum for each particle size and V_S up to the voltage beyond which the radiation from the bombarded side is poorly transmitted through the screen. With commercial CRT, writing speeds up to about 10^6 cm sec^{-1} may be used, and with special high-voltage high-current CR oscilloscopes writing speeds in excess of 10^8 cm sec^{-1} may be used for photographic or visual detection.[508-510] For accelerating voltages near 5 kv, the minimum number of electrons (in one pulse) striking a square centimeter to produce a trace which is just visible to the scotopic eye is about 10^9 electrons per square centimeter (10^{-10} coulomb per square centimeter) for the short-persistent screens, such as cub.-$MgS:Sb$, P15, P5, P11, and P1, and about 10^{10} electrons per square centimeter for long-persistent screens such as the P14 and P7. About 10^{12} electrons per square centimeter are required in the case of the tenebrescent P10 screen operated at 10 kv.[58, 63]

2. Cathodoluminescence Excited by Medium- and High-voltage Electrons (Kinescopes, Image Tubes, and Electron Microscopes). (a) *Television kinescopes.* Image reproduction by means of cathodoluminescent screens has become a valuable growing art during the past two decades since

TABLE 22

PROPERTIES OF SOME USEFUL CATHODOLUMINESCENT MATERIALS

The Relative Efficiency Data Were Obtained at 8 kv and about 1 $\mu a\ cm^{-2}$, with Measurements Made on the Bombarded Sides of Thick Screens

Data from Austin E. Hardy

No.	Phosphor	RMA Code	λ_{pk}, Å	ICI Coordinates x	y	Relative Efficiency Peak Output	Total Output	Visual
1.	Rhbdl.-$Zn_2SiO_4 \cdot 0.012MnSiO_3$, 1250°C	P1	5250	0.206	0.700	100*	100*	100*
2.	Rhbdl.-$Zn_8BeSi_5O_{19} \cdot 0.25MnSiO_3$, 1250°C	P3	5300	0.469	0.525	36	87	85
3.	Rhbdl.-$Zn_8BeSi_5O_{19} \cdot 0.45MnSiO_3$, 1250°C	(P3* special)	6100	0.527	0.470	44	94	69
4.	Tetr.-$CaWO_4$:[W], 1100°C	P5	4300	0.171	0.149	12	25	11
5.	Monocl.-Mg_2WO_5:[W], 1100°C		4930	0.234	0.311	10	33	31
6.	Monocl.-$CaMg(SiO_3)_2$:$0.02TiO_2$, 1200°C		4150	0.164	0.120	35	95	28
7.	Hex.-ZnO:[Zn], 1000°C	P15	5050	0.233	0.436	48	102	81
8.	Tetr.-$ZnF_2 \cdot 0.01MnF_2$, 900°C	P12	5870	0.539	0.461	50	69	71
9.	Cub.-ZnS:Ag(0.003), 850°C	P11	4550	0.146	0.127	204	320	82
10.	Hex.-ZnS:Ag(0.015), 1250°C	P7/1 and P14/1	4350	0.151	0.032	206	234	23
11.	Hex.-ZnS(95)CdS(5):Ag(0.015), 1100°C		4490	0.146	0.046	134		22
12.	Hex.-ZnS(85)CdS(15):Ag(0.015), 1100°C		4850	0.128	0.184	87		37
13.	Hex.-ZnS(75)CdS(25):Ag(0.015), 1100°C		4980	0.156	0.391	112		77
14.	Hex.-ZnS(65)CdS(35):Ag(0.015), 1100°C		5260	0.229	0.572	147		146
15.	Hex.-ZnS(60)CdS(40):Ag(0.015), 1100°C		5350	0.324	0.601	135		163
16.	Hex.-ZnS(55)CdS(45):Ag(0.015), 1100°C		5600	0.385	0.585	120		188
17.	Hex.-ZnS(45)CdS(55):Ag(0.015), 1100°C		5720	0.462	0.527	122		210
18.	Hex.-ZnS(35)CdS(65):Ag(0.015), 1100°C		6040	0.548	0.450	96		149
19.	Hex.-ZnS(25)CdS(75):Ag(0.015), 1100°C		6420	0.624	0.376	85		80
20.	Hex.-ZnS(15)CdS(85):Ag(0.015), 1100°C		6780	0.692	0.307	70		37
21.	Hex.-ZnS:Cu(0.006), 1200°C		5160	0.221	0.551	107		140
22.	Hex.-ZnS(96)CdS(4):Cu(0.006), 1200°C		5250	0.255	0.556	64		107
23.	Hex.-ZnS(92)CdS(8):Cu(0.006), 1200°C		5400	0.320	0.559	57		107
24.	Hex.-ZnS(88)CdS(12):Cu(0.007), 1200°C	P7/2	5500	0.364	0.547	54		109
25.	Hex.-ZnS(86)CdS(14):Cu(0.007), 1200°C		5600	0.357	0.537	58	109	110
26.	Hex.-ZnS(80)CdS(20):Cu(0.006), 1200°C		5700	0.458	0.494	45		106
27.	Hex.-ZnS(76)CdS(24):Cu(0.006), 1200°C		5800	0.501	0.471	46		99
28.	Hex.-ZnS(75)CdS(25):Cu(0.006), 1200°C	P14/2	6000	0.516	0.455	69	141	94
29.	Hex.-ZnS:Ag(0.02):Cu(0.01), 1250°C	P2	5200	0.207	0.406	90	183	120
30.	Tricl.-$Ca_2MgSi_2O_7 \cdot 0.05Ce_2O_3$, 1250°C		3700			≈200		

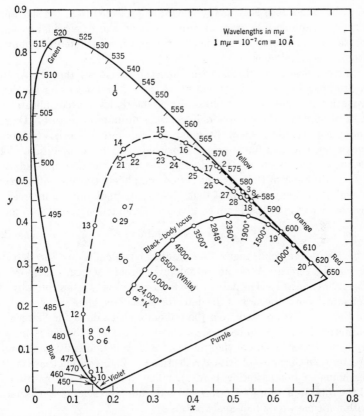

Fig. 128. ICI-coordinate plot of some of the luminescence-emission colors of phosphors listed in Table 22. The bounded area is the domain of real colors perceptible to the average human eye. Saturated colors are plotted around the periphery, and the colors become less saturated (paler) on proceeding in toward the white region near 6500°K on the black-body locus. The saturated colors are monochromatic, except for the dichromatic purples.

the introduction of the electronic-television kinescope.[72, 423, 567]   A **kinescope** is essentially a CR oscilloscope with a control grid for modulating the CR-beam current; hence, a kinescope can provide shades of light and dark (that is, halftones) in the patterns or images traced on its screen.   Conventional direct-viewing kinescopes, whose images are viewed directly on the luminescing screen, are operated at about 4 to 20 kv, whereas conventional projection kinescopes, whose images are projected onto larger viewing screens, are operated at about 15 to over 100 kv.   In both cases, the luminescing screen is brighter on the scanned

side, but the screen is generally viewed or projected from the side opposite the scanning CR beam in order to avoid the trapezoidal scanning
patterns which result when the electron gun is offcentered to remove it
from the field of view.

Screen structures. It has been remarked previously that the scattering produced by fine-crystal phosphors is disadvantageous for the determination of some of their physical characteristics and complicates the
task of applying the phosphors in coatings of uniform texture. Despite
these complexities, small irregular crystals are essential to obtain
maximum effective luminescence efficiency with high resolution. This
becomes obvious by considering that the total luminescence-radiation
flux F_t (in lumens) from a point P in a completely transparent "screen"
is emitted equally in all directions; hence, the radiation F_d intercepted
by the detector in the assumed absence of refraction and absorption is

$$F_d = F_t\omega/4\pi \text{ lumens} \qquad (125)$$

where ω is the solid angle (in steradians) subtended by the detector
facing the screen. With an actual transparent "screen," as shown in
Fig. 129A, only a small part of the total radiation is intercepted by the
detector, even if the area of the detector be made so large that essentially
all the radiation emitted from the unbombarded side of the "screen" is
intercepted. This is so because an equal amount of radiation is emitted
from the opposite side of the "screen," and radiation emitted at angles
greater than the angle of critical reflection θ_c remains within the "screen"
(assuming a nonscattering border). The approximate proportion F_e
of the total radiation F_t which is emitted from both sides of the "screen"
within a cone whose half-angle is θ_c is

$$F_e \text{ (both sides)} = (1 - \cos\theta_c)F_t \qquad (126)$$

or, taking into account the radiation escaping from only one side,

$$F_e \text{ (one side)} = (1 - \cos\theta_c)F_t/2 \qquad (127)$$

Since $\sin\theta_c = \mu^{-1}$, where μ is the index of refraction of the "screen"
(crystal), *in vacuo*,

$$\cos\theta_c = (1 - \sin^2\theta_c)^{1/2} = (1 - \mu^{-2})^{1/2} \qquad (128)$$

and eq. 127 becomes

$$F_e \text{ (one side)} = [1 - (1 - \mu^{-2})^{1/2}]F_t/2 \qquad (129)$$

From Table 19 it is apparent that the index of refraction μ of most
phosphors is about 2, and so only about 13 per cent of the total internal
radiation would be emitted from one side of a transparent single-crystal

"screen." The distribution in angle of the radiation emitted from such a transparent "screen" would be approximately hemispherical; that is, the radiant flux from a point on the surface would be the same at all angles away from the surface. The effective luminescence efficiency of a large flat single crystal can be greatly increased by roughening the side to be observed or detected. This roughening also provides a cosine-law flux distribution (eq. 130), but the resolution is not as fine as with fine-crystal screens unless the screen thickness is very small.

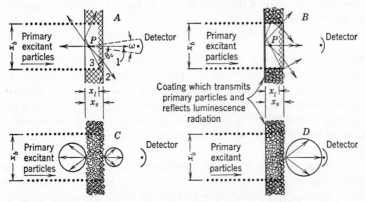

FIG. 129. Illustrative cross-section examples of phosphor screens of thickness x_s excited to a depth x_l by a primary beam of diameter x_b. A = single-crystal phosphor plate. B = large single phosphor crystal bordered by small phosphor crystals. C and D = conventional small-crystal phosphor screens (substrate omitted for simplicity). A and C = unaluminized, B and D = aluminized.

If as shown in Fig. 129B, a smaller portion of a single-crystal "screen" be bordered by ideal nonabsorbing scattering crystals, then the bulk of the radiation which normally remains within the screen by total internal reflection, until absorbed, is scattered out of the screen. This is shown merely as an intermediate step leading to the conventional fine-crystal screen shown in Fig. 129C. If there were *no* absorption of such a scattering screen for its own radiation, then, assuming that all the crystals are sufficiently irregular in shape to allow all of the radiation to escape from each, all the internally produced luminescence radiation would escape from the screen. In the absence of reflecting coatings, the proportion of radiation escaping from the bombarded side of the screen would be higher than that escaping from the opposite side, because (1) the excitation density decreases as the primary beam penetrates farther into the screen (Fig. 22), and (2) photons emitted inside such scattering screens have a probability of escape from one or the other screen surface which is inversely related to the distance from the particular surface. An *ideal* nonabsorbing uniformly excited scattering screen, therefore, would emit $F_t/2$ from each side of the screen, and an *ideal* nonabsorbing screen with a *perfect* reflecting coating (Fig. 129D) would emit all of the internally produced radiation F_t from the unbombarded side, regardless of the distribution of excitation density in the screen. Also, a scattering screen affords a cosine-law distribution of emission, where the radiant flux dF through the small solid angle $d\omega$ varies with the angle θ from the normal to the screen surface according to Lambert's law,

$$dF = (F/\pi) \cos \theta \, d\omega \text{ lumens} \tag{130}$$

where $F =$ lamberts $\times$ centimeter^2, and the apparent luminance L of the screen is independent of θ. Fine-crystal phosphor screens obey Lambert's law quite closely,[58] and so they have practically the same apparent luminance, regardless of the angle of view, although the total radiation emitted from a screen in a given direction decreases as the apparent area of the screen is reduced on viewing at increasing angles to the surface normal.

Phosphor screens made of irregularly shaped fine crystals are, therefore, generally more efficient than transparent regular single-crystal screens because (1) radiation retention by total internal reflection is largely eliminated in the scattering screen, and (2) the cosine-law distribution of radiation from a scattering screen increases the intensity of radiation in the direction normal to the surface, that is, in the direction of the usual detector. In actual fine-crystal phosphor screens, however, there is always a finite absorption of the material for its own luminescence radiation, and so it is usually necessary to increase the crystal size to reduce scattering as the energy (penetration) of the primary particle is increased. Generally, the optimum average crystal size $\bar{x}_c$ is of the order of the penetration "limit" x_l, and the optimum screen thickness x_s is made approximately four times x_l, that is,

$$4x_l \approx x_s \approx 4\bar{x}_c \tag{131}$$

This approximate empirical relation obtains for screens detected on the unbombarded side, because x_s may be made much larger than x_l when the detector is placed on the bombarded side of the screen. Use of the foregoing relationship is complicated by (1) the difficulties encountered in growing phosphor crystals with a wide range of average crystal sizes, especially when very large crystals are needed, as in the case of gamma-ray excitation where x_l may exceed 0.1 cm (Fig. 127), and (2) the increase in graininess (decrease in resolution) of the screen as the average crystal size is increased. Most phosphor screens, especially CRT screens, are detected (viewed) on the unbombarded side, and it has been found to be preferable, on occasion, to disobey eq. 131; for example, (a) by using a screen of hex.-ZnO:[Zn] according to eq. 131, optimum efficiency is obtained for emission of the visible band which is little absorbed by the phosphor (Fig. 50), whereas by using $x_s \approx x_l \approx \bar{x}_c$ optimum efficiency is obtained for emission of the UV band which is strongly absorbed by the phosphor,[354] (b) by using $\bar{x}_c \ll x_s$ it is possible to decrease the spreading of the luminous spot in CRT screens, thereby improving the resolution and detail contrast of the image; and (c) by using $\bar{x}_c \ll x_s$ it is possible to decrease the transmission of extraneous radiation by the screen, for example, this may be done to minimize transmission of illumination from the incandescent cathodes in CRT.

Secondary emission. During luminescence excited by primary photons, the photons which enter and leave the phosphor crystals are, of course, uncharged, and so the insulator crystals do not gain or lose charge during luminescence, except for a small number of photoelectrons which may be emitted until the crystals become charged sufficiently positive to deter emission of further photoelectrons (whose escape energies are usually less than 10 ev).[511, 514] During cathodoluminescence, however, the absorbed primary electrons tend to charge the crystals negatively, and so the crystals must lose by conduction or emission at least an equal number of electrons in order to prevent becoming charged so strongly negative that further primary electrons are electrostatically repelled (deflected) away from the crystals. (The average energies of escaping secondary electrons are, in general, less than 10 ev.[99, 130, 134]) Therefore, an important factor in conventional unmetallized cathodoluminescent phosphor screens deposited on insulating substrates (for example, glass) is the secondary-emission ratio of the phosphor screen.[72] The **secondary-emission ratio** is the ratio of the number of emitted secondary electrons to the number of *absorbed* primary electrons.[124-134, 568] This neglects those primary electrons which are simply reflected. As shown in Fig. 130, the secondary-emission ratios of phosphor screens (or insulators in general) rise to a **first unity intercept (first crossover voltage)** at about 50 to 200 volts applied potential, and fall to a **second unity intercept (second crossover voltage)** at about 1000 to over 60,000 volts. The first slow rise of the secondary-emission-ratio curve occurs in the low-voltage region in which most of the input CR energy is dissipated as heat in the imperfect and irregular surface layers of the crystals; the second greater-than-unity portion of the curve is in the medium-voltage region where most of the input CR energy is absorbed in the more regular volumes of the crystals where internal "free" secondary electrons are readily produced and may travel considerable distances without encountering an undue number of imperfections which decrease their energies by inelastic collisions; and the final decline of the curve is in the high-voltage region where *deep* internal secondary electrons have a low probability of escaping without encountering sufficient imperfections (including traps and thermal scattering) to reduce their kinetic energies to values below that required for escape from the crystals (screens).

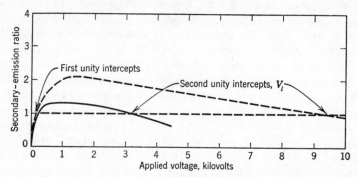

Fɪɢ. 130. Typical secondary-emission characteristics for two hypothetical insulators (phosphors).

It is not possible to present a quantitative account of the mechanism of secondary emission, accounting in particular for emission of electrons in the backward direction, that is, in the direction opposite to the direction of motion of the primary particle. Qualitatively, however, one may picture the *modus operandi* of secondary emission in the backward direction in the two following ways, assuming an isotropic (cubic) solid for simplicity: (1) the primary electron imparts some of its energy to an atom in the solid, whereupon the atom has nearly equal probability of emitting an internal secondary electron ("internal ionization") in all directions, and (to a lesser extent) (2) the primary electron interacts with a specific bound electron in the solid, imparting sufficient energy to that electron to free it from its parent atom and raising it in energy through the forbidden region of energy levels where it satisfies the Bragg relation (eq. 45) and may be diffracted back along the direction opposite to the motion of the primary electron. In both of these examples, momentum is conserved by the solid acting as a third body in addition to the primary and secondary electrons. For anisotropic solids, of course, the emission probability may be considerably influenced by variation in direction relative to the different crystal axes.

Specific intercept values and secondary-emission ratios at given voltages depend strongly on the chemical compositions and crystallinities of the phosphors, and so the secondary-emission characteristics of phosphors with a given formula often differ considerably from batch to batch; also, the secondary-emission characteristics are affected by the crystal sizes, methods of application, screen structures, and conditions of operation. The second unity intercepts (also called **limiting potentials**) of metals are usually about 2 kv, and phosphors with relatively high conductivities generally have low limiting potentials V_l; for example, Knoll reports that the V_l of hex.-ZnO:[Zn] (P15) is about 2.5 kv, whereas rbhdl.-Zn_2SiO_4:Mn (P1), tetr.-$CaWO_4$:[W] (P5), and cub.-ZnS:[Zn] have V_l's higher than 60 kv.[132] (It should be noted that these values are not representative of the phosphors alone, because the screens were applied with a binder of potassium silicate.) Also, the V_l values of several (Zn:Cd)(S:Se)-type phosphors were reported to range from 15 to over 60 kv, and it has been found that all-sulphide P2, P4, P6, P7, and P14 screens in commercial CRT often have V_l values less than 20 kv, whereas the oxygen-dominated P1, P3, and P5 screens usually have V_l values well above 20 kv. Incidentally, the relatively low secondary-emission ratios of pure metals are practically independent of temperature up to the melting point of the metal, whereas the much higher secondary-emission ratios obtained with insulator crystals tend to decrease to the values obtained from metals as the temperature of the insulator crystal is increased. This temperature dependence is reflected in the results of Nelson, who measured the secondary-emission characteristics of phosphor screens on glass which was heated high enough to conduct.[124] By this method, Nelson found that samples from a given batch of rbhdl.-Zn_2SiO_4:Mn (P1) applied in three different CRT had V_l values of 3, 11, and 18 kv, and found that the V_l of a P3 screen decreased from about 12 to 8 kv in 765 hr excitation by a scanning CR beam with an *average* current density of about 0.6 μa cm^{-2} for the screen as a whole. Nelson's results indicate that the secondary-emission characteristics of phosphor screens are not determined by the phosphors alone, and the characteristics of a given screen may alter greatly during operation.

At applied voltages below the first unity intercept (V_s tends toward cathode potential), and above the second unity intercept (V_s tends toward V_l), phosphor screens usually require auxiliary aids, such as conducting coatings or admixed conducting particles connected to the positive electrode, to prevent deceleration of the electrons by the lower potential of the screen relative to the applied potential.[424, 566] In the applied-voltage region between the two unity intercepts, however, a phosphor screen may be used directly, because it maintains itself at the applied potential by emitting at least one secondary electron, on the average, for each incident primary electron. In practice, (1) at applied voltages below about 500 volts, conducting particles are usually admixed with phosphor particles deposited on a conducting anode, (2) at applied voltages from approximately 500 to 15,000 volts, phosphor screens are used directly, or with alkali-compound admixtures, such as K_2SiO_3 or Cs_2O, and (3) at applied voltages above about 10,000 volts, phosphor screens are often coated on the CR-beam side with thin electron-permeable light-reflecting conducting coatings, such as a 0.1-micron-thick reflecting coating of aluminum.[58, 63, 100, 569, 570] The reflecting coating is usually applied by evaporating aluminum onto a taut film of water-floated organic material (for example, nitrocellulose) covering the phosphor screen. In cases 1 and 3 electric contact is established between the conducting particles or conducting coating and the anode, whereas in case 2 no electric connection is necessary.

Phosphor screens for television kinescopes. As indicated in Table 21, television CRT screens emit *white* light, although the first screens used in practical kinescopes in this country were green-emitting rbhdl.-Zn_2SiO_4:Mn and hex.-ZnS:Cu.[72] These two phosphors exemplify the parent phosphor systems from which have been developed the present useful white-emitting combination screens indicated in Table 21. It is possible to make a single-component white-emitting phosphor from the general O-dominated system $(Zn:Be)O:SiO_2:(Ti:Zr:Th)O_2:-MnO$,[429] but it has been found that better results are obtained by (*a*) mixing a material from the most efficient blue-emitting part of this system, for example, rbhdl.-Zn_2SiO_4:Ti, with the most efficient yellow-emitting material, for example, rbhdl.-$8ZnO\cdot BeO\cdot 5SiO_2$:Mn(1.4), or (*b*) mixing the blue-emitting rbhdl.-Zn_2SiO_4:Ti (or monocl.-CaMg-$(SiO_3)_2$:Ti) with green-emitting rbhdl.-Zn_2SiO_4:Mn and orange–red-emitting rbhdl.-$8ZnO\cdot BeO\cdot 5SiO_2$:Mn(2.4) to make white-emitting combination screens. In the S(Se)-dominated system, the single white-emitting cub.-ZnS:[Zn]:P phosphor[698] has about one tenth of the peak output and half the luminance of the more efficient blue-, green-, yellow-, and red-emitting phosphors now available in the general system $(Zn:Cd)(S:Se)$:Ag (Figs. 33, 36, 39[170]).[575] Combination phosphor screens for television kinescopes are designated as **P4** if the kinescope is intended for monochrome (black-and-white) television, and as **P6** if the kinescope is intended for color television. P4 screens are usually two-component screens, comprising a blue-emitting and a yellow-emitting phosphor, whereas P6 screens are usually three-component screens, comprising a blue-emitting, a green-emitting, and a red-emitting phosphor (in either case, the components may be (1) mixed, or (2) in separate small adjacent areas, or (3) in superimposed layers).[58, 63, 72, 312, 539, 571–578]

At present, most unaluminized P4 screens, for direct-viewing kine-scopes, comprise admixed hex.-ZnS:Ag (blue component) and hex.-ZnS(47)CdS(53):Ag (yellow component).[579] These two sulphide phos-phors have closely matched emission characteristics as a function of CR current density, and screens made of the two phosphors give uniform white emission in image halftones and high lights. Coating densities are usually in the range from 3 to 8 mg cm^{-2}, with crystals averaging 4 to 10 microns in size, for operation at 6 to 12 kv. Each different voltage requires a different proportion of the blue and yellow components for constant color, and a different coating density for maximum efficiency.

To emphasize the latter point in a different manner, consider a conventional kinescope screen of fixed thickness excited by a CR beam of *fixed power* as the energy (penetration) of the primary electrons is increased (and the intensity of the beam is decreased to keep the power constant). As the voltage accel-erating the primary electrons is increased from zero, (1) at voltages below the first unity intercept the screen is at first practically nonluminescent because (a) the screen tends to charge negative and fall to cathode potential, and (b) most of the absorbed CR energy is spent in the imperfect and inefficient surface layers of the phosphor crystals; (2) at voltages between the first and second unity intercepts the screen exhibits a maximum cathodoluminescence output at a given voltage where most of the CR energy is absorbed in the efficient vol-umes of the phosphor crystals, and eq. 131 is approximately satisfied; (3) at high voltages the cathodoluminescence emission falls to vanishingly low values because (a) a decreasingly small portion of the primary-beam *power* is absorbed in the fixed-thickness screen as the energy (penetration) of the primary elec-trons is increased (this effect occurs even when the beam *current* is kept constant, because $x_l \propto V_0^2$), and (b) the screen tends to remain near V_l unless it is alu-minized.

Incidentally, even at optimum thickness, statistical variations in the texture of fine-particle screens cause some decrease in effective cathodoluminescence efficiency. The losses occur where (a) the screen is thin enough to transmit appreciable CR beam power and (b) the screen is too thick and efficiency is lowered by undue scattering and absorption of luminescence emission.

When carefully applied in conventional kinescopes, such as the 10BP4 and 16AP4,[697] unaluminized all-sulphide screens give effective luminescence efficiencies of about 2 to 3 candles per watt at voltages between 8 and 10 kv, and high-light luminances between 40 and 60 mL, at *average* current densities of about 1 μa cm^{-2}. At about 0.1 μa cm^{-2}, the luminance is about 8 mL, and the effective cathodoluminescence efficiency is about 4.5 candles per watt (compare Fig. 113). Somewhat higher efficiency may be obtained by applying the cited sulphide phos-phors in separate superimposed layers, with the blue-component hex.-ZnS:Ag next to the glass substrate so that there is less absorption of the blue light by the yellow-emitting (and blue-absorbing) hex.-(Zn:Cd)S:Ag phosphor.[312, 539] In this case, however, the particle sizes and layer thicknesses must be adjusted to optimum values for maximum efficiency and correct color balance for each particular voltage, and the layer screens do not give such constant resultant white emission color over a range of voltages as do the mixed-phosphor screens. When P4

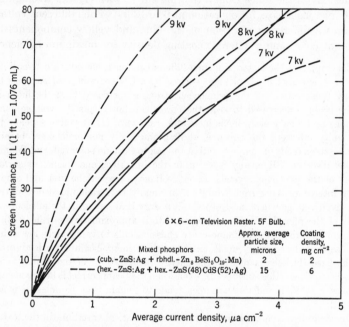

FIG. 131. Cathodoluminescence emissions of two unaluminized P4 screens as a function of applied voltage and *average* current density. Measurements on unbombarded side. (A. E. Hardy)

screens are aluminized, selective absorption effects are accentuated, in either mixed or layer screens, and the proportions of the different phosphors must be adjusted to obtain the same satisfactory (6500°K) white obtained from unaluminized screens. With aluminizing, however, increased effective cathodoluminescence efficiencies may be obtained at operating voltages above about 5 kv. With an aluminized all-sulphide P4 kinescope screen operated at 15 to 20 kv and an average current density of about 1 μa cm^{-2}, it is possible to obtain an effective luminescence efficiency of about 5 candles per watt at a high-light luminance in excess of 250 mL. In all cases, luminance increases and effective luminescence efficiency decreases as the average current density is increased, although at current densities high enough to raise the temperature of the screen above T_B even the luminance decreases (compare Figs. 111–113).

Figure 131 shows plots of the luminances of some unaluminized P4 screens as a function of average current density at three different voltages for (1) an all-sulphide phosphor mix (dotted lines), and (2) a sulphide–silicate phosphor mix (solid lines), both detected on the *unbombarded* side of the screen.[58] At any of the given voltages, the all-sulphide P4 screen has higher effective luminescence efficiency at the lower current densities, but the sulphide–silicate P4 screen has higher efficiency at high current densities, and an all-silicate P4 screen (not shown) is superior to the sulphide–silicate screen at high current densities. As shown in Fig. 131, the efficiency-crossover points for the two given screens occur at increasingly higher current densities as the applied voltage is raised. The given sulphide phosphors saturate at lower current densities than the silicate phosphors (Fig. 112), and an increase in accelerating voltage, V_0 ($= V_s$), at constant beam current lowers the average power density in the excited portion of the screen because the beam power increases as V_0, while the volume in which the primary beam power is expended increases as V_0^2 (eqs. 76 and 77). At high current densities, then, the higher luminance of the sulphide–silicate P4 screen is due to the high-activator-content silicate component which saturates at a much higher current density than the low-activator-content sulphide component.[34, 58] The sulphide–silicate screen, therefore, usually becomes yellower in emission color at higher current densities. This color shift may be eliminated by using, instead of the sulphide component, a complementary high-activator-content O-dominated blue-emitting phosphor, such as rbhdl.-Zn_2SiO_4:Ti or monocl.-$CaMg(SiO_3)_2$:Ti whose cathodoluminescence-versus-current-density characteristic matches that of the yellow-emitting rbhdl.-$Zn_8BeSi_5O_{19}$:Mn phosphor.

Figure 132a shows a plot of the cathodoluminescence emissions L from the *bombarded* sides of thick screens of cub.-ZnS:Ag and rbhdl.-$Zn_8BeSi_5O_{19}$:Mn(1.4). These luminances of phosphors were measured *separately* as a function of accelerating voltage, V_0 ($= V_s$), at several different current densities over the voltage range from 0.5 to 10 kv, using a stationary unmodulated CR beam.[126, 399] The slopes of the L vs. V_0 curves at constant current density in the beam may vary considerably with the method of preparing the phosphor and screen, even when L is measured on the bombarded side of very thick screens. For example, Martin and Headrick report that the slopes of the L vs. V_0 curves of different thick screens of rbhdl.-Zn_2SiO_4:Mn, measured at 20 to 50 μa cm^{-2} over the voltage range from 0.5 to 10 kv, are (*a*) 2.5 for a fused layer, (*b*) 1.6 for a screen made of particles obtained by com-

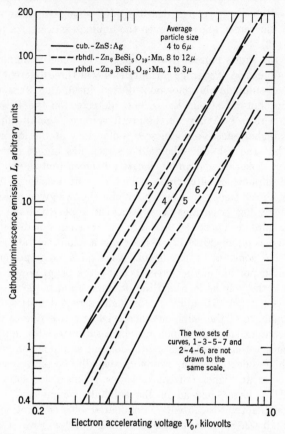

FIG. 132a. Log–log plot of cathodoluminescence emission (*bombarded* side of screen) as a function of accelerating voltage for two representative phosphors. (S. T. Martin and L. B. Headrick)

Curve	Current Density, $\mu a\ cm^{-2}$	Slope
1	200	1.77
2	200	1.6
3	150	1.79
4	200	1.5
5	50	1.84
6	10	2.0 and 1.5
7	10	2.0

minuting the fused material, and (c) 1.5 for a screen made of particles synthesized by the carbonate process.[126] This increase in L according to

$$L \propto V_0^n \tag{132}$$

where $n \approx 2$, does not agree with the expected linear relationship (from eqs. 1, 75, 110, and 116):

$$L = N_e h\nu = N_0 e V_0 \tag{133}$$

Because h, ν, e, and N_0 are constant, it is not possible for L (N_e) to increase indefinitely as $V_0{}^n$, with $n > 1$, without violating the principle of conservation of energy. Therefore, L vs. $V_0{}^n$ curves which start with $n > 1$ at low V_0 values must change to $n \leq 1$ at high V_0 values. The observed greater-than-linear increases of L with V_0, from 0.5 to 10 kv,

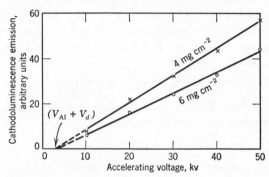

Fig. 132b. Linear plot of cathodoluminescence emission from the unbombarded side of an aluminized screen of unground hex.-ZnS:Ag(0.01), 1100°C. Defocussed 16-μa CR beam scanning a 3 × 5-cm raster. (S. Larach and R. E. Shrader)

are attributed to a combination of (1) increasing the penetration of the primary electrons $(x_l \propto V_0{}^2$, eq. 77), and thereby minimizing saturation by (a) decreasing the average excitation density per unit volume in the penetrated portions of the phosphor crystals, and (b) decreasing the average current density per unit volume by increasing the scattering of primary electrons out of the projected path of the primary beam (larger x_l increases the scattering probability per primary particle); and (2) increasing the proportion of CR-beam energy absorbed in the efficient volumes, rather than in the inefficient surface layers, of the phosphor crystals (compare Fig. 111). With large phosphor crystals and very high V_0, L vs. V_0 should approach the linear relationship expressed in eq. 133, but this has not been determined experimentally. With screens having finite absorption of their own radiation, the externally measured L should eventually increase at a rate less than linear with V_0, that is, n becomes less than unity in eq. 132. The V_0 at which n becomes less than unity is lower for small-crystal screens than for large-crystal screens, because absorption of internal radiation is increased by scattering.

Figure 132*b* shows a plot of the cathodoluminescence emissions from the *unbombarded* sides of aluminized screens of washed and unground hex.-ZnS:Ag(0.01), [NaCl(6)], 1100°C. Here, the expected linear relationship between L and V_0, at constant current, is obtained for both screen densities over the range of V_0 from 10 to 50 kv. A too thin 2 mg cm^{-2} screen gives $L \propto V_0^{n<1}$ in this range, because the amount of CR beam power lost by penetration through thin portions of the screen increases as the penetration increases (eq. 77). Also, a too thick screen gives $L \propto V_0^{n>1}$, because more luminescence photons escape out of the observed side of the screen as the CR penetration is increased.

With unaluminized phosphor screens, the effective accelerating voltage V_0 available for excitation of the phosphor crystals varies with the secondary-emission ratio of the screen at the particular applied voltage, V_A (see Fig. 130), that is,

$$V_s = V_0 \leq V_A \qquad (134)$$

whereas, with aluminized screens (connected to the anode) the V_0 available for excitation of luminescence is

$$V_0 = V_A - V_{Al} = V_s - V_{Al} \qquad (135)$$

where V_{Al} is a constant voltage decrement determined by the thickness of the particular aluminum layer. For 1000-Å-thick reflecting coatings of aluminum on P4 screens, the **break-even voltage,** for L observed on the *unbombarded* side of the screen, is about 5 kv; that is, the aluminized and unaluminized screens have nearly the same cathodoluminescence efficiency at about 5 kv. Below the break-even voltage the unaluminized screens are the more efficient, and above the break-even voltage the aluminized screens are the more efficient. A study of the variation of L with V_0, for a given screen composition, when L is measured on the *unbombarded* side, requires the preparation of different screens with optimum coating densities (milligrams per square centimeter) for each voltage (and particle size). This is a lengthy procedure, and data on the variation of L (*unbombarded* side) versus V_0 are difficult to obtain for a wide range of V_0 (compare comments on Fig. 134).

When optimum efficiency is required, and L is observed on the *unbombarded* side of a screen, it is advantageous to have the CR-beam penetration x_l the same order of magnitude as the average particle (crystal) diameter $\bar{x}_c$ (Fig. 7). In this respect, it may be noted that a screen composed of a range of different particle sizes about a given $\bar{x}_c$ (Figs. 7 and 10) may be made more compact than a screen composed of just one particle size, $x_c = \bar{x}_c$, because the smaller particles can fill the voids between the larger particles. When x_l is small relative to $\bar{x}_c$, the coating density is generally made a minimum for coverage of the substrate. When x_l is large relative to $\bar{x}_c$, the optimum coating density is a complex function of x_l (that is, V_0), $\bar{x}_c$, the particle-size distribution of the phosphor (Fig. 7), and the particle-size distribution in the screen (taking into account any selective deposition of different sizes and densities of particles).[72] Figure 133 shows an example of the variation in white-light transmission, as a function of coating density, of a phosphor which has very little light absorption.[58] The transmission is a function of intrinsic absorption, particle size, index of refraction, and thickness.[580-582] Maximum opacity to white light is obtained with particles

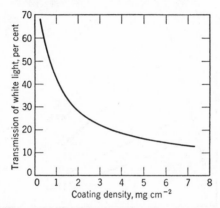

Fig. 133. Transmission of white light versus coating density of a sprayed screen of a rbhdl.-$Zn_8BeSi_5O_{19}$:Mn(1.4) phosphor with an average particle size near 2 microns. (L. E. Swedlund)

having (uniform) diameters from 0.3 to 0.4 microns, when the index of refraction of the material is 2.3 to 2.8.[581] Figure 133 indicates that the luminescence emission from within the surface layers on the *bombarded* side on an unaluminized screen makes a decreasing contribution to L observed on the *unbombarded* side as the coating density is increased. It is advantageous, in this case, to have the average screen thickness x_s the same order of magnitude as $\bar{x}_c$, as indicated in eq. 131, because the excitation density (and L) is always greater on the bombarded side of the screen. Figure 134 shows the variation of cathodoluminescence output from the unbombarded side of an aluminized kinescope screen, made of cub.-ZnS:Ag, as a function of coating density at various applied voltages.[582] With the given phosphor, whose average crystal size was about 3 microns, the optimum coating density varies from about 1.4 mg cm^{-2} at 17 kv to about 3.6 mg cm^{-2} at 47 kv. With a mixed [hex.-ZnS:Ag + hex.-(Zn:Cd)S:Ag] P4 screen, with average crystal size near 8 microns, the optimum coating density varies from about 5 mg cm^{-2} at 17 kv to about 10 mg cm^{-2} at 50 kv. The decreasing light output at coating densities above the optimum is caused by the absorption of the screen for its own emission, and by the less than 100 per cent reflectivity of the aluminum layer. For a *completely* nonabsorbing transparent screen, the light output as a function of increasing coating density would rise to a maximum value for each voltage and remain at the maximum as the coating density was increased beyond the penetration limit of the excitant. It may be noted that the data in Fig. 134,

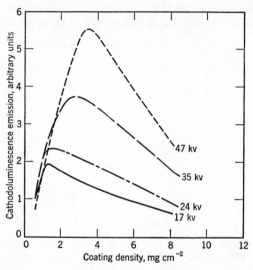

FIG. 134. Relative cathodoluminescence emission (unbombarded side of screens) as a function of accelerating voltage and coating density for settled aluminized screens of cub.-ZnS:Ag having an average crystal size near 3 microns. Stationary unmodulated CR beam delivering about 20 μa cm^{-2}. (F. H. Nicoll)

when replotted, show that the cathodoluminescence output of the screens *at optimum coating density* increases almost linearly with increasing V_0 at constant current density.

It is apparent from the foregoing discussion that there are many difficulties encountered in the preparation, application, and use of fine-crystal phosphors, and yet these screens afford the highest efficiencies. It would be useful to have *efficient* transparent screens, perhaps made in the form of a honeycomb of flat-ended transparent hexagonal phosphor prisms surrounded at their contiguous sides with nonabsorbing electrically conducting sheaths of metal (perhaps roughened to minimize retention of radiation by total internal reflection), so that the screens could be viewed in rooms with normal illumination without losing contrast unduly by reflection of the room light. These screens, however, could not be aluminized without destroying the nonreflecting quality. Attempts have been made to produce transparent screens of phosphors by evaporation or crystallization in large sheets,[58, 262, 292, 583] but none has given effective cathodoluminescence efficiencies above about 10 per cent of the same phosphors prepared by conventional solid-state reaction.

In general, low-activator-content all-sulphide (that is, S(Se)-dominated) P4 or P6 screens give highest effective luminescence efficiencies for use in direct-viewing kinescopes operated near or below 10 kv at average current densities below about 3 μa cm^{-2}, whereas high-activator-content all-silicate (that is, O-dominated) P4 or P6 screens give highest efficiencies for use in projection kinescopes operated above 10 kv at average current densities above about 3 μa cm^{-2}. If highest visually effective luminescence efficiency is required, *irrespective of color*, then a green–yellow-emitting phosphor with a composition near hex.-6ZnS·4CdS:Ag (Fig. 36) is superior below about 300 mL, whereas the green-emitting rbhdl.-Zn_2SiO_4:Mn phosphor is superior from about 300 to over 30,000 mL. With aluminized screens of these phosphors, operated at 20 to 50 kv with average current densities below 0.1 μa cm^{-2}, it is possible to obtain efficiencies of about 12 candles per watt (compare efficiency data on the phosphors in Figs. 107–109).

It is fortunate that several of the oxygen-dominated screen materials, which afford the highest cathodoluminescences, also have ϵ^{-at} decays with values of a near 80. These intermediate-duration exponential decays are more effective than any t^{-n} decays, or any shorter decays in general, in filling the "dark" interval between successive excitations (every $\frac{1}{30}$ sec) during television image reproduction. Longer persistences, for example, exponential decays with $a < 40$, give excessive carry-over beyond the frame interval and cause moving objects in television images to appear blurred and have comet-like "tails." The intermediate-duration exponential decays minimize flicker, without blurring moving objects in the image and are particularly beneficial at high image luminances when the eye is most susceptible to flicker.[58, 72]

One of the important qualities of a television image is **contrast,** which broadly connotes the ratio of light to dark in different image areas **(area contrast)** and more specifically includes the sharpness of demarcation between the light and dark areas **(detail contrast).** In kinescopes, contrast is determined by many factors, including (1) the diameter and electron-intensity distribution of the CR beam, (2) the scattering of the electrons and the luminescence emission in the screen, (3) the halation produced by multiple reflections in the transparent substrate, (4) the general background luminance of the screen, produced by stray electrons or ions exciting random luminescence or by luminescence emission reflected back from the interior of the kinescope, and (5) the amount of extraneous (ambient) illumination falling on the screen. The phosphor screen, and its application are involved in items 2 and 3, and the latter part of 4. The scattering of electrons and emitted photons in conventional small-crystal screens broadens the projected CR-beam diameter by an amount comparable with the thickness of the screen; halation may be reduced by reducing the optical contact of the crystals with the substrate (larger crystals afford less optical contact than small crystals, and settled screens usually have less optical contact than sprayed screens because the larger crystals tend to deposit first during settling—compare eq. 123); [72, 584] and reflected general illumination from the kinescope interior may be minimized by coating the interior with light-absorbing materials (such as graphite) or eliminated by aluminizing the screen.[570]

Another important quality of a television image is its **resolution (definition)**, that is, the fineness of discernible detail. The graininess of the phosphor screen, which is determined by its texture and the sizes of its crystals, is usually negligible. This is so, because P4- and P6-screen crystals average less than 10^{-3} cm in diameter, whereas CR beams are seldom less than 10^{-2} cm in diameter. For some uses of cathodoluminescent screens, such as in small projection CRT, large crystals and nonuniformities in texture of the order of 10^{-2} cm are objectionable, and special pains must be taken to obtain uniform small-crystal screens to avoid "grainy" images. At high operating voltages with small projection kinescopes, an engineering compromise must be made between the conflicting requirements of (1) small crystals for maximum resolution, and (2) larger crystals (to satisfy eq. 131) for maximum effective cathodoluminescence efficiency.

The problem of change in effective luminescence efficiency during operation of cathodoluminescent screens in CRT is usually more acute than during operation of photoluminescent screens in "fluorescent" lamps, because much higher excitation densities are used in CRT screens, and the screens are not excited uniformly over their areas (especially in oscilloscopes). As shown in Fig. 135, phosphor screens in CRT sometimes exhibit a small increase in effective luminescence efficiency during the first few minutes of operation (**"bright burn"**) followed by a slow decrease in efficiency at a rate depending on the intrinsic stability of the phosphor, the chemical and mechanical treatment of the phosphor during application (Fig. 115), the screen temperature during operation, and the power input per unit volume penetrated by the CR beam.[58, 72]

When the efficiency of a bombarded screen area has decreased below that of an unbombarded area it is called a **"dark burn."** Figure 136 shows some accelerated life test made on thick layers of several phosphors excited by an intense unmodulated stationary CR beam, using the recording spectroradiometer [93] to obtain a continuous record of efficiency as a function of time during excitation. It may be noted that in Figs. 135 and 136, as in Figs. 124 and 125, the P3 phosphor is somewhat more stable than the P1 phosphor, and the phosphate phosphors are much

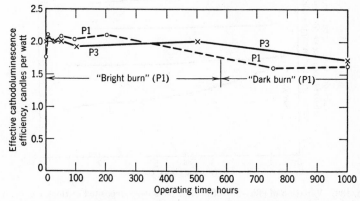

FIG. 135. Variation of efficiency of two (non-standard) phosphor screens in experimental CRT operated under television scanning conditions at 6 kv and an average current density of 1 μa cm^{-2}. Measurements made on the unbombarded sides of unaluminized screens. (L. E. Swedlund and L. B. Headrick)

less stable than the other tested materials. In the case of unaluminized phosphor screens, it is difficult to obtain a direct measure of the change in luminescence efficiency, *per se*, because changes in the *apparent* effective luminescence efficiencies may be caused largely by changes in the secondary-emission ratios of the screens.[124-134] This difficulty is eliminated by using aluminized screens operated at voltages above about 5 kv. The "bright burns" of phosphor screens are probably caused by temporary increases in the secondary-emission ratio, or by volatilization of inert surface layers (for example, organic materials) which may have formed during screen application and tube processing. With respect to "dark burn," it is found that, in general, large crystals are more stable than small crystals, O-dominated phosphors are more stable than S-dominated phosphors, and phosphors with high bonding energies are more stable than phosphors with low bonding energies (for example, low-melting cub.-Ba(NO$_3$)$_2$ is readily decomposed by fast electrons,[464] whereas high-melting rhomb.-$BaSO_4$:Pb is relatively stable, although not so stable as the silicates and aluminates).

Qualitative observations reported by R. B. Head indicate that for a number of oxide-type phosphors with a given activator the host crystals with the highest heats of formation are the least susceptible to electron or ion "burn" (see next paragraph), and the susceptibility to "burning"

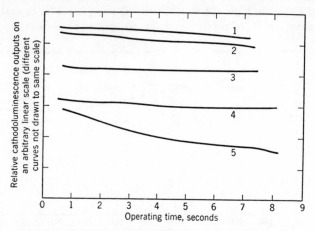

FIG. 136. Variation of efficiency of several phosphors deposited in thick layers and measured on the bombarded side during excitation by an intense stationary unmodulated CR beam at 6 kv and about 10 μa cm^{-2}.

1. rbhdl.-Zn_2SiO_4:Mn(0.3) (P1)
2. hex.-ZnS:Ag(0.015) (P7/1)
3. hex.-$9ZnS \cdot CdS$:Cu(0.0073) (P7/2)
4. hex.-ZnS:Ag(0.02):Cu(0.01) (P2), also rbhdl.-$Zn_8BeSi_5O_{19}$:Mn(1.4) (P3)
5. $Ca_2P_2O_7$:Dy(0.1) (P9)

is least for phosphors containing activators whose oxides also have high heats of formation.[555a] The rate of efficiency decrease of a phosphor during operation is accelerated by excessive grinding of the phosphor before application (Fig. 115) and by increasing the average current density at constant accelerating voltage. Increasing the accelerating voltage at constant current density usually reduces the rate of decrease of cathodoluminescence efficiency by decreasing the power dissipated per unit of penetrated volume in the phosphor crystals. Closely packed phosphor screens with a minimum of voids are desirable to minimize "burning," because most of the heat dissipation from an excited crystal is by conduction to neighboring crystals and the substrate.

Another detrimental factor during CRT operation is **"ion burn,"** which usually appears as a centrally located dark spot **(ion spot)** on the phosphor screen and is generally caused by bombardment of the screen by negative ions (usually O^{2-}) from the cathode.[585] The heavy ions cause dissociation of the crystals and atomic dislocations and tend to lodge in or on the surface layers of the screen as inert material which wastefully absorbs electron energy and luminescence radiation. High vacua in well-degassed tubes, or magnetic ion traps, serve to eliminate ion spot, and aluminized CRT screens are relatively free from the effect when the CRT are operated at voltages well above the break-even point.

In addition to the cited discolorations or efficiency losses which may be suffered by the phosphor crystals during operation, there are sometimes discolorations of the glass substrate and detrimental acquisition of alkali which may diffuse from the glass.[407, 556] These effects are most pronounced at high voltages, when residual primary electrons may penetrate into the substrate to cause decomposition and electrolysis, especially at high current densities when the screen and substrate become heated sufficiently to allow appreciable quantities of the more mobile alkali elements or their compounds to migrate into the screen from the glass. The glass substrates of heavily operated projection kinescopes are often darkened to less than 50 per cent transmission in an area corresponding to the scanned pattern (this may be seen by opening the kinescope and removing the phosphor screen, which is generally less discolored than the glass). In addition to the darkening caused by direct bombardment of the glass substrate by primary electrons, there is often a strong darkening produced by x rays emitted from the phosphor crystals and absorbed by the CRT walls. This x-ray-produced darkening extends along the glass sides of the CRT as well as in the faceplate on which the phosphor is deposited.

During 1000 hr of television operation of a kinescope screen, a given screen element the size of the cross-sectional area of the CR beam is excited for about 1.5×10^{-7} sec, 30 times a second, for a total duration of only 16 sec of actual operation time per area element. On this basis alone, it seems odd that the efficiency of the screen should change appreciably during 1000 hr elapsed (16 sec actual) time of operation. During the actual operation time, however, the *instantaneous* power input into a screen element is about 10,000 volts $\times 2 \times 10^{-4}$ amp/0.001 cm^2 = 2000 watts per square centimeter. This high value is considerably higher than the 250 watts per square centimeter absorbed power input (and radiated power output) of a tungsten filament in an ordinary incandescent lamp operated at 2850°K; hence, even 16 sec of operation at an input of 2000 watts per square centimeter is considerable for a sensitive phosphor. If the screen elements were not excited intermittently, the phosphor would become incandescent and volatilize in a second or so under a stationary unmodulated beam delivering 2000 watts per square centimeter at 10 kv. (As an example of the potency of CR beams, powdered diamonds may be made to incandesce and are converted into graphite during a few seconds of bombardment by a 10-kv CR beam delivering about 1000 watts per square centimeter, or about 10^7 watts per cubic centimeter of volume penetrated by the primary electrons). The saving feature, in scanning operation for television, is the rest period between successive excitation pulses. The rest period allows an excited crystal to dispose of most of its momentary excess of heat by radiation and by conduction to the (preferably cooled) substrate. In general, the *average* power input, rather than the *instantaneous* power input, per unit of excited volume appears to be the chief factor determining efficiency changes during scanning- or pulse-type operation with short periods of excitation.

It has been proposed to produce monochrome (white) television images by projecting blue luminescence emission from a kinescope onto a screen coated with a cascade-excited complementary yellow-emitting phosphor.[586] The yellow emission plus any transmitted blue light, when carefully matched, should produce a white image. This system is optically inefficient, because only a small percentage of the blue emission from the kinescope is directed onto the yellow-emitting screen, even with reflective optics, and much of the blue light incident on the yellow-emitting screen is lost by reflection. Much higher optical efficiency may be obtained by making the blue- and yellow-emitting layers contiguous, and viewing or projecting the resultant transmitted and cascade luminescence.[58, 63, 312, 539, 662-664]

The P6 screen was originally devised for use in a semimechanical sequential color-television system in which blue, green, and red images are picked up and reproduced in sequence on a single kinescope emitting all three primary colors.[587] By using synchronized blue, green, and red optical filters, the separate primary-color images are reproduced in time sequence, the resultant image being considerably less than one-third the luminance of the directly viewed kinescope screen. A more satisfactory all-electronic simultaneous color–television system has been introduced in which all three primary-color images are picked up and reproduced simultaneously.[588, 589] The reproduced image is usually produced by projecting and superimposing separate images from three kinescopes (the **trinoscope**), whereby each kinescope emits a different primary-color image which is optically focused and superimposed on the other images on a common viewing screen. Suitable phosphors for the trinoscope projection kinescopes are: (1) blue-emitting kinescope, hex.-ZnS:Ag, rbhdl.-Zn_2SiO_4:Ti, monocl.-(Ca:Mg)SiO_3:Ti, or $CaSiO_3$:Ti; (2) green-emitting kinescope, rbhdl.-Zn_2SiO_4:Mn; and (3) red-emitting kinescope, rbhdl.-$Zn_8BeSi_5O_{19}$:Mn(2.4) (with red optical filter) or $Cd_2B_2O_5$:Mn. Other color–television systems have been proposed with various geometrical arrangements of kinescopes with different screens, or with mosaic or line-structure screens in a common kinescope, but these have not yet achieved practical significance.[572–574, 590]

(b) *Radar kinescopes.* Most radar kinescopes have been essentially identical with television kinescopes, except that (1) they were operated at relatively low voltages (about 5 kv) to minimize electrical difficulties during air-borne operation; and (2) they were operated at many different image repetition frequencies (**IRF**), extending from above the critical flicker frequency (about 60 repetitions per second) to as low as 1 repetition in 30 sec.[58, 63, 520–522, 591–603] The low applied voltages precluded aluminizing, and the many different IRF's (usually dictated by different mechanical inertias of the oscillating radar antennas) necessitated the use of several different screens having sufficient persistences to provide visible image traces during the different intervals between successive excitations.

At IRF's in the **nonflicker region,** above about 60 repetitions per second, the fast- and intermediate-decay P1, P3, and P4 screens have proved adequate for most uses, although considerably higher visually effective cathodoluminescence efficiency may be obtained by using a special *nonstandard* screen with a composition near hex.-$6ZnS \cdot 4CdS$:Ag (Fig. 36). At IRF's in the objectionable **flicker region,** from about 2 to 60 repetitions per second, really satisfactory screens have not been devised, although P12 screens provide useful images with relatively little flicker at IRF's over most of the flicker region. By aluminizing P12 screens and operating them at about 20 kv, it is possible to obtain high-light luminances of about 50 mL, which is usually adequate for observation in well-lighted rooms or aircraft cockpits. A disadvantage of the P12 screen is its relatively rapid loss of efficiency and increasing rate of decay during operation (this follows from the low bonding energy, for example, low hardness and low melting point, of the phosphor host crystal). At IRF's in the less objectionable **flacker region,** below 2 repetitions per second, the cascade-type blue-luminescing yellow-phosphorescing P14 and P7 screens have been particularly useful for operation at about 1 repetition per second and at about 1 repetition in 5 to 10 sec, respectively. The single-layer white-emitting **P9** screen ($Ca_2P_2O_7$:Dy) approximates the persistence of the P7 screen, except that the P9 screen affords lower luminance during phosphorescence after a given excitation and loses efficiency more rapidly during operation. The operating luminance of a P7-screen trace 3 sec after an average excitation is about

0.03 mL, whereas the luminances of P14- and P9-screen traces 2 sec after an average excitation are about 0.003 mL. These low luminances necessitate darkened operating booths and mesopic vision. For IRF's lower than about 1 repetition in 10 sec, the tenebrescent P10 screen (cub.-KCl) affords sufficient persistence of dark image traces on a very bright background, but the P10 screen has been found to be generally less satisfactory than long-persistent phosphor screens because (1) the CR-beam energy required to produce a barely visible trace on a P10 screen is about 100 times higher than that required for phosphor screens, (2) the contrast in P10-screen images is usually about $1.2/1$ compared with about $20/1$ for phosphor screens, (3) a strong source of heat and auxiliary illumination must be provided to bleach the image, and (4) even with high intensity of irradiation strongly excited image traces on P10 screens often persist for several minutes or even days.[58, 63, 215, 591, 596, 598, 604] In this respect, it is noteworthy that long-persistent phosphor screens (for example, P7) and tenebrescent screens (for example, P10) have opposite variations of decay rate with increasing degree of excitation; that is, the phosphor screen decays at a *faster* rate with increasing degree of excitation, whereas the tenebrescent screen decays at a *slower* rate with increasing degree of excitation.

It is possible to conceive of a hypothetical stratified screen of cascading phosphors, such that the CR beam *briefly* excites one persistent phosphor whose emission efficiently excites a second persistent phosphor, whereby the light output from the second phosphor rises and then falls (concave-downward decay) after the short excitation pulse.[58, 63] In the most favorable case, when both phosphors have temperature-independent ϵ^{-at} decays, with constants a_1 and a_2, the light output from the final layer is given by

$$L_{\epsilon_1\epsilon_2}(t) = C_{\epsilon_1\epsilon_2} \frac{\epsilon^{-a_1t} - \epsilon^{-a_2t}}{a_2 - a_1} \tag{136}$$

where $C_{\epsilon_1\epsilon_2}$ is proportional to the CR energy input and the efficiencies of the cascaded phosphors. From the symmetry of the constants in eq. 136 (and in eq. 138), it may be seen that the sequence of the two phosphors in the cascade screen is immaterial, it being assumed that each will excite the other; so the phosphor with the shortest persistence (largest a in eq. 136) dominates the persistence of the cascade screen. When $a_1 = a_2$, eq. 136 becomes, by L'Hospital's rule,

$$L_{\epsilon_1\epsilon_2}(t) = C_{\epsilon_1\epsilon_2} t \epsilon^{-a_1t} \tag{137}$$

If two ϵ^{-at} decay phosphors with both a_1 and a_2 equal to 10 sec^{-1} could be made to cascade, the maximum $L(t)$ value would occur at 0.1

sec, whereas, *if* a_1 and a_2 equaled 1 sec^{-1}, the maximum $L(t)$ value would occur at 1 sec (Fig. 137). These *hypothetical* cascade screens would be very useful in providing practically flickerless radar images, but the rare occurrence of ϵ^{-at} decay phosphors with a less than 10 sec^{-1} and

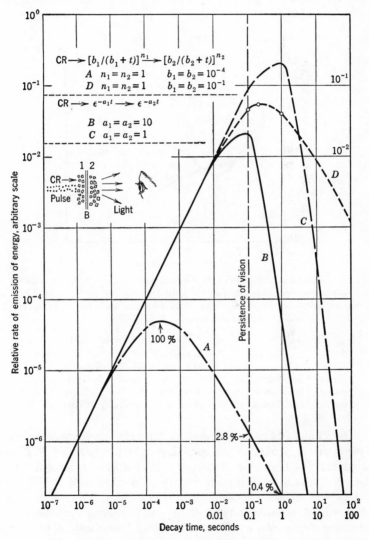

Fig. 137. Phosphorescence emission curves for the second layers of some hypothetical cascade screens excited by a primary pulse lasting less than 10^{-7} sec.

the wide separation between the excitation and emission spectra of efficient ϵ^{-at} decay phosphors have thus far precluded attainment of practical cascaded screens of this type, even with both a values near 60 sec^{-1} (this would be especially useful in eliminating flicker in television).

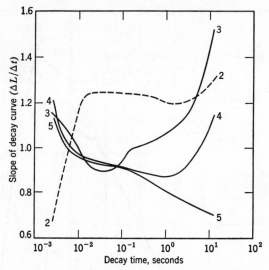

FIG. 138. Variations of slopes n of the decay curves of some t^{-n} decay unaluminized phosphor screens excited by one 4-kv CR pulse carrying 2×10^{-8} coulomb. Screens previously de-excited by red light, and measurements made on the unbombarded side of the screen. (A. B. White)

	First Layer	Second Layer
1.	. . . . . .	10 mg cm^{-2} hex.-ZnS(86)CdS(14):Cu (P7/2)
2.	. . . . .	10 mg cm^{-2} hex.-ZnS(93)CdS(7):Ag:Cu (AB)
3.	6 mg cm^{-2} hex.-ZnS:Ag on	10 mg cm^{-2} hex.-ZnS(70)CdS(30):Cu
4.	6 mg cm^{-2} hex.-ZnS:Ag on	10 mg cm^{-2} hex.-ZnS(75)CdS(25):Cu (P14)
5.	10 mg cm^{-2} hex.-ZnS:Ag on	12 mg cm^{-2} hex.-ZnS(86)CdS(14):Cu (P7)

The **P7** and **P14** cascade screens used in radar kinescopes comprise a layer of about 12 mg cm^{-2} of hex.-9ZnS·CdS:Cu(0.007) (= **P7/2**), or 12 mg cm^{-2} of hex.-8ZnS·2CdS:Cu(0.005) (= **P14/2**), covered with a layer of about 8 to 10 mg cm^{-2} of hex.-ZnS:Ag(0.015) (= P7/1, or P14/1).[662–664] These are temperature-dependent t^{-n} decay phosphors which exhibit increasing rates of decay with increasing temperature and increasing degree of excitation. The decay "constants" n of several previously de-excited single-layer and double-layer screens, made of t^{-n} decay phosphors, are shown in Figs. 138 and 139 as a function of time after (a) one 2×10^{-8} coulomb pulse, and (b) 100 2×10^{-8} coulomb pulses spaced 1 sec apart.[58, 600, 601] None of these screens afford maximum $L(t)$ near or after 0.1 sec. For a *hypothetical* cascading pair of t^{-n} decay phosphors with $n_1 = n_2 = 1$ (this hyperbolic decay violates

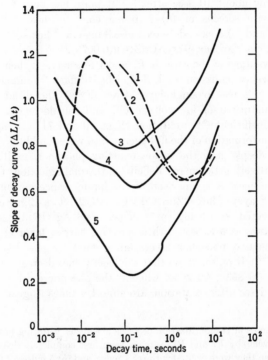

FIG. 139. Variations of slopes n of decay curves of t^{-n} decay unaluminized phosphor screens excited by 100 4-kv CR pulses, each carrying 2×10^{-8} coulomb and spaced 1 sec apart. Screens previously de-excited by red light, and measurements made from unbombarded side of screen. Same legend as Fig. 138. (A. B. White)

the law of conservation of energy for $t \to \infty$ but *approximates* the exponent for most of the phosphor screens in the interval shown in Figs. 138 and 139; compare Figs. 73 and 80) and "constants" b, according to eq. 99, the light output from the final layer is given by

$$L_{h_1 h_2}(t) = C_{h_1 h_2} \frac{b_1 b_2 \log_\epsilon \dfrac{(b_1 + t)(b_2 + t)}{b_1 b_2}}{b_1 + b_2 + t} \tag{138}$$

where the phosphorescence intensity, $L(t)$, is a maximum when $t \approx (b_1 + b_2)$. In this hypothetical case, the "constants" b are less than about 10^{-4} for known cascadable phosphors under practical operating conditions; hence, maximum $L(t)$ does not occur at decay times longer than about 10^{-3} sec (Fig. 137). A brief concave-downward decay lasting about 10 μsec after a CR pulse has been found experimentally with screens of (a) 14.6 mg cm^{-2} of hex.-ZnS:Ag(0.01)-Cu(0.005), and (b) a cascade screen of 8.3 mg cm^{-2} hex.-ZnS:Ag(0.015) on 11.9 mg cm^{-2} of hex.-9ZnS·CdS:Cu(0.0073).[58]

The advantage of the cascade P7 and P14 screens is their intensified phosphorescence emission and diminished **"flash"** (emission during excitation) relative to single-layer screens.[58,63] The disadvantageous shallow penetration of low-voltage CR and the decreased t^{-n} decay caused by inefficient CR excitation (Figs. 70 and 71) are obviated by cascade transformation of the CR energy into violet and blue light which penetrates deeply into the yellow-emitting phosphor, exciting it and leaving relatively little energy deficit (as thermal agitation) to decrease phosphorescence. A comparison of the luminescence outputs of (a) a 10-mg cm^{-2} layer of hex.-9ZnS·CdS:Cu (P7/2), (b) a 12-mg cm^{-2} layer of P7/2 covered with 8 mg cm^{-2} of hex.-ZnS:Ag (P7/1), and (c) the same structure as b except with a selective **barrier layer** of thin glass between the two phosphor layers, all excited by one $\frac{1}{720}$-sec 6-kv 220-μa/cm^{-2} CR pulse, shows that the flash values decrease in the order (a) 50,000, (b) 8530, (c) 2650, whereas the phosphorescence intensities of b and c, 1 sec after excitation, are almost 4 times as great as that of a.[58,63]

For purposes of standardization of cascade screens, a testing procedure was established whereby P7 and P14 screens were first completely de-excited with red light, then they were excited with (a) 1 pulse, and (b) 5 pulses of CR of given intensity, duration, and spacing, and the light output 1 sec after a and b was measured as B_1 and B_5.[58,592,593,599] The ratio B_5/B_1 was called "buildup," and P7 screens with B_5/B_1 less than 5 were rejected. In practical use, of course, the screens are always partially excited, that is, the deep traps are filled, and so the superproportional "buildups" obtainable with completely de-excited screens (compare Fig. 74) are not realized in operation. Therefore, the B_5/B_1 ratio has no direct significance unless it is measured on screens which are partially excited to an extent corresponding to average operating conditions. Perhaps the simplest meaningful test of persistent screens is to excite them under simulated operating conditions and measure the phosphorescence intensity at two instants of time after excitation, for example, at $0.3t_i$ and t_i, where t_i is the time interval between successive images. The value at, for example, $0.3t_i$ should be made a maximum to facilitate observation, and the value at t_i should be made a minimum to minimize carry-over into the next image.

When a cascade screen comprises stratified layers of an exponential-decay phosphor with decay constant a, and an (approximately) hyperbolic-decay phosphor with decay constant b, and $n = 1$ as in eq. 138, then the resultant decay of the cascade screen after a brief excitation may be expressed by

$$L_{h\epsilon}(t) = C_{h\epsilon}\epsilon^{-a(b+t)} \int_b^{b+t} \frac{\epsilon^{a\xi}}{\xi} \, d\xi \qquad (139)$$

where $\xi = b + t$ is introduced to put the expression in a form capable of solution with the aid of tables of exponential integrals.[58] Again, the sequence of the two phosphors is immaterial insofar as the decay alone is concerned (it being assumed that each excites the other; which is not the case with actual phosphors).

If a cascade screen be excited to *equilibrium* with the (unmodulated) source of excitation, then the growth and decay processes differ from those of the previous examples which were based on very brief excitation pulses, that is, pulses short relative to the half-life of the fastest-decay phosphor component of the screen. As an example, given two cascading ϵ^{-at} decay phosphor layers, with decay constants a_1 and a_2, the *growth* of luminescence emission L during attainment of equilibrium is expressed by

$$L_{\text{growth}} = C_{\epsilon_1\epsilon_2}[1 - (a_2\epsilon^{-a_1t} - a_1\epsilon^{-a_2t})/(a_2 - a_1)] \qquad (140)$$

where the phosphor layer with the decay constant a_1 is the layer excited by the primary excitation energy. Under these circumstances, a plot of L vs. t gives an S-shaped growth curve similar to that shown in Fig. 74a for a single (deep-trap) phosphor. The *decay* of luminescence emission after cessation of excitation of such a cascade screen to equilibrium is given by

$$L_{\text{decay}} = 1 - L_{\text{growth}} = C_{\epsilon_1\epsilon_2}(a_2\epsilon^{-a_1t} - a_1\epsilon^{-a_2t})/(a_2 - a_1) \qquad (141)$$

In the case when $a_2 = a_1 \equiv a$, eq. 141 becomes

$$L_{\text{decay}} = C_{\epsilon_1\epsilon_2}(1 + at)\epsilon^{-at} \qquad (142)$$

where the cascade factor, $at\epsilon^{-at}$, provides a plateau of practically constant L for a time determined by the magnitude of a, immediately after excitation, and thereafter L decays in an increasingly exponential manner. Comparison of eqs. 137 and 142 shows the difference between the decay of a cascade screen following a brief excitation pulse and following excitation to equilibrium.

In addition to the usual *positive modulation* of phosphor outputs, that is, the normal increase of luminescence emission with increasing excitation density at temperatures below the temperature break point T_B (Figs. 18 and 103–106), there is the possibility of using *negative modulation*, that is, decreasing the luminescence emission by increasing T beyond T_B.[58, 63, 605-607] This may be accomplished by uniform scanning or flooding of a cathodoluminescent screen with a CR beam of sufficient power to bring the excited area to a high luminance and to a temperature near T_B, whereupon a dark-trace image may be traced by supplying additional CR energy with a modulated CR beam which effectively "pushes the phosphor over the brink" of its L vs. T curve (Fig. 18). The dark traces, thus formed on an intense luminescent background, decay at a rate determined by the rate of dissipation of the excess heat imparted by the tracing CR beam. Dark traces on experimental tubes have been made to persist for times up to 7 sec.[58] This brute-force method is relatively insensitive and requires considerable CR-beam power, but it provides exceptionally high-contrast dark-trace images with background luminances in excess of 10,000 mL.

One of the major lacks in radar CRT is a screen with an intense long-persistent *red* phosphorescence emission.[58] Such a screen would be of great advantage in allowing radar operators to preserve their dark adaptations. It is remarkable that no efficient long-persistent red-emitting phosphors have thus far been devised. At present, intense long-persistent red radar images are best produced by using electronic storage tubes [608, 609] which actuate kinescopes having conventional red-emitting phosphor screens, for example, hex.-$3ZnS \cdot 7CdS:Ag$ at moderate current densities, and rbhdl.-$Zn_8BeSi_5O_{19}:Mn(2.4)$ or $2CdO \cdot B_2O_3:Mn$ at high current densities (with suitable optical filters).

(c) *Image tubes.* An image tube is an evacuated glass bulb with a photoelectron-emitting surface at one end and a cathodoluminescent screen at the other.[610-612] By focusing an optical image on the photosurface, a corresponding pattern of electrons is emitted, whereupon the electrons are accelerated and the electron pattern is focused on the phosphor screen which reproduces the original optical image in the color characteristic of the phosphor(s). With an image tube, such as the 1P25A, it is possible to "see" with infrared and ultraviolet radiations which are invisible to the human eye but eject photoelectrons from suitable photosurfaces (Fig. 120). The image tube is a vital component of the snooperscopes and sniperscopes used to conduct nocturnal military operations with the aid of infrared. From the phosphor standpoint, the chief difficulties in image tubes comprise producing short-persistent (nonblurring) phosphor screens which (1) can withstand cesium vapor, from the photosurface, at tube temperatures up to about 200°C without undue reduction and loss of cathodoluminescence efficiency, and (2) have very uniform and fine texture to allow resolution of detail down to about 1000 image elements per centimeter. Although all the phosphors thus far investigated darken and lose efficiency to some extent under the reducing action of hot cesium vapor, the high-melting oxygen-dominated phosphors, such as rbhdl.-Zn_2SiO_4:Mn, are more resistant than the higher-efficiency sulphur-dominated phosphors, such as hex.-$6ZnS\cdot 4CdS$:Ag. Unfortunately, phosphors with small crystals are usually more susceptible to the action of hot cesium vapor than those with larger crystals. The impasse is best eliminated by using image-tube voltages above the break-even voltage (about 5 kv) and carefully aluminizing the screens. A pinhole-free coating of aluminum protects the screen from the cesium vapor and allows the use of hex.-$6ZnS\cdot 4CdS$:Ag or cub.-$7ZnS\cdot 3ZnSe$:Ag (Figs. 33, 36, and 39) which provides higher effective cathodoluminescence efficiency at low current densities than rbhdl.-Zn_2SiO_4:Mn. (It is noteworthy that the incorporation of Se in these sulphide-type phosphors apparently increases the stability of the resultant material towards photolysis or the reducing action of hot cesium vapor.)

High-resolution screens may be obtained by careful deposition of phosphors which have been synthesized with average crystal sizes less than 10^{-4} cm, or by deposition of phosphor "fines" which are obtained by segregating the smaller crystals from ordinary phosphor batches. Synthesis of small-crystal batches is usually accomplished by starting with very finely divided ingredients which are reacted and crystallized at as low a temperature and for as short a time as is needed to obtain adequate luminescence efficiency. Segregation of small crystals from a batch with ordinary size distribution (Figs. 7 and 10) may be done by taking advantage of the selective settling of particles of different sizes in a tall column of liquid [the largest and densest particles settle most rapidly (eq. 123), leaving the "fines" fraction in the upper part of the column]. Very small crystals are usually applied by centrifugation, coating densities ranging from about 0.5 to 4 mg cm^{-2} being used, depending on the average size of the phosphor crystals and on the operating voltage.

(d) *Electron microscopes.*[101] The phosphor screens in electron microscopes, as in image tubes, should have (1) visible persistences shorter than 0.1 sec to avoid blurring of moving images, and (2) very fine and uniform texture. A satisfactory fine-crystal phosphor for electron microscopes is yellow-emitting hex.-$6ZnS\cdot 4CdS$:Ag (800°C, Fig. 33), although white-emitting P4 screens may be used to advantage, especially when the image is to be photographed. In both image tubes and electron microscopes, the CR current densities are usually less than 0.1

μa cm^{-2}; hence, the image luminances are usually less than 0.1 mL, and the observer must be partially dark-adapted. In the electron microscope luminescence efficiency is not a pressing problem, and so the phosphor screen may be made thinner than the optimum coating density in order to minimize loss of resolution due to scattering in the thicker screens. For electron microscopes operated in the voltage range from 50 to 200 kv, coating densities from 2 to 4 mg cm^{-2} afford screens with resolutions as high as 1000 lines per centimeter when the average particle size of the phosphor does not exceed about 3 microns.

EXCITATION BY PRIMARY IONS. (IONOLUMINESCENCE). As shown in Fig. 23, ions, such as alpha particles (He^{2+} ions), penetrate much shorter distances into solids than electrons of the same energy (in ev). For this reason, heavy ions used as primary excitants must have energies in excess of about 10 kev in order to penetrate sufficiently into the efficient volumes of phosphor crystals to produce readily detectable luminescence. Further complications which are encountered in the use of ions to excite luminescence in solids are (1) the strong local disruptions (atomic and ionic dislocations) produced in the crystals along the paths of these massive and often highly charged particles, and (2) the additional impurity incorporated in the phosphor when the absorbed primary ion remains, at least temporarily, in or on the phosphor crystal. Both these effects generally reduce luminescence efficiency during operation.[34, 102, 240–242, 360, 362, 364]

The chief commercial use of phosphors excited by ions is in radioactive luminescent coatings for watch dials, instrument dials, and the like, to be viewed in the dark. By admixing approximately 2×10^{-6} g of alpha-particle-emitting RaBr$_2$ or RaSO$_4$ per gram of hex.-ZnS:Cu, a surface luminance of about 0.02 mL may be obtained without further excitation as long as 2 months after making the mixture.[211]

One gram of radium, freed of its disintegration products (for example, radon, and radium A, B, and C), produces about 3.7×10^{10} alpha particles per second; the average alpha particle having an energy of about 4.7 million electron volts (**mev**). (*Note*. It has become customary to refer to 3.7×10^{10} radioactive disintegration particles per second as one **curie**, regardless of the nature of the particles, that is, regardless of whether the particles are gamma-ray photons, beta-ray electrons, alpha particles, or other particles). When phosphors and radioactive compounds are freshly mixed, they do not have such high luminescence outputs as they do several weeks after mixing, because the radium must re-establish equilibrium with its shorter-lived disintegration products, such as radon (gas), RaC, and RaF (polonium). According to Rutherford, Chadwick, and Ellis,[499] the energy emitted by 1 g of radium in equilibrium with its disintegration products is distributed as follows: 25.2 g cal hr^{-1} of 4.7-mev alpha particles from Ra, 99.2 g cal hr^{-1} of 5.4- to 7.7-mev alpha particles from the disintegration products, 6.3 g cal hr^{-1} of 0.03- to 2.5-mev beta particles (average energy about 0.5 mev) from RaB and RaC, and 9.4 g cal hr^{-1} of 0.05- to 2.2-mev gamma rays from RaB and RaC.

Higher luminances than 0.02 mL may be obtained by using higher radium (or radiothorium, or mesothorium) contents, but the rate of decrease of luminescence output from the phosphor during operation is approximately proportional to the proportion of radioactive material. A binderless radioactive luminescent coating with an initial maximum luminance (2 to 4 weeks after compounding) of 0.03 mL reportedly decreases to 0.015 mL in about 9 months, whereas a coating with an initial maximum luminance of only 0.015 mL decreases to 0.008 mL in about 18 months.[102] In practical use, the luminance of a radioactive luminescent coating usually decreases more rapidly than the rate indicated, because of (1) darkening of the organic binders (used to affix the coarse-crystal mixture to the substrate) under alpha-particle bombardment, (2) darkening (decreased reflectivity) of the substrate caused by the radioactive disintegration products and particles, and (3) photolytic darkening of the hex.-ZnS:Cu phosphor when exposed to sunlight. In the latter respect, superior performance may be obtained by using the stable green–yellow-emitting rbhdl.-Zn_2GeO_4:Mn phosphor, which has been found to have excellent efficiency under excitation by alpha particles and is not susceptible to photolysis.[192] The most economical coating density for a hex.-ZnS:Cu phosphor, with an average crystal size of about 10 microns, admixed with radioactive salt, is about 40 mg cm^{-2} (0.2 mm thick), although the luminance may be increased by increasing the coating density up to a maximum value of about 400 mg cm^{-2} (no further increase in luminance is obtained when the thickness is increased beyond about 2 mm). Recently, it has been reported that 3-cm crystals of potassium uranyl sulphate have fairly constant self-luminances of the order of 10^{-5} mL,[613] whereas Kabakjian found that the self-luminescence of radium sulphate crystals decreased to 50 per cent of the initial value in 40 min and became practically extinct in 10 hr.[244] Reheating the crystals to 650°C restored the luminescence efficiency to its initial value.

Objective Detectors (for Example, Phototubes, Photographic Emulsions, and Phosphors)

The foregoing outline of several uses of phosphors, involving the human eye as the detector, presents most of the salient features of the procedure for selecting a phosphor for a particular purpose. From the standpoint of efficiency alone, a phosphor would be chosen to give a maximum effective luminescence efficiency, ε (compare eq. 107),

$$\varepsilon = 100 \iint N_e E_e \cdot X_d(E_e) \cdot dt dE_e / \iint N_0 E_a dt dE_a \quad \text{in per cent} \quad (143)$$

where N_e is the number of luminescence photons of energy E_e emitted per unit time, $X_d(E_e)$ is the sensitivity of the detector to photons of energy E_e, and N_0 is the number of primary excitant particles, of energy E_a, absorbed per unit time by the phosphor screen. In a practical case, assuming the average photopic human eye as the detector, the maximum ε for excitation by 3650-Å UV is given by hex.-6ZnS·4CdS:Ag, whereas maximum ε for excitation by 2537-Å UV is given by rbhdl.-Zn$_2$SiO$_4$:Mn. Similarly, maximum ε for excitation by television-type scanning with a 15-kv CR beam delivering an average of less than 3 μa cm^{-2} is given by hex.-6ZnS·4CdS:Ag(0.01), whereas maximum ε for the same excitation, except at an average of well over 3 μa cm^{-2}, is given by rbhdl.-Zn$_2$SiO$_4$: Mn(1) (or, preferably, with as high a proportion of *effective* Mn as possible).

Other factors, such as color, persistence, resolution, and contrast, sometimes force the selection of phosphors which have relatively low ε. For example, if intense visible phosphorescence emission is required 1 min after cessation of excitation, the lower-efficiency (during excitation) hex.-ZnS:Cu may be chosen for use with 3650-Å UV excitation, rbhdl.-Zn$_9$BeSi$_6$SnO$_{24}$:Mn [or rbhdl.-(Zn:Be)$_2$SiO$_4$:Mn:As] may be chosen for excitation by 2537-Å UV, and a P7 or tetr.-(Zn:Mg)F$_2$:Mn (P12) screen may be chosen for excitation by CR. Also, if extremely high resolution is required, it is necessary to use fine-crystal phosphors (or thin plates of large single-crystal phosphors, preferably roughened on the detector side) which may be synthesized with some sacrifice in ε. Similarly, if the eye is to be the detector, but dark adaptation must be preserved, then red-emitting phosphors are required, with a large loss in ε owing to the insensitivity of the scotopic eye to red light (Fig. 120).[58]

It is not feasible to list all the engineering compromises which are necessary in selecting (or devising) and applying the best phosphor for a specific purpose. In the following brief discussion, only certain outstanding examples of presently known phosphors and their uses are cited to provide some orientation and aid to researchers who wish to use phosphors to transform various primary particles into radiations which are detectable by objective photosensitive devices. For many uses, the best available phosphor catering to the eye will provide adequate luminescence emission to actuate phototubes or photographic emulsions whose sensitivities overlap the spectral sensitivity curve of the eye (Fig. 120); hence, it is often sufficient to refer to the preceding section and use the phosphor recommended for production of visible emission by a particular excitant. One of the gratifying aspects of providing phosphors for use with *objective* detectors, such as phototubes, is the absence of certain intangible *subjective* factors, such as the psychological and physiological effects of different colors and intensities on the rather whimsical human eye (and brain).

EXCITATION BY PHOTONS. Inspection of Figs. 32–45 and 51–56, 60, 90–96 shows that, in general, (1) the highest peak-energy output under excitation by UV (or CR) is obtained from the simplest progenitors in each of the cited phosphor families (for example, hex.-ZnS:Ag and rbhdl.-Zn_2SiO_4:Mn), and (2) these simple progenitor phosphors usually have their emissions in the violet–green portion of the spectrum. Hence, it is particularly advantageous to use blue-sensitive panchromatic photographic emulsions or phototubes (especially phototubes with the S4 or S5 photosurfaces) coupled with a good hex.-ZnS:Ag phosphor to detect CR or 3650-Å UV. For excitation by 2537-Å UV, monocl.-Mg_2WO_5:[W] or tetr.-$CaWO_4$:[W] may be substituted for the hex.-ZnS:Ag phosphor, and, for excitation by x rays and gamma rays, hex.-ZnO:[Zn], rhomb.-$BaSO_4$:Pb, cub.-NaI:Tl(1), monocl.-stilbene, or monocl.-naphthalene (see page 250) are also advantageous, especially when microsecond persistence is required.[58, 290–293, 369, 502–505, 674, 675] When green emissions, such as from rbhdl.-Zn_2SiO_4:Mn, are to be recorded photographically, orthochromatic emulsions are preferred.[508–510, 525]

EXCITATION BY CHARGED MATERIAL PARTICLES. Of particular practical importance is the **flying-spot CRT,** which is used as a television pickup device.[440, 588, 589] In one use of flying-spot CRT, an unmodulated CR beam scans a normal television pattern **(raster)** on a fast-decay phosphor screen, and the resultant flying spot of light on the screen is projected by a lens system through a movie film onto a phototube. The shades of light and dark in the film are converted into corresponding high- and low-output currents from the phototube, thereby providing a point-by-point analysis of the photographic image on the film. Ultrashort persistence of the phosphor is required because the scanning CR beam travels at about 10^5 cm sec^{-1} while exploring over 10^6 image elements, so the luminescence emission from a given image element should decay in less than 10^{-6} sec to a negligibly low value. Particularly suitable ultrafast-decay phosphors for flying-spot CRT are hex.-ZnO: [Zn] (P15), tricl.-$Ca_2MgSi_2O_7$:Ce, and cub.-MgS:Sb (Tables 5, 21, and 22). If the microsecond-decay *visible* emission band of hex.-ZnO: [Zn] (Fig. 50) is to be used in an aluminized screen for operation at 20 kv, the optimum coating density is about 6 mg cm^{-2} for crystals averaging about one micron. If the faster-decay UV emission band is to be used, however, the optimum coating density is practically a monolayer of crystals, that is, about 1 mg cm^{-2} for an aluminized screen operated at 20 kv. The UV output from an aluminized hex.-ZnO:[Zn] screen, operated at 20 kv and 500 μa in a well-focused spot, has been found to increase about a hundredfold when the coating density is decreased from 6 to 1 mg cm^{-2}. The microsecond-decay cub.-MgS:Sb phosphor is advantageous for color-television pickup, because this phosphor emits more red than the hex.-ZnO:[Zn] phosphor, and the most efficient (S4 and S5) photosurfaces are relatively insensitive in the red end of the visible spectrum (Fig. 120). A disadvantage of the conventional hex.-ZnO:[Zn] phosphor is its low-temperature break point T_B (Fig. 104),[58, 63] and so it is not always suitable for operation at high excitation densities or at temperatures much above room temperature. By skillful variations in processing it is possible to prepare hex.-ZnO:[Zn] phosphors which have T_B values about 100°C higher than normal, but the fast-decay cub.-MgS:Sb and rhomb.-BaSO$_4$:Pb phosphors are less easily saturated and operate efficiently at considerably higher temperatures (up to about 400°C in the case of rhomb.-BaSO$_4$:Pb). The hex.-ZnO:[Zn] phosphor, however, is outstanding for excitation by low-voltage primary electrons, giving detectable luminescence emission under bombardment by primary electrons with energies as low as 5 ev.[58, 701]

For general use, with blue-sensitive detectors and charged material-particle excitants, such as beta particles, protons, and alpha particles, the hex.-ZnS:Ag (or hex.-ZnS:[Zn]) phosphor is outstanding when its weak t^{-n} decay is not objectionable. Under alpha-particle excitation, for example, most of the luminescence (scintillation) produced by a single alpha particle is emitted within about 10^{-6} sec, whereafter there is a feeble t^{-n} decay phosphorescence. Oscilloscope recordings of scintillations excited by successive single alpha particles often differ considerably in magnitude. The difference is caused in part by the distribution in energy of the primary particles and in the excitations produced in an assumedly homogeneous medium,[614] and in part by the random points and angles of entry of the excitant particles into the jumble of phosphor crystals comprising a screen. Of two primary particles of equal energy, one may penetrate through a larger proportion of inefficient edges and sides of the phosphor crystals and, therefore, produce less luminescence emission than the particle which penetrates more into the efficient volumes of the phosphor crystals. In other words, even when all the energy of the primary particle is absorbed in the screen, different equal-energy primary particles entering a conventional multicrystal phosphor screen may give different N_e's in eq. 143. A coating density of about 30 mg cm^{-2} of hex.-ZnS:Ag ($\bar{x}_c \approx$ 15 microns) on the outer surface of a blue-sensitive multiplier phototube, such as the 931A, gives optimum results with 6-mev alpha particles emanated from radium. As previously mentioned, the luminescence efficiencies *reported* under excitation by alpha particles range up to 100 per cent, with the efficiency under excitation by beta particles somewhat lower

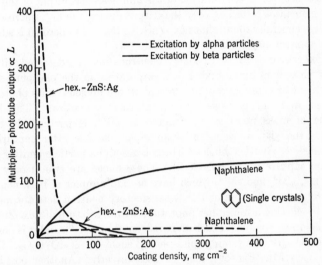

FIG. 140. Relative microsecond outputs of different thicknesses of a fine-crystal hex.-ZnS:Ag phosphor and of a series of large single-crystal "screens" of monocl.-naphthalene excited by alpha particles and by beta particles. Measurement on the unbombarded side of the screen, with a UV + blue-sensitive multiplier phototube (Compare Figs. 126 and 127). (H. Kallmann)

than under alpha particles.[102, 399, 615] A more conservative value for the maximum effective luminescence efficiency under high-energy photons or charged material particles is of the order of 10 per cent; that is, about 10 per cent of the available energy of the primary photons or charged material particles is converted into emitted (conventional) luminescence radiation.[34, 400] In addition to the conventional luminescence, x rays may be emitted with a maximum efficiency of the order of 1 per cent.[395-397]

The same general considerations which applied to the choice of phosphors and sizes of phosphor crystals for excitation by x rays and gamma rays (Figs. 126 and 127) apply also for excitation by charged material particles. As shown in Fig. 140, Kallmann reports maximum microsecond output from a hex.-ZnS:Ag phosphor under excitation by alpha particles which penetrate a coating density of about 6 mg cm^{-2}, whereas maximum microsecond output was obtained with large transparent crystals of monocl.-naphthalene under excitation by beta particles which penetrate a coating density of over 50 mg cm^{-2} in hex.-ZnS:Ag, tetr.-CaWO$_4$:[W], and rbhdl.-Zn$_2$SiO$_4$:Mn.[292] When the detector integrates over most of the duration of the luminescence output of the phosphor, instead of recording only the output during the first microsecond, then it is generally found that hex.-ZnS:Ag ($\bar{x}_c \approx 15$ microns) is superior to the monocl.-naphthalene.

"EXCITATION" BY NEUTRONS. Neutrons may be detected with the aid of phosphors by admixing or incorporating in the phosphor screen an element or compound of an element whose nucleus emits photons or charged material particles after capturing a neutron. The boron isotope of mass number 10, denoted as $_5B^{10}$, is particularly useful because the $_5B^{10}$ nucleus emits an alpha particle after capturing a neutron.[68, 505] Another nucleus which is an alpha-particle emitter after neutron capture is $_3Li^6$ (Table 2). These nuclei are two of only four, $_1H^2$, $_3Li^6$, $_5B^{10}$, and $_7N^{14}$, which have an odd number of both neutrons and protons (and a spin of 1). A satisfactory neutron detector may be made by coating a multiplier phototube with a mixture of hex.-ZnS:Ag and a colorless compound such as B$_2$O$_3$ containing $_5B^{10}$, and it is possible that some $_5B^{10}$-containing phosphors, such as 2CdO·B$_2$O$_3$:Mn or 2ZnO·B$_2$O$_3$:Mn (Fig. 59), may be used directly. Another possibility is to embed the hex.-ZnS:Ag crystals partially or completely in a B$_2$O$_3$-rich glass (using $_5B^{10}$).

Phosphors as Detectors and Indicators. Phosphors are used mostly, at temperatures near room temperature, as (1) direct transformers of the free energies of primary particles (photons and charged material particles having energies in excess of about 2 ev) into photons with energies ranging from about 1 to 7 ev, and (2) means for temporarily storing excitation energy (with decreasing retention during the storage interval) for time intervals lasting from about 10^{-7} sec to many months. Primary particles with energies from about 2 to over 10^7 ev excite the more efficient phosphors to readily detectable luminescence emission, and the luminances of phosphor screens are limited more by the available sources of excitation energy than by the phosphors themselves. Indeed, instantaneous (pulsed) luminances in excess of 10^7 mL, and sustained average luminances of the order of 3 × 10^4 mL have already been obtained under CR excitation. Higher luminances are possible with refrigerated phosphor screens excited by more penetrating (higher-energy) particles with higher beam densities (see eq. 117). *Initial* temporary storage of excitation energy may be accomplished up to a maximum of about 10^{21} potential photons per cubic centimeter for some ϵ^{-at} decay phosphors, such as rbhdl.-$(Zn:Be)_2SiO_4:Mn$, and up to about 10^{19} potential photons per cubic centimeter for some t^{-n} decay phosphors, such as hex.-$ZnS:Cu$. The rate of dissipation of the stored energy depends on the nature of the phosphor and, in the t^{-n} decay phosphors, on the temperature and other operating conditions. The luminescence outputs of some of the ultrafast-decay phosphors, such as hex.-$ZnO:[Zn]$ (UV component) and rhomb.-$BaSO_4:Pb$, may be detectably modulated at rates up to about 10^7 cycles per second, whereas the potential phosphorescence emission after excitation of some phosphors, such as cub.-$Sr(S:Se):Sm:Eu$, may be "frozen in" at low temperatures to be released several years later by raising the temperature or irradiating the phosphor with infrared.

In addition to the uses of phosphors which have been cited already, there are innumerable other special uses taking advantage of the unique capabilities of phosphors.[471-481, 616, 617] Some of the more interesting special uses include:

1. *Temperature indicators*, especially in inaccessible locations. A phosphor whose efficiency is a known function of temperature (Figs. 103–106) may be coated on a remote object and excited by a steady source of UV, so that the detector (multiplier phototube) output may be calibrated in terms of the temperature of the substrate. This method is especially useful for detecting temperature changes over extended areas where (*a*) the entire surface may be coated with one or more phosphors which are heated by conduction, or (*b*) the thermal radiation from the surface may be optically projected onto a remote phosphor screen which is heated in proportion to the infrared intensity. According to Urbach *et al.*, temperature differences as low as 4°C can be detected by using $(Zn:Cd)S:Ag:Ni$ phosphors excited by intense 3650-Å UV and "quenched" by heat conduction or thermal radiation.[172] Some of these phosphors have very precipitous decreases in ε with increasing T (Fig. 103) and may be used to detect temperature differences with radiation intensities as low as 500 μw cm^{-2}.

2. *Flow indicators*. Phosphorescing or steadily excited phosphors with different emission colors and correspondingly different particle sizes may be injected into fluids to facilitate the study of flow in hydrodynamics and aerodynamics. (In the laboratory, finely divided phosphors may be used to demonstrate how certain liquids, for example, mercury, vaporize [618]).

3. *Infrared detectors*. In addition to the method described under temperature indicators, infrared radiations, up to about 14,000 Å, may

be detected by utilizing at room temperature (a) previously excited infrared-stimulable phosphors (Fig. 26b), such as Nos. 12 and 20–22 in Table 5, or (b) phosphors whose dielectric constants change on irradiation with infrared, for example, hex.-Cd(S:Se)Cu. The infrared-stimulable phosphors were used in **metascopes** as direct receivers and indicators of infrared during nocturnal signaling and observation, and the dielectric-change phosphors were used in **dielectric-cell "bolometers"** for infrared signaling.[65, 117, 169-180, 221, 237, 238, 483, 484, 488, 501, 619] The phosphorescence emission of a hex.-(Zn:Cd)S:Cu phosphor at 90°K is reportedly quenched by 27,000-Å photons.[611, 712, 713]

4. *Cascade transformers of the luminescence emissions of other phosphors.* When a certain phosphor has satisfactory characteristics except that its emission spectrum is inefficiently located relative to the sensitivity spectrum of the detector, it is sometimes possible to use (a) a second cascade-excited phosphor to transform the primary phosphor's emission directly into more suitable *longer-wavelength* radiation, or (b) a second previously excited cascade-stimulated phosphor to transform the primary phosphor's emission into more suitable *shorter-wavelength* radiation. Similarly, when the detectable phosphorescence of the first phosphor is too short, a second cascade-excited longer-persistent phosphor may be used; also, when the detectable phosphorescence of the first phosphor is too long and feeble, a second cascade-excited infrared-stimulable phosphor may be used, whereby the energy accumulated in the infrared-stimulable phosphor during a long time interval may be released in short high-intensity bursts by pulses of intense infrared.

5. *Contrast compression.* As shown in Fig. 81, the initial rate of decay of t^{-n} decay phosphors *increases* as the initial luminance increases. Therefore, the ratio L_1/L_2 ($L_1 > L_2$), between the luminances of two different excited areas of a t^{-n} decay phosphor screen, decreases during the initial period of phosphorescence. This effect has been used in the **icaroscope,** a device incorporating a rotating phosphor-coated disk which allows one to make a slightly delayed observation of objects viewed near the line of sight of the sun, by selectively reducing the intensity of the sun image relative to that of the object to be observed.[221, 547] (*Contrast expansion* may be obtained by using certain scotophors whose rates of decay *decrease* with increasing intensity of excitation [58, 63]).

6. *Time delays.* Information may be stored as potential phosphorescence in phosphor screens, tapes, etc.,[620-624] for later detection, and the information may be stimulated (released) or quenched (erased) by proper choice of applied heat or radiation (usually infrared). One such use is in the production of electronically simulated echoes produced by recording

sound as a variable luminescent area or intensity on a moving phosphorescent screen whose output is detected by phototubes located at various time-displaced intervals along the screen.[622] The resultant output from all the phototubes, when used to drive sound reproducers, simulates the acoustics of an "electronically adjustable" room. When it is necessary to erase phosphorescence quickly,[623] it is advantageous to use an infrared-quenchable phosphor, such as hex.-ZnS:Cu:Fe (No. 13, Table 5).

For a given phosphor, the ratio of the phosphorescence intensity to the intensity of luminescence emission at the instant of cessation of excitation, $L(t)/L_0$, may be calibrated as a function of time and used as a timing device with the aid of phototubes driving electronic circuits which perform certain operations according to the value of $L(t)/L_0$. The timing accuracy of such a device is relatively independent of the normal operating conditions in the case of temperature-independent ϵ^{-at} decay phosphors operated below their temperature break points, whereas the timing accuracy is strongly dependent on the degree of standardization of excitation and the control of operating conditions in the case of the temperature-dependent t^{-n} decay phosphors.

Recent developments in the production of transparent conducting coatings on glass,[625, 626] make possible controlled heating of transparent substrates for phosphor screens. These semiconductor coatings have surface resistivities of 10^2 to 10^5 ohm cm^{-2}, and they and the glass may be heated to about 260°C by passing current through the surface coating. By depositing phosphors (or scotophors) on such transparent conducting coatings, and coating the screens on the excited side with thin conducting and reflecting films of aluminum (compare Fig. 129), it is possible to apply electric fields across the screens and thereby influence the motion of "free" electrons in the screen crystals. This procedure makes it possible to combine semiconduction (including photoconduction, Figs. 83–85, 116–118) with luminescence (compare Figs. 72, 74, and 75, and Table 15) for special electronic uses.

Conclusion. In addition to the practical progress which has resulted from empirical research on phosphors used as detectors of invisible particles, this research has greatly increased our knowledge of electronically active solids in general. Luminescence is such a convenient and sensitive indicator of changes of composition, structure, and atomic interactions in many solids that the practical consequences of this broad aspect of luminescence research may eventually overshadow the tangible results already obtained.

APPENDIX 1

PREPARATION OF PURE ZnS, CdS, AND ZnSe

These important host-crystal ingredients are often used with 0.01 weight per cent, or less, of activator, and so they must contain *less* than 10^{-5} weight per cent of metallic-element impurities in order to insure that the resultant phosphors will not be poisoned or contain unknown activators which may obscure the results. Suitable purification procedures for the listed substances are outlined in the following, with the assumption of (1) scrupulously clean working conditions, (2) an adequate supply of water which has been (slowly) triple-distilled in insoluble glass or silica stills, and (3) laboratory ware which has been cleaned in hot mineral acids and heated in triple-distilled water for several hours before removing, rinsing with triple-distilled water, and allowing to dry on clean glass racks in a dust-free chamber.[58, 89, 221]

Purification of ZnS and CdS by the Alkali Process

Dissolve cp ZnO (added in excess) in dilute cp HCl, stir well, and filter through a fine fritted-glass filter. Heat the solution of $ZnCl_2$ to about 90°C, and pass through a long heated glass column containing sp vacuum-distilled zinc metal (obtainable from the New Jersey Zinc Company, Palmerton, Pa.). Repeat the passage over zinc metal several times, cleaning the zinc with dilute HNO_3 occasionally, or let the warm solution stand for several weeks in contact with zinc granules or dust. Filter the solution. Add concentrated cp H_2O_2 or bromine water to oxidize iron, chromium, etc., and add cp NH_4OH to resolution of the precipitate. Filter again to remove insoluble hydroxides, such as $Fe(OH)_3$. Fractionally precipitate with purified tank H_2S (pass through several gas-washing bottles containing HCl, $Ba(OH)_2$, and finally distilled water). Decant the liquid (discard the precipitate) through a fine fritted-glass filter, and pass in H_2S to complete precipitation. Wash the precipitate well by decantation, and dry at about 120°C. Where *pure* $(NH_4)_2S$ is obtainable, this may be substituted for the H_2S. The purification process is similar for CdS. The reactions are

$$ZnO + 2HCl \rightleftharpoons ZnCl_2 + H_2O \tag{144}$$

$$Fe^{2+} + Br^0 \rightarrow Fe^{3+} + Br^- \tag{145}$$

$$Fe^{3+} + 3(OH)^- \rightarrow Fe(OH)_3 \downarrow \tag{146}$$

$$ZnCl_2 + 2NH_4OH \rightarrow Zn(OH)_2 \downarrow + 2NH_4Cl \qquad (147)$$

$$2Zn(OH)_2 + 4NH_4OH + 4NH_4Cl \rightarrow 2Zn(NH_3)_4Cl_2 + 8H_2O \qquad (148)$$
$$\text{(resolution of ppt.)}$$

$$Zn(NH_3)_4Cl_2 + H_2S + 2H_2O \rightarrow ZnS \downarrow + 2NH_4Cl + 2NH_4OH \qquad (149)$$

The step involving contact of the $ZnCl_2$ solution with metallic zinc is an electrochemical deposition in which ions lying below zinc in the electromotive-force series are precipitated out of the solution by replacement with zinc; for example:

$$Fe^{2+} + Zn^0 \rightarrow Fe^0 \downarrow + Zn^{2+} \qquad (150)$$

$$Cu^{2+} + Zn^0 \rightarrow Cu^0 \downarrow + Zn^{2+} \qquad (151)$$

The alkaline solution is necessary when H_2S is used as a precipitant, because ZnS will not precipitate in the presence of HCl. The main new ingredients required to "feed" this process are ZnO, HCl, NH_4OH, and H_2S [or $(NH_4)_2S$].

Purification of ZnS and CdS by the "Acid" Process

Dissolve CP ZnO in CP H_2SO_4. Filter through a fritted-glass filter. Pass the hot solution of $ZnSO_4$ over SP zinc several times. Add 30 per cent CP H_2O_2, and add just enough dilute CP NH_4OH to obtain a *small* precipitate. Let the precipitate settle, and decant the supernatant solution carefully, discarding the precipitate in the bottom of the vessel. Electrolyze the hot stirred solution with clean *pure* platinum electrodes at 2.2 volts for several days. Fractionally precipitate twice with purified tank H_2S, decanting the supernatant $ZnSO_4$ solution carefully off the discarded ZnS precipitate (plus occluded low-solubility sulphides) each time. Saturate the remaining $ZnSO_4$ solution with pure H_2S, decant, and discard the liquid. Wash the precipitated ZnS at least four times with hot distilled water, and dry at about 120°C. The purification process is again similar for CdS. The reactions are

$$ZnO + H_2SO_4 \rightarrow ZnSO_4 + H_2O \qquad (152)$$

$$2Fe^{2+} + O^0 \rightarrow 2Fe^{3+} + O^{2-} \qquad (153)$$

$$Fe_2(SO_4)_3 + 6NH_4OH \rightarrow 2Fe(OH)_3 \downarrow + 3(NH_4)_2SO_4 \qquad (154)$$

$$ZnSO_4 + H_2S \rightarrow ZnS \downarrow + H_2SO_4 \qquad (155)$$

The treatment with zinc metal and the electrolysis accomplish the same result, except that peroxides of manganese and lead are deposited on the anode during electrolysis (these peroxides produce a coppery sheen which may be mistaken for the copper which deposits on the cathode). For high yields it is necessary to use fairly dilute solutions of $ZnSO_4$ at the final precipitation; otherwise, the increasing high concentration of hydrogen ions, from

$$H_2SO_4 \rightleftharpoons 2H^+ + SO_4^{2-} \qquad (156)$$

on the right side of eq. 155, eventually prevents progress of the reaction toward the right. Introduction of the pure H_2S under pressure increases the yield of ZnS by driving eq. 155 toward the right. By leaving some zinc (cadmium) unprecipitated, the final ZnS or CdS represents a middle fraction between the first discarded fractional precipitates and the unprecipitated substance. The by-product H_2SO_4 from eq. 155 may be re-used at the beginning of the process, thereby reducing raw-material requirements to only ZnO and H_2S as the major ingredients.

The pure ZnS made by the "acid" process is generally superior to that made by the alkali process, because the latter process introduces variable amounts of oxygen-containing compounds (for example, SiO_2, B_2O_3, and Na_2O) dissolved from the glass vessels, and perhaps some occluded $Zn(OH)_2$, in the final ZnS. It has been found that cub.-ZnS:Ag(0.01) phosphors produced at temperatures below 1000°C from the alkali and "acid" processes have substantially the same luminescence efficiency, although the phosphor made from the alkali-process ZnS has about half the average crystal size and is more aggregated than the phosphor made from the "acid"-process ZnS, and the alkali-process phosphor has a *green* afterglow. On the other hand, hex.-ZnS:Ag(0.01) phosphors produced at about 1360°C from "acid"-process ZnS have about ten times the luminescence efficiency of those prepared from alkali-process ZnS. Again, the product made from "acid"-process ZnS has a blue afterglow, whereas that made from the alkali-process ZnS has a green afterglow.[58] Luminescence-pure ZnS (33-Z-19) and CdS (33-C-165) prepared by the "acid" process may be obtained from the RCA, Victor Division, Lancaster, Pa., with the aid of the indicated code numbers.

Purification of ZnSe

Dissolve 99.5 per cent selenium (obtainable from the Foote Mineral Company, Philadelphia, Pa.) in CP HNO_3, and evaporate to leave crude SeO_2 according to

$$3Se + 4HNO_3 \rightarrow 3SeO_2 + 2H_2O \uparrow + 4NO(\text{dilute } HNO_3) \uparrow \qquad (157)$$

The SeO_2 may then be purified by repeated fractional sublimations in a stream of anhydrous NO_2 (from HNO_3) to sweep the vaporized SeO_2 from the hot to the cold end of the silica or hard-glass tube in which the sublimation is carried out.[221] After about three fractional sublimations, the purified SeO_2 may be reduced to pure Se metal by treatment with SO_2 or a hydrazine salt, according to

$$SeO_2 + 2H_2O + 2SO_2 \rightarrow Se \downarrow + 2H_2SO_4 \qquad (158)$$

$$SeO_2 + H_2O + N_2H_4 \cdot H_2SO_4 \rightarrow Se + N_2 \uparrow + 3H_2O + H_2SO_4 \qquad (159)$$

The pure Se may be dissolved in purified Na_2SO_3, in aqueous solution, and from this solution ZnSe may be precipitated by the addition of purified ("acid"-process) $ZnSO_4$, according to

$$2Na_2SO_3 + Se \rightarrow (Na_2SO_3)_2 \cdot Se \qquad (160)$$

$$(Na_2SO_3)_2 \cdot Se + ZnSO_4 + H_2O \rightarrow ZnSe \downarrow + 2NaHSO_4 + Na_2SO_3 \quad (161)$$
$$\text{(in acid solution)}$$

Zinc selenide of high purity (ultrapure ZnSe) may be obtained from the Mallinckrodt Chemical Company, St. Louis, Mo.

As incidental information on purification procedures: (1) Flux solutions may generally be purified by a process similar to the "acid" process; for example, to an aqueous solution of CP NaCl add CP H_2O_2, make slightly alkaline with CP NH_4OH, and saturate with purified H_2S. Let stand until all the precipitate has settled, and carefully decant the purified solution. (2) Fused-silica crucibles are usually cast in iron molds and contain some dissolved iron oxide on their surfaces. According to A. L. Smith,[253] the iron can be removed by heating the crucibles to red heat in a stream of HCl which removes the iron as volatile $FeCl_3$ (boiling point = 315°C), according to

$$Fe_2O_3 + 6HCl \rightarrow 2FeCl_3 \uparrow + 3H_2O \uparrow \quad (162)$$

An alternative method is to dissolve away some of the crucible by a brief treatment with hydrofluoric acid solution.

When precipitates of ZnS, CdS, and ZnSe are dried at temperatures below 120°C, their particles average much less than 0.1 micron in size, and these particles are made up of loosely aggregated solids with structure finer than 100 Å. A typical 33-Z-19 ZnS batch dried at about 115°C has particles averaging approximately 250 Å in diameter. When these particles are heated with 6 per cent NaCl flux at 1250°C for about 15 min, the resultant 15-micron hex.-ZnS:[Zn] phosphor crystals are on the average 10^8 times larger in volume and weight than the original 250-Å ingredient particles.[678]

Crystallization of small research lots of ZnS-type phosphors at high temperatures is readily carried out by placing the covered silica crucible and its contents in a cold furnace whose temperature is raised quickly (30 to 60 min) to the indicated highest temperature. The highest temperature should be held *without fluctuation* for about 15 min, or until the batch is uniformly heated (larger batches take longer times). By this method of rising to the highest temperature, atmospheric oxygen is kept outside because the gas in the crucible is steadily expanding according to the ideal gas law (eq. 16),

$$PV = NRT \quad (163)$$

where P is the pressure, V is the volume, and N is the number of moles of gas. Fluctuation of temperature T at constant (atmospheric) pressure is undesirable because there is a corresponding fluctuation of volume per mole of gas and this produces an alternating inflow and outflow of air which oxidizes the phosphor batch.

PROPORTIONS OF ISOLATED AND STRUCTURALLY ADJACENT IMPURITY ATOMS, ASSUMING A RANDOM DISTRIBUTION

Derivation of General Probability Distribution Formula

Consider a host crystal with a total number N_T of available sites, of which N_C "empty" sites are occupied by normal host-crystal atoms, N_I "filled" sites are occupied by a given species of impurity atom, and N_V sites are really vacant, so that

$$N_T = N_C + N_I + N_V \tag{164}$$

For a random distribution of impurity atoms, the probability that an available site is occupied by an impurity atom is represented by

$$\mathbf{P} = N_I/N_T \tag{165}$$

Consider next one of the N_I sites known to be occupied by an impurity atom, and consider $(N_L - 1)$ nearest available "filled" sites satisfying a group-N_L condition (this condition may involve special geometrical arrangements, but N_L is just the number of impurity atoms in the group being investigated). Consider one such configuration, the rth, which may be produced in r equivalent ways, each of which provides M_r available sites adjacent to at least one of the N_L filled sites. Then, given a site occupied by an impurity atom, and assuming $N_I \ll N_T$, the probability of finding the rth configuration is

$$\mathbf{P}_r = r\mathbf{P}^{N_L-1}(1 - \mathbf{P})^{M_r} \tag{166}$$

and the probability of finding an N_L group around this site is

$$\mathbf{P}_{N_L} = \sum_r \mathbf{P}_r = \sum_r r\mathbf{P}^{N_L-1}(1 - \mathbf{P})^{M_r} \tag{167}$$

This is the fraction of N_I impurity atoms which lie in groups of N_L, assuming a random distribution of the impurity atoms.

Examples for Two General Types of Phosphor Crystals

(a) For substitutional impurities in cation sites of structures of the type $Fm3m$-KCl, $Fm3m$-CaF$_2$, $F\bar{4}3m$-ZnS (cubic), and $C6mc$-ZnS (hexagonal), or interstitial impurities in structures of the type $F\bar{4}3m$-ZnS, the number of *isolated*

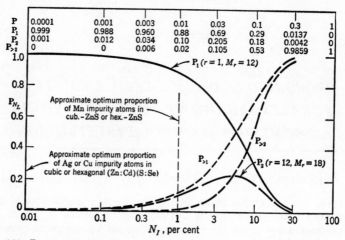

FIG. 141. Proportions of isolated impurity atoms, P_1, atoms in isolated pairs of impurity atoms P_2, and nonisolated impurity atoms $P_{>1} = 1 - P_1$, in certain structures with 12-coordinated equivalent sites.

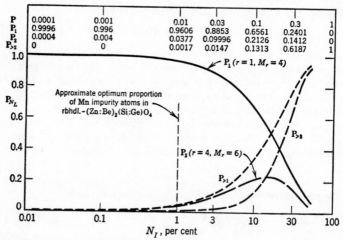

FIG. 142. Proportions of isolated impurity atoms, P_1, atoms in isolated pairs of impurity atoms P_2, and nonisolated impurity atoms $P_{>1}$, in certain structures with 4-coordinated equivalent sites.

individual impurity atoms (that is, lone impurity atoms having no impurity atoms in next-adjacent available sites of the type being considered) is given by setting $N_L = 1$, so

$$\mathbf{P}_1 = \sum_r r(1 - \mathbf{P})^{M_r} = (1 - \mathbf{P})^{12} \tag{168}$$

because this N_L group can be produced in only one way ($r = 1$), and the number of nearest available sites is $M_1 = 12$ (assuming that the impurity atoms go into only one type of available site, for example, into *either* substitutional *or* interstitial sites in cub.-ZnS, but not into both). Then, the proportion of atoms in *isolated pairs* of impurity atoms is given by setting $N_L = 2$, and so

$$\mathbf{P}_2 = \sum_r r\mathbf{P}(1 - \mathbf{P})^{M_r} = 12\mathbf{P}(1 - \mathbf{P})^{18} \tag{169}$$

because $r = 12$; that is, there are 12 nearest available sites around an impurity atom in the given structures, and $M_{12} = 18$; that is, there are 18 nearest available sites around a pair of adjacent impurity atoms in available sites. Figure 141 shows plots of the proportions of isolated individual impurity atoms, $\mathbf{P}_1$; isolated pairs of impurity atoms, $\mathbf{P}_2$; nonisolated impurity atoms, $\mathbf{P}_{>1} = 1 - \mathbf{P}_1$; and groups of adjacent impurity atoms with three or more atoms in a group, $\mathbf{P}_{>2} = 1 - (\mathbf{P}_1 + \mathbf{P}_2)$.

(b) Similar calculations based on substitutional impurities in cation sites in the $R\bar{3}$-Zn_2SiO_4 structure lead to

$$\mathbf{P}_1 = \sum_r r(1 - \mathbf{P})^{M_r} = (1 - \mathbf{P})^4 \tag{170}$$

and

$$\mathbf{P}_2 = \sum_r r\mathbf{P}(1 - \mathbf{P})^{M_r} = 4\mathbf{P}(1 - \mathbf{P})^6 \tag{171}$$

From these equations, the incidence of isolated impurity atoms $\mathbf{P}_1$; atoms in isolated pairs of impurity atoms $\mathbf{P}_2$; $1 - \mathbf{P}_1 = \mathbf{P}_{>1}$; and $1 - (\mathbf{P}_1 - \mathbf{P}_2) = \mathbf{P}_{>2}$ were calculated and are shown plotted in Fig. 142.

Average Proximity of Activator Atoms

It is of interest to calculate the approximate average internuclear spacing between activator atoms in the representative phosphors cub.-ZnS:Ag(0.01) and rbhdl.-$Zn_2SiO_4 \cdot 0.02Mn$. In these cases, it is possible to correlate the average spacing between activator impurity atoms with the spacing between certain host-crystal atoms, because the number of suitable interstitial sites in cub.-ZnS is roughly equal to the number of Zn atoms, and the number of available substitutional sites in rbhdl.-Zn_2SiO_4 is equal to the number of Zn atoms. If one assumes, as a first approximation, a uniform distribution of the Zn host-crystal atoms in both cases, the average center-to-center spacings between Zn atoms is given by

$$\bar{x}_{Zn-Zn} \approx (\sigma N_V N_A / M_W)^{-1/3} \text{ cm} \tag{172}$$

where σ is the density of the host crystal in g cm^{-3} ($\sigma = 4.1$ for cub.-ZnS, and 4.24 for rbhdl.-Zn_2SiO_4), N_V is the number of Zn's per simple molecule ($N_V = 1$ in ZnS, and 2 in Zn_2SiO_4), $N_A = 6.02 \times 10^{23}$, and M_W is the gram-molecular weight per simple molecule ($M_W = 97.44$ for ZnS, and 222.82 for Zn_2SiO_4). Substituting the above values in the formula, one obtains

$$\bar{x}_{Zn-Zn} \approx 3.4 \times 10^{-8}\ cm = 3.4\ \text{Å in cub.-ZnS (actually, according to Fig. 12,}$$

$$\bar{x}_{Zn-Zn} = 5.43\ \text{Å} \times 2^{-\frac{1}{2}} = 3.84\ \text{Å})$$

$$\bar{x}_{Zn-Zn} \approx 3.5 \times 10^{-8}\ cm = 3.5\ \text{Å in rbhdl.-}Zn_2SiO_4$$

The average distance between activator atoms, assuming a uniform distribution, is

$$\bar{x}_{Zn-Zn} R_{Zn/act.}{}^{\frac{1}{3}} \tag{173}$$

where $R_{Zn/act.}$ is the ratio of Zn to activator atoms in the particular phosphor. From this expression, it is found that

$$\bar{x}_{Ag-Ag} \approx 3.84\ \text{Å} \times \left(\frac{97.44 \times 10^4}{107.88} \right)^{\frac{1}{3}} \approx 83\ \text{Å in cub.-ZnS:Ag(0.01)}$$

$$\bar{x}_{Mn-Mn} \approx 3.5\ \text{Å} \times 100^{\frac{1}{3}} \approx 16\ \text{Å in rbhdl.-}Zn_2SiO_4 \cdot 0.02Mn$$

APPENDIX 3

INTERCONVERSION OF SOME UNITS OF LUMINANCE (SURFACE BRIGHTNESS)

1 candle cm^{-2} = 1 stilb = 6.452 candles $in.^{-2}$ = 929 candles ft^{-2} = 10^4 candle m^{-2} = π lamberts = 3142 millilamberts (mL) = $\pi \times 10^6$ microlamberts = 2919 ft-lamberts (or apparent ft-candles, or equivalent ft-candles) = 31,416 meter-lamberts = 31,416 effective lux = 31,416 apostilbs = 31,416 blondels = 1.11 Hefnerkerze cm^{-2} = π lumens emitted per square centimeter from a cosine-law luminator (eq. 130).

1 millilambert (mL) = 0.0003183 candle cm^{-2} = 0.0003183 stilb = 0.002054 candle $in.^{-2}$ = 0.2957 candle ft^{-2} = 3.183 candle m^{-2} = 10^{-3} lambert = 10^3 microlambert = 0.929 ft-lambert (or apparent ft-candle, or equivalent ft-candle) = 10 meter-lamberts = 10 effective lux = 10 apostilbs = 10 blondels = 10^{-3} lumen emitted per square centimeter from a cosine-law luminator (eq. 130).

Using the latest values of the fundamental physical constants,[106, 518, 552] 685 lumens = 1 watt = 1 joule sec^{-1} = 10^7 ergs sec^{-1} = 6.28×10^{18} ev sec^{-1} = 0.239 cal sec^{-1} for 100 per cent conversion of power into emitted 2.22-ev photons, that is, into monochromatic radiation at 5560 Å, which corresponds to the peak of the sensitivity curve of the average photopic human eye (Fig. 120).

GENERAL SPECTRAL CHARACTERISTICS OF LUMINESCENCE (RADIESCENCE)

This book is concerned with chiefly the efficient *conventional luminescence* of solids in the photon energy range from about 1 to 10 ev (12,400 to 1240 Å). Conventional luminescence generally comprises transitions of excited electrons to the very uppermost normally filled energy levels of a solid and may provide efficient external emission when the emission spectrum lies in a region where the solid has low absorption. Radiative transitions to the numerous lower-lying energy levels are possible and sometimes afford appreciable externally detectable luminescence emission when the solid does not reabsorb all its own emission, but there is usually a high probability of reabsorbing the high-energy photons so produced. In general, then, the efficiency of nonconventional luminescence is quite low, being of the order of 1 per cent or less for x-ray fluorescence.

Although conventional luminescence emission is usually the most efficient and useful, its relation to the other higher-energy luminescence emissions is important in obtaining a broad perspective of the general phenomenon. For this reason, Fig. 143 is presented to provide a rough pictorial correlation of the principal absorption and emission characteristics of a *hypothetical* typical phosphor solid, such as hex.-ZnO:[Zn] or hex.-ZnS:Ag, over the energy range from 10^{-3} to 10^7 ev. The structural details are not intended to be accurate, but the *approximate* spectral and efficiency relationships of conventional luminescence to general luminescence (radiescence) and thermal radiation are presented in representative form.

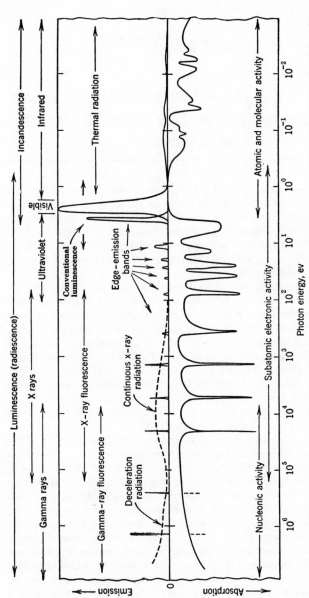

FIG. 143. A simplified schematic of the general absorption and emission spectra of a *hypothetical* "typical" phosphor solid at room temperature. Absorption and emission lines may be present in the conventional luminescence region when there are unfilled inner electron shells in some of the atoms in the solid. The line and narrow-band spectra of luminescence radiation are characteristic of the composition and structure of the luminescent material. The broad-band spectra of deceleration radiation and continuous x-ray radiation, however, are independent of the nature of the emitting material.[2,5]

APPENDIX 5

SUMMARY OF SOME MAGNITUDES AND STRUC-
TURES INVOLVED IN THE LUMINESCENCE
OF SOLIDS

Table 23 gives a condensed and pictorial presentation of some of the chief features in the evolution of solids which may be made luminescent. As may be seen from the table, enormous ranges of size, mass, and energy are encountered on going from the elementary particles of radiation and matter to some typical solids which have been used as phosphor host crystals. Of particular interest in the table is a graphic representation of the *structures* and *shapes* of atoms, molecules, and crystals.

The upper drawings of isolated hydrogen *atoms* show the simple spherical shapes of s-state electron "clouds," with the electron-cloud density $\psi\psi^* \cdot 4\pi r^2$ plotted as a function of atomic radius r below each cross-section view. On the other hand, the p-state electron clouds, as shown, have knobby spatial distributions, and d- and f-state distributions (not shown) are even more complex. It is strongly recommended that the reader study H. E. White, "Pictorial Representation of the Electron Cloud for Hydrogen-Like Atoms," *Phys. Rev.*, Vol. 37, No. 11, 1416–1424, June 1, 1931 (or White's *Introduction to Atomic Spectra*, McGraw-Hill Book Co., New York, 1934). When both s and p states are occupied, as in nitrogen or oxygen atoms, the p-state electron clouds penetrate through and may project beyond the s-state clouds.

The drawings of electron clouds of several *molecules* illustrate some of the changes in spatial distributions that occur when atoms combine. At one extreme, the originally spherical s-state electron clouds of the two H atoms in covalent H_2 are changed and merged into a single prolate ellipsoid enveloping both nuclei. At the other extreme, the near-spherical $3s^2 3p^6$-state electron clouds of K^+ and Cl^- ions remain distinct and nearly spherical [except for repulsion of filled shells (not shown)] when these ions combine to form an ionic KCl molecule. An intermediate example is shown for the H_2O molecule, where the $1s$ electrons of the two H atoms are partly shared by and partly transferred to the two unfilled $2p$ states of the O atom. Here, the H nuclei are embedded in the O electron cloud. This embedding does not occur generally, because nuclei that are surrounded by filled states are repelled by the filled $2s^2$ (or $2p^6$) shells of O. Thus, the $Zn^{(n+)}$ $3s$-shell remains well outside the $O^{(n-)}$ $2s$-shell in a ZnO molecule. In ZnO, the bonding is semi-ionic by partial transfer of some electron(s) from Zn to occupy the unfilled $2p$ states of O, and semicovalent by some sharing of electron(s) with opposite spins in a given s or p state. Where two atoms in a molecule share p-state electrons, there is an extended prolate-ellipsoid electron-cloud "bridge" between the two atomic nuclei (this obtains also for shared d- and f-state electrons).

The lower drawings of atomic arrangements in *solids (crystals)* are greatly simplified, because detailed drawings of the electron clouds of the atoms would unduly complicate the sketches. It should be kept firmly in mind, however, that *the dots or circles used to denote atoms or ions in crystals represent atoms which generally have nonspherical shapes and complex electron-cloud structures.* In the diamond structure, for example, each tetrahedrally arranged covalent bond along the *dotted* line between two carbon atoms is an extended prolate-ellipsoid electron cloud containing two electrons (of opposite spin), each atom contributing one electron. These tetrahedral bonds are formed by hybridization of the states occupied by the two $2s$ and two $2p$ electrons of each carbon atom. The lowermost drawings are idealized sketches of some typical shapes of crystals having the given atomic arrangements. Further information on common crystal forms may be obtained by consulting Dana, *The System of Mineralogy*, John Wiley & Sons, New York, 1944. Here, again, the fine structures of atoms and the normal imperfections, such as mosaic structure, omission defects, impurities, and interstitial atoms, are omitted for simplicity. Sketches of crystals, such as naphthalene, which contain distinct structural units weakly bonded together by dipole-type van-der-Waals' energies are not shown, but may be found in standard texts.[47]

In closing, it is worth noting that a complex activator atom, such as Mn, which has outer electrons in $l = 0$ (s) states *and* $l \geq 1$ ($p, d, f, \cdots$) states with nearly equal energies can have the spatial arrangements and distributions of electrons in these states altered greatly when the atom is (a) incorporated as an impurity in different host crystals, and (b) excited to one or another of several allowed states above the ground state (Fig. 16c). Detailed diagrams and quantitative solutions of such configurations challenge the serious researcher on luminescence of solids.

TABLE

	Number of Species	Name	Charge, e	Mass, g	Size, cm	Binding Energy, ev
Radiation	1	Photon	0	$h\nu c^{-2} (\approx 10^{-32}$ for light$)$	?	(free)
Matter	1	Electron	-1	10^{-27}	$\approx 10^{-13}$	(free)
	1	Proton	$+1$	10^{-24}	$\approx 10^{-13}$	(free)
	1	Neutron	$e - e = 0$	10^{-24}	$\approx 10^{-13}$	(free)
	> 900	Atomic nuclei	$+Z$	10^{-24} to 4×10^{-22}	$\approx 10^{-12}$	$\approx 10^7$ per nucleon
	103	Isolated atoms	$Ze - Ze = 0$	10^{-24} to 4×10^{-22}	$\approx 3 \times 10^{-8}$	Binding to atomic nucleus: Innermost (1s) electron = 13.5 (H) to $> 10^5$ (Cm) Outermost electron = 3.9 (Cs) to 24.5 (He)
	$\approx 10^6$	Isolated elementary (simple) molecules	0	3×10^{-24} to $\approx 10^{-20}$	$\approx 10^{-7}$ to $\approx 10^{-6}$	Atom to atom ≈ 1 to 15 Outermost electron to nearest atomic nucleus ≈ 0.1 to 20
	∞	Solids (crystals)	0	$\approx 10^{-18}$ to $> 10^3$	$\approx 10^{-6}$ to $> 10^2$	Atom to atom ≈ 1 to 20 Simple molecule to simple molecule ≈ 0 to 10 Outermost electron to nearest atomic nucleus ≈ 0.01 to 30

23
FEATURES OF MATTER AND RADIATION

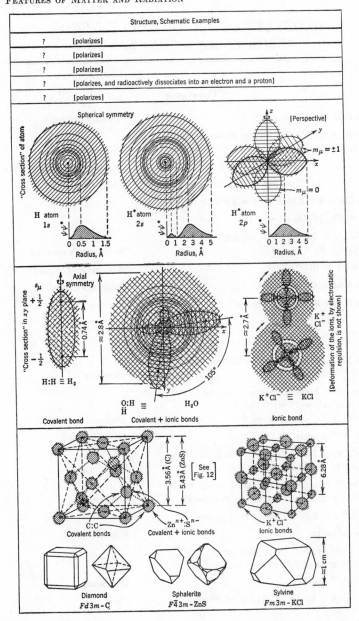

	Structure, Schematic Examples
?	[polarizes]
?	[polarizes]
?	[polarizes]
?	[polarizes, and radioactively dissociates into an electron and a proton]
?	[polarizes]

Spherical symmetry

[Perspective]

"Cross section" of atom

$m_\mu = \pm 1$

$m_\mu = 0$

H atom
$1s$

H* atom
$2s$

H* atom
$2p$

$\psi \cdot \psi^*$

Radius, Å

0 0.5 1 1.5

0 1 2 3 4 5

0 1 2 3 4 5

"Cross section" in xy plane

Axial symmetry

s_μ

$+\frac{1}{2}$

$-\frac{1}{2}$

0.74 Å

≈ 2.8 Å

≈ 2.7 Å

[Deformation of the ions, by electrostatic repulsion, is not shown]

K⁺
Cl⁻

H:H ≡ H_2

105°

$\overset{O:H}{\underset{H}{}}$ ≡ H_2O

K⁺Cl⁻ ≡ KCl

Covalent bond

Covalent + ionic bonds

Ionic bond

3.56 Å (C)
5.43 Å (ZnS)

[See Fig. 12]

6.28 Å

C:C
Covalent bonds

$Zn^{n+}:S^{n-}$
Covalent + ionic bonds

K⁺Cl⁻
Ionic bonds

≈ 1 cm

Diamond
$Fd3m$ - C

Sphalerite
$F\bar{4}3m$ - ZnS

Sylvine
$Fm3m$ - KCl

GLOSSARY OF FREQUENTLY USED SYMBOLS AND CONSTANTS

Constants

$c = 2.9978 \times 10^{10}$ cm sec^{-1} — Velocity of light

$e = 4.80 \times 10^{-10}$ esu of charge (statcoulomb)
$= 1.60 \times 10^{-20}$ emu of charge — Electronic charge

$\epsilon = \lim_{N \to \infty} (1 + N^{-1})^N = 2.71828 \cdots$ — Base of natural logarithms

$h = 6.624 \times 10^{-27}$ erg sec $= 4.14 \times 10^{-15}$ ev sec — Planck's constant

$\hbar \equiv h/2\pi = h/2(3.14159\cdots) = 1.053 \times 10^{-27}$ erg sec

$i = (-1)^{1/2}$ (Note: $\epsilon^{\pi i} = -1$)

$k = R/N_A = 1.38 \times 10^{-16}$ erg deg^{-1}
$= 8.62 \times 10^{-5}$ ev deg^{-1} — Boltzmann's constant

$m_0^- = 9.1 \times 10^{-28}$ g — Rest mass of electron

$N_A = 6.02 \times 10^{23}$ mole^{-1} — Avogadro's constant

$R = 8.314 \times 10^7$ erg deg^{-1} mole^{-1} — Universal gas constant

Units

1 Å $= 10^{-8}$ centimeter (cm) $= 10^{-4}$ micron (μ)
$= 10^{-1}$ millimicron (mμ) — Angstrom

1 a $= c/10 = 2.998 \times 10^9$ esu of current
$= 1$ coulomb sec^{-1} $= 10^{-1}ce^{-1}$
$= 6.24 \times 10^{18}$ electrons sec^{-1} — Ampere

1 μa $= 10^{-6}$ ampere — Microampere

1 ev $= 4.80 \times 10^{-10}/299.8 = 1.602 \times 10^{-12}$ erg
$= 3.82 \times 10^{-20}$ cal $= 23.05$ kcal mole^{-1} — Electron volt

1 mL $= 10^{-3}$ lambert $= 0.0003183$ candle cm^{-2}
$= 10^{-3}$ lumen cm^{-2} (cosine-law luminator) — Millilambert

1 v $= 10^8/c = 3.336 \times 10^{-3}$ esu of potential — Volt

1 kv $= 10^3$ volts — Kilovolt

1 watt $= 10^7$ erg sec^{-1} $= 1$ joule sec^{-1}
$= 685$ lumens of 2.22-ev photons — Watt

1 μw $= 10^{-6}$ watt — Microwatt

1 Ω $= 1$v/1a $= 10^9/c^2 = 1.113 \times 10^{-12}$ esu — Ohm

ABBREVIATIONS

a-c	alternating current	cub.	cubic
CP	chemically pure	hex.	hexagonal
CR	cathode ray	monocl.	monoclinic
CRT	cathode-ray tube	rbhdl.	rhombohedral
d-c	direct current	rhomb.	rhombic
i	interstitial	tetr.	tetragonal
IR	infrared	tricl.	triclinic
IRF	image repetition frequency		
LP	luminescence pure		
RCA	Radio Corporation of America		
s	substitutional		
SP	spectroscopically pure		
UV	ultraviolet		

SYMBOLS AND COMMON UNITS

a, b	Usually constants, as the decay constant in ϵ^{-at}
δ	Number of active luminescence centers per cubic centimeter
ϵ_D	Dielectric constant
E	Energy, in ergs (or electron volts or calories)
E_a	Absorbed energy
E_e	Emitted (radiated) energy
E_k	Kinetic energy
E_p	Potential energy
E_0	Ground-state energy level
E^{**}	Excited-state energy level *before* atomic readjustment
E^*	Excited-state energy level *after* atomic readjustment
ΔE^* (or E^*)	Activation energy
$\overline{\Delta E}$	Magnitude of average energy bit in a solid
$\overline{\Delta E^*}$	Width (indeterminacy), in energy, of a spectral line
ε	Efficiency, in per cent
F	Electric field, in volts per centimeter
i	Electric current, in amperes
J	Total angular momentum (resultant of L and S_e)
$\mathbf{K}$	Wave-number vector of a free particle
$\mathbf{k}$	Reduced wave-number vector of a particle in a crystal
$\mathbf{k_r}$	Reduced wave-number vector of an electron in a crystal
$\mathbf{K}$	Wave-number vector in the reciprocal lattice of a crystal
l	Angular-momentum quantum number; $l = 0, 1, 2, 3, \cdots = s, p, d, f, \cdots$
L	Total resultant l; $L = 0, 1, 2, 3, \cdots = S, P, D, F, \cdots$
L	Radiance or luminance, in millilamberts (or in per cent of L_0)
L_0	Radiance or luminance at time $t = 0$
λ	Wavelength, in centimeters (or angstroms)
λ_d	Diffraction wavelength
λ_E	Radiation conversion wavelength (see Table 1)
λ_{pk}	Wavelength at peak (maximum intensity) of spectral-distribution curve
m	Mass, in grams (g)

m_μ	Magnetic-moment quantum number; $m_\mu = -l, -l + 1, \cdots 0 \cdots l - 1, l$
μ	Index of refraction (also used to indicate microns or micro-)
n	Principal quantum number; $n = 1, 2, 3, \cdots = K, L, M, \cdots$
n	A variable which may be a noninteger, as in t^{-n}
N	An integer, 1, 2, 3, $\cdots \infty$
N_n	Mass number
$\mathrm{Mn}(N)$	Indicates 0.006 mole of Mn per mole of host-crystal cation(s)
ν	Frequency of oscillation (vibration), in $\sec^{-1}$
ν_a	Atomic vibration frequency, or thermally induced attempt-to-escape frequency of a trapped electron (or positive hole)
ν_c	Characteristic frequency
$\bar{\nu}$	Wave number, in cm^{-1}
p_m	Momentum, in gram centimeters per second
P	Power, in ergs per second (or watts, or lumens)
$\mathbf{P}$	Probability (usually per unit time)
r	Radius, in centimeters (or angstroms)
s_μ	Spin-moment quantum number; $s_\mu = \pm \frac{1}{2}$
S_s	Resultant angular momentum of electron spin
σ	Density, in grams per cubic centimeter for volume density, grams per square centimeter for area density, and current per square centimeter for electric-current density
t	Time, in seconds (sec)
T	Temperature, in degrees Kelvin (or in degrees centigrade, where $0\,^{\circ}\mathrm{C} = 273.1\,^{\circ}\mathrm{K}$)
T_B	Temperature breakpoint beyond which ε decreases rapidly in plot of ε vs T
T_{50}	Temperature at which ε is 50 per cent of $\varepsilon_{\max}$
τ	Lifetime of an excited state, in seconds
τ_F	Natural (fluorescence) lifetime of an unconstrained excited state
v	Velocity, in centimeters per second
V	Voltage
V_0	Initial voltage
V_d	Extrapolated "dead voltage"
V_l	Limiting voltage (second crossover)
V_s	Effective electron-accelerating voltage of a screen
W	Atomic weight (weight of N_A atoms of a given species), in grams
x	Distance (length), in centimeters (or in angstroms)
$\bar{x}$	Averaged distance and configuration coordinate
$\bar{x}_c$	Average crystal diameter
x_l	Depth of penetration "limit"
x_s	Screen thickness
x_τ	Unit identity translation distance
ψ	Wave function
z	Charge
Z	Atomic number = nuclear charge = number of extranuclear electrons of an atom

REFERENCES

Chapter 1

1. C. MORLEY, *Travels in Philadelphia*, J. B. Lippincott and Co., Philadelphia, 1920.
2. W. HEITLER, *The Quantum Theory of Radiation*, Oxford, The Clarendon Press, 1944; P. A. M. DIRAC, *The Principles of Quantum Mechanics*, Oxford, 1947; N. F. MOTT and I. N. SNEDDON, *Wave Mechanics and Its Applications*, Oxford, 1948. (Dover reprint).
3. E. SCHRÖDINGER, *Collected Papers on Wave Mechanics*, Blackie and Son, London, 1929; H. A. BETHE et al, *Quantentheorie, Handbuch der Physik*, Vol. 24, Part 1, J. Springer, Berlin, 1933; L. I. SCHIFF, *Quantum Mechanics*, McGraw-Hill Book Co., New York, 1949.
4. C. J. ELIEZER, "The Interaction of Electrons and an Electromagnetic Field," *Revs. Modern Phys.*, Vol. 19, 147–184, July 1947.
5. F. K. RICHTMYER and E. H. KENNARD, *Introduction to Modern Physics*, McGraw-Hill Book Co., New York, 1947; E. G. RAMBERG and F. K. RICHTMYER, "Radiation Probabilities," *Phys. Rev.*, Vol. 51, No. 11, 913–929, June 1, 1937.
6. N. ROSEN, "Statistical Geometry and Fundamental Particles," *Phys. Rev.*, Vol. 72, No. 4, 298–303, Aug. 15, 1947; *ibid.*, Vol. 72, No. 12, 1253, Dec. 15, 1947.
7. W. E. SHOUPP, "The Structure of the Nucleus," *Westinghouse Engr.*, Vol. 6, No. 4, 118–125, July 1946.
8. H. A. BETHE, *Elementary Nuclear Theory*, John Wiley & Sons, New York, 1947.
9. E. U. CONDON, "Foundations of Nuclear Physics," *Nucleonics*, Vol. 1, No. 1, 3–12, Sept. 1947.
10. J. A. WHEELER, "Elementary Particle Physics," *Am. Scientist*, Vol. 35, No. 2, 177–195, Apr. 1947; *ibid.*, Vol. 37, 202–218, 444 ff., 1949.
11. G. GAMOW, *Atomic Energy in Cosmic and Human Life*, Macmillan Co., New York, 1946; *One, Two, Three · · · Infinity*, Viking Press, New York, 1947.
12. L. PAULING, "Atomic Radii and Interatomic Distances in Metals," *J. Am. Chem. Soc.*, Vol. 69, No. 3, 542–553, Mar. 1947 (*cf. Revs. Modern Phys.*, Vol. 20, No. 1, 112–122, Jan. 1948).
13. H. W. LEVERENZ, "A Convenient Periodic Chart of the Elements," *Foote Prints*, Vol. 12, No. 1, 22–24, June 1939.
14. H. SEMAT, *Introduction to Atomic Physics*, Rinehart and Co., New York, 1946.
15. K. K. DARROW, *Introduction to Contemporary Physics*, D. Van Nostrand Co., New York, 1939.
16. R. TOMASCHEK, "Physics of the Atom," Vol. 5 of E. GRIMSEHL'S, *A Textbook of Physics*, Blackie and Son, London, 1935.
17a. C. H. D. CLARK, *The Electronic Structure and Properties of Matter*, Chapman & Hall, London, 1934.
17b. C. H. D. CLARK, *The Fine Structure of Matter*, Part III, John Wiley & Sons, New York, 1938.

18. H. FROEHLICH, *Elektronentheorie der Metalle*, J. Springer, Berlin, 1936; E. E. SCHNEIDER, "Electronic Structure of Solids," *Sci. Progress*, Vol. 36, 614–632, 1948.

19. F. SEITZ, *The Physics of Metals*, McGraw-Hill Book Co., New York, 1943; G. V. RAYNOR, *An Introduction to the Electron Theory of Metals*, Inst. Metals, London, 1947.

20. R. M. BOZORTH, "Magnetism," *Revs. Modern Phys.*, Vol. 19, 29–86, Jan. 1947; *Elec. Eng.*, 471–476, June 1949.

21. J. L. SNOEK, *New Developments in Ferromagnetic Materials*, Elsevier Pub. Co., Amsterdam, 1947 (*cf. Physica*, Vol. 3, No. 6, 463–483, June 1936; *Philips Tech. Rev.*, Vol. 8, No. 12, 353–360, Dec. 1946).

22. P. W. SELWOOD, *Magnetochemistry*, Interscience Pub., New York, 1943; "Valence Inductivity," *J. Am. Chem. Soc.*, Vol. 70, No. 2, 883, Feb. 1948; "Valence Inductivity and Catalytic Action," *Pittsburgh International Conf. on Surface Reactions*, 49–51, Corrosion Publ. Co., Pittsburgh, 1948.

23. M. POLÁNYI, *Atomic Reactions*, Williams and Norgate, London, 1932 (*cf. Uspekhi Khim.*, Vol. 2, 412–444, 1933).

24. L. PAULING, *The Nature of the Chemical Bond and the Structure of Molecules and Crystals*, Cornell Univ. Press, Ithaca, 1945; "The Modern Theory of Valency," *J. Chem. Soc.*, 1461–1467, 1948; *Z. Naturforsch.*, Vol. 3a, 438–447, 1948; *Proc. Roy. Soc. (London)*, A, Vol. 196, No. 1046, 343–362, 1949.

24a. G. W. WHELAND, *The Theory of Resonance*, John Wiley & Sons, New York, 1944; "Molecular Constants and Chemical Theories," *J. Chem. Phys.*, Vol. 13, 239–248, 1945; R. SAMUEL, "Molecular Constants and Chemical Theories," *J. Chem. Phys.*, Vol. 12, 167–202, 380–390, 521–522, 1944, *ibid.*, Vol. 13, 251, 572–584, 1945; W. C. FERNELIUS, C. P. SMYTH, H. S. TAYLOR, B. E. WARREN, *et al.*, *Chemical Architecture*, Interscience Pub., 1948.

25. S. GLASSTONE, *Theoretical Chemistry*, 1945, and *Textbook of Physical Chemistry*, 1946, D. Van Nostrand Co., New York; "Oxidation Numbers and Valence," *J. Chem. Education*, Vol. 25, No. 5, 278–279, May 1948.

26. E. U. CONDON, "The Franck–Condon Principle and Related Topics," *Am. J. Phys.*, Vol. 15, No. 5, 365–375, Sept.–Oct. 1947.

27. O. K. RICE, *Electronic Structure and Chemical Binding*, McGraw-Hill Book Co., New York, 1940.

28. J. C. SLATER, *Introduction to Chemical Physics*, McGraw-Hill Book Co., New York, 1939.

29. J. FRENKEL, *Kinetic Theory of Liquids*, Oxford, The Clarendon Press, 1946.

30. J. FRENKEL, "Viscous Flow of Crystalline Bodies under the Action of Surface Tension," and "On the Surface Motion of Particles in Crystals and the Natural Roughness of Crystalline Faces," *J. Phys.*, (*USSR*), Vol. 9, No. 5, 385–398, 1945.

31. N. S. GINGRICH, "The Diffraction of X Rays by Liquid Elements," *Revs. Modern Phys.*, Vol. 15, No. 1, 90–110, Jan. 1943.

32. J. T. McCARTNEY and R. B. ANDERSON, "Crystalline Aggregation of Cobalt Powder," *J. Applied Phys.*, Vol. 18, No. 10, 902–903, Oct. 1947.

33. A. VLASOV, "On the Theory of the Solid State," *J. Phys.* (*USSR*), Vol. 9, No. 2, 130–138, 1945 (*cf. J. Exp. Theoret. Phys.* (*USSR*), Vol. 18, 368–373, 449–456, 1948).

34. H. W. LEVERENZ, "Excitation and Emission Phenomena in Phosphors," section in *Preparation and Characteristics of Luminescent Materials*, John Wiley & Sons, New York, 1948.

35. N. F. MOTT and R. W. GURNEY, *Electronic Processes in Ionic Crystals*, Oxford, The Clarendon Press, 1948. (Dover reprint).

36. F. SEITZ, *The Modern Theory of Solids*, McGraw-Hill Book Co., New York, 1940; "Basic Principles of Semi-Conductors," *J. Applied Phys.*, Vol. 16, 553–563, Oct. 1945.

37. W. SHOCKLEY, "The Quantum Physics of Solids, I. The Energies of Electrons in Crystals," *Bell System Tech. J.*, Vol. 18, No. 4, 645–723, Oct. 1939.

38. R. W. POHL and F. STÖCKMANN, "Die Rolle sekundärer Elektronen bei der lichtelektrischen Leitung," *Ann. Physik*, Ser. 6, Vol. 1, No. 6, 275–284, 1947; H. H. HAUSNER, "Semiconducting Ceramic Materials," *J. Am. Ceram. Soc.*, Vol. 30, No. 9, 290–296, 1947; *Electronics*, Jan. 1948.

Chapter 2

39. A. F. WELLS, *Structural Inorganic Chemistry*, Oxford, The Clarendon Press, 1945; "Crystal Habit and Internal Structure," *Phil. Mag.* [7], Vol. 37, 184–199, 217–236, 605–630, 1946; "The Structures of Metallic Oxides," *Quart. Revs. (London)*, Vol. 2, 185–202, 1948.

40. L. BRAGG and J. F. NYE, "A Dynamical Model of a Crystal Structure," *Proc. Roy. Soc. (London)*, A, Vol. 190, No. 1023, 474–482, Sept. 9, 1947; *Science*, Vol. 108, 455–463, 1948; J. C. SLATER, "The Physics of Metals," *Physics Today*, Vol. 2, No. 1, 6–15, Jan. 1949.

41. H. W. FAIRBAIRN, "Packing in Ionic Minerals," *Bull. Geol. Soc. Am.*, Vol. 54, 1305–1374, Sept. 1, 1943; S. O. MORGAN, et al., "Crystal Chemistry Symposium," *Ceramic Age*, Nov. 1948.

42. W. P. DAVEY, *A Study of Crystal Structure and Its Applications*, McGraw-Hill Book Co., New York, 1934; R. W. G. WYCKOFF, *Crystal Structures*, Interscience Pub., New York, 1948.

43. P. NIGGLI, *Grundlagen der Stereochemie*, Birkhäuser, Basel, 1945; E. BRANDENBERGER, *Röntgenographisch-Analytische Chemie*, Birkhäuser, Basel, 1945.

44. H. W. LEVERENZ, *A Chart of the Geometry of Crystal Lattices*, Radio Corp. Am., RCA Laboratories Div., Princeton, N. J., 1947.

45. R. D. LOWDE, "The Crystallographic Theory of Space-Groups," *Sci. J. Roy. Coll. Sci.*, Vol. 16, 130–140, 1946.

46. C. HERMANN, Editor, *International Tables for the Determination of Crystal Structures*, Vols. I and II, Edwards Bros., Ann Arbor, Mich., 1944.

47. C. W. BUNN, *Chemical Crystallography*, Oxford, The Clarendon Press, 1945; J. M. BIJVOET, N. H. KOLKMEIJER, and C. H. MACGILLAVRY, *Roentgenanalyse von Krystallen*, J. Springer, Berlin, 1940.

48. F. C. PHILLIPS, *An Introduction to Crystallography*, Longmans Green and Co., London, 1946.

49. W. H. ZACHARIASEN, *Theory of X-Ray Diffraction in Crystals*, John Wiley & Sons, New York, 1945. (Dover reprint).

50. M. J. BUERGER, *X-Ray Crystallography*, John Wiley & Sons, New York, 1943; "The Photography of the Reciprocal Lattice," *Am. Soc. X-Ray and Electron Diffraction Monograph No. 1*, Aug. 1944; "Derivative Crystal Structures," *J. Chem. Phys.*, Vol. 15, No. 1, 1–16, Jan. 1947.

50a. H. M. JAMES and V. A. JOHNSON, "Electron Distribution in ZnO Crystals," *Phys. Rev.*, Vol. 56, 119, July 1, 1939; C. H. EHRHARDT and K. LARK-HOROVITZ, "Intensity Distribution in X-Ray and Electron Diffraction Patterns," *Phys. Rev.*, Vol. 57, 603–613, Apr. 1, 1940.

51. H. A. BETHE, M. BORN, K. F. HERZFELD, A. SMEKAL, et al., "Aufbau der zusammenhängenden Materie," *Handbuch der Physik*, Vol. 24, Part 2, J. Springer, Berlin, 1933; D. C. STOCKBARGER, "Production of Large Single Crystals of Alkali Halides," *J. Optical Soc. Am.*, Vol. 27, 416–419, 1937; *Rev. Sci. Instr.*, Vol. 10, 205–211, 1939; G. P. THOMSON, "The Growth of Crystals," *Proc. Phys. Soc.*, Vol. 61, Part 5, No. 347, 403–416, Nov. 1948; J. C. FISHER, J. H. HOLLOMON, and D. TURNBULL, "Rate of Nucleation of Solid Particles in a Subcooled Liquid," *Science*, Vol. 109, No. 2825, 168–169, Feb. 18, 1949; S. V. STARODUBTSEV and N. I. TIMOKHINA, "Nature of the Conglomeration of Crystals," *Doklady Akad. Nauk SSSR*, Vol. 62, 619–621, 1948; S. ZERFOSS et al., "The Properties of Synthetic Rutile Single Crystals," *J. Chem. Phys.*, Vol. 16, 1166, 1948; W. K. BURTON, "Role of Dislocations in Crystal Growth," *Nature*, Vol. 163, No. 4141, 398–399, Mar. 12, 1949; P. H. EGLI, "Crystal Research," *Sci. Monthly*, Vol. 58, No. 4, 270–278, Apr. 1949.

52. N. W. H. ADDINK, "Complete and Incomplete Crystals," *Nature*, Vol. 157, 764, June 8, 1946.

53. R. C. WILLIAMS and R. W. G. WYCKOFF, "Applications of Metallic Shadow-Casting to Microscopy," *J. Applied Phys.*, Vol. 17, No. 1, 23–33, Jan. 1946; also S. TOLANSKY and W. L. WILCOCK, "Interference Studies of Diamond Faces," *Proc. Roy. Soc. (London) A.*, Vol. 191, No. 1025, 182–195, Nov. 11, 1947.

54. B. W. ANDERSON, "Absorption and Luminescence in Diamonds," *Goldsmiths J. Gemnologist*, Vol. 47, No. 290, Suppl., 33–35, 1943.

55. V. CHANDRASEKHARAN, C. V. RAMAN, K. G. RAMANATHAN, S. RAMESHAN, et al., "Luminescence, Phosphorescence, and Photoconductivity of Diamonds," *Proc. Ind. Acad. Sci.*, A, Vol. 24, 1–197, 1946; *ibid*, Vol. 26A, 479–480, 1947; *ibid*, Vol. 27A, 171–175, 316–320, 1948; *Current Sci. (India)*, Vol. 15, 205–213, 1946.

56. D. E. WOOLDRIDGE, A. J. AHEARN, and J. A. BURTON, "Conductivity Pulses Induced in Diamond by Alpha-Particles," *Phys. Rev.*, Vol. 71, No. 12, 913, June 15, 1947; L. F. CURTISS and B. W. BROWN, "Diamond as a Gamma-Ray Counter," *Phys. Rev.*, Vol. 72, No. 7, 643, Oct. 1, 1947; *Chem. Eng. News*, Vol. 25, No. 42, 3100–3102, Oct. 20, 1947; H. FRIEDMAN et al., "Ultraviolet Transmission of 'Counting' Diamonds," *Phys. Rev.*, Vol. 73, No. 2, 186, Jan. 15, 1948; A. J. AHEARN, "Conductivity Pulses Induced in Single Crystals of Zinc Sulphide by Alpha-Particle Bombardment," *Phys. Rev.*, Vol. 73, No. 5, 524, Mar. 1, 1948; "Conductivity Induced in Diamond by Alpha-Particle Bombardment and Its Variations among Specimens," *Phys. Rev.*, Vol. 73, No. 9, 1113, May 1, 1948; R. HOFSTADTER, "Remarks on Diamond Crystal Counters," *Phys. Rev.*, Vol. 73, No. 6, 631, Mar. 15, 1948; F. SEITZ, "On the Mobility of Electrons in Pure Non-Polar Insulators," *Phys. Rev.*, Vol. 73, No. 6, 547–564, Mar. 15, 1948; K. G. MCKAY, "Electron Bombardment Conductivity in Diamond," *Phys. Rev.*, Vol. 74, No. 11, 1606–1621, Dec. 1, 1948; *cf.*, Ref. 676 and L. PENSAK, "Conductivity Induced by Electron Bombardment in Thin Insulating Films," *Phys. Rev.*, Vol. 75, No. 3, 472–478, Feb. 1, 1949.

57. K. LONSDALE, "The Structure of Real Crystals," *Science Progress*, Vol. 35, No. 137, 1–11, Jan. 1947; "Remarks on Diamond Crystal Counters," *Phys. Rev.*, Vol. 73, No. 12, 1467, June 15, 1948; "Vibration Amplitudes of Atoms

in Cubic Crystals," *Acta Cryst.*, Vol. 1, 142–149, 1948; *Crystals and X Rays*, G. Bell & Sons, London, 1948.

58. H. W. LEVERENZ, *Final Report on Research and Development Leading to New and Improved Radar Indicators*, PB25481, Office Pub. Board, Washington, D. C., 1945 (*cf.* PB32546, PB23326, PB23337, and PB25732).

59. W. EITEL, *Synthesis of Fluorine Mica*, PB20530–20532, and PB32545, Office Pub. Board, Washington, D. C., 1946; *The Synthetic Mica Program*, Office of Naval Research, Washington, D. C., 12-1-48; G. MAYR, "Effects of a Magnetic Field on Crystallization of Some Paramagnetic Salts," *Rend. ist. lombardo sci.*, Vol. 78, No. 1, 459–471, 511–519, 1944–1945; J. T. KENDALL and D. YEO, "Magnetic Susceptibility of Mica," *Nature*, Vol. 161, 476–477, Mar. 27, 1948.

60. J. A. HEDVALL, *Reaktionsfähigkeit fester Stoffe*, J. A. Barth, Leipzig, 1938; *Trans. Chalmers Univ. Technol. Gothenberg*, No. 4, 25 pp., 1942 (*cf.* H. L. RILEY, "Reactions in the Solid Phase," *J. Oil & Colour Chemists' Assoc.*, Vol. 29, No. 308, 25–36, Feb. 1946; G. COHN, "Reactions in the Solid State," *Chem. Rev.*, Vol. 42, No. 3, 527–579, June 1948.)

61. H. W. LEVERENZ and F. SEITZ, "Luminescent Materials," *J. Applied Phys.*, Vol. 10, No. 7, 479–493, July 1939.

62. R. NAGY and C. K. LUI, "Thermal and X-Ray Analysis of Some Common Phosphors," *J. Optical Soc. Am.*, Vol. 37, No. 1, 37–41, Jan. 1947; A. P. RICE, "Differential Thermal Analysis Studies in Some Silicate Systems," *J. Electrochem. Soc.*, Vol. 96, 114–122, 1949.

63. H. W. LEVERENZ, "Luminescence and Tenebrescence as Applied in Radar," *RCA Rev.*, Vol. 7, No. 2, 199–239, June 1946.

64a. E. WAINER, "High Titania Dielectrics," *Electrochem. Soc. Trans.*, Vol. 89, 331–356, 1946.

64b. B. MATTHIAS and A. VON HIPPEL, "Structure, Electrical, and Optical Properties of Barium Titanate," *Phys. Rev.*, Vol. 73, No. 3, 268, Feb. 1, 1948; *ibid.*, No. 11, 1378–1385, June 1, 1948; *ibid.*, Vol. 74, No. 11, 1622–1639, Dec. 1, 1948.

64c. W. SHOCKLEY, "Electronic Polarizability of O^{2-} in Presence of Ti^{4+}," *Bull. Am. Phys. Soc.*, Vol. 23, No. 2, 46, Jan. 29, 1948.

65. R. TOMASCHECK, Editor, *Leuchten und Struktur fester Stoffe*, R. Oldenburg, Munich, 1943; PBL38183, Office Pub. Board, Washington, D. C., 1947.

66. F. SEITZ, "Color Centers in Alkali Halide Crystals," *Revs. Modern Phys.*, Vol. 18, 384–409, July 1946; I. ESTERMANN, W. J. LEIVO, and O. STERN, "Change in Density of Potassium Chloride upon Irradiation with X-rays," *Phys. Rev.*, Vol. 75, No. 4, 627–633, Feb. 15, 1949.

67. H. F. IVEY, "Spectral Location of the Absorption Due to Color Centers in Alkali Halide Crystals," *Phys. Rev.*, Vol. 72, No. 4, 341–343, Aug. 15, 1947, *ibid.*, Vol. 72, No. 9, 873, Nov. 1, 1947.

68. D. M. YOST, H. RUSSELL, and C. S. GARNER, *The Rare-Earth Elements and Their Compounds*, J. Wiley & Sons, New York, 1947.

69. P. F. A. KLINKENBERG, "On the Spectra of the Rare-Earth Elements," *Physica*, Vol. 13, No. 1–3, 1–6, Mar. 1947.

70. W. F. MEGGERS, "Electron Configurations of 'Rare-Earth Elements'," *Science*, Vol. 105, No. 2733, 514–516, May 16, 1947.

70a. S. FREED and S. KATCOFF, "The Absorption and Fluorescence Spectra of Bivalent Europium Ion in Crystals," *Physica*, Vol. 14, No. 1, 17–28, Jan.

1948; F. D. S. BUTEMENT, "Absorption and Fluorescence Spectra of Bivalent Sm, Eu, and Yb," *Trans. Faraday Soc.*, Vol. 44, Part 9, No. 309, 617–626, Sept. 1948; K. PRZIBRAM, "The Light Emitted by Europium Compounds," *Nature*, Vol. 163, No. 4156, 989, June 25, 1949; B. RINCK, "Transition Probabilities for Radiating and Radiationless Processes in Crystalline $Eu_2(SO_4)_3 \cdot 8H_2O$," *Ž. Naturforschung*, Vol. 3a, 406–412, July 1948.

70b. J. HOOGSCHAGEN and C. J. GORTER, "Absorption of Light in Solutions of the Rare Earth Salts," *Physica*, Vol. 14, 197–206, 1948.

Chapter 3

71a. H. G. DEMING, *General Chemistry*, 5th Ed., John Wiley & Sons, New York, 1947; L. PAULING, *General Chemistry*, W. H. Freeman and Co., San Francisco, 1947.

71b. F. P. HALL and H. INSLEY, *Phase Diagrams for Ceramists*, Am. Ceram. Soc., Columbus, Ohio, 1947.

72. H. W. LEVERENZ, "Cathodoluminescence as Applied to Television," *RCA Rev.*, Vol. 5, No. 2, 131–175, Oct. 1940.

73. A. S. EDDINGTON, *The Nature of the Physical World*, Cambridge Univ. Press, 1929.

74. *Luminescence, a General Discussion*, published for Faraday Soc. by Gurney and Jackson, London, 1938.

75. P. PRINGSHEIM and M. VOGEL, *Luminescence of Liquids and Solids*, Interscience Pub., New York, 1943; P. PRINGSHEIM, *Fluorescence and Phosphorescence*, Interscience Pub., New York, 1949.

76. P. W. DANCKWORTT, "Lumineszenz-analyse im filtrierten ultravioletten Licht," *Akad. Verlag.*, Leipzig, 1940.

77. G. N. LEWIS, D. LIPKIN, and T. T. MAGEL, "Reversible Photochemical Processes in Rigid Media. A Study of the Phosphorescent State," *J. Am. Chem. Soc.*, Vol. 63, 3005–3018, 1941; *ibid.* Vol. 64, 1774–1782, 1942 (see B. Y. SVESHIKOV and P. P. DIKUN, *Doklady Akad. Nauk SSSR*, Vol. 59, No. 1, 37–40, 1948); M. KASHA, "Phosphorescence and the Role of the Triplet State in the Electronic Excitation of Complex Molecules," *Chem. Rev.*, Vol. 41, No. 2, 401–419, Oct. 1947; "Fabrication of Boric Acid Glass for Luminescence Studies," *J. Optical Soc. Am.*, Vol. 38, No. 12, 1068–1073, Dec. 1948; H. CHOMSE and G. HEINRICH, "Über die Lumineszenz des Phosphorpentoxyds," *Naturwissenschaften*, Vol. 34, No. 4, 122–123, Jan. 1948.

77a. C. A. COULSON, "Excited Electronic Levels in Conjugated Molecules: I. Long Wavelength Ultraviolet Absorption of Naphthalene, Anthracene, and Homologs," *Proc. Phys. Soc.*, Vol. 60, Part 3, No. 339, 257–281, Mar. 1, 1948; "Wave Mechanics," *Sci. Progress*, Vol. 36, 436–449, 1948.

78. I. V. OBREIMOV and C. G. SHABALDAS, "Fluorescence of Naphthalene Crystals," *J. Phys. (USSR)*, Vol. 7, No. 4, 168–178, 1943; C. E. FEAZEL and C. D. SMITH, "Production of Large Crystals of Naphthalene and Anthracene," *Rev. Sci. Instr.*, Vol. 19, No. 11, 817–818, Nov. 1948.

79. A. KUHN, "Fluoreszenz fester Stoffe im kompakten und dispergierten Zustand," *Kolloid-Z.*, Vol. 100, No. 1, 126–135, 1942 (*cf.* W. E. KASKAN and A. B. F. DUNCAN, "The Fluorescence of Solid Acetone," *J. Chem. Phys.*, Vol. 16, No. 3, 223–224, Mar. 1948; *ibid.*, 407–410, 1948).

80. M. Déribéré, *Les applications pratiques de la luminescence*, Dunod, Paris, 1938 (Edwards Bros., Ann Arbor, Mich., 1943). M. Curie, *Fluorescence et phosphorescence*, Hermann et Cie., Paris, 1946.

81. E. N. Harvey, *Living Light*, Princeton Univ. Press, 1940; "The Luminescence of Living Things," *Sci. American*, Vol. 178, No. 5, 46–50, May 1948; "Bioluminescence," *Ann. N. Y. Acad. Sci.*, Vol. 49, Art. 3, 327–482, Apr. 15, 1948.

82. H. E. Millson *et al.*, "Fluorescent Dyes in Coatings and Plastics," *Calco Tech. Bull.*, Nos. 571 and 755, Calco Chemical Co., Bound Brook, N. J.; "The Phosphorescence of Textile Fibers and Other Substances," *Textile Colorist*, Vol. 65, 495, Nov. 1943; *ibid.*, Vol. 66, 109, Mar. 1944; "A Study of Fluorescent Dyes," (to be published).

82a. H. E. Millson and H. E. Millson, Jr., "Observations on Mineral Luminescence," (to be published); H. Yagoda, "Luminescent Phenomena as Aids in the Localization of Minerals in Polished Sections," *Econ. Geol.*, Vol. 41, 813–819, 1946.

83. N. Campbell, "The Fluorescence of Organic Compounds," *Endeavour*, Vol. 5, 155–159, 1946.

84. R. Tomaschek, "The Inner Molecular Sensitivity of Luminous Substances," *Reichber. Physik.*, Vol. 1, 139–140, 1945; Th. Förster, "Zwischenmolekulare Energiewanderung und Fluoreszenz," *Ann. Physik*, Vol. 2, No. 1–2, 55–75, 1948.

85. E. H. Archibald, *The Preparation of Pure Inorganic Substances*, John Wiley & Sons, New York, 1932.

86. R. C. Hughes, "Preparation of Purified Inorganic Compounds for Use in Spectrographic Standards," *J. Optical Soc. Am.*, Vol. 33, No. 1, 49–60, Jan. 1943.

87. R. E. Shrader and E. J. Wood, "Automatic Control of Stills," *Electronics*, 98ff., Sept. 1944.

88. Carborundum Co., Globar Div., Niagara Falls, N. Y., or Harper Electric Furnace Corp., Niagara Falls, N. Y. Harper furnaces HL7610 and HS81011-special have been found satisfactory for phosphor research. (The latter furnace, which was designed by the writer, has a rotating stage to assure uniform heat treatment of several samples heated at one time.)

89. H. W. Leverenz, "Phosphors Brighten Radio Future," *Radio Age*, Vol. 3, No. 1, 7–11, Oct. 1943.

90. E. W. Beggs, "Activating Light Sources for Luminescent Materials," *J. Optical Soc. Am.*, Vol. 33, No. 2, 61–70, Feb. 1943; "New Developments in Mercury Lamps and Their Applications," *Illum. Eng.*, Vol. 42, No. 4, 435–462, Apr. 1947; "Progress with Mercury Lamps," *Westinghouse Engr.*, Vol. 7, No. 6, 179–184, Nov. 1947 (*cf.* Ref. 518); P. Schulz, "A High-intensity 40–Atmosphere Xenon Source of Continuous UV and Visible Radiation," *Z. Naturforsch.*, Vol. 2a, 583–584, Oct. 1947.

91. A. H. Compton and S. K. Allison, *X-Rays in Theory and Experiment*, D. Van Nostrand Co., New York, 1935.

92. G. L. Clark, *Applied X-Rays*, McGraw-Hill Book Co., New York, 1940.

93. V. K. Zworykin, *et al.*, "An Automatic Recording Spectroradiometer for Cathodoluminescent Materials," *J. Optical Soc. Am.*, Vol. 29, 84–91, Feb. 1939.

94. F. E. E. GERMANN and R. WOODRIFF, "Cross-Prism Investigation of Fluorescence," *Rev. Sci. Instruments*, Vol. 15, No. 6, 145–149, June 1944.

95. A. E. HARDY, "A Combination Phosphorometer and Spectroradiometer for Luminescent Materials," *J. Electrochem. Soc.* Vol. 91, 221–240, 1947.

96. F. J. STUDER, "Method for Measuring the Spectral Energy Distribution of Low Brightness Light Sources," *J. Optical Soc. Am.*, Vol. 37, No. 4, 288–292, Apr. 1947; "Method for Automatically Plotting Spectral Energy Distribution of Luminescent Materials," *J. Optical Soc. Am.*, Vol. 38, No. 5, 467–470, May 1948.

97. W. S. PLYMALE, JR., "A Sensitive Photoelectric Method for Determining the Chromaticity of Phosphorescent and Fluorescent Materials," *J. Optical Soc. Am.*, Vol. 37, No. 5, 399–402, May 1947.

98. R. A. BURDETT and L. C. JONES, "The Photoelectric Measurement of Fluorescence Spectra," *J. Optical Soc. Am.*, Vol. 37, No. 7, 554–557, July 1947.

99. O. KLEMPERER, *Einführung in die Elektronik*, J. Springer, Berlin, 1933.

100a. J. STRONG, *Procedures in Experimental Physics*, Prentice-Hall, New York, 1938.

100b. C. H. BACHMAN, *Techniques in Experimental Electronics*, John Wiley & Sons, New York, 1948.

101. V. K. ZWORYKIN, G. A. MORTON, E. G. RAMBERG, J. HILLIER, and A. W. VANCE, *Electron Optics and the Electron Microscope*, John Wiley & Sons, New York, 1945.

102. N. RIEHL, *Physik und technische Anwendungen der Lumineszenz*, J. Springer, Berlin, 1941 (Edwards Bros., Ann Arbor, Mich., 1944).

103. MANHATTAN PROJECT, "Availability of Radioactive Isotopes," *Science*, Vol. 103, No. 2685, 697–705, June 14, 1946; G. T. SEABORG and I. PERLMAN, "Table of Isotopes," *Revs. Modern Phys.*, Vol. 20, No. 4, 585–667, Oct. 1948.

104. PLUTONIUM PROJECT, "Nuclei Formed in Fission: Decay Characteristics, Fission Yields, and Chain Relationships," *Revs. Modern Phys.*, Vol. 18, No. 4, 513–544, Oct. 1946.

105. W. E. FORSYTHE, *Measurement of Radiant Energy*, McGraw-Hill Book Co., New York, 1937; D. W. EPSTEIN, "Photometry in Television Engineering," *Electronics*, 110–113, July 1948.

106. W. B. NOTTINGHAM, *Notes on Photometry, Colorimetry, and an Explanation of the Centibel Scale*, Report 804, Contract OEMsr-262 (NDRC Div. 14), Dec. 17, 1945.

107. R. W. WOOD, *Physical Optics*, Macmillan Co., New York, 1934; F. A. JENKINS and H. E. WHITE, *Fundamentals of Physical Optics*, McGraw-Hill Book Co., New York, 1937. (Dover reprint).

108. A. E. H. MEYER and E. O. SEITZ, *Ultraviolette Strahlen*, W. de Gruyter and Co., Berlin, 1942 (Edwards Bros., Ann Arbor, Mich., 1944).

109. C. SCHAEFER and F. MATOSSI, *Das Ultrarote Spektrum*, J. Springer, Berlin, 1930 (Edwards Bros., Ann Arbor, Mich., 1943).

110. W. DE GROOT, "Luminescence Decay and Related Phenomena," *Physica*, Vol. 6, 275–289, 1939; *ibid.*, Vol. 7, 432–446, 1940; *ibid.*, Vol. 8, 789–795, 1941; *ibid.*, Vol. 12, 402–404, 1946.

111. J. H. GISOLF and W. DE GROOT, "Fluorescentie en Fosforescentie," *Philips Tech. Tijdschr.*, Vol. 3, No. 8, 245, Aug. 1938; "The Absorption Spectra of Zinc Sulphide and Willemite," *Physica 's Grav.*, Vol. 8, 805–809, July 1941.

112. R. B. NELSON, R. P. JOHNSON, and W. B. NOTTINGHAM, "Luminescence during Intermittent Electron Bombardment," *J. Applied Phys.*, Vol. 10, No. 5, 335–342, May 1939.

113. G. R. FONDA, "Phosphorescence of Zinc Silicate Phosphors," *J. Applied Phys.*, Vol. 10, No. 6, 408–420, June 1939.

114. R. P. JOHNSON and W. L. DAVIS, "Luminescence during Intermittent Optical Excitation," *J. Optical Soc. Am.*, Vol. 29, No. 7, 283–290, July 1939.

115. J. SADDY, "Sur l'étude et l'interpretation du déclin de la phosphorescence des sulfures," *Compt. rend.*, Vol. 219, 314–315, 1944; *ibid.*, Vol. 222, 1002–1004, 1946.

116a. R. DELORME and F. PERRIN, "Durées de fluorescence des sels d'uranyle solides et de leurs solutions," *J. phys. radium*, Vol. 10, No. 5, 177–187, May 1929; *cf.* L. BRUNINGHAUS, *Ann. physik*, [12], Vol. 3, 199–274, 1948.

116b. J. A. KHVOSTIKOV, "Fluoreszenz der Platinocyanide," *Physik. Z. Sowjetunion* Vol. 9, 210–236, 1936.

116c. W. KIRCHHOFF, "Messung von Abklingzeiten bei der Fluoreszenz," *Z. Physik*, Vol. 116, No. 1/2, 115–122, 1940.

117. G. WOLLWEBER, "Über eine neue Methode zur Messung der Abklingung der Phosphoreszenz," *Ann. Physik*, Vol. 34, 29–40, 1939.

118. H. B. BRIGGS, "A Supersonic Cell Fluorometer," *J. Optical Soc. Am.*, Vol. 31, No. 8, 543–549, Aug. 1941.

119. J. T. RANDALL and M. H. F. WILKINS, "The Phosphorescence of Various Solids," *Proc. Roy. Soc. (London) A*, Vol. 184, No. 999, 347–408, 1945 (*cf. Nature*, Vol. 143, No. 3632. 978–979, June 10, 1939).

120. G. F. J. GARLICK and M. H. F. WILKINS, "Short Period Phosphorescence and Electron Traps," *Proc. Roy. Soc. (London) A*, Vol. 184, No. 999, 408–434, 1945.

121. L. BERGMANN and F. RONGE, "Über lichtelektrische Untersuchungen an Leuchtstoffen," *Physik. Z.*, Vol. 41, No. 15, 349–356, Aug. 1, 1940.

122. R. C. HERMAN and R. HOFSTADTER, "Photoconductivity of a Natural Willemite Crystal," *Phys. Rev.*, Vol. 59, 78–84, 1941.

123. A. E. HARDY, "The Photoconductivity of Zinc–Cadmium Sulphide as Measured with the Cathode-Ray Oscillograph," *Trans. Electrochem. Soc.*, Vol. 87, 355–367, 1945.

123a. C. C. KLICK and J. H. SCHULMAN, "Absence of Photoconductivity in Tungstate Phosphors," *Phys. Rev.*, Vol. 75, No. 10, 1606–1607, May 15, 1949.

123b. N. S. BAYLISS and J. C. RIVIERE, "Photoconductivity of Naphthalene and Anthracene," *Nature*, Vol. 163, No. 4150, 765, May 14, 1949.

124. H. NELSON, "Method of Measuring Luminescent Screen Potential," *J. Applied Phys.*, Vol. 9, 592–599, Sept. 1938.

125. W. B. NOTTINGHAM, "Electrical and Luminescent Properties of Phosphors under Electron Bombardment," *J. Applied Phys.*, Vol. 10, No. 1, 73–83, Jan. 1939.

126. S. T. MARTIN and L. B. HEADRICK, "Light Output and Secondary Emission Characteristics of Luminescent Materials," *J. Applied Phys.*, Vol. 10, No. 2, 116–127, Feb. 1939.

127. R. FRERICHS and E. KRAUTZ, "Messung der Aufladepotentiale Elektronenbestrahlter Leuchtstoffschichten," *Physik. Z.*, Vol. 40, 229–230, 1939.

128. C. HAGEN, "Aufladungs- und Ermüdungserscheinungen von Leuchtsubstanzen," *Physik. Z.*, Vol. 40, 621–640, 1939; *Fernseh A. G.*, Vol. 1, No. 5, 187–193, 1939.

129. B. KÜHNREICH, *Secondary Emission of Insulators and Luminescent Materials at Electron Velocities up to 60 kv*, PBL73874 (4554–4587), Office Pub. Board, Washington, D. C., 1939.

130. H. BRUINING, *Die Sekundär-Elektronen-Emission fester Körper*, J. Springer, Berlin, 1942; H. W. LEVERENZ, J. E. RUEDY, and V. K. ZWORYKIN, *MgO-Mg:Ag Dynode*, U. S. Pat. No. 2,233,276, 2-25-41.

131. K. H. GEYER, "Beobachtungen an der Sekundärelektronenstrahlung aus Nichtleitern," *Ann. Physik*, Vol. 41, No. 2, 117–144, 1942; *ibid.*, Vol. 42, No. 4, 241–254, 1942.

132. M. KNOLL, "Zum Verhalten von Leuchtstoffen bei intermittierender Elektronenbestrahlung," *Z. Physik*, Vol. 116, 385–414, 1940; M. KNOLL, *et al.*, "Zum Mechanismus der Sekundäremission im Inneren von Ionenkristallen," *Z. Physik*, Vol. 122, No. 1–4, 137–163, Mar. 1944; *Reichsber. Physik*, Vol. I, 108–110, 1944.

133. L. R. KOLLER, "Secondary Emission," *G. E. Rev.*, Vol. 51, 33–40, 50–52, Apr.–May 1948.

134. K. G. McKAY, "Secondary Electron Emission," Chapter in *Advances in Electronics*, Vol. 1, Academic Press, New York, 1948.

135. J. H. HELLMERS *et al.*, *Die Mineralogischen Grundlagen der Silikosefrage*, J. A. Barth, Leipzig, 1942; Staub, 5–19, 1943.

136. A. J. ARMOR, *The Chemical Aspects of Silicosis*, H. K. Kewis, London, 1943.

137. P. P. GAUTHIER, "Toxic Gases," *J. Hôtel-Dieu Montréal*, Vol. 13, 322–334, 1944.

138. L. T. FAIRHALL, "Inorganic Industrial Hazards," *Physiol. Revs.*, Vol. 25, 182–202, 1945.

139. H. S. VAN ORDSTRAND *et al.*, "Beryllium Poisoning," *J. Am. Med. Assoc.*, Vol. 129, 1084–1090, 1945 (*cf.* H. L. HARDY, "Delayed Chemical Pneumonitis Occurring in Workers Exposed to Beryllium Compounds," *Bull. New England Med. Center*, Vol. 9, 16, 1947; *Brit. J. Ind. Med.*, Vol. 5, 35, 1948; J. N. AGATE, "Delayed Pneumonitis in a Beryllium Worker," *Lancet*, Vol. 255, 530–533, 1948; R. J. HASTERLIK, "Beryllium Poisoning," *Physics Today*, Vol. 2, No. 6, 14–17, June 1949).

140. G. G. JENNER and J. A. K. CUNNINGHAM, "An Outbreak of Cadmium Poisoning," *New Zealand Med. J.*, Vol. 43, 282–283, 1944.

141. A. J. ARMOR and P. PRINGLE, "A Review of Selenium as an Industrial Hazard," *Bull. Hyg.*, Vol. 20, 239–241, 1945.

142. A. W. FREIREICH, "Hydrogen Sulphide Poisoning," *Am. J. Path.*, Vol. 22, 147–155, 1946.

142a. P. M. VAN ARSDELL, "Health Hazards of Common Laboratory Reagents," *Ind. Eng. Chem., News Ed.*, Vol. 26, No. 5, 304–309, Feb. 2, 1948.

143. G. N. LEWIS and M. RANDALL, *Thermodynamics and the Free Energy of Chemical Reactions*, McGraw-Hill Book Co., New York, 1923.

144. R. TOMASCHEK, "Strukturerforschung fester und flüssiger Körper mit Hilfe der Linien Fluoreszenzspektren," *Ergeb. exakt. Naturw.*, Vol. 20, 268–303, 1942.

145. P. LENARD, F. SCHMIDT, and R. TOMASCHEK, "Phosphoreszenz und Fluoreszenz," *Handbuch Experimentalphysik*, Vol. 23, Parts 1 and 2, Akad. Verlag., Leipzig, 1928.

146. H. RUPP, *Die Leuchtmassen und ihre Verwendung*, G. Borntraeger, Berlin, 1937.

147. F. FRITZ, *Leuchtfarben*, G. Bodenbender, Berlin, 1940 (Edwards Bros., Ann Arbor, Mich., 1946).

148. H. T. LEFTLEY *et al.*, *Boule Manufacture*, PB48441, Office Pub. Board, Washington, D. C., 1946.

149. A. SCHLOEMER, "Beiträge zur Kenntnis lumineszierender Verbindungen," *J. prakt. Chem.*, Vol. 133, 51–59, 1932.

150. E. L. Nichols et al., "Studies in Luminescence," *Carnegie Inst. Wash. Pub.* Nos. 152, 1912; 298, 1919; 384, 1928; and *Bull. Natl. Research Council*, Vol. 5, part 5, No. 30, Natl. Acad. Sci., Washington, D. C. 1923.

151. V. M. Kudryavtzeva, "Temperature Luminescence of ZnO and CaO Powders," *Compt. rend. sci. URSS*, Vol. 52, No. 7, 581–583, 1946.

152. P. Pringsheim et al., "Quantenhafte Ausstrahlung," *Handbuch Physik*, Vols. 19, 21, and 23/1, J. Springer, Berlin, 1928–1933.

153. M. Swanson, *Infrared-Stimulable ZnS:Mn Phosphor*, U. S. Pat. 2,396,298, 3-12-46.

154. F. A. Kröger, *Luminescence in Solids Containing Manganese*, Dissertation, J. van Campen, Amsterdam, 1940; *Some Aspects of the Luminescence of Solids*, Elsevier Pub. Co., Amsterdam, 1948.

155. F. A. Kröger, "Fluorescence of Tungstates and Molybdates," *Nature*, Vol. 159, No. 4046, 674–675, May 17, 1947; *Philips Research Repts.*, Vol. 2, 177–189, 340–348, 1947.

156. F. A. Kröger, "Tetravalent Manganese as an Activator in Luminescence," *Nature*, Vol. 159, No. 4047, 706–707, May 24, 1947; "Applications of Luminescent Substances," *Philips Tech. Rev.*, Vol. 9, No. 7, 215–221, 1947; "The Incorporation of Uranium in Calcium Fluoride," *Physica*, Vol. 14, No. 7, 488, Sept. 1948; *Luminescent MgO:TiO₂:Mn*, Dutch Pat. 62,462, 2-15-49.

156a. F. A. Kröger, J. M. Stevels, and Th. P. J. Botden, "The Fluorescence of Hexavalent Uranium in Glass," *Philips Research Repts.*, Vol. 3, 46–48, 1948.

156b. F. A. Kröger et al., "The Influence of Temperature Quenching on the Decay of Fluorescence," *Physica*, Vol. 14, No. 2–3, 81–96, Apr. 1948; "Decay and Quenching in Willemite," *Physica*, Vol. 14, No. 7, 425–442, Sept. 1948.

156c. F. A. Kröger and J. E. Hellingman, "Luminescence of Zinc Sulphide," *J. Electrochem. Soc.*, Vol. 93, No. 5, 156–171, May 1948; *ibid.*, Vol. 95, No. 2, 68–69, Feb. 1949.

157. H. W. Leverenz, "Phosphors *versus* the Periodic System of the Elements," *Proc. IRE*, 256–263, May 1944.

158. E. A. Ab, V. M. Gugel, V. P. Levshin, A. V. Moskvin, et al., "Symposium on Luminescence" (in Russian), *Bull. acad. sci. URSS, Sér. phys.*, Vol. 9, No. 4/5, 269–578, 1945; *J. Tech. Phys. (USSR)* Vol. 13, No. 4/5, 182–187, 1943; *Uspekhi Phys. Nauk SSSR*, Vol. 36, No. 4, 557–566, 1948.

159. H. C. Froelich, *Silicate Phosphors with Ti Activator*, U. S. Pat. 2,415,129, 2-4-47 (*cf.* Ref. 429 and 430, and K. G. Zimmer, Ger. Pat. App. T51,782, 2-10-39; Fr. Pat. 869,448, 2-2-42); K. Iwase and U. Nisioka, "Equilibrium Diagrams of the Systems CaO:SiO₂:TiO₂, CaO:SiO₂:TiO₂:MnO, and CaO:-MgO:SiO₂:TiO₂," *Sci. Repts. Tôhoku Imp. Univ.*, Ser. 1, Honda Anniv. Vol. 1–10, 455–456, 1936; *ibid.*, Vol. 24, 715, 1935–1936; *ibid.*, Vol. 25, 504–509, 1936–1937; *ibid.*, Vol. 26, 593, 1937–1938; S. S. Cole and W. K. Nelson, "The System ZnO–TiO₂. Zinc Orthotitanate and Solid Solutions with Titanium Dioxide," *J. Phys. Chem.*, Vol. 42, 245–251, 1938.

160. R. Coustal and F. Prevet, "Sur un Nouveau Procédé de préparation du sulfure de zinc phosphorescent," *Compt. rend.*, Vol. 188, No. 10, 703–705, Mar. 1929.

161. L. Wesch, *Method of Incorporating an Activator in a Pre-formed Crystal by Electrolysis*, U. S. Pat. 2,245,843, 6-17-41.

162. W. Bünger and W. Flechsig, "Über die Abklingung eines KCl-Phosphors mit TlCl-Zusatz und ihre Temperaturabhängigkeit," *Z. Physik*, Vol. 67, No. 1/2, 42–53, 1931.

163. E. Wasser, "Sur la décadence de la luminescence du phosphore KCl:Tl en état pulverisé," *Cahiers phys.*, Vol. 2, No. 11, 6–16, 1942.

164. F. Seitz, "Interpretation of the Properties of Alkali Halide–Thallium Phosphors," *J. Chem. Phys.*, Vol. 6, No. 3, 150–163, Mar. 1938.

165. P. Pringsheim, "Fluorescence and Phosphorescence of Thallium-Activated Potassium–Halide Phosphors," *Revs. Modern Phys.*, Vol. 14, No. 2/3, 132–138, Apr.–June 1942; *J. Chem. Phys.*, Vol. 16, No. 3, 241–246, Mar. 1948.

166. V. V. Antonov-Romanovskii, "Mechanism of Luminescence of Phosphors," *J. Phys. USSR*, Vol. 6, 120–140, 1942; *ibid.*, Vol. 7, No. 4, 153–167, 1943.

167. H. W. Leverenz, *ZnS:Ag:Cu Phosphors*, U. S. Pat. 2,402,757, 6-25-46.

168. A. L. Landau and H. Meyer, *ZnS-type Phosphors with U or Pb in Addition to the Normal Activators, such as Ag and Cu*, Brit. Pat. 571,690, 9-5-45.

169. G. R. Fonda, "Preparation and Characteristics of Zinc Sulphide Phosphors Sensitive to Infrared," *J. Optical Soc. Am.*, Vol. 36, No. 7, 382–390, July 1946; U. S. Pat. 2,447,322, 8-17-48 (*cf.* A. A. Cherepnev, *Compt. rend. acad. sci. URSS*, Vol. 56, No. 8, 807–810, 1947).

170. H. W. Leverenz, E. J. Wood, R. E. Shrader, and S. Lasof, "Zinc:Cadmium–Sulphide:Selenide Phosphors," in *Preparation and Characteristics of Luminescent Materials*, John Wiley & Sons, New York, 1948.

171. B. O'Brien, "Development of Infrared-Sensitive Phosphors," *J. Optical Soc. Am.*, Vol. 36, No. 7, 369–371, July 1946.

172. F. Urbach, D. Pearlman, and H. Hemmendinger, "On Infrared-Sensitive Phosphors," *J. Optical Soc. Am.*, Vol. 36, No. 7, 372–381, July 1946; PB-33250, Office Pub. Board, Washington, D. C., 1945; *Bull. Am. Optical Soc.*, 3–5, Mar. 10, 1949; U. S. Pat. 2,482,813–5, 9-27-49.

173. W. Primak, R. K. Osterheld, and R. Ward, "The Function of Fluxes in the Preparation of Infrared-Sensitive Phosphors of the Alkaline-Earth Sulphides and Selenides," *J. Am. Chem. Soc.*, Vol. 69, No. 6, 1283–1287, June 1947; PB16094 and PB33341, Office Pub. Board, Washington, D. C. (*cf. J. Am. Chem. Soc.*, Vol. 70, No. 6, 2043–2046, June 1948).

174. A. L. Smith, R. D. Rosenstein, and R. Ward, "The Preparation of Strontium Selenide and Its Properties as a Base Material for Infrared-Sensitive Phosphors," *J. Am. Chem. Soc.*, Vol. 69, No. 7, 1725–1729, July 1947.

175. J. J. Pitha, A. L. Smith, and R. Ward, "The Preparation and Properties of Lanthanum Oxysulphide," *J. Am. Chem. Soc.*, Vol. 69, No. 8, 1870–1871, Aug. 1947; U. S. Pat. 2,462,547, 2-22-49.

176. K. F. Stripp and R. Ward, "The Effect of Varying Composition on the Properties of the Infrared-Sensitive Phosphor Strontium Sulphide Activated with Samarium and Europium," *J. Am. Chem. Soc.*, Vol. 70, No. 1, 401–406, Jan. 1948.

177. V. J. Russo, *The Preparation of Pure Magnesium Sulphide and Its Use as a Base Material for Infrared Phosphors*, Thesis, Polytechnic Inst. Brooklyn, Sept. 1946.

177a. R. Ward, "Infrared-Sensitive Phosphors of the Alkaline-Earth Sulphides and Selenides," *J. Electrochem. Soc.*, Vol. 93, No. 5, 171–176, May 1948; *Inorganic Syntheses*, Vol. 3, McGraw-Hill Book Co., 1949; *Final Report for Contract NObsr 39045*, June 30, 1949.

178. P. Brauer, "On a Group of Complex Phosphors with Mixed Activators," *Z. Naturforsch.*, Vol. 1, 70–78, Feb. 1946; *ibid.*, Vol. 2a, 238–239, Apr. 1947; PBL6478 and PBL19204, Office Pub. Board, Washington, D. C.; "Luminescent Materials which Change Color," *Optik*, Vol. 2, 399–414, Nov. 1947.

179. V. V. Antonov-Romanovskii *et al.*, "Phosphors Sensitive to Red Light," *Compt. rend. acad. sci. URSS*, Vol. 54, 19–22, 775–778, 1946.

180. R. E. Shrader, E. J. Wood, and H. W. Leverenz, "Some Recent Results from Experimentation with Infrared-Sensitive Phosphors," *J. Optical Soc. Am.*, Vol. 36, No. 6, 353, June 1946; also Part 2 of final report on Contract OEMsr-440, Sept. 22, 1945.

181. M. Schilling, "Spektralphotometrische Untersuchungen an einer technischen Leuchtfarbe," *Z. Physik*, Vol. 21, No. 10, 232–239, 1940.

182. J. N. Bowtells and E. E. Miles, "Phosphorescence and its Applications," *GEC Journal*, Vol. 7, 256–265, Aug. 1941.

183. A. H. Taylor, "Luminous Characteristics of two Phosphorescent Materials," *J. Optical Soc. Am.*, Vol. 32, No. 9, 506–509, Sept. 1942.

184. G. F. A. Stutz, "Fluorescent and Phosphorescent Paints," *Illum. Eng.*, Vol. 37, No. 10, 798–806, Dec. 1942; *Chem. Ind.*, Vol. 54, 520–523, 1944; *Paper Trade J.*, 1–3, March 2, 1944.

185. C. E. Barnett, "Properties of Luminescent Materials," *Chem. Eng. News*, Vol. 20, 1006–1009, 1942; *Mining and Met.*, Vol. 29, 603–605, 1948; *Bull. Am. Optical Soc.*, 3, Mar. 10, 1949.

186. R. P. Teele, "Photometer for Luminescent Materials," *J. Optical Soc. Am.*, Vol. 35, 373–378, 1945.

187. G. R. Fonda, "The Magnesium–Tungstate Phosphor," *J. Phys. Chem.*, Vol. 48, 303–307, 1944; T. E. Foulke, U. S. Pats. 2,203,682, 6-11-40, and 2,232,-780, 2-25-41; W. A. Roberts, *Calcium Tungstate Phosphors*, U. S. Pat. 2,312,267, 2-23-43; H. R. Schoenfeldt, *Calcium Tungstate Phosphors*, U. S. Pat. 2,363,090, 11-31-44.

188a. C. G. A. Hill, "The Fluorescence of Pure Magnesium Tungstate," *Trans. Faraday Soc.*, Vol. 42, 685–689, 1946 (*cf. ibid.*, Vol. 42, 705–708, 1946).

188b. N. V. Philips' Gloeilampenfabrieken, *Mg and Zn Tungstate Phosphors with 0.05 to 0.5 percent Cd, Pb, Bi, Th, or Ce*, Dutch Pat. 55,143, 9-15-43; *Alkaline-earth Tungstate Phosphors with WO_3 Deficiency*, Dutch Pat., 66,610, 9-15-48.

189. H. G. Jenkins and J. W. Ryde, *Tungstate Phosphors with U(0.1) or Bi(2) Activator*, U. S. Pat. 2,295,040, 9-8-42.

190. R. C. Hultgren and S. Isenberg, *Magnesium Tungstate Phosphor*, U. S. Pat. 2,324,843, 7-20-44.

191. H. W. Leverenz, *Zinc–Beryllium–Silicate:Mn and/or Zinc–Beryllium–Germanate:Mn Phosphors*, U. S. Pats. 2,118,091, 5-24-38, and 2,274,272, 2-24-42.

192. H. W. Leverenz, *Zinc–Germanate:Mn and Magnesium–Germanate:Mn Phosphors*, U. S. Pat. 2,066,044, 12-29-36; F. E. Williams, U. S. Pats. 2,447,448 and 2,447,449, 8-17-48.

193. H. Gobrecht, "Über die Absorptions- und Fluoreszenzspektren der Ionen der Seltenen Erden in festen Körpern, insbesondere in Ultrarot," *Ann. Physik*, Vol. 28, No. 8, 673–700, 1937; *ibid.*, Vol. 31, 181–186, 600–608, 755–760, 1938.

194. O. Deutschbein, "Über die Spektren des Chroms in Kristallen," *Ann. Physik*, Vol. 14, No. 6/7, 712–754, Aug.–Sept. 1932; *ibid.*, Vol. 20, 828–842, Sept. 1934; *Physik. Z.*, Vol. 33, 874–878, 1932; Z. Physik, Vol. 77, No. 7/8, 489–

504, Aug. 1932; *Naturwissenschaften*, Vol. 30, 228, 1942 (*cf.* Brit. Pat. 548,997, 11-2-42).

195. B. V. THOSAR, "The Fluorescent Chromium Ion in Ruby," *Phil. Mag.*, Ser. 7, Vol. 26, 380–389, 878–887, Sept. and Nov. 1938; "Crystal Luminescence–Paramagnetic Ions as Centers," *J. Chem. Phys.*, Vol. 12, No. 10, 424, Oct. 1944.

196. R. S. KRISHNAN, "Raman Spectrum of Alumina and the Luminescence and Absorption Spectra of Ruby," *Nature*, Vol. 160, No. 4053, 26, July 5, 1947; *Proc. Indian Acad. Sci.*, Vol. 26A, 450–468, 1947.

197. H. G. JENKINS, A. H. McKEAG, and P. W. RANBY, *Alkaline-Earth Pyrophosphate with 0.1 percent-Dy*, U. S. Pat. 2,427,728, 9-23-47.

198. W. A. ROBERTS, *Calcium–Phosphate:Ce*, U. S. Pat. 2,306,567, 12-29-42.

199. F. A. KRÖGER and J. BAKKER, "Luminescence of Cerium Compounds," *Physica*, Vol. 8, 628–646, July 1941; Dutch Pat. 61,327, 7-15-48; U. S. Pat. 2,450,548, 10-5-48; 2,476,681, 7-19-49 (*cf.* U. S. Pat. 2,254,956, 9-2-41).

200. H. C. FROELICH, "Spectral Distribution Curve of Calcium–Phosphate:Ce Phosphor," *Trans. Electrochem. Soc.*, Vol. 87, 395–396, 1945; "New Ultraviolet Phosphors," *Trans. Electrochem. Soc.* Vol. 91, 241–263, 1947.

201. H. C. FROELICH, *Strontium–Aluminate:Mn*, U. S. Pat. 2,392,814, 1-15-46.

202. W. A. WEYL, *Zinc–Vanadate Phosphor*, U. S. Pat. 2,322,265, 6-22-43.

203. A. H. McKEAG and P. W. RANBY, *Alkaline-Earth-Silicate:Eu Phosphors*, Brit. Pat. 544,160, 3-31-42; U. S. Pat. 2,297,108, 9-29-42.

204. H. W. LEVERENZ, *Decomposition Method of Preparing Silicate-Type Phosphors*, U. S. Pat. 2,210,087, 8-6-40.

205. J. EINIG, "Eine neue Methode zur Darstellung von leuchtenden Zinksulfid," *Chem. Zt.*, No. 3, 31, 1931; *ibid.*, No. 19, 185–186, 1932.

206. E. STRECK, "Über eine Präparationsmethode für Phosphore," *Z. physik. Chem.*, A186, 19–26, 1940.

207. L. WESCH, *Herstellung von Leuchtstoffen*, PB9010, PB14529, Office Pub. Board, Washington, D. C., Mar. 4, 1942; Ger. Pat. 682,660, 9-28-49.

208. BRIT. COMBINED INTELLIGENCE OBJECTIVES SUBCOMMITTEE (CIOS), *German Fluorescent Lamp Industry and Phosphor Chemical Manufacture*, PB34017, PB6674, Office Pub. Board, Washington, D. C., 1945.

209. W. HARTMANN, *Chemical Production of the Siemens-Reiniger Works*, PB2143, Office Pub. Board, Washington, D. C., May 15, 1945.

210a. AUERGESELLSCHAFT A. G., *Manufacturing Instructions for Luminous Paints, etc.*, PB73285, Microfilm Reel B257, Frames 45–94, 1940–1946; PBL74079, Frames 1960–1979, 1941, Office Pub. Board, Washington, D. C.; C. H. LOVE, *Important Inorganic Pigments*, PB40350, Jan. 1947; also PBL87366 and PBL74289, Frames 1763–1768.

210b. E. R. OWEN, *Production of Luminous Compounds at the Works of Auergesellschaft A. G.*, Brit. Intelligence Objectives Rept., PB44680, Office Pub. Board, Washington, D. C., 1945.

210c. W. HANLE and A. SCHMILLEN, "Phosphorescence and Luminescence of Solids," *FIAT Rev. German Sci.* 1939–1946, *Physics of Solids*, Part 2, 185–209, 1948; *cf.* "Kaltes Licht durch Leuchtstoffe," *Chem. Ber.* Prüf. Nr. 15 (PB52016, Office Pub. Board, Washington, D. C.), 478–688, 1942.

211. SCHAD, TESKE, *et al.*, *Production of Luminescent Materials by the I. G. Farbenindustrie A. G.*, PBL70358, Frames 1951–1975, 1999–2007, 2130–2136, 2203–2209, 2510–2520, 2524–2530, Office Pub. Board, Washington, D. C.

212. A. Schleede, R. Tomaschek, L. Wesch, et al., "Luminescent Materials," Reichsamt Wirtschaftsausbau, Chem. Ber., Prüf-Nr. 15, 487–688, 1942; PB52016, PBL86685, Office Pub. Board, Washington, D. C.

213a. J. R. Spraul, "Manufacture of Phosphors," The Frontier, Vol. 9, No. 3, 12–15, 1946.

213b. P. N. Campbell, "Silicate Phosphors for Cathode-Ray Tubes," J. Soc. Chem. Ind., Vol. 66, No. 6, 191–194, June 1947.

214. H. W. Leverenz, Three Fluxes Used in Preparation of ZnS-Type Phosphors, U. S. Pat. 2,402,759, 6-25-46.

215a. E. D. Reed, German Research and Development Work on CRT, Television Tubes, and Television Systems, PB52345, 1946, Office Pub. Board, Washington, D. C.

215b. E. S. Henning, "Special Developments in German CRT," Air Tech. Intelligence, Tech. Data Digest, Vol. 13, No. 3, 7–17, Feb. 1, 1948; also CIOS Rept. Item No. 1 and 9, File No. 32–95, "Telefunken CRT Laboratories and Leuchtstoffe."

216. V. E. Lashkarev and K. M. Kossonogova, "Infrared Luminescence of Cuprous Oxide," Compt. rend. acad. sci. URSS, Vol. 54, 125–126, 1946.

217. J. M. Dunoyer, "Sur la réduction par l'hydrogène gazeux des oxydes de plomb et des oxydes d'argent," Compt. rend., Vol. 214, 556–557, Mar. 16, 1942; Z. Szabo et al., "Zur Reaktion zwischen Kohlenoxyd und Silberoxyd," Z. anorg. Chem., Vol. 252, 201–204, 1944; B. N. Sen, "Some Physicochemical Evidence of the Bivalency of Silver," Current Sci. (India), Vol. 17, 182–183, 1948.

218. F. Ephraim et al., Inorganic Chemistry, 5th Ed., Interscience Pub., New York, 1948.

Chapter 4

219. H. W. Leverenz, Research on High Intensity Phosphor Light Source, PB33347, Office Pub. Board, Washington, D. C.; U. S. Pat. 2,298,947, 10-13-42 (cf. C. G. Suits, G. R. Harrison, and L. Jordan, Science in World War II. Applied Physics, Little, Brown and Co., Boston, 1948).

220. E. Ingerson, O. F. Tuttle, et al., Studies and Investigations in Connection with the Development of a New Light Source Consisting of a Phosphor Rod Irradiated by High Voltage Electrons, PB33342, Office Pub. Board, Washington, D. C., 1945; "Artificial Willemite Needles," Am. J. Sci., Vol. 245, No. 5, 313–319, May 1947; "The Systems K_2O–ZnO–SiO_2, ZnO–B_2O_3–SiO_2, and Zn_2SiO_4–Zn_2GeO_4," Am. J. Sci., Vol. 246, No. 1, 31–40, Jan. 1948.

221. G. R. Fonda and F. Seitz, Editors, Preparation and Characteristics of Luminescent Materials, John Wiley & Sons, New York, 1948.

222. K. Birus, "Kristallphosphore," Ergeb. exakt. Naturw., Vol. 20, 183–268, 1942.

223. G. R. Fonda, "Characteristics of Silicate Phosphors," J. Phys. Chem., Vol. 43, No. 5, 561–577, May 1939; "The Preparation of Fluorescent Calcite," ibid., Vol. 44, No. 4, 435–439, Apr. 1940; "The Yellow and Red Zinc Silicate Phosphors," ibid., Vol. 44, No. 7, 851–861, Oct. 1940; "The Constitution of Zinc Beryllium Silicate Phosphors," ibid., Vol. 45, No. 2, 282–288, Feb. 1941; "Double Activation of Zinc Fluoride Phosphors," J. Optical Soc. Am., Vol. 38, No. 12, 1007–1014, Dec. 1948; "The Constitution of Zinc-Beryllium-Silicate Phosphors," J. Electrochem. Soc., Vol. 95, 304–315, June 1949.

224a. H. G. JENKINS, A. H. McKEAG, and H. P. ROOKSBY, "Position Occupied by the Activator in Impurity-Activated Phosphors," *Nature*, Vol. 143, No. 3632, 978, June 10, 1939.

224b. H. P. ROOKSBY, "Some Applications of X-Ray Powder Method in Industrial Laboratory Problems," *J. Sci. Instruments*, Vol. 18, 84–90, 1941; *J. Roy. Soc. Arts*. Vol. 90, 673ff, 1942; *J. Electrochem. Soc.*, Vol. 96, 1949.

225. H. C. FROELICH, "The Low-Temperature Yellow Zinc Silicate Phosphor," *J. Phys. Chem.*, Vol. 47, No. 9, 669–677, Dec. 1943 (*cf.* U. S. Pats. 2,238,026, 4-8-41 and 2,247,192, 6-24-41).

226. K. J. MURATA and R. L. SMITH, "Manganese and Lead as Coactivators of Red Fluorescence in Halite," *Am. Mineral.*, Vol. 31, 527–538, 1946.

227. J. H. SCHULMAN, L. W. EVANS, R. J. GINTHER, and K. J. MURATA, "Sensitized Luminescence of Manganese-Activated Calcite," *J. Applied Phys.*, Vol. 18, No. 8, 732–740, Aug. 1947.

228. J. H. SCHULMAN, E. BURSTEIN, L. W. EVANS, and R. J. GINTHER, "The Sensitized Luminescence of NaCl:Mn," *Bull. Am. Phys. Soc.*, Vol. 24, No. 4, 25, April 29, 1949.

229. J. B. MERRILL and J. H. SCHULMAN, "The CaSiO₃:(Pb + Mn) Phosphor," *J. Optical Soc. Am.*, Vol. 38, No. 5, 471–479, May 1948; *ibid.*, Vol. 38, No. 9, 817–818, Sept. 1948 (*cf.* Brit. Pat. 577,694, 5-28-46).

229a. TH. P. J. BOTDEN and F. A. KRÖGER, "Energy Transfer in Sensitized Ca₃-(PO₄)₂:Ce:Mn and CaSiO₃:Pb:Mn," *Physica*, Vol. 14, No. 8, 553–566, Dec. 1948.

229b. F. J. STUDER and G. R. FONDA, "Optical Properties and Constitution of Calcium–Silicate:Pb Phosphors," *Bull. Am. Phys. Soc.*, Vol. 23, 48, Apr. 29, 1948; G. R. FONDA and H. C. FROELICH, "The Structure of Calcium Silicate Phosphors," *J. Electrochem. Soc.*, Vol. 93, No. 4, 114–122, Apr. 1948; H. C. FROELICH, "Manganese Activated Calcium Silicate Phosphors," *J. Electrochem. Soc.*, Vol. 93, No. 4, 101–113, Apr. 1948; H. C. FROELICH, *Calcium–Cadmium Molybdate Phosphor with Pb and Sm Activators*, U. S. Pat. 2,434,-764, 1-20-48 (*cf.* French Pat. 856,607, 7-29-40); *UV-emitting Al-Ca-Mg-phosphate:Ce:Th Phosphors*, U. S. Pats. 2,455,413–2,455,415, 12-7-48; "Double and Triple Activated Magnesium Pyrophosphate Phosphors," *J. Electrochem. Soc.*, Vol. 95, No. 5, 254–266, May 1949; J. T. G. OVERBEEK, *Lanthanum-silicate:Ce Phosphor*, U. S. Pat. 2,467,689, 4-19-49; R. NAGY, R. G. WOLLENTIN, and C. K. LUI, "Ultraviolet-emitting Phosphor," *J. Electrochem. Soc.*, Vol. 96, 1949.

230. N. RIEHL and H. ORTMANN, "Über Einbau und Diffusion von Fremdatomen im Zinksulfidgitter," *Z. physik. Chem.*, Vol. 188, 109–126, Mar. 1941.

231. G. R. FONDA, "Factors Affecting Phosphorescence Decay of the Zinc Sulfide Phosphors," *Trans. Electrochem. Soc.*, Vol. 87, 339–355, 1945.

232. M. CURIE, "L'excitation de la luminescence des sulfures par transferts d'activation moléculaires," *Cahiers phys.*, Vol. 4, No. 21, 12–18, 1944.

233. H. A. KLASENS *et al.*, "Transfer of Energy between Centres in Zinc Sulphide Phosphors," *Nature*, Vol. 158, 306–307, Aug. 31, 1946; "The Relation between Efficiency and Exciting Intensity for Zinc–Sulphide Phosphors," *J. Optical Soc. Am.*, Vol. 38, No. 1, 60–65, Jan. 1948; "Fluorescence Efficiency and Hole Migration in Zinc Sulphides," *J. Optical Soc. Am.*, Vol. 38, No. 3, 226–231, Mar. 1948; "Erratum," *ibid.*, Vol. 38, No. 7, 649, July 1948.

234. H. A. KLASENS and M. E. WISE, "Decay of Zinc–Sulphide-Type Phosphors," *Nature*, Vol. 158, 483–484, Oct. 5, 1946.

235. M. P. LORD and A. L. G. REES, "Note on the Behaviour of Zinc Sulphide Phosphors under Conditions of Periodic Excitation," *Proc. Phys. Soc.*, Vol. 58, 280–291, 1946.

236. M. P. LORD, A. L. G. REES, and M. E. WISE, "The Short-Period Time Variation of the Luminescence of a Zinc Sulphide Phosphor under Ultraviolet Excitation," *Proc. Phys. Soc.*, Vol. 59, 473–502, 1947.

237. G. F. J. GARLICK, *Luminescent Materials*, Oxford Univ. Press, 1949; "Cathodoluminescence," chapter in *Advances in Electronics*, Academic Press, New York, 1949.

238. G. F. J. GARLICK and A. F. GIBSON, "Luminescence Processes in Zinc Sulphide Phosphors," *Nature*, Vol. 158, 704–705, Nov. 16, 1946; "Electron Traps and Dielectric Changes in Phosphorescent Solids," *Proc. Roy. Soc. (London) A*, Vol. 188, No. 1015, 485–509, Feb. 25, 1947; "Luminescence Efficiency Changes in Zinc Sulphide Phosphors below Room Temperature," *Nature*, Vol. 161, No. 4088, 359, Mar. 6, 1948; "The Electron Trap Mechanism of Luminescence in Sulphide and Silicate Phosphors," *Proc. Phys. Soc.*, Vol. 60, No. 342, 574–590, June 1, 1948; *ibid.*, Vol. 62, No. 353A, 317–319, May 1, 1949.

238a. M. H. F. WILKINS and G. F. J. GARLICK, "Relation between Photoconduction and Luminescence in Zinc Sulphide," *Nature*, Vol. 161, No. 4093, 565–566, Apr. 10, 1948.

238b. R. C. HERMAN, C. F. MEYER, and H. S. HOPFIELD, "Theory of Electron Retrapping in Infrared Phosphors," *Bull. Am. Phys. Soc.*, Vol. 23, No. 3, 48, Apr. 29, 1948; *J. Optical Soc. Am.*, Vol. 38, No. 12, 999–1006, Dec., 1948; W. L. PARKER and J. J. DROPKIN, "Stimulated Phosphorescence and Photoconductivity of Infrared Phosphors," *Bull. Am. Phys. Soc.*, Vol. 23, No. 3, 48–49, Apr. 29, 1948.

238c. H. KAHLER, "The Photoelectrical Properties of the Phosphors," *Phys. Rev.*, Vol. 21, 214–215, 1923.

239. V. V. ANTONOV-ROMANOVSKII, "Recombination Phosphorescence," *Bull. acad. sci. URSS, Sér. phys.*, Vol. 10, 504–508, 1946 (*cf. ibid.*, pp. 465–475).

240. A. BECKER and I. SCHAPER, "Über die Phosphoreszenzzerstörung," *Ann. Phys.*, Vol. 42, 297–336, 487–500, 1943; F. FREY, *ibid.*, [6], Vol. 2, No. 3/4, 147–157, 1948.

241. A. BECKER and F. BECKER, "Über die Phosphoreszenztilgung," *Ann. Physik*, Vol. 43, 598–607, 1943; *Reichsber. Physik*, Vol. 1, 169–179, 1945; *Z. Naturforsch.*, Vol. 2a, 100–108, 1947; *Z. Physik*, Vol. 125, 475–486, 694–706, 1949.

242. F. WECKER, "Roentgenographische Strukturuntersuchungen an Zinksulfid–Kupfer–Phosphoren nach der Pulvermethode," *Ann. Physik*, Vol. 43, 561–573, 1943; "Lumineszenzrückgang und Teilchengrösse beim druckzerstörten ZnS-Phosphore," *ibid.*, Vol. 43, 607–616, 1943.

243. J. H. SCHULMAN, "Luminescence of $(Zn:Be)_2SiO_4:Mn$ and Other Manganese-Activated Phosphors," *J. Applied Phys.*, Vol. 17, No. 11, 902–908, Nov. 1946; *J. Electrochem. Soc.*, Vol. 96, 57–74, Aug. 1949.

244. D. H. KABAKJIAN, "Dependence of Luminescence on Physical Structure in Zinc Borate Compounds," *Phys. Rev.*, Vol. 51, 365–368, Mar. 1, 1937; "Luminescence of Pure Radium and Barium Compounds," *ibid.*, Vol. 57, 700–705, Apr. 15, 1940.

245. H. FRIEDRICH, *Luminescence Studies on Crystalline and Vitreous Compounds Derived from the System Zinc Oxide–Boric Acid*, PBL74110 (432–460), Office Pub. Board, Washington, D. C., 1938.

246. H. S. TASKER, *Alkaline-Earth Sulphates Activated with PbSO₄*, Brit. Pats. 540,252; 557,841, 12-8-43; 571,324, 8-20-45; 574,494, 1-8-46; and U. S. Pat. 2,336,815, 12-14-43 (*cf. Chem. Eng. News*, Vol. 26, No. 24, 1794–1797, June 14, 1948).

247. K. H. BUTLER, "Barium Silicate Phosphors," *Electrochem. Soc.* Vol. 91, 265–278, 1947; "Fluorescence of Silicate Phosphors," *J. Optical Soc. Am.*, Vol. 37, No. 7, 566–571, July 1947; "Spectra of Silicate Phosphors," *J. Electrochem. Soc.*, Vol. 93, No. 5, 143–156, May 1948; *ibid.*, Vol. 95, No. 5, 267–281, May 1949; U. S. Pat. 2,467,810, 4-19-49.

248. R. H. CLAPP and R. J. GINTHER, "Ultraviolet Phosphors and Fluorescent Sun Tan Lamps," *J. Optical Soc. Am.*, Vol. 37, No. 5, 355–362, May 1947; W. A. ROBERTS, *Calcium-Orthophosphate:Tl Phosphor*, U. S. Pat. 2,447,210, 8-17-48 (*cf.* Brit. Pat. 577,693, 3-28-46, and Ref. 229*b*).

249. M. GUILLOT and J. FERRAND, "Sur la fluorescence du chlorure de plomb cristallisé," *Compt. rend.*, Vol. 218, 874–876, 1944.

250. L. A. LEVY and D. W. WEST, *Alkali- or Alkaline-Earth Silicate Phosphor with Lead Activator*, Brit. Pat. 569,273, 5-16-45.

251. W. SCHOTTKY, *Zusammenfassende Bearbeitung der bisherigen experimentellen Ergebnisse zu einer Theorie der Phosphore*, PB56266, PBL18221, PBL61875, Office Pub. Board, Washington, D. C., 1944.

252. A. TOPOREC, "Luminescence Spectra of Phosphors Activated by Silver," *Mem. Phys.* (*Kiev*), Vol. 8, 161–166, 1940.

253*a*. H. SCHLEGEL, "Eine Bemerkung über die Wirkungsweise der Schmelzmittel bei ZnS-, CdS-Leuchtstoffen," *Naturwissenschaften*, Vol. 30, 242, 1942.

253*b*. A. L. SMITH, "Influence of Fluxes on the Cathodoluminescence of Zinc Sulfide Phosphors," *J. Electrochem. Soc.*, Vol. 96, 75–84, Aug. 1949.

253*c*. A. L. SMITH, "Influence of Atmospheres on the Cathodoluminescence of Zinc-Sulphide:Ag-Type Phosphors," *J. Electrochem. Soc.*, Vol. 93, No. 6, 324–333, June 1948.

254. K. KAMM, "Die Zink-Cadmium-Sulfid Phosphoren," *Ann. Physik*, Vol. 30, 332–353, 1937.

255. S. ROTHSCHILD, "The Luminescent Spectra of Zinc Sulphide and Zinc Cadmium Sulphide Phosphors," *Trans. Faraday Soc.*, Vol. 42, No. 290, 635–642, Sept.–Oct. 1946 (*cf.* Brit. Pat. 560,484, 4-5-44).

256. C. A. PIERCE, "Studies in Thermoluminescence," *Am. Historical Rev.*, Vol. 13, No. 4, 312–330, 1908; *Phys. Rev.*, Vol. 26, No. 6, 454–469, 1908.

257. M. DÉRIBÉRÉ, "La thermoluminescence et ses applications," *Rev. sci.*, Vol. 76, 382–384, 1938.

258. R. C. HERMAN and C. F. MEYER, "Thermoluminescence and Conductivity of Phosphors," *J. Applied Phys.*, Vol. 17, No. 9, 743–749, Sept. 1946; *ibid.*, Vol. 18, No. 2, 258–260, Feb. 1947.

259. M. PIRANI and A. RÜTTENAUER, "Lichterzeugung durch Strahlungsumwandlung," *Licht*, Vol. 5, 93–98, May 1935; *Naturwissenschaften*, Vol. 23, No. 25, 393–405, June 21, 1935.

260. R. P. JOHNSON, "Solid Fluorescent Materials," *Am. J. Phys.*, Vol. 8, No. 3, 143–153, June 1940.

261*a*. F. A. VICK, "Recent Advances in Science: Physics," *Science Progress*, Vol. 34, 562–569, 1946.

261*b*. F. D. KLEMENT, "Some Basic Phenomena of Photoluminescence in Solid Bodies," *Vestnik Leningrad. Univ.*, Vol. 1, No. 4/5, 38–52, 1946.

262. F. E. WILLIAMS, "Some New Aspects of Germanate and Fluoride Phosphors," *J. Optical Soc. Am.*, Vol. 37, No. 4, 302–307, Apr. 1947; *Bull. Am. Phys. Soc.*, Vol. 23, No. 3, 48, Apr. 29, 1948; U. S. Pat. 2,447,447, 8-17-48; *Bull. Optical Soc. Am.*, 2–5, Mar. 10, 1949; *J. Chem. Phys.*, Vol. 17, No. 6, 583–584, June 1949; *cf. Phys. Rev.*, Vol. 76, No. 1, 144–145, July 1, 1949.

263. F. E. WILLIAMS and H. EYRING, "The Mechanism of the Luminescence of Solids," *J. Chem. Phys.*, Vol. 15, No. 5, 289–305, May 1947; F. E. WILLIAMS, "Theory of the Luminescence of Ionic Phosphors," *Bull. Am. Phys. Soc.*, Vol. 23, No. 2, 27, Jan. 29, 1948.

264. E. IWASE, "Studies on the Luminescence of Inorganic Solids," *Sci. Papers, Inst. Phys. Chem. Research (Tokyo)*, Vol. 34, 173–181, 487–503, 761–774, 1938; *ibid.*, Vol. 35, 426–436, 1939; *ibid.*, Vol. 38, 67–80, 1940.

265. S. MAKASHIMA, "Photochemische Untersuchungen über Metallsulfide, Kristallstruktur und Lumineszenzfähigkeit der Sulfid–Phosphore," *J. Soc. Chem. Ind. Japan*, Vol. 41, No. 6, 203a–205b, 1938; *ibid.*, Vol. 46, 1045–1053, 1943; *ibid.*, Vol. 47, 757–762, 1944.

266. T. MUTA, "On the Quantum Theory of the Phosphorescence of Crystal Phosphors," *Inst. Phys. Chem. Research (Tokyo)*, Vol. 28–29, No. 613, 171–206, 1935; *ibid.*, Vol. 32, 5–25, 1937.

267. S. IIMORI, "On the Constitution of Phosphorescence Centres in Fluorite," *Inst. Phys. Chem. Research (Tokyo)*, Vol. 20, 189–200, 1933; *ibid.*, 79–110, 1936; B. MUKHERJEE, "Cathodoluminescence Spectra of Indian Fluorites," *Indian J. Phys.*, Vol. 22, 221–228, 1948.

268. Y. UEHARA, "Studies on Luminescent Materials," *Bull. Chem. Soc. Japan*, Vol. 14, 539–546, 1939; Vol. 15, 214–223, 1940; *ibid.*, Vol. 16, 99–105, 1941; *J. Chem. Soc. Japan*, Vol. 60, 133–138, 1939.

269. F. BANDOW, "Über die Phosphoreszenzzentren," *Ann. Physik*, Vol. 41, No. 2, 172–176, 1942; *ibid.*, Folge 6, Vol. 1, No. 7/8, 399–404, 1947; *Z. Naturforsch.*, Vol. 3a, 16–20, 1948.

270a. A. SHECHTER *et al.*, "An Electron-Microscopic Investigation of Freshly Formed Precipitates from Solutions," *Acta Physicochim. URSS*, Vol. 21, 849–852, 1946.

270b. W. O. MILLIGAN, "The Color and Crystal Structure of Precipitated CdS," *J. Phys. Chem.*, Vol. 36, No. 6, 797–800, June 1934

271. J. A. HEDVALL *et al.*, "Fotoaktivierung av kristaller och des inverkan på adsorptionsprocesser," *Svensk Kem. Tid.*, Vol. 47, 156–161, 1935; *ibid.*, Vol. 51, 163–173, 1939; *Arkiv Kemi, Mineral. Geol.*, Vol. 17A, No. 11, 11 pp, 1943.

272. J. ROCKSTROH, *Verfahren zur Herstellung polymorpher Metallverbindungen*, Ger. Pat. No. 633,973, 7-23-36; W. O. MILLIGAN and L. MERTEN, "X-Ray Diffraction Studies in the System $Fe_2O_3–Cr_2O_3$," *J. Phys. & Colloid Chem.*, Vol. 51, No. 7, 521–528, Mar. 1947.

273. W. KOCH, "Über Absorption und Dispersion von Alkalihalogenidphosphoren von bekanntem Fremddionengehalt," *Z. Physik*, Vol. 59, No. 5/7, 378–386, 1930.

274a. W. BORCHERT, "Gitterumwandlungen im System $Cu_{2-x}Se$," *Z. Krist.*, Vol. 106, 5–24, 1945.

274b. G. AMINOFF and B. BROOMÉ, "Strukturtheoretische Studien über Zwillinge," *Z. Krist.*, Vol. 80, No. 5/6, 355–376, 1931.

275. F. D. KLEMENT, "Effect of Oxygen on Diffusion Processes during the Formation of Alkali–Halide Phosphors," *J. Phys. Chem. (USSR)*, Vol. 21, 563–568, 1947.

276. K. PRZIBRAM, "Über die Rekristallisation des Steinsalzes," *Z. Physik*, Vol. 67, No. 1/2, 89–106, 1931; "On the Fluorescence of Fluorite and the Divalent Rare Earths," *Compt. rend. acad. sci. URSS*, Vol. 56, No. 1, 31–33, 1947.

277. M. V. CLASSEN-NEKLUDOVA, "Spinel-Type Mechanical Twinnings of Rock Salt Crystals," *J. Phys. (USSR)*, Vol. 7, No. 6, 272–278, 1943.

278. J. WHITNEY and N. DAVIDSON, "Spectrophotometric Investigation of the Interaction between Ions of Different Oxidation States of an Element," *J. Am. Chem. Soc.*, Vol. 69, No. 8, 2076–2077, Aug. 1947.

279. L. SIBAIYA and H. S. VENKATARAMIAH, "Magnetism of Phosphors," *Current Sci. (India)*, Vol. 9, No. 5, 224–225, May 1940.

280. A. BERTON, "Absorption ultraviolette de corps solides polymorphes," *Compt. rend.*, Vol. 208, 1898–1900, 1939.

281. J. N. FERGUSON, JR., "The Photoconductivity of NaCl in the Far Ultraviolet," *Phys. Rev.*, Vol. 66, 220–223, 1944; also, E. FISHER, "Energy Levels of Sodium Chloride," *Phys. Rev.*, Vol. 73, No. 1, 36–42, Jan. 1, 1948.

282. E. MOLLWO, "Das ultraviolette Absorptionsspektrum der Sulfide und Oxyde von Zink und Kadmium," *Reichsber. Physik*, Vol. 1, No. 1, 1–4, Apr. 1944.

283. S. RAMESHAN, "The Faraday Effect in Diamond," *Proc. Indian Acad. Sci.*, Vol. 24a, 104–113, 1946; "The Faraday Effect in Some Cubic Crystals," *Proc. Indian Acad. Sci.*, Vol. 25a, 459–466, 1947.

284. H. B. FRIEDMAN and K. E. SHULER, "Evidence for the Ionic Nature of Certain Crystal Lattices," *J. Chem. Education*, Vol. 24, 11–18, Jan. 1947.

285. E. J. W. VERWEY et al., "Physical Properties and Cation Arrangements of Oxides with Spinel Structures," *J. Chem. Phys.*, Vol. 15, No. 4, 174–187, Apr. 1947; *ibid.*, Vol. 16, No. 12, 1091–1092, Dec. 1948; *Philips Tech. Rev.*, Vol. 9, No. 6, 185–190, 1947; *ibid.*, Vol. 9, No. 8, 239–248, 1947.

286. E. T. GOODWIN, "Electronic States at the Surfaces of Crystals," *Proc. Cambridge Phil. Soc.*, Vol. 35, 205–241, 474–484, 1939; W. SHOCKLEY, "On the Surface States Associated with a Periodic Potential," *Phys. Rev.*, Vol. 56, No. 4, 317–323, Aug. 15, 1939.

287. J. BARDEEN, "Surface States and Rectification at a Metal Semi-Conductor Contact," *Phys. Rev.*, Vol. 71, No. 10, 717–727, May 15, 1947 (*cf. ibid.*, Vol. 72, No. 4, 345, Aug. 15, 1947; *ibid.*, Vol. 74, No. 2, 230–233, July 15, 1948; *ibid.*, Vol. 75, No. 4, 689–691, Feb. 15, 1949; also, *Elec. Eng.*, Vol. 68, No. 3, 215–223, Mar. 1949, and *Phys. Rev.*, Vol. 75, No. 8, 1181–1182, 1208–1225, Apr. 15, 1949).

288. N. BLOEMBERGEN, "Note on the Internal Secondary Emission and the Influence of Surface States," *Physica*, Vol. 11, No. 4, 343–344, Dec. 1945.

289. B. GUDDEN and R. W. POHL, "Das Quantenäquivalent bei der lichtelektrischen Leitung," *Z. Physik*, Vol. 17, 331–346, 1923.

290. P. J. VAN HEERDEN, *The Crystalcounter*, Dissertation, Utrecht, 1945.

291. R. FRERICHS and R. WARMINSKY, "Die Messung von Beta- und Gamma-Strahlen durch inneren Photoeffekt in Kristallphosphoren," *Naturwissenschaften*, Vol. 33, 251, Oct. 30, 1946; R. FRERICHS, "Photoconductivity of 'Incomplete Phosphors'," *Phys. Rev.*, Vol. 72, No. 7, 594–602, Oct. 1, 1947; *ibid.*, Vol. 74, No. 12, 1875, Dec. 15, 1948; *Research (London)*, Vol. 1, 208–211, Feb. 1948; PB73630, 1-42 (ZWB-FB-1874), PB56304, and PBL75946,

Office Pub. Board, Washington, D. C.; C. J. Humphreys, "Spectral Response of a Cadmium Sulphide Crystal," *J. Optical Soc. Am.*, Paper 34, Mar. meeting, 1948; J. Fassbender, "Zu den optischen und elektrischen Eigenschaften von Kadmiumsulfid," *Naturwissenschaften*, Vol. 34, No. 7, 212–213, 1947; K. Weiss, "The Absorption, Sensitivity to Light, and Electrical Conductivity of Cadmium Sulphide Layers," *Z. Naturforsch.*, Vol. 2a, 650–652, 1947; J. C. M. Brentano and D. H. Davis, "The Change of Photoconductivity of Cadmium Sulphide between Room Temperature and the Temperature of Liquid Air," *Phys. Rev.*, Vol. 74, No. 6, 711–712, Sept. 15, 1948; G. J. Goldsmith and K. Lark-Horovitz, "Cadmium Sulphide as a Crystal Counter," *Phys. Rev.*, Vol. 75, No. 3, 526–527, Feb. 1, 1949.

292. H. Kallmann, "Luminescent Naphthalene Crystals and Photomultipliers as Beta- and Gamma-Ray Counters," *Natur u. Tech.*, July 1947; M. Deutsch, "High Efficiency, High Speed Scintillation Counters for Beta- and Gamma-Rays," *Bull. Am. Phys. Soc.*, Vol. 23, No. 2, 7–8, Jan. 29, 1948; G. B. Collins and R. Hoyt, "Detection of Beta-Rays by Scintillations," *Bull. Am. Phys. Soc.*, Vol. 23, No. 2, 29–30, Jan. 29, 1948; "Decay Times of Scintillations," *Phys. Rev.*, Vol. 74, No. 10, 1543–1544, Nov. 15, 1948; H. N. Bose, "Luminescence of Some Organic Compounds under X-Ray Excitation," *Proc. Natl. Inst. Sci. (India)*, Vol. 11, No. 2, 152–157, 1945; P. R. Bell, "Use of Anthracene as a Scintillation Counter," *Phys. Rev.*, Vol. 73, No. 11, 1405–1406, June 1, 1948; S. C. Curran and W. R. Baker, "Photoelectric Alpha-Particle Detector," *Rev. Sci. Instruments*, Vol. 19, 116, Feb. 1948; H. Kallmann, "Neutron Radiography," *Research (London)*, Vol. 1, 254–260, Mar. 1948; "Scintillation and Crystal Counters," *ONR European Sci. Notes*, Vol. 2, No. 21, 299, Nov. 1, 1948; I. Broser and H. Kallmann, "Über den Elementarprozess der Lichtanregung in Leuchtstoffen durch α-Teilchen, schnelle Electronen, und γ-Quanten," *Z. Naturforsch.*, Vol. 2a, 439–440, 642–650, 1947; *ibid.*, Vol. 3a, 6–15, 1948; "Measurements of Alpha-particle Energies with the Crystal Fluorescence Counter," *Nature*, Vol. 163, No. 4131, 20–21, Jan. 1, 1949; *Ann. Phys.*, [6], Vol. 3, 317–321, 1948; L. F. Wouters, "Pulse Characteristics of Anthracene Scintillation Counters," *Phys. Rev.*, Vol. 74, 489–490, 1948; R. Hofstadter *et al.*, "Scintillation Counters," *Bull. Am. Phys. Soc.*, Vol. 24, No. 1, 16–17, Jan. 26, 1949; *cf. ibid;* Vol. 24, No. 2, 9–10, Feb. 3, 1949; H. T. Gittings *et al.*, "Relative Sensitivities of Some Organic Compounds for Scintillation Counters," *Phys. Rev.*, Vol. 75, No. 1, 205–206, Jan. 1, 1949; H. Kallmann, "Quantitative Measurements with Scintillation Counters," *Phys. Rev.*, Vol. 75, No. 4, 623–626, Feb. 15, 1949; E. J. Schillinger *et al.*, "Scintillation Counting with Chrysene," *Phys. Rev.*, Vol. 75, No. 5, 900–901, Mar. 1, 1949; L. Roth, "Fluorescence of Anthracene Excited by High Energy Radiation," *Phys. Rev.*, Vol. 75, No. 6, 983, Mar. 15, 1949; R. Hofstadter, "Crystal Counters," *Nucleonics*, Vol. 4, No. 4, 2–27, Apr. 1949; R. Hofstadter and J. A. McIntyre, "Simultaneity in the Compton Effect," *Bull. Am. Phys. Soc.*, Vol. 24, No. 4, 18, Apr. 28, 1949; G. T. Reynolds and F. B. Harrison, "Measurements of Meson Decay by Scintillation Counters," *Bull. Am. Phys. Soc.*, Vol. 24, No. 4, 15, Apr. 28, 1949; G. F. J. Garlick and R. A. Fatehally, "Measurement of Particle Energies with Scintillation Counters," *Phys. Rev.*, Vol. 75, No. 9, 1446, May 1, 1949; W. A. MacIntyre,

"Decay of Scintillations in Calcium Fluoride Crystals," *Phys. Rev.*, Vol. 75, No. 9, 1439–1440, May 1, 1949; J. PINE and R. SHERR, "Constant Sensitivity Gamma-ray Detector," *Bull. Am. Phys. Soc.*, Vol. 24, No. 5, 7, June 16, 1949; E. C. FARMER, H. B. MOORE, and C. GOODMAN, "Studies on Synthetic LiF as a Scintillation Detector of Heavy Particles and Quanta," *Bull. Am. Phys. Soc.*, Vol. 24, No. 5, 7, June 16, 1949; G. E. KOCH and J. D. GRAVES, "Temperature Dependence of a Scintillation Counter Using Several Organic and Inorganic Phosphors," *Bull. Am. Phys. Soc.*, Vol. 24, No. 6, 11, June 27, 1949; R. H. DAVIS and J. D. GRAVES, "The Energy Dependence of Several Organic and Inorganic Counters," *Bull. Am. Phys. Soc.*, Vol. 24, No. 6, 11, June 27, 1949; G. N. HARDING *et al.*, "Scintillation Counters Using Organic Compounds," *Nature*, Vol. 163, No. 4156, 990, June 25, 1949; H. KALLMANN *et al.*, "Luminescent Counters," *Research (London)*, Vol. 2, 62–68, 87–89, 1949; J. H. SCHULMAN *et al.*, "Effect of X-rays on the Absorption and Luminescence of Alkali-halide Phosphors," *Bull. Am. Phys. Soc.*, Vol. 24, No. 5, 13, 1949; W. S. KOSKI and C. O. THOMAS, "Scintillations Produced by Alpha Particles in a Series of Structurally Related Organic Crystals," *Phys. Rev.*, Vol. 76, No. 2, 308–309, July 15, 1949.

293. R. HOFSTADTER *et al.*, "Behaviour of Silver Chloride Crystal Counters," *Phys. Rev.*, Vol. 72, No. 10, 977–978, Nov. 15, 1947; "Thallium Halide Crystal Counter," *Phys. Rev.*, Vol. 72, No. 11, 1120–1121, Dec. 1, 1947; Vol. 75, No. 5, 796–810, Mar. 1, 1949 (cf. *ibid.*, Vol. 72, No. 11, 1127–1128, Dec. 1, 1947; Brit. Pat. No. 492,722, 2-10-38); "Alkali Halide Scintillation Counters," *Phys. Rev.*, Vol. 74, No. 1, 100–101, July 1, 1948; "Temperature Dependence of Scintillations in Sodium–Iodide:Tl Crystals," *Phys. Rev.*, Vol. 75, No. 1, 203–204, Jan. 1, 1949; S. I. GOLUB, "Luminescence of Silver Halides," *Doklady Akad. Nauk SSSR*, Vol. 60, 1153–1155, 1948.

294. H. LENZ, "Elektronenleitung in Kristallen," *Ann. Physik*, Vol. 77, No. 13, 449–476, 1925; *ibid.*, Vol. 82, No. 6, 775–793, 1927; *ibid.*, Vol. 83, No. 15, 941–956, 1927.

295. P. PRINGSHEIM, "Zwei Bemerkungen über den Unterschied von Lumineszenz- und Temperaturstrahlung," *Z. Physik*, Vol. 57, No. 11/12, 739–747, 1929.

296. S. VAVILOV, "Some Remarks on the Stokes Law," *J. Phys. (USSR)*, Vol. 9, No. 2, 68–72, 1945.

297. P. PRINGSHEIM, "Some Remarks concerning the Difference between Luminescence and Temperature Radiation. Anti-Stokes Fluorescence," *J. Phys. (USSR)*, Vol. 10, No. 6, 495–499, 1946.

298. S. VAVILOV, "Photoluminescence and Thermodynamics," *J. Phys. (USSR)*, Vol. 10, No. 6, 499–502, 1946.

299. L. LANDAU, "On the Thermodynamics of Photoluminescence," *J. Phys. (USSR)*, Vol. 10, No. 6, 503–506, 1946.

300. L. BRILLOUIN, *Wave Propagation in Periodic Structures*, McGraw-Hill Book Co., New York, 1946. (Dover reprint).

301. F. C. VON DER LAGE and H. A. BETHE, "Method for Obtaining Electronic Eigenfunctions and Eigenvalues in Solids with an Application to Sodium," *Phys. Rev.*, Vol. 71, No. 9, 612–622, May 1, 1947.

302. H. FROELICH, "On the Theory of Dielectric Breakdown in Solids," *Proc. Roy. Soc. (London) A*, Vol. 188, No. 1015, 521–541, Feb. 25, 1947.

303. J. FRENKEL, "On the Transformation of Light into Heat in Solids," *Phys. Rev.*, Vol. 37, Nos. 1 and 10, 17–45 and 1276–1295, 1931.

304. J. FRANCK and E. TELLER, "Migration and Photochemical Action of Excitation Energy in Crystals," *J. Chem. Phys.*, Vol. 6, 861–872, Dec. 1938; TH. FORSTER, "Energy Migration and Fluorescence," *Naturwissenschaften*, Vol. 33, 166–175, 220–221, 1946; E. J. BOWEN *et al.*, "Resonance Transfer of Electronic Energy in Organic Crystals," *Proc. Phys. Soc.*, Vol. 62, Part 1, No. 349A, 26–31, Jan. 1, 1949.

305. F. MÖGLICH and R. ROMPE, "Zur Statistik der Vielfachstösse," *Z. Physik*, Vol. 117, No. 1/2, 119–124, 1940; "Über Energieumwandlung im Festkörper," *ibid.*, Vol. 117, No. 11/12, 707–729, 1940; Über die Energieumwandlung in Kristallphosphoren," *Physik. Z.*, Vol. 41, No. 9/10, 236–242, May 10, 1940; "Über die Anregung von Kristallphosphoren durch Korpuskularstrahlen," *ibid.*, Vol. 41, No. 23/24, 552–555, Dec. 15, 1940; "Zur Theorie fester Isolatoren," *Naturwissenschaften*, Vol. 29, No. 8, 105–113, 129–134, Feb. 21, 1941; "Zur Hedvallschen Störstellenwanderung," *ibid.*, Vol. 31, No. 5/6, 69, Jan. 29, 1943.

306. K. BIRUS, F. MÖGLICH, and R. ROMPE, "Über einige die Kristallphosphore und Isolatoren betreffende Fragen," *Physik. Z.*, Vol. 44, No. 6, 122–129, Mar. 1943 (*cf.*, A. RADKOWSKY, "Temperature Dependence of Electron Energy Levels in Solids," *Phys. Rev.*, Vol. 73, No. 7, 749–761, Apr. 1, 1948).

307. M. SCHÖN, "Über die Temperaturabhängigkeit der Helligkeit von Kristallphosphoren bei monochromatischer Anregung," *Naturwissenschaften*, Vol. 31, No. 14/15, 169, Apr. 2, 1943; "Eine Bemerkung zur Abklingung der Phosphoreszenz von Kristallphosphoren," *ibid.*, Vol. 31, No. 16/18, 203–204, Apr. 16, 1943.

308. P. BRAUER, "Über die obere Temperaturgrenze des Leuchtens von Phosphoren," *Naturwissenschaften*, Vol. 32, No. 1/4, 32, Jan. 1944; PB19206, Office Pub. Board, Washington, D. C., 1944.

309. H. M. O'BRYAN and H. W. B. SKINNER, "The Soft X-ray Spectroscopy of Solids," *Proc. Roy. Soc. (London) A*, Vol. 176, No. 965, 229–262, Oct. 1940; *Phys. Rev.*, Vol. 57, No. 11, 995–997, June 1, 1940 (*cf.* D. COSTER and S. KIESTRA, "On the Empty Electron Bands of Lowest Energy of the Transition Metals Mn and Fe and their Oxides," *Physica*, Vol. 14, No. 2–3, 175–188, Apr. 1948).

Chapter 5

310. G. SZIGETI and E. NAGY, "Dielectric Losses and Fluorescence of Zinc Silicate," *Nature*, Vol. 160, No. 4071, 641–642, Nov. 8, 1947; "Luminescent Materials," *Elektrotechnika*, Vol. 39, 70–73, 81–86, 1947; *Müegyetemi Közlemények*, No. 1, 115–130, 1948; E. NAGY, "Optical and Electrical Properties of Zinc Silicate Phosphors," *J. Optical Soc. Am.*, Vol. 39, No. 1, 42–49, Jan. 1949; E. MAKAI, "High Valent Manganese as Activator of Luminescence," *J. Electrochem. Soc.*, Vol. 95, No. 3, 107–111, Mar. 1949.

311. F. URBACH, A. URBACH, and M. SCHWARTZ, "The Brightness of Apparent Fluorescence as a Function of the Exciting Intensities," *J. Optical Soc. Am.*, Vol. 37, No. 2, 122, 1947.

312. H. W. LEVERENZ, "Optimum Efficiency Conditions for White Luminescent Screens in Kinescopes," *J. Optical Soc. Am.*, Vol. 30, No. 7, 309–315, July 1940.

313. E. A. ENGLE and L. SCHULTZ, "The Crystalloluminescence of Barium Bromate," *J. Colo.–Wyo. Acad. Sci.*, Vol. 1, No. 3, 22, 1931.

314. V. D. KUZETZOV and V. N. KOTLER, "The Crystalloluminescence of NaCl," *J. Phys. Chem.* (*USSR*), Vol. 4, 871–872, 1933; *Physik. Z. Sowjetunion*, Vol. 5, 40–56, 1934.

315. R. T. DUFFORD, "Luminescence Associated with Electrolysis," *J. Optical Soc. Am.*, Vol. 18, 17–28, 1929; R. T. DUFFORD and R. R. SULLIVAN, "Galvanoluminescence," *J. Optical Soc. Am.*, Vol. 21, 513–523, 1931.

316. A. GUNTHERSCHULZE and H. BETZ, "Kaltes Temperaturleuchten," *Z. Physik*, Vol. 74, No. 9/10, 681–691, Mar. 3, 1932.

317. A. O. HAUGEN, "Lumaforer," *Tek. Ukeblad*, Vol. 88, No. 4, 35–39, Jan. 23, 1941.

318. J. DeMENT, *Fluorochemistry*, Chemical Pub. Co., Brooklyn, 1945 (*Note:* Consult first the conscientious review by G. R. FONDA, *J. Am. Chem. Soc.*, Vol. 68, No. 2, 347, Feb. 19, 1946, and note that DeMent's "first law" was rightfully considered "a commonplace in physics" much earlier by T. Lyman in "Notes on the Luminescence of Glass and Fluorite," *Phys. Rev.*, Vol. 40, 578–582, May 15, 1932).

319. T. C. POULTER and H. McCOMB, "A Study of Phosphorescent Zinc Sulphide Screens and Radioactivity under Extremely High Pressure," *Proc. Iowa Acad. Sci.*, Vol. 37, 311–312, 1930.

320a. H. E. WHITE, *Introduction to Atomic Spectra*, McGraw-Hill Book Co., New York, 1934.

320b. G. HERZBERG, *Atomic Spectra and Atomic Structure*, Dover Publications, New York, 1944.

320c. G. HERZBERG, *Molecular Spectra and Molecular Structure*, Prentice-Hall, New York, 1939.

321. G. RUTHEMANN, "Diskrete Energieverluste schneller Elektronen in Festkörpern," *Naturwissenschaften*, Vol. 29, 648, 1941; *ibid.*, Vol. 30, 145, 1942; *ibid.*, Vol. 34, No. 6, 187, 1947; W. LANG, "Geschwindigkeitsverluste mittelschneller Elektronen," *Optik*, Vol. 3, No. 3, 233–246, Mar. 1948; *cf. Ann. Physik*, [6], Vol. 2, No. 3/4, 113–146, 1948.

322. J. HILLIER and R. F. BAKER, "Microanalysis by Means of Electrons," *J. Applied Phys.*, Vol. 15, No. 9, 663–675, Sept. 1944; E. G. RAMBERG and J. HILLIER, "Chromatic Aberration and Resolving Power in Electron Microscopy," *J. Applied Phys.*, Vol. 19, No. 7, 678–682, July 1948.

323. U. FANO, "On the Theory of Ionization Yield of Radiations in Different Substances," *Phys. Rev.*, Vol. 70, No. 1/2, 44–52, July 1 and 15, 1946.

324. H. W. LEVERENZ, "Problems Concerning the Production of Cathode-Ray Tube Screens," *J. Optical Soc. Am.*, Vol. 27, No. 1, 25–35, Jan. 1937; M. SADOWSKY, "Preparation of Luminescent Screens," *J. Electrochem. Soc. Am.*, Vol. 95, No. 3, 112–128, Mar. 1949.

325. B. MUKHOPADHYAY, "Transmission of Light through Suspensions of Powdered Crystals," *Indian J. Phys.*, Vol. 7, 307–315, Sept. 1, 1932; S. BHAGAVANTAM, *Scattering of Light and the Raman Effect*, Chemical Pub. Co., Brooklyn, 1942; G. OSTER, "The Scattering of Light and its Applications to Chemistry," *Chem. Rev.*, Vol. 43, No. 2, 319–365, Oct. 1948.

326. G. HEYNE and M. PIRANI, "Eine spektrographische Methode zur Feststellung der Lumineszenzerregung," *Z. tech. Physik*, No. 1, 31–33, 1933.

327. E. M. PELL and D. H. TOMBOULIAN, "The Response of Certain Inorganic Phosphors in the Extreme Ultraviolet Region," *Bull. Am. Phys. Soc.*, Vol. 23, No. 3, 49, Apr. 29, 1948; *J. Appl. Phys.*, Vol. 20, No. 3, 263–267, Mar. 1949.

328. N. C. BEESE, "The Response of Several Fluorescent Materials to Short-Wavelength Ultraviolet Radiations," *J. Optical Soc. Am.*, Vol. 29, 278–282, July 1939.

329. C. K. LUI, "Absorption and Excitation of Zinc-Silicate Phosphors," *J. Optical Soc. Am.*, Vol. 35, No. 7, 492–494, July 1945.

330. M. KATO, "Spectroscopic Investigations of Optically Homogeneous Luminescent Substances," *Sci. Papers Inst. Phys. Chem. Research (Tokyo)*, Vol. 41, 113–160, 1943; *ibid.*, Vol. 42, 35–48, 95–107, 1944.

331. H. N. BOSE, "On the Afterglow of Sodium Chloride and its Decay," *Indian J. Phys.*, Vol. 21, 29–42, Feb. 1947.

332. W. L. LEMCKE, "Excitation of Luminescence of Mercury," *Science*, Vol. 69, 75–78, 1929; T. S. LOGAN and R. K. TAYLOR, "Triboelectric Luminescence," *Science*, Vol. 72, 89–90, 1930; R. SCHNURMANN, "Contact Electrification of Solid Particles," *Proc. Phys. Soc.*, Vol. 53, 547–553, 1941; *ibid.*, Vol. 54, 14–26, 1942; J. L. COX, *Mechanically Actuated "Fluorescent" Lamp*, U. S. Pat. 2,449,880, 9-21-48.

333. T. INOUE et al., "Triboluminescence," *J. Chem. Soc. Japan*, Vol. 60, 149–156, 1939; M. CURIE and M. PROST, "Sur une des causes possibles de la triboluminescence," *Compt. rend.*, Vol. 223, No. 26, 1125–1126, Dec. 23, 1946.

334. E. H. HUNTRESS, L. HARRIS, A. S. PARKER, and L. N. STANLEY, "The Preparation and Chemiluminescence of 3-aminophthalhydrazide," *J. Am. Chem. Soc.*, Vol. 56, 241–242, 1934; *ibid.*, Vol. 57, 1939–1942, 1935.

335. B. SVESNIKOV, "On the Mechanism of the Chemiluminescence of 3-aminophthalhydrazide," *Acta Physicochim. URSS*, Vol. 8, No. 4, 441–461, 1938.

336. H. KAUTSKY et al., "Phosphoreszenzumwandlung durch Sauerstoff," *Naturwissenschaften*, Vol. 29, 150–151, 1941; *ibid.*, Vol. 30, 148, 1942; Z. *Naturforsch.*, Vol. 2a, 167–172, Mar. 1947.

337. R. T. DUFFORD et al., "Efficiency of Chemiluminescence," *Phys. Rev.*, Vol. 44, 315, Aug. 15, 1933.

338. B. TAMAMUSHI, "Quantum Yield of Chemiluminescence Reaction in Solution," *Sci. Papers Inst. Phys. Chem. Research (Tokyo)*, Vol. 41, 166–176, 1943.

339. H. SCHÜLER and A. WOLLDIKE, "Über den Anregungsmechanismus der Fluoreszenz bei organischen Molekülen," *Physik. Z.*, Vol. 45, No. 3/4, 61–65, Aug. 1944.

339a. R. KLING, "Sonoluminescence and Its Relation to Certain Chemical Effects of Ultrasonic Radiation," *Rev. Sci.*, Vol. 85, 364–366, 1947.

340. E. J. BOWEN, *The Chemical Aspects of Light*, Oxford Univ. Press, New York, 1946; "Fluorescence Quenching in Solutions," *Analyst*, Vol. 72, 377–379, 1947; *Quart. Revs. (London)*, Vol. 1, No. 1, 1–15, 1947.

341. S. PEKAR, "Local Quantum States of an Electron in an Ideal Ionic Crystal (Autolocalization of an Electron)," *J. Phys. (USSR)*, Vol. 10, No. 4, 341–350, 1946; J. J. MARKHAM and F. SEITZ, "Binding Energy for a Self-trapped Electron in NaCl," *Phys. Rev.*, Vol. 74, No. 9, 1014–1024, Nov. 1, 1948.

342. R. T. ELLICKSON and W. L. PARKER, "Effect of Absorption on Decay of Infrared-Sensitive Phosphors," *Phys. Rev.*, Vol. 69, No. 9/10, 534, May 1946; "Decay in Brightness of Infrared-Sensitive Phosphors," *Phys. Rev.*, Vol. 70, No. 5/6, 290–300, Sept. 1946; R. T. ELLICKSON et al., "Inertia Effects in Infrared-Sensitive Phosphors," *J. Optical Soc. Am.*, Vol. 39, No. 1, 64–67, Jan. 1949.

343. R. T. ELLICKSON, "The Effect of Wavelength Distribution on the Brightness of Phosphors," *J. Optical Soc. Am.*, Vol. 36, No. 5, 261–264, May 1946; "Light Sum of Phosphors under Thermal and Infrared Stimulation," *J. Optical Soc. Am.*, Vol. 36, No. 5, 264–269, May 1946; "Light Storage in Infrared-Sensitive Phosphors," *J. Optical Soc. Am.*, Vol. 36, No. 9, 501–502, Sept. 1946; "Excitation of Infrared Phosphors by Alpha and Beta Particles," *Phys. Rev.*, Vol. 73, No. 2, 185–186, Jan. 15, 1948.

344. Z. L. MORGENSTERN, "Light Sum of Flash and Phosphorescence in Calcium Sulphide–Strontium Sulphide + Cerium, Samarium Phosphors," *Compt. rend. acad. sci. URSS*, Vol. 54, 783–785, 1946.

345. H. KORTE and R. BÜNNAGEL, "Über die spektrale Energieverteilung der Lumineszenz einiger Luminophore," *Physik. Z.*, Vol. 43, No. 21, 437–439, 1942.

346a. L. DE BOISBAUDRAN, "Luminescence of SrSO₄ under Cathode Rays," *Compt. rend.*, Vol. 103, 468, 629, 1886; *J. Chem. Soc.*, Vol. 52, 3–4, 1887; F. A. KRÖGER et al., "Bismuth as an Activator in Fluorescent Solids," *J. Electrochem. Soc.*, Vol. 96, 132–141, 1949.

346b. H. W. LEVERENZ, *Silicate Phosphors with Cu, Ag, and Re Activators*, U. S. Pats. 2,110,161 and 2,110,162, 3-8-38.

347. J. T. RANDALL, "Luminescence of Solids at Low Temperatures," *Nature*, Vol. 142, 113–114, July 16, 1938.

348. J. EWLES, "Resolution and Interpretation of the Luminescent Spectra of Some Solids at Low Temperatures," *Proc. Roy. Soc. (London) A*, Vol. 167, No. 928, 34–52, July 1938; Y. UCHIDA and K. FUKUDA, "The Fine Structure of Some Emission Bands of Crystalline Phosphors," *Bull. Inst. Phys. Chem. Research, Ser. Phys. (Tokyo)*, Vol. 22, 577–582, 1943; *ibid.*, Vol. 23, 156–158, 1944.

349. S. T. HENDERSON, "Band Spectra of Cathodoluminescence," *Proc. Roy. Soc. (London) A*, Vol. 173, 323–338, 1939.

350. T. TANAKA, "On the Cathodoluminescence of Solid Solutions of 42 Metals," *J. Optical Soc. Am.*, Vol. 8, 287–319, 411–419, 659–669, 1924.

351. W. H. BYLER, "Multiband Emission Spectra of Some Zinc Sulphide and Zinc–Cadmium Sulphide Phosphors," *J. Optical Soc. Am.*, Vol. 37, No. 11, 920–923, Nov. 1947; *ibid.*, Vol. 39, No. 1, 91–92, Jan. 1949 (*cf.* F. J. STUDER and D. A. LARSEN, "Emission Spectra of Zinc Cadmium Sulphides," *J. Optical Soc. Am.*, Vol. 38, No. 5, 480–481, May 1948).

352. F. J. STUDER and L. GAUS, "Effect of Temperature on the Spectral Energy Distribution of Several Phosphors," Section in the book cited in Ref. 221.

353. F. E. SWINDELLS, "The Luminescence of Alkaline Earth Tungstates Containing Lead," *J. Optical Soc. Am.*, Vol. 23, 129–132, 1933.

354. R. E. SHRADER and H. W. LEVERENZ, "Cathodoluminescence Spectra of Zinc-Oxide Phosphors," *J. Optical Soc. Am.*, Vol. 37, No. 11, 939–940, Nov. 1947; F. H. NICOLL, "Temperature Dependence of the Emission Bands of Zinc Oxide Phosphors," *J. Optical Soc. Am.*, Vol. 38, No. 9, 817, Sept. 1948; F. I. VERGUNAS and F. F. GAVRILOV, "Luminescence Spectra of Zinc Oxide," *J. Exptl. Theor. Phys. USSR*, Vol. 18, 224–227, 1948; "Dependence of the Relative Quantum Efficiency of Luminescent Zinc Oxide on the Exciting Wavelength," *Doklady Akad. Nauk SSSR*, Vol. 59, 1273–1275, 1948.

355. W. KUTZNER, "Phosphoreszenz- und Szintillationsspektren," *Z. Physik*, Vol. 117, No. 9/10, 575–589, 1941.

356. V. L. LEVSHIN, "Interaction of Zinc and Manganese Emission in Zinc Sulphide–Manganese Phosphors; Effect of Wavelength of Exciting Light," *Compt. rend. acad. sci. URSS*, Vol. 54, 127–129, 215–218, 1946; *J. Exp. Theoret. Phys. (USSR)*, Vol. 17, 675–688, 1947; *J. Phys. (USSR)*, Vol. 11, No. 5, 1947.

357. A. LEVIALDI and V. LUZZATI, "Sur une transformation du silicate de cadmium phosphorescent produite par les radiations ultraviolettes," *Compt. rend.*, Vol. 223, 328–329, Aug. 12, 1946; *J. Phys. radium*, Vol. 8, 306–312, 341–344, 1947; J. SADDY, "Increase of the Photoluminescence of Zinc Sulphide Following a Constant Excitation," *Compt. rend.*, Vol. 226, 896–898, 1948.

358. K. BRÜCHERSTEINKUHL, "Über das Nachleuchten von Phosphoren und seine Bedeutung für den Lichtstrahlabtaster mit Braunscher Röhre," *Fernseh A.G.*, Vol. 1, No. 5, 179–186, Aug. 1939.

359. H. M. JAMES, "Decay of Phosphorescence in Cu-activated ZnS," *Phys. Rev.*, Vol. 71, No. 2, 137, Jan. 15, 1947.

360. H. HERZFINKIEL and L. WERTENSTEIN, "Phosphorescence du sulfure de zinc sous l'action des rayons alpha," *J. phys. radium*, Sér. 6, Vol. 2, 31–32, 1921.

361. S. J. VAVILOV and V. L. LEVSCHIN, "Studien zur Kenntnis der Natur der Photolumineszenz von Uranylsalzen," *Z. Physik*, Vol. 148, No. 11/12, 397–425, 1928; M. FREEMAN, "Classification of the Fluorescence Bands of Various Uranyl Salts," *Compt. rend.*, Vol. 225, 529–531, 1947.

362. B. KARLIK, "Versuch zur Lumineszenz von Zinksulfid und Diamant unter Einwirkung radioaktiver Strahlung," *Sitzber. Akad. Wiss. Wien, Abt. IIa*, Vol. 139, No. 7/8, 509–519, 1930.

363. J. STALONY-DOBRZANSKI, "O przebiegu zjawiska scyntylacji w czasie," *Roczniki Chem.*, Vol. 12, 299–310, 1932.

364. G. DESTRIAU et al., "Influence de la taille des cristaux phosphorescents sur le rayon d'action des particules alpha," *J. chim. phys.*, Vol. 33, 587–625, 1936; *ibid.*, Vol. 36, 161–163, 1939.

365. L. A. TUMERMAN, "On the Law of Decay of Luminescence of Complex Molecules," *J. Phys. (USSR)*, Vol. 4, No. 1/2, 151–166, 1941.

366. J. W. STRANGE and S. T. HENDERSON, "Cathodoluminescence," *Proc. Phys. Soc.*, Vol. 58, Part 4, No. 328, 369–401, July 1946; M. D. GALANIN, "The Duration of the Initial Emission Process of Phosphors," *Doklady Akad. Nauk SSSR*, Vol. 60, No. 5, 783–784, 1948.

367. M. CALVIN and G. D. DOROUGH, "The Phosphorescence of Chlorophyll and Some Chlorin Derivatives," *Science*, Vol. 105, No. 2730, 433–434, 1947; *Science News Letter*, 117, Feb. 22, 1947; *J. Am. Chem. Soc.*, Vol. 70, No. 2, 699–706, Feb. 1948; M. KASHA and R. E. POWELL, "The Correlation of the Spectroscopic and Thermal Energy Differences between the Fluorescence and Phosphorescence Levels of Dye Molecules," *J. Am. Chem. Soc.*, Vol. 69, 2909–2910, Nov. 1947.

368. O. YADOFF, "Sur la représentation fonctionnelle du déclin de la phosphorescence d'un produit composé de plusieurs centres luminogènes," *Compt. rend.*, Vol. 224, No. 2, 111–113, 1947.

369. F.-H. MARSHALL, "Microsecond Measurement of the Phosphorescence of X-Ray Fluorescent Screens," *J. Applied Phys.*, Vol. 18, No. 6, 512–519, June 1947.

370. W. H. BYLER, "Studies on Phosphorescent Zinc Sulfide," *J. Am. Chem. Soc.*, Vol. 60, 632–639, Mar. 1938.

371. S. C. Hsin *et al.*, "Phosphorescence Produced by Mechanical Means," *Chinese J. Phys.*, Vol. 7, 53–55, 1947.

372. See Fig. 164, page 786, of "Phosphoreszenz und Fluoreszenz," Part 2, Ref. 145.

373. V. V. Antonov-Romanovskii, "Abklingung von Zinkphosphoren in einzelnen Kristallen," *Physik. Z. (USSR)*, Vol. 7, 366–379, 1935.

374. H. W. Leverenz, *Fluorides in ZnS-Type Phosphors*, U. S. Pat. 2,409,574, 10-15-46.

375. G. F. J. Garlick and A. F. Gibson, "Decay of Luminescence Due to Forbidden Optical Transitions," *Nature*, Vol. 160, No. 4061, 303, Aug. 30, 1947.

376. R. W. Engstrom, "Multiplier Phototube Characteristics: Application to Low Light Levels," *J. Optical Soc. Am.*, Vol. 37, No. 6, 420–432, June 1947; *Rev. Sci. Instruments*, Vol. 18, No. 8, 587–588, Aug. 1947.

377. J. Saddy, "The Photoluminescence of Zinc Sulphide," *Ann. phys.* [12], Vol. 2, 414–455, 1947.

378. G. G. Blake, "Infrared Radiations with Special Reference to Their Quenching Effects upon Zinc–Sulphide Phosphors," *J. Proc. Roy. Soc. N. S. Wales*, Vol. 73, 112–124, 1939.

379. S. A. Popok and F. D. Klement, "Effect of Infrared Rays on the Excitation of the Luminescence of ZnS:Cu Phosphors," *J. Exp. Theoret. Phys. (USSR)*, Vol. 10, 800–807, 1940.

380. H. W. Leverenz, *Alkali-Halide Scotophors with Multivalent Impurities to Accelerate Bleaching*, U. S. Pats., 2,432,908, 12-16-47; 2,451,292, 10-12-48; G. R. Fonda, U. S. Pat. 2,435,435, 2-3-48.

381. W. D. Wright, *Researches on Normal and Defective Colour Vision*, Henry Kimpton, London, 1946; also E. N. Willmer, *Retinal Structure and Colour Vision*, Cambridge Univ. Press, 1946.

382. A. Rose, "The Sensitivity Performance of the Human Eye on an Absolute Scale," *J. Optical Soc. Am.*, Vol. 38, No. 2, 196–208, Feb. 1948.

383. A. Rüttenauer, "Technische Lichterzeugung durch Leuchtstoffe," *Elektrotech. Z.*, Vol. 59, No. 43, 158–160, Oct. 27, 1938; *Z. tech. Physik*, Vol. 17, No. 11, 384–387, 1936; *ibid.*, Vol. 19, 148–150, 1938.

384. J. W. Marden, N. C. Beese, and G. Meister, "Effect of Temperature on Fluorescent Lamps," *Trans. Illum. Eng. Soc.*, Vol. 34, No. 1, 55–64, Jan. 1939.

385. G. E. Inman, "Characteristics of Fluorescent Lamps," *Trans. Illum. Eng. Soc.*, Vol. 34, No. 1, 65–87, Jan. 1939.

386. R. N. Thayer and B. T. Barnes, "The Basis for High Efficiency in Fluorescent Lamps," *J. Optical Soc. Am.*, Vol. 29, 131–134, 1939; R. N. Thayer, "Some Physical Properties of Fluorescent-Lamp Coatings," *Trans. Electrochem. Soc.*, Vol. 87, 413–429, 1945.

387. D. A. Schklover, "Quantum Luminescence Efficiency of Certain Silicates, Tungstates, and Borates," *J. Tech. Phys. (USSR)*, Vol. 17, No. 11, 1239–1252, Nov. 1947.

388. H. G. Jenkins, "Mains Voltage Fluorescent Lighting," *Textile Mfr.*, 193–201, May 1947.

389. C. C. Paterson, "Electric Lighting by Luminescence," *Endeavour*, Vol. 5, 18–23, 1946; A. H. McKeag and P. W. Ranby, "Phosphors for Fluorescent Discharge Lamps," *Ind. Chemist*, Vol. 23, 513–520, 597–601, 1947; H. G. Jenkins, A. H. McKeag, and P. W. Ranby, "The Halophosphates: A New Class of Phosphors for Fluorescent Lamps," *J. Electrochem. Soc.*, Vol. 96,

Oct.–Nov., 1949; R. NAGY, R. W. WOLLENTIN, and C. K. LUI, "Calcium Halophosphate Phosphors," *J. Electrochem. Soc.*, Vol. 95, 187–193, Apr. 1949; K. H. BUTLER, "Design of Fluorescent Lamp Phosphors," *Illum. Eng.*, Vol. 44, 267–277, May 1949.

390. TH. P. J. BOTDEN and F. A. KRÖGER, "Fluorescence of Cadmium Borates Activated by Manganese," *Physica*, Vol. 13, No. 4/5, 216–225, May 1947.

391. O. GAERTNER, "Nutzeffekt des Röntgenstrahlen-Leuchtschirmes Absolut Gemessen," *Z. tech. Physik*, Vol. 16, No. 1, 9–10, 1935.

392. H. A. KLASENS, "The Light Emission from Fluorescent Screens Irradiated by X-Rays," *Philips Research Repts.*, Vol. 2, 68–78, Feb. 1947; *ibid.*, Vol. 9, No. 11, 321–329, 1947/1948; *ibid.*, Vol. 9, No. 12, 364–369, 1947/1948.

393. V. PETRESEN and E. LUCA, "Maximum Efficiency of Fluorescent Screens under X-Rays," *Bull. école polytech. Jassy*, Vol. 2, 193–202, Jan.–June 1947.

394. J. W. COLTMAN, E. G. EBBIGHAUSEN, and W. ALTAR, "Physical Properties of Calcium Tungstate X-Ray Screens," *J. Applied Phys.*, Vol. 18, No. 6, 530–545, June 1947.

395. H. W. PAEHR, "Über das Auftreten von Röntgenstrahlen bei Braunschen Röhren," *Fernseh A.G.*, Vol. 1, No. 2, 23–27, Dec. 1938.

396. W. FERRANT, "Über den Wirkungsgrad bei der Umwandlung von Kathodenstrahl- in Röntgenstrahlenergie und die Energieabgabe des Radiums durch die RaC-γ-Strahlung," *Z. Physik*, Vol. 115, Nos. 11/12, 747–755, 1940.

397. P. KIRKPATRICK and A. V. BAEZ, "Absolute Energies of Kα-Radiation from Thick Targets of Silver," *Phys. Rev.*, Vol. 71, No. 8, 521–529, Apr. 15, 1947; "Formation of Optical Images by X rays," *J. Optical Soc. Am.*, Vol. 38, No. 9, 766–774, Sept. 1948; *Sci. American*, Vol. 180, No. 3, 44–47, Mar. 1949.

398. G. I. FINCH and S. FORDHAM, "The Effect of Crystal-Size on Lattice Dimensions," *Proc. Phys. Soc. (London)*, Vol. 48, No. 264, 85–95, Jan. 1, 1936.

399. U. FANO, "A Theory on Cathode Luminescence," *Phys. Rev.*, Vol. 58, No. 6, 544–553, Sept. 15, 1940 (*cf.* W. MAURER, *Production of Light by Ionic and Atomic Bombardment*, PBL74867, Frames 2414–2425).

400. H. A. KLASENS, "The Light Output of Zinc Sulfide on Irradiation with α-Rays," *Trans. Faraday Soc.*, Vol. 42, 666–668, 1946.

401. H. W. LEVERENZ, *Multiple-Penetration Method for Obtaining High Efficiency from CR-Excited Phosphor Screens*, U. S. Pat. 2,404,077, 7-16-46.

402. A. SCHLOEMER, "Zur Frage der Struktur phosphoreszierender Gläser," *Glastech. Ber.*, Vol. 11, 128, 1933; *ibid.*, Vol. 13, 424, 1935.

403. D. DOBISCHEK, *Darstellung und Eigenschaften luminezenzfähiger Gläser, unter bes. berücksichtigung von Zinksilikatphosphoren*, Dissertation, Berlin, 1934 (*cf.* H. WIEHR, *Dissertation on Devitrification of Phosphorescent Glasses*, Dissertation, Halle, 1936).

404. O. DEUTSCHBEIN, "Zur Fluoreszenz seltener Erden in Gläsern," *Z. Physik*, Vol. 102, Nos. 11/12, 772–781, Oct. 6, 1936.

405. W. HÄNLEIN, *Luminescent Glass Containing Phosphor Crystals*, U. S. Pat. 2,219,895, 10-29-40.

406. S. H. LINWOOD and W. A. WEYL, "The Fluorescence of Manganese in Glasses and Crystals," *J. Optical Soc. Am.*, Vol. 32, No. 8, 443–453, Aug. 1942.

407. W. A. WEYL, "Fluorescence and Solarization of Glass," *Ind. Eng. Chem.*, Vol. 34, 1035–1041, Sept. 1942; *J. Soc. Glass Technol.*, Vol. 27, 133–206, 265–295, 1943; *ibid.*, Vol. 30, No. 138, 90–173, April–June 1946; *Glass Science Bull. (Penn. State Coll.,)* Vols. 5 and 6, 1947; *J. Am. Ceram. Soc.*, Vol. 30,

314–319, 1947; *Chem. & Eng. News*, Vol. 27, No. 15, 1048–1049, Apr. 11, 1949.

408. A. R. RODRIGUEZ, C. W. PARMALEE, and A. E. BADGER, "Study of Photoluminescence in Glass," *J. Am. Chem. Soc.*, Vol. 26, 137–150, 1943.

409. N. J. KREIDL, "Recent Studies on the Fluorescence of Glass," *J. Optical Soc. Am.*, Vol. 35, 249–258, 1945.

410. B. E. COHN and S. C. LIND, "Luminescence and Color Excited by Radium in Zinc–Borate Glasses Which Contain Manganese," *J. Phys. Chem.*, Vol. 42, No. 3, 441–453, Mar. 1938.

411. B. E. COHN and W. D. HARKINS, "Thermoluminescence in Glasses Which Contain Two Activators,"*J. Am. Chem. Soc.*, Vol. 52, 5146–5154, Dec. 1930.

412. S. ROTHSCHILD, "Über Sensibilisierung von Phosphoren," *Physik. Z.*, Vol. 35, Nos. 14/15, 557–560, 1934; *ibid.*, Vol. 37, No. 21, 757–763, 1936.

413. P. BOUROFF, "Confirmations expérimentales de l'augmentation de l'intensité lumineuse d'un produit phosphorescent par l'adjonction successive de plusieurs activateurs," *Compt. rend.*, Vol. 218, 317–318, 970–972, 1944.

414. H. M. FERNBERGER, *CdBe–tungstate:Bi:Sm Phosphor*, U. S. Pat. 2,361,467, 10-31-44.

415. W. O. GRAFF, *Cd–borate (or silicate):Mn:Bi (or Mn:Sn) Phosphor*, U. S. Pat. 2,400,925, 5-28-46.

416. J. W. MARDEN and G. MEISTER, "Effects of Impurities on Fluorescent Compounds," *Trans. Illum. Eng. Soc.*, Vol. 34, No. 5, 503–513, May 1939; *Chem. & Met. Eng.*, 80ff, Aug. 1941.

417. L. A. LEVY and D. W. WEST, "New and Rapid Intensifying Screen," *Brit. J. Radiology*, Vol. 6, 85–107, Feb. 1933; (*Zn:Cd*)*S:Ag:Ni Phosphors*, Brit. Pats. 424,195, 2-11-35 and 440,818, 1-7-36; U. S. Pat. 2,075,399, 3-30-37; "Fluorescent Screens for Cathode-Ray Tubes for Television and Other Purposes," *J. Inst. Elec. Engrs. (London)*, Vol. 79, 11–19, July 1936; "Self-Luminous Materials and Their Utilization," *Chem. Age (London)*, 169–172, Mar. 23, 1940; *Endeavour*, 22–25, Jan. 1943.

418. H. W. LEVERENZ, "General Correlations between the Efficiency Characteristics and Constitutions of Phosphors," *J. Optical Soc. Am.*, Vol. 37, No. 6, 520, June 1947.

419. A. V. HIPPEL, "Einige prinzipielle Gesichtspunkte zur Spektroskopie der Ionenkristalle und ihre Anwendung auf die Alkalihalogenide," *Z. Physik*, Vol. 101, No. 11/12, 680–720, 1936.

420. W. HEIMANN and F. SCHUH, "Lichtausbeute der ZnS-ZnSe Phosphoren bei schwachstrom Kathodenstrahlerregung," *Naturwissenschaften*, Vol. 33, 157, 1946.

421. S. LASOF, "The Efficiency of Cathodoluminescence as a Function of Current Density," *Phys. Rev.*, Vol. 72, 165, July 15, 1947.

422. N. RIEHL, "Über einen neuen Effekt an lumineszierendem Zinksulfid," *Z. tech. Physik*, Vol. 20, No. 5, 152–155, 1939.

423. V. K. ZWORYKIN and G. A. MORTON, *Television, the Electronics of Image Transmission*, John Wiley & Sons, New York, 1940.

424. H. W. LEVERENZ, *Phosphor Screen with Fibrillar Electrode Structure*, U. S. Pat. 2,392,161, 1-1-46.

Chapter 6

425. E. BEUTEL and A. KUTZELNIGG, "Zur Kenntnis der Fluoreszenz des Zinkoxydes," *Monatsh.*, Vol. 61, 437–454, 1932; *Wien. Ber.*, Vol. 146, 297, 1937 (*cf.* Brit. Pat. 558,312, 12-24-43).

426. H. C. FROELICH and G. R. FONDA, "Exaggerated Phosphorescence of Zinc Silicate Phosphors," *J. Phys. Chem.*, Vol. 46, No. 8, 1–8, 1942.

427. H. C. FROELICH, *Silicate Phosphors with As to Impart Long Persistence*, U. S. Pat. 2,206,280, 7-20-40.

428. G. N. RAMACHANDRAN, "Thermo-optic Behaviour of Solids. IV–VI. Zinc Blende, Alkali Halides, and Optical Glasses," *Proc. Indian Acad. Sci.*, Vol. 25a, 375, 481–514, 1947.

429. H. W. LEVERENZ, *Silicate and Germanate Phosphors Containing Ti, Zr, Hf, and Th, and Having Mn Activator*, U. S. Pats. 2,171,145, 8-29-39; 2,182,087, 12-5-39; and 2,212,209, 8-20-40.

430. H. W. LEVERENZ, *Silicate and Germanate Phosphors Containing Ti, Zr, Hf, and Th, without Mn Activator*, U. S. Pat. 2,402,760, 6-25-46.

430a. H. W. LEVERENZ, "Relative Emission Spectra of Zinc Silicates and Other Cathodoluminescent Materials," *Phys. Rev.*, Vol. 53, No. 11, 919–920, June 1, 1938.

431. K. BUTKOW and IR. WOJCIECHOWSKA, "Elementary Photochemical Process in Halides of Bivalent Metals," *Nature*, Vol. 159, No. 4043, 570–571, Apr. 26, 1947.

432. J. R. MOURELO, "Nuevas observaciones acerca de la fototropica de los sistemas inorganicos," *Anales soc. espan. fis. y quim.*, Vol. 1, 346, 1903; Vol. 3, 40, 1905; Vol. 20, 601–605, 1922; Vol. 28, 572–579, 1930; *Compt. rend.*, Vol. 158, 122–125, 1914; Vol. 160, 174–176, 1915; Vol. 161, 172–175, 1915; *Arch. sci. phys. nat.*, Vol. 4, 25, 1908.

433. A. SCHLEEDE, "Über die Schwärzung des Zinksulfids durch Licht," *Z. physik. Chem.*, Vol. 106, No. 5/6, 386–399, 1923.

434. S. ROTHSCHILD, "Über die Verfärbung von Calciumsulfidphosphoren durch Licht," *Z. physik. Chem.*, Vol. 172, No. 3, 188–196, 1935.

435. H. PLATZ and P. W. SCHENCK, "Über die Verfärbung des Zinksulfids in Licht," *Angew. Chem.*, Vol. 49, 822–826, 1936 (*cf.* E. PODSCHUS et al., PB70066, 74892, and 74886, Frames 2267–2296, Office Pub. Board, Washington, D. C.); *Z. anorg. Chem.*, Vol. 255, 45–64, 1947.

436. K. GLOOR, "Photolysen mit Zinksulfid," *Helv. Chim. Acta*, Vol. 20, 853–877, 1937.

437. N. T. GORDON, F. SEITZ, and F. QUINLAN, "The Blackening of Zinc Sulphide Phosphors," *J. Chem. Phys.*, Vol. 7, 4–7, 1939; *J. Applied Phys.*, Vol. 13, 639, 1942.

438. J. A. KITCHENER, "The Photochemistry of Solids," *Sci. J. Roy. Coll. Sci.*, Vol. 16, 1–14, 1946.

438a. C. F. GOODEVE and M. R. TAYLOR, "The Effect of Temperature and Impurities on Certain Photochemical Reactions in Solids," *J. Phys. & Colloid Chem.*, Vol. 52, No. 5, 828–836, May 1948.

439. E. TIEDE and A. SCHLEEDE, "Über mit seltenen Erdmetallen aktivierte Magnesiumsulfidphosphore," *Ann. Physik* [4], Vol. 62, 573–580, 1922.

440. F. SCHRÖTER, "Über Grenzgebiete der Fernsehforschung," *Telefunken Mitt.*, Vol. 21, No. 85, 7–23, Dec. 1940.

441. H. W. LEVERENZ, *Electrolyte in Liquids for Settling Phosphors*, U. S. Pat. 2,108,683, 2-15-38.

442. H. W. LEVERENZ, *Lithium Compounds in Liquids for Settling Phosphors*, U. S. Pat. 2,376,437, 5-22-45.

442a. M. SADOWSKY, *Settling Phosphors Through Colloidal Silica*, U. S. Pat. 2,412,654, 12-17-46 (see also U. S. Pat. 2,278,742, 4-7-42, and 2,451,590, 10-19-48).

443. R. B. HEAD, "The Application of Chemically Unstable Phosphors to Cathode Ray Tubes," *Electronic Eng.*, Vol. 29, No. 237, 363–365, Nov. 1947.

444. H. W. LEVERENZ, *Washing Sulphide-Type Phosphors in Water Containing S⁻⁻ Ions to Obtain Free-Flowing Products*, U. S. Pat. 2,421,207, 5-27-47.

445. H. W. LEVERENZ, *Anhydrous Organic Media with Inorganic Binders for Settling Hydrophobic Phosphors*, U. S. Pat. 2,421,208, 5-27-47.

446. L. WESCH, "Verfärbung und Nachleuchten der Carbonat- und Oxydphosphore," *Ann. Physik* [5], Vol. 12, 730–743, 1932.

447. H. NAGAOKA, Y. SUGIURA, and T. MISHIMA, "Coloration of Fluorides by Kathode Rays," *Proc. Imp. Acad. (Tokyo)*, Vol. 9, 486–489, Nov. 1933.

448. R. HERMAN and S. SILVERMAN, "The Decoloration of Natural Fluorite Crystals," *J. Optical Soc. Am.*, Vol. 47, No. 10, 871–872, Oct. 1947.

449. W. KUDRJAWZEWA, "Über die ultraviolette Fluoreszenz der röntgenisierten Steinsalzkristalle," *Z. Physik*, Vol. 90, Nos. 7/8, 489–504, 1934.

450. M. SCHEIN and M. L. KATZ, "Ultraviolet Luminescence of NaCl," *Nature*, Vol. 138, 883, 1936.

451. M. L. KATZ, "Luminescence of X-Rayed Rock Salt," *Physik. Z. Sowzetunion*, Vol. 12, 373–382, 1937.

452. M. N. DIATCHENKO, "The Ultraviolet Phosphorescence of X-Rayed Rock Salt," *Physik. Z. Sowzetunion*, Vol. 13, No. 1, 55–65, 1938.

453. P. M. WOLF and N. RIEHL, "Über die Zerstörung von ZnS-phosphoren durch Alpha-strahlen," *Ann. Physik* [5], Vol. 17, No. 5, 581–587, July 1933.

454. N. RIEHL, "Über die Stabilität des Lenardschen Leuchtzentrums bei Zinksulfid," *Ann. Physik* [5], Vol. 24, No. 6, 536–543, Nov. 1935.

455. W. MEIDINGER, "Fluoreszenz und Empfindlichkeit photographischer Halogensilbergelatineschichten bei tiefen Temperaturen," *Physik. Z.*, Vol. 41, Nos. 11/12, 277–285, June 1940.

456. F. GOOS, "Lichtelektrische Eigenschaften zerstörter Zinksulfid-kupferphosphore," *Ann. Physik* [5], Vol. 34, 77–95, 1939; Vol. 37, 76–88, 1940.

457. V. A. YASTREBOV, "Peculiarities in the Luminescence of the Zinc–Cadmium Sulphide Phosphors," *Compt. rend. acad. sci. URSS*, Vol. 53, 605–606, 1946.

458. N. C. BEESE and J. W. MARDEN, "The Fatigue Effect in Luminescent Materials," *J. Optical Soc. Am.*, Vol. 32, No. 6, 317–323, June 1942.

459. H. C. FROELICH, "Chemical and Physical Stability of Silicate Phosphors," *Trans. Electrochem. Soc.*, Vol. 87, 429–441, 1945; U. S. Pat. 2,443,728, 6-22-48.

460. H. C. FROELICH, "Depreciation of Fluorescent Lamps without Passage of Current," *J. Applied Phys.*, Vol. 17, No. 7, 573–579, July 1946.

461. H. C. FROELICH, "Note on the Stability of Phosphors under Ultraviolet Irradiation," *J. Optical Soc. Am.*, Vol. 37, No. 4, 308, Apr. 1947.

462. G. MEISTER and R. NAGY, "Notes on the Stability of Phosphors under Ultraviolet Irradiation," *J. Optical Soc. Am.*, Vol. 37, No. 5, 403, May 1947.

462a. E. F. LOWRY, "The Long-term Deterioration in Certain Phosphors Exposed to Mercury Resonance Radiation," *J. Electrochem. Soc.*, Vol. 95, No. 5, 242–253, May 1949.

463. M. E. NORDBERG, "Ultraviolet-Transmitting Glasses for Mercury-Vapour Lamps," *J. Am. Ceram. Soc.*, Vol. 30, No. 6, 174–179, June 1947.

464. A. O. ALLEN and J. A. GHORMLEY, "Decomposition of Solid Barium Nitrate by Fast Electrons," *J. Chem. Phys.*, Vol. 15, No. 4, 208–209, Apr. 1947.

464a. E. KINDER, "Damage of Crystals by Electron Beams," *Naturwissenschaften*, Vol. 34, No. 1, 23–24, 1947.

465. F. H. POUGH and T. H. ROGERS, "Experiments in X-Ray Irradiation of Gem Stones," *Am. Mineral.*, Vol. 32, 31–43, 1947.

466. T. ALPER, "Fluorescence Fatigue," *Nature*, Vol. 158, 451, 1946.

467. A. J. HENRY, "Fluorescence 'Fatigue'," *Nature*, Vol. 160, No. 4057, 163, 1947.

468. G. F. ROLLEFSON and M. BURTON, *Photochemistry and the Mechanism of Chemical Reactions*, Prentice-Hall, New York, 1939; M. BURTON *et al.*, "Radiation Chemistry," *J. Phys. & Colloid Chem.*, Vol. 51, 611–625, 1947; *ibid.*, Vol. 52, No. 3, 437–611, Mar. 1948; *ibid.*, Vol. 52, No. 5, 810–819, May 1948; *Chem. Eng. News*, Vol. 26, No. 4, 1764–1766, June 14, 1948.

469. L. B. LOEB and L. SCHMIEDESKAMP, "The Destruction of Phosphorescent Zinc Sulphides by Ultraviolet Light," *Proc. Natl. Acad. Sci. U. S.*, Vol. 7, 202–207, July 1921.

470. H. W. LEVERENZ, *Method for Improving Luminescence Efficiency by Removing the Dead-Voltage Layer from Phosphor Crystals*, U. S. Pat. 2,164,533, 7-4-39.

471. C. E. BARNETT, *Final Report on Luminescent Materials for Military Blackout Uses*, PB33365, Office Pub. Board, Washington, D. C., 6-30-45.

472. BRITISH STANDARDS INST., "Fluorescent and Phosphorescent Materials (excluding Radioactive Materials)," *Brit. Standard* 1316, 1946.

473. J. L. SWITZER and R. C. SWITZER, *Signal Flag with Fluorescent Organic Dye*, U. S. Pats. 2,417,383, 3-14-47, and 2,417,384, 3-14-47.

474. B. G. KODJBANOFF, *Luminescent Material Incorporated in Inorganic Enamel*, U. S. Pat. 899,873, 9-29-08.

475. E. O'HARA, *Radioactive Luminescent Coating Covered with Inorganic Enamel*, U. S. Pats. 1,364,950, 1-11-21, and 1,364,951, 1-11-21.

476. L. VIELHABER, "Lumineszierende Emaillie," *Emaillewaren-Ind.*, Vol. 8, 227, 1931.

477. H. FISCHER, *Luminescent Glass*, U. S. Pat. 2,049,765, 7-4-36; H. P. HOOD, U. S. Pat. 2,440,048, 4-20-48.

478. W. KERSTAN and H. DIEHL, *Transparent Inorganic Enamel for Suspending ZnS-type Phosphors*, U. S. Pat. 2,224,516, 12-10-40.

479. N. V. PHILIPS' GLOEILAMPENFABRIEKEN, *Borate or Phosphate Glass for Suspending ZnS-Type Phosphors*, Dutch Pat. 51,911, 2-16-42.

480. M. L. MACHT and M. M. RENFREW, *Luminescent Sulphides Incorporated in Cast Synthetic Resins*, U. S. Pat. 2,383,067, 8-21-45.

481. A. STROBL, "Luminescent Paints, Pigments, and Inks," *Colloid Chemistry*, Vol. 6, 735–741, 1946, Reinhold Pub. Co., New York; R. MANSELL, "Vehicles for Luminescent Pigments," *Organic Finishing*, Vol. 9, 11–19, 23, 26–40, 1948.

482. V. M. KRAVCHENKO, "Melting Temperatures of Organic Crystals," *J. Applied Chem. (USSR)*, Vol. 19, 1241–1250, 1946; *Acta Physicochim. URSS*, Vol. 22, 187–197, 1947.

483. E. BREUNIG, "Die Spektrale Verteilung der durch Licht hervorgerufenen Dielektrizitätskonstantenänderung des ZnS:Cu-Phosphors an dünnen Schichten," *Ann. Physik* [5], Vol. 11, No. 7, 863–885, 1931.

484. L. WEBER, "Lichtelektrische Leitung und Absorption der Lenardphosphore im roten und ultraroten Spektralgebiet," *Ann. Physik* [5], Vol. 16, 821–843, 1933.

485. G. DÉCHÊNE, "Phosphorescence. Influence du passage d'un courant électrique sur la phosphorescence du sulfure de zinc," *Compt. rend.*, Vol. 201, 139–142, July 8, 1935.

486. G. DESTRIAU, "Recherches expérimentales sur les actions du champ électrique sur les sulfures phosphorescents," *J. chim. phys.*, Vol. 34, 117–124, 1937;

J. phys. radium, Vol. 4, No. 8, 77–80, 1943; Vol. 7, 259–265, 1946; Vol. 9, No. 10, 258–264, Oct. 1948.

487. E. VOYATZAKIS, "Phosphorescence. Sur l'effet photoélectrique et la photoconductibilité des sulfures phosphorescents et des fluorines," *Compt. rend.*, Vol. 209, 31–33, July 3, 1939; *cf.* F. R. LAPPE, *Ann. Physik*, [5], Vol. 39, 604*ff.*, 1941.

488. J. H. GISOLF, "Dielectric Losses of Irradiated Zinc Sulphide Phosphors," *Physica*, Vol. 6, 918–928, 1939.

489. K. BIRUS, "Zur Erklärung der Dielektrizitätskonstantenerhöhung durch Belichtung bei Kristallphosphore," *Naturwissenschaften*, Vol. 52, 779–780, Dec. 26, 1941.

490. O. FRITSCH, "Elektrisches und Optisches Verhalten von Halbleitern. X. Electrische Messungen an Zinkoxyd," *Ann. Physik*, Vol. 22, [5], 375–401, 1935; E. MOLLWO and F. STÖCKMANN, "Über die elektrische Leitfähigkeit von Zinkoxyd," *Ann. Physik*, [6], Vol. 3, 223–229, 1948.

491. G. DESTRIAU *et al.*, "Electrophotoluminescence," *Ann. phys.*, Vol. 17, 318–323, June–Aug. 1942; *J. phys. radium*, Vol. 6, 12–16, 1945; *Phil Mag.*, Vol. 38, 700–739, 774–793, 880–887, Oct.–Dec. 1947; *J. phys. radium*, Vol. 9, 258–264, 1948.

492. G. DESTRIAU and J. MATTLER, "Photoluminescence. Influence de l'orientation des cristaux en électrophotoluminescence," *Compt. rend.*, Vol. 223, 894–896, 1946.

493. P. PRINGSHEIM and S. J. VAVILOV, "Polarisierte und unpolarisierte Phosphoreszenz fester Farbstofflösungen," *Z. Physik*, Vol. 37, No. 10/11, 705–713, 1926.

494. W. L. LEWSCHIN, "The Possibility of Interpreting Phenomena of Polarized Luminescence Using Linear Oscillator Model," *J. Phys. (USSR)*, Vol. 7, No. 4, 265–275, 1939.

495. A. N. SEVCHENKO, "Polarization of Photoluminescence of Uranium Glasses," *J. Phys. (USSR)*, Vol. 8, No. 3, 163–170, 1940; *J. Exp. & Theoret. Phys. (USSR)*, Vol. 17, No. 11, 1063–1070, Nov. 1947.

496. J. STARK, "Die Axialität der Atomstruktur in der polarisierten Fluoreszenz von Kristallen," *Physik. Z.*, Vol. 44, Nos. 9/10, 215–216, 1943.

496a. S. V. CHERBYNTSEV, "Optical Anisotropy of Phosphors with Organic Dyes," *J. Exp. Theoret. Phys. (USSR)*, Vol. 18, No. 4, 352–367, 1948.

497. P. P. FEOFILOV, "Nature of Elementary Emitters and the Polarization of Photoluminescence," *Compt. rend. acad. sci. URSS*, Vol. 55, 403–406, 1947; *J. Acad. Sci. (USSR)*, Vol. 55, 343–346, 1947 (*cf. ibid.*, 883–886, 1947).

498. A. HOCH, "Die Absorptionsspektra einiger Zinksulfidphosphore im angeregten Zustand," *Ann. Physik* [5], Vol. 38, 486–494, 1940.

Chapter 7

499. E. RUTHERFORD, J. CHADWICK, and C. D. ELLIS, *Radiations from Radioactive Substances*, Macmillan Co., Cambridge, England, 1930.

500. O. E. BERG and H. F. KAISER, "The X-Ray Storage Properties of the Infrared Storage Phosphor and Application to Radiography," *J. Applied Phys.*, Vol. 18, No. 4, 343–347, Apr. 1947.

501. R. T. ELLICKSON, "Recent Developments in the Detection of Infrared Radiation," *Am. J. Phys.*, Vol. 15, No. 3, 199–202, May–June 1947.

502. M. BLAU and B. DREYFUS, "The Multiplier Phototube in Radioactive Measurements," *Rev. Sci. Instruments*, Vol. 16, 245–248, 1945.

503. J. A. RAJCHMAN and W. H. CHERRY, "Electron Mechanics of Induction Acceleration," *J. Franklin Inst.*, Vol. 243, 261–285, 1947.

504. F.-H. MARSHALL, J. W. COLTMAN, and L. P. HUNTER, "The Photomultiplier X-Ray Detector," *Rev. Sci. Instruments*, Vol. 18, No. 7, 504–513, July 1947; *ibid.*, Vol. 19, No. 11, 744–770, Nov. 1948.

505. F.-H. MARSHALL and J. W. COLTMAN, "The Photomultiplier Radiation Detector," *Phys. Rev.*, Vol. 72, No. 6, 528, Sept. 15, 1947; W. E. SHOUPP and K.-H. SUN, "Neutron Detection by Uranium Fission," *Westinghouse Engr.*, Vol. 8, No. 6, 185, Nov. 1948; J. W. COLTMAN, "The Scintillation Counter," *Proc. IRE*, Vol. 37, No. 6, 671–682, June 1949.

506. R. SHERR, "Scintillation Counter for the Detection of Alpha-Particles," *Rev. Sci. Instruments*, Vol. 18, No. 10, 767–771, Oct. 1947.

507. CORNING GLASS WORKS, "Photosensitive Glass Containing Ag or Cu(0.1) with or without CeO$_2$(0.05)," *Rev. Sci. Instruments*, Vol. 18, No. 8, 588–589, Aug. 1947; R. H. DALTON, U. S. Pat. 2,422,472, 6-17-47; W. H. ARMISTEAD, Can. Pat. 442,272, 6-17-47; S. D. STOKEY, Can. Pats. 442,273 and 442,274, 6-17-47.

507a. P. SELENYI, "Photography on Selenium," *Nature*, Vol. 161, No. 4092, 522, Apr. 3, 1948.

508. R. FELDT, "Photographing Patterns on Cathode-Ray Tubes," *Electronics*, 130–268, Feb. 1944.

509. R. J. HERCOCK, R. G. HOPKINSON, W. F. BERG, and W. NETHERCOT, "Photographic Recording of Cathode-ray Tube Screen Traces," *Phot. J.*, 86B, 138–164, 1946; R. J. HERCOCK, *The Photographic Recording of Cathode-ray Tube Traces*, Ilford, London, 1947.

510. H. GOLDSTEIN and P. D. BALES, "High Speed Photography of the Cathode-Ray Tube," *Rev. Sci. Instruments*, Vol. 17, 89–96, 1946.

511. A. L. HUGHES and L. A. DuBRIDGE, *Photoelectric Phenomena*, McGraw-Hill Book Co., New York, 1932.

512. A. SOMMER, "Photo-electric Cells," *Sci. J. Roy. Coll. Sci.*, Vol. 16, 82–94, 1946.

513. G. P. KERR, "New Meters Employing Light-Sensitive Cells for the Measurement of Erythemal Energy," *Rev. Sci. Instruments*, Vol. 18, No. 7, 472–473, July 1947.

514. V. K. ZWORYKIN and E. G. RAMBERG, *Photoelectricity and Its Application*, John Wiley & Sons, New York, 1949.

515. W. J. TENNANT, *Improvements in Lighting by Gas and Vapour Discharge Tubes with Fluorescence; Ultraviolet-Emitting ZnS:Gd Phosphor*, Brit. Pat. 474,907, 11-9-37.

516. P. PRINGSHEIM, "Lumineszenzlichtquellen," *Handbuch Physik*, Vol. 19, J. Springer, Berlin, 1928.

517. L. J. BUTTOLPH and L. B. JOHNSON, "Ultraviolet Radiation and Fluorescence," *Trans. Illum. Eng. Soc.*, Vol. 30, 21–40, Jan. 1936.

518. *IES Lighting Handbook*, Illum. Eng. Soc., New York, 1947.

519. H. COTTON, *Electric Discharge Lamps*, Chapman & Hall, London, 1946.

520. G. BRADFIELD, J. G. BARTLETT, and D. J. WATSON, "A Survey of Cathode-Ray-Tube Problems in Service Applications, with Special Reference to Radar," *J. Inst. Elec. Engrs. (London)*, Vol. 93, 3A, No. 1, 128–148, 1946; *Electronic Eng.*, Vol. 8, No. 219, 143–148, May 1946.

521. G. F. J. GARLICK, S. T. HENDERSON, and R. PULESTON, "Cathode-Ray-Tube Screens for Radar," *J. Inst. Elec. Engrs. (London),* Vol. 93, 3A, No. 5, 815–821, 1946; L. C. JESTY, H. MOSS, and R. PULESTON, "War-Time Developments in Cathode-Ray Tubes for Radar," *J. Inst. Elec. Engrs. (London),* Vol. 93, Part 3A, No. 1, 149–166, 1946.

522. W. B. NOTTINGHAM, *Cathode Ray Tube Displays,* MIT Radiation Laboratory Ser., Vol. 22, Chap. 18, McGraw-Hill Book Co., New York, 1948.

523. C. C. HEIN, "Quantum Efficiency of Certain Light Sensitive Devices," *J. Optical Soc. Am.,* Vol. 25, No. 7, 203–206, July 1935.

524. *RCA Tube Handbook,* Radio Corp. Am., Victor Div., Harrison, N. J.

525. T. H. JAMES and G. C. HIGGINS, *Photographic Theory,* John Wiley & Sons, New York, 1948; C. E. K. MEES, *The Theory of the Photographic Process,* Macmillan Co., New York, 1942; *Photographic Plates for Use in Spectroscopy and Astronomy,* Eastman Kodak Co., Rochester, N. Y., 1948.

526. W. P. LEUCH, *Production of Photographic Diazotype Prints,* U. S. Pat. 2,113,944, 4-12-38 (*cf.* U. S. Pat. 1,821,281, 9-1-31); H. G. GREIG, "The Chemistry of High-Speed Facsimile Recording," *Proc. IRE,* 1224–1235, Oct. 1948.

527. C. SHEARD, "Dark Adaptation: Some Physical, Physiological, Clinical, and Aeromedical Considerations," *J. Optical Soc. Am.,* Vol. 34, 464–508, 1944.

528. W. S. PLYMALE, JR., "Filters for Spectral Corrections of Multiplier Phototubes Used from Scotopic to Photopic Brightness Levels," *Rev. Sci. Instruments,* Vol. 18, No. 8, 535–539, Aug. 1947.

529. S. DUSHMAN, "Search for High Efficiency Light Sources," *J. Optical Soc. Am.,* Vol. 27, No. 1, 1–25, Jan. 1937.

530. F. G. BROCKMAN, "The Nature of the Light Emitter in Photoflash Lamps," *J. Optical Soc. Am.,* Vol. 37, No. 8, 652–659, Aug. 1947.

531. C. F. KUTSCHER, "Ocular Effects of Radiant Energy," *Ind. Med.,* Vol. 15, No. 5, 311–316, May 1946.

532. T. W. J. TAYLOR and W. BAKER, *Sidgwick's Organic Chemistry of Nitrogen,* Oxford Univ. Press, 1942.

533. F. J. WELCHER, *Organic Analytical Reagents,* D. Van Nostrand Co., New York, 1947.

534. L. S. PRATT, *The Chemistry and Physics of Organic Pigments,* John Wiley & Sons, New York, 1947; L. N. FERGUSON, "Relationship between Absorption Spectra and Chemical Constitution of Organic Molecules," *Chem. Rev.,* Vol. 43, No. 3, 385–446, Dec. 1948.

535. E. J. BOWEN and E. MIKIEWICZ, "Fluorescence of Solid Anthracene," *Nature,* Vol. 159, No. 4047, 706, May 24, 1947.

536. J. MIGLARESE, *Method of Marking Fabrics with Luminescent Hydrochloride Salt of Alpha Phenyl Meta Amino Benzimidazole,* U. S. Pat. 2,333,329, 11-2-43.

537. W. F. ROESER, "Fluorescent Materials," *Rev. Sci. Instruments,* Vol. 13, 467, Oct. 1942.

538. R. C. SWITZER, *Detection of Surface Cracks in Solids by Incorporation of Luminescent Materials and Observation under UV,* U. S. Pat. 2,259,400, 10-14-41; R. A. WARD, U. S. Pat. 2,405,078, 7-30-46.

539. H. W. LEVERENZ, *Stratified Phosphor Screens according to the Absorption and Emission Characteristics of the Component Phosphors,* U. S. Pat. 2,243,828, 5-27-41.

540. L. T. WEAGLE, *Luminescent Marking Material,* U. S. Pat. 2,396,219, 3-5-46.

541. M. A. WEIL, *Luminescent Plastic Crayons,* U. S. Pat. 2,317,159, 4-20-43.

542. F. A. BUTAEVA, "Excitation of Luminophors in Low Pressure Mercury Lamps," *Compt. rend. acad. sci. URSS*, Vol. 27, No. 7, 654–657, 1940.

543. J. N. ALDINGTON, "Fluorescent Light Sources and Their Applications," *Trans. Illum. Eng. Soc. (London)*, Vol. 7, 57–75, 1942.

544. O. C. RALSTON and A. G. STERN, "Fluorescent Minerals Used in Lighting and Elsewhere," *U. S. Bur. Mines Inform. Circ.* 7276, Apr. 1944.

544a. O. C. SMITH, *Identification and Qualitative Chemical Analysis of Minerals*, D. Van Nostrand Co., 1946 (Excellent color photographs of luminescing minerals).

545. P. S. MILLAR, "Recent Developments in Light Sources," *Elec. Eng.*, 1-13, Apr. 1944.

546. W. E. FORSYTHE and E. M. WATSON, "Brightnesses of Light Sources," *G. E. Rev.*, Vol. 44, No: 9, 489–491, Sept. 1941.

547. W. E. FORSYTHE and E. Q. ADAMS, *Fluorescent and Other Gaseous Discharge Lamps*, Murray Hill Books, New York, 1948.

548. A. L. MARSHALL, "Hydrogen Peroxide Formation Photosynthesized by Mercury Vapour," *J. Am. Chem. Soc.*, Vol. 49, No. 11, 2763–2772, Nov. 1927.

549. H. KREFFT and M. PIRANI, "Quantitative Messungen im Gesamtspektrum technischer Strahlungsquellen," *Z. tech. Physik*, Vol. 14, 242–246, 392–411, 1933.

550. A. C. HARDY, *Handbook of Colorimetry*, Technology Press, Mass. Inst. Tech., Cambridge, Mass., 1936.

551. E. G. BORING *et al.*, "A Symposium on Color Tolerance," *Am. J. Psychol.*, Vol. 52, 383–448, July 1939; P. J. BOUMA, *Physical Aspects of Colour*, Elsevier Book Co., New York, 1947.

552. COMMITTEE ON COLORIMETRY, "Radiant Energy and Colorimetry," *J. Optical Soc. Am.*, Vol. 33, 534–554, Oct. 1943; *ibid.*, Vol. 34, 183–218, 245–266, 601–668, April–Nov. 1944; *ibid.*, Vol. 35, 1–25, Jan. 1945.

553. F. TARNAY, "Definition of the Color Temperature of Fluorescent Light Sources," *Bull. soc. électriciens*, Vol. 6, No. 61, 168–175, Aug.–Sept. 1946.

554. A. H. MCKEAG *et al.*, *Luminescent hex.-Cadmium-Chlorophosphate:Mn*, U. S. Pats. 2,191,351, 2-20-40; 2,201,698, 5-21-40; 2,214,643, 9-10-40; 2,216,407, 12-24-40; Brit. Pat. 577,089, 5-3-46 (*cf.* D. MCCONNELL, "A Structural Investigation of the Isomorphism of the Apatite Group," *Am. Mineral.*, Vol. 23, 1–19, 1938; H. STRUNZ, "Isomorphie zwischen Tilasit, Durangit, und Cryphiolith," *Zentr. Mineral. Geol.*, (A), 59–60, 1938)

555. M. SERVIGNE, *Luminescent Tungstates or Molybdates with Sm(1) Activator*, Brit. Pat. 490,663, 8-18-38.

555a. R. H. HEAD, "The Luminescence of Simple Oxide Phosphors," *Electronic Eng.*, Vol. 20, No. 245, 219–226, July 1948.

556. J. E. STANWORTH, *Soda-free Glass to Improve the Stability of Phosphor Coatings during Operation of Fluorescent Lamps*, U. S. Pat. 2,418,202, 4-1-47.

557. J. H. CAMPBELL, "Fluorescent Lamp Efficiency Increased," *Science News Letter* 228, Oct. 11, 1947.

558. G. R. FONDA and H. HUTHSTEINER, "The Fluorescence of Phosphors in the Rare Gases," *J. Optical Soc. Am.*, Vol. 32, No. 3, 156–160, Mar. 1942.

559. A. ROSE, P. K. WEIMER, and H. B. LAW, "The Image Orthicon—A Sensitive Television Pickup Tube," *Proc. IRE*, Vol. 34, No. 7, 424–432, July 1946.

560. F. F. RENWICK, *Roentgenoluminescent Phosphor Screen with a Thin Reflecting Coating of Aluminum*, Brit. Pat. 449,244, 6-22-36.

561. O. UHLE, *Roentgenoluminescent UV-emitting BaFCl Phosphor*, U. S. Pat. 2,303,963, 12-1-42.

562. R. P. JOHNSON and F. B. QUINLAN, *Roentgenoluminescent Screen Prepared by Compressing Alkali–Halide Phosphors into Translucent Sheets*, U. S. Pat. 2,248,630, 7-8-41.

563. H. S. NEWCOMER, "The X-Ray Fluorescence of Certain Organic Compounds," *J. Am. Chem. Soc.*, Vol. 42, 1997–2007, 1920; H. N. BOSE, "Luminescence of Some Organic Compounds Under X-ray Excitation," *Indian J. Phys.*, Vol. 22, 316–318, 1948.

564. R. G. FRANKLIN and A. J. ALLEN, "Excitation and Detection of Ultraviolet Fluorescence," *J. Franklin Inst.*, Vol. 230, No. 2, 263–268, Aug. 1940.

565. G. MILEY, "Ultraviolet Blood Irradiation Therapy (Knott Technic) in Acute Pyogenic Infections," *Am. J. Surg.*, 493–507, Sept. 1942.

566. J. D. LEVAN, *Carbon-Admixture in Phosphor Coating for Low-Voltage Operation*, U. S. Pat. 2,226,567, 12-31-40 (*cf.* N. V. PHILIPS' Co., Dutch Pat. 53,618, 12-15-42).

567. V. K. ZWORYKIN, *Television Kinescopes*, U. S. Pats. 1,988,469, 1-22-35; 2,084,-364, 6-22-37; 2,109,245, 2-22-38.

568. YU M. KUSHNIR *et al.*, "Dependence of the Secondary-Electron Emission of Semiconductors on the Angle of Incidence of the Primary Beam," *J. Tech. Phys. (USSR)*, Vol. 16, 1105–1110, 1946.

569. J. H. DEBOER, *Coating Phosphor Screens with Cesium Oxide to Improve Their Secondary-Emission Characteristics*, U. S. Pat. 2,242,644, 5-20-41.

570. D. W. EPSTEIN and L. PENSAK, "Improved Cathode-Ray Tubes with Metal-Backed Luminescent Screens," *RCA Rev.*, Vol. 7, No. 1, 5–11, Mar. 1946.

571. J. B. JOHNSON, *White-Light-Emitting CRT Screen Comprising a Blue- and a Yellow-Emitting Phosphor*, U. S. Pat. 1,603,284, 10-19-26.

572. V. K. ZWORYKIN, *Mosaic Color-Television Phosphor Screen for CRT*, U. S. Pat. 1,691,324, 11-13-28.

573. R. RÜDENBERG, *Ruled Line-Structure Phosphor Screen for Color-Television CRT*, U. S. Pat. 1,934,821, 11-14-33.

574. H. W. LEVERENZ, *Method of Producing Line-Structure Phosphor Screens*, U. S. Pat. 2,310,863, 2-9-43.

575. F. MICHELSSEN, *Cathodoluminescent Screen Comprising Zinc-Sulphide and Zinc-Cadmium-Sulphide Phosphors*, Ger. Pat. 640,929, 8-21-31; U. S. Pat. 1,988,-605, 1-22-35.

576. M. V. ARDENNE, *Stratified CRT Screen with the Phosphors Arranged in Order of Their Efficiency*, U. S. Pat. 2,096,986, 10-26-37.

577. A. SCHLEEDE and F. SCHRÖTER, *Composite CRT Screen with Large Particles Partially Embedded in the Glass Substrate*, U. S. Pat. 2,137,118, 11-15-38 (*cf.* H. W. LEVERENZ, U. S. Pat. 2,072,115, 3-2-37).

578. H. W. KAUFMANN, *White-Emitting CRT Screen Comprising Blue-Emitting ZnS and Yellow-Emitting (Zn:Be)Silicate:Mn Phosphors*, U. S. Pat. 2,219,929, 10-29-40.

579. A. E. HARDY, "Application of the I. C. I. Color System to the Development of the All-Sulphide White Television Screen," *RCA Rev.*, Vol. 8, No. 3, 554–563, Sept. 1947.

580. H. C. HAMAKER, "Radiation and Heat Conduction in Light-Scattering Material," *Philips Research Repts.*, Vol. 2, 55–67, 103–125, 1947.

581. J. R. DEVORE and A. H. PFUND, "Optical Scattering by Dielectric Powders of Uniform Particle Size," *J. Optical Soc. Am.*, Vol. 37, No. 10, 826–832,

Oct. 1947; R. L. HENRY, "The Transmission of Powder Films in the Infrared," *J. Optical Soc. Am.*, Vol. 38, No. 9, 775–789, Sept. 1948.

582. F. H. NICOLL, "Some Observations on Light Absorption of Powdered Luminescent Materials," Chapter in Ref. 221; "Note on Ultraviolet Absorption of Zinc–Oxide Phosphors," *J. Optical Soc. Am.*, Vol. 38, No. 4, 417, Apr. 1948.

583. J. H. DEBOER and C. J. DIPPEL, *Method of Evaporating Phosphors to Produce Roentgenoluminescent Screens*, U. S. Pat. 1,954,691, 4-10-34.

584. R. R. LAW, "Contrast in Kinescopes," *Proc. IRE*, Vol. 27, 511–522, Aug. 1939.

585. G. LIEBMANN, "Ion Burn in Cathode-Ray Tubes," *Nature*, Vol. 157, No. 3987, 228, 2-23-46; *Electronic Eng.*, Vol. 18, 289–290, Sept. 1946; R. M. BOWIE, "The Negative-Ion Blemish in a Cathode-Ray Tube and its Elimination," *Proc. IRE*, Vol. 36, No. 12, 1482–1486, Dec. 1948.

586. J. H. DEBOER, *Television Projection System with Separated Complementary Phosphor Screens in Cascade*, U. S. Pat. 2,227,070, 12-31-40 (*cf.* W. J. SCOTT, Brit. Pat. 490,029, 8-4-38).

587. P. C. GOLDMARK, J. N. DYER, E. R. PIORE, and J. M. HOLLYWOOD, "Color Television," *Proc. IRE*, Vol. 30, No. 4, 162–182, Apr. 1942; Vol. 31, No. 9, 465–478, Sept. 1943.

588. V. K. ZWORYKIN, "Television," *J. Franklin Inst.*, Vol. 244, No. 2, 131–147, Aug. 1947.

589. R. D. KELL et al., "Simultaneous Color Television," *Proc. IRE*, Vol. 35, No. 9, 861–875, Sept. 1947.

590. F. ROCKETT, "Experimental C-R Tubes for Television," *Electronics*, 112–115, Mar. 1947.

591. L. N. RIDENOUR, L. J. HAWORTH, T. SOLLER, et al., *Radar System Engineering* (Vol. 1), *Cathode Ray Tube Displays* (Vol. 22), McGraw-Hill Book Co., New York, 1947.

592. W. B. NOTTINGHAM, *Excitation and Decay of Luminescence under Electron Bombardment*, PB 8070, Office Pub. Board, Washington, D. C., Jan. 22, 1942.

593. R. E. JOHNSON and A. E. HARDY, "Performance Characteristics of Long-Persistence CRT Screens; Their Measurement and Control," *RCA Rev.*, Vol. 8, No. 4, 660–681, Dec. 1947; F. HAMBURGER and E. J. KING, "A Recording Photometer and its Use in Studies of Cathode-Ray Screen Displays," *J. Optical Soc. Am.*, Vol. 38, No. 10, 875–879, Oct. 1948.

594. H. W. LEVERENZ, *Summary of the RCA Research on Radar Indicator Screens*, PB 32564, Office Pub. Board, Washington, D. C., Apr. 11, 1942.

595. S. DUSHMAN, *Progress Report of Work on Slow Phosphors*, PB 20267, Office Pub. Board, Washington, D. C., May 18, 1942.

596. S. DUSHMAN, *Report of Progress of Work on Dark-Trace Tube*, PB 20245, Office Pub. Board, Washington, D. C., Mar. 1, 1943.

597. W. B. NOTTINGHAM, *Studies of British Phosphors of the Type C, H, K, and M*, PB 2724, Office Pub. Board, Washington, D. C., Aug. 2, 1943.

598. H. W. LEVERENZ, *Dark-Trace Radar Indicator Screens*, PB 23326, Office Pub. Board, Washington, D. C., Feb. 18, 1944.

599. W. B. NOTTINGHAM, *Comparison of P7-Screen Test Methods*, PB32744, Office Pub. Board, Washington, D. C., Mar. 14, 1944.

600. A. B. WHITE, *Intermediate Persistence CRT Screens*, Report 62-2/20-45, Contract OEMsr-262, Feb. 20, 1945.

601. A. B. WHITE, *Tabulation of CRT Screen Properties*, Rept. S-48, Contract OEMsr-262, Radiation Laboratory (MIT), May 1, 1945.

602. W. G. WHITE, "Visual Photometer for After-Glow Tests on Cathode-Ray Tube Screens," *J. Sci. Instruments Phys. Ind.*, Vol. 25, No. 1, 1–4, Jan. 1948.

603. R. PAYNE-SCOTT, "The Visibility of Small Echoes on Radar PPI Displays," *Proc. IRE*, Vol. 36, No. 2, 180–196, Feb. 1948 (*cf.* A. L. SWEET and N. R. BARTLETT, "Visibility on Cathode-Ray Tube Screens: Signals on a P7 Screen Seen at Different Intervals after Excitation," *J. Optical Soc. Am.*, Vol. 38, No. 4, 329–337, Apr. 1948; also, R. M. HANES and S. B. WILLIAMS, "Visibility on Cathode-Ray Tube Screens: The Effect of Light Adaptation," *J. Optical Soc. Am.*, Vol. 38, No. 4, 363–377, Apr. 1948; *ibid.*, Vol. 39, No. 6, 463–473, 1948; *J. Psychol.*, Vol. 25, 401–417, 1948; *ibid.*, Vol. 27, 231–244, 1949).

604. U. S. NAVAL TECHNICAL MISSION IN EUROPE, *Blauschrift: A German Dark-Trace CRT with Method of Erasure and Illumination*, PB23089, Tech. Rept. No. 544–545, Office Pub. Board, Washington, D. C., Oct. 1945.

605. V. A. JONES, *Modulating a UV-excited Television Screen by Local Variation of Temperature*, Brit. Pat. 517,483, 1-31-40.

606. J. C. BATCHELOR, *Modulating a UV-excited CRT Screen by Local Variation of Temperature*, U. S. Pat. 2,247,112, 6-24-41.

607. H. W. LEVERENZ, *Modulating a CR-excited Screen by a Second CR Beam*, U. S. Pat. 2,402,761, 6-25-46.

608. A. V. HAEFF, "A Memory Tube," *Electronics*, Vol. 20, 80–83, Sept. 1947.

609. R. A. McCONNELL, "Video Storage by Secondary Emission from Simple Mosaics," *Proc. IRE*, Vol. 35, No. 11, 1258–1264, Nov. 1947; H. KLEMPERER and J. T. DE BETTENCOURT, "Repeller Storage Tube," *Electronics*, 104–106, Aug. 1948; A. S. JENSEN, J. P. SMITH, M. H. MESNER, and L. E. FLORY, "Barrier Grid Storage Tube and Its Operation," *RCA Rev.*, Vol. 9, No. 1, 112–135, Mar. 1948; L. PENSAK, "The Graphecon, a Picture Storage Tube," *RCA Rev.*, Vol. 10, No. 1, 59–73, Mar. 1949.

610. F. COETERIER and M. C. TEVES, "An Apparatus for the Transformation of Light of Long Wavelength into Light of Short Wavelength," *Physica*, Vol. 3, No. 9, 968–976, Nov. 1936.

611. T. H. PRATT, "An Infrared Image-Converter Tube," *J. Sci. Instruments*, Vol. 24, No. 12, 312–314, Dec. 1947; G. B. B. M. SUTHERLAND and E. LEE, "Developments in the Infrared Region of the Spectrum," *Reports on Progress in Physics*, Vol. 11, 144–147, 1948.

612. G. A. MORTON and L. E. FLORY, "An Infrared Image Tube and Its Military Applications," *RCA Rev.*, Vol. 7, No. 7, 385–413, Sept. 1946.

612a. I. LANGMUIR, *Image Intensifier*, U. S. Pat. 2,198,479, 4-23-40; G. A. MORTON, J. E. RUEDY, and G. L. KRIEGER, "The Brightness Intensifier," *RCA Rev.*, Vol. 9, No. 3, 419–432, Sept. 1948; J. W. COLTMAN, "Fluoroscopic Image Brightening by Electronic Means," *Radiology*, Vol. 51, 359–367, Sept. 1948.

613. T. V. TIMOFEEVA, "Radioluminescence of Potassium Uranyl Sulphate as the Basis for a Low-Brightness Standard," *Compt. rend. acad. sci. URSS*, Vol. 47, 554–557, 1945; *Doklady Akad. Nauk SSSR*, Vol. 47, 575–578, 1945; M. BLAU and I. FEUER, "Radioactive Light Sources," *J. Optical Soc. Am.*, Vol. 36, No. 10, 576–580, Oct. 1946.

614. U. FANO, "Ionization Yield of Radiations," *Phys. Rev.*, Vol. 72, 26–29, Jan. 1947.

615. H. J. Born, N. Riehl, and K. G. Zimmer, "Light Yield on Irradiation of Luminescent Zinc Sulphide by Beta-Rays," *Reichsber. Physik*, Vol. 1, 154–158, 1945; *Doklady Akad. Nauk SSSR*, Vol. 59, 1269–1272, 1948; W. Heimann and F. Schuh, "Die Lichtausbeute elektronenerregten ZnS–ZnSe–Phosphore im Bereich geringer Stromdichten," *Naturwissenschaften*, Vol. 33, No. 5, 157, Sept. 15, 1946; R. F. Taschek and H. T. Gittings, "Observations of Naphthalene Scintillations Caused by Tritium Beta Rays," *Phys. Rev.*, Vol. 74, No. 10, 1553–1554, Nov. 15, 1948; R. L. Longini, "The Intrinsic Efficiency of Fluorescence Excited by Electron Bombardment," *Bull. Am. Phys. Soc.*, Vol. 23, No. 7, 13, Nov. 26, 1948.

616. W. Sommer, "The Industrial Applications of Luminescence," *Electronic Eng.*, Vol. 18, No. 226, 361–369, Dec. 1946.

617. L. H. Dawson, "Luminous Materials. Properties and Typical Applications," *Products Eng.*, Vol. 18, No. 7, 144–149, 1947; *cf.* R. Mansell, "Fluorescence," *Western Paint Rev.*, No. 10, 29A–30A, 46A–48A; No. 11, 25A–26A, 28A–32A, 1947.

618. H. C. Froelich, "Mercury Vapour Made Visible," *J. Chem. Education*, Vol. 19, 314, 1942.

619. V. Krizek and V. Vand, "The Development of Infrared Technique in Germany," *Electronic Eng.*, Vol. 18, No. 224, 316ff, Oct. 1946.

620. C. H. Corwin, *Luminescent Adhesive Tape*, U. S. Pat. 2,333,641, 11-9-43.

621. J. B. Leavy, *Luminescent Adhesive Tape*, U. S. Pat. 2,310,740, 2-9-43.

622. P. C. Goldmark, *Echo Simulation by Means of a Moving Phosphorescent Sound Image*, U. S. Pat. 2,203,352, 6-4-40.

623. H. Dudley and O. O. Gruenz, Jr., "Visible Speech Translators with External Phosphors," *J. Acoust. Soc. Am.*, Vol. 18, No. 1, 62–73, July 1946.

624. J. B. Johnson, "Cathode-Ray Tube for Viewing Continuous Patterns," *J. Applied Phys.*, Vol. 17, No. 11, 891–894, Nov. 1946.

625. Corning Glass Works, *EH Coated Glass*; *cf.* J. T. Littleton, U. S. Pat. 2,118,795, 5-24-38.

626. Pittsburgh Plate Glass Co., "NESA Coated Glass," *Tech. Glass Bull.*, No. 15, Product Development Dept.

Supplemental List of Recent References

627. C. A. Popok and F. D. Klement, "Influence of Infrared Rays on the Excitation of Luminescence of a CaS:Pr Phosphor," *J. Exp. Theoret. Phys. (USSR)*, Vol. 17, No. 10, 914–924, Oct. 1947.

628. V. L. Levshin *et al.*, "Investigation of Alkaline-Earth Phosphors Having High Sensitivity toward Infrared," *J. Exp. Theoret. Phys. (USSR)*, Vol. 17, No. 11, 949–964, Nov. 1947.

629. V. V. Antonov-Romanovskii, "Particular Properties of Alkaline-Earth Phosphors Sensitive to Infrared Rays," *Doklady Acad. Sci. SSSR*, Vol. 58, 771–775, 1947.

630. Z. A. Trapeznikova, "Direct Evidence for the Induced Emission of Light for the Phosphor SrS:Sm:Eu," *Doklady Acad. Sci. SSSR*, Vol. 58, 791–794, 1947.

631. Z. L. Morgenstern, "The Storage of Light in Phosphors Sensitive to Infrared," *Doklady Acad. Sci. SSSR*, Vol. 58, 783–787, 1947.

632. V. L. LEVSHIN, "The Origin and Composition of Different Forms of Luminescence of CaS–SrS:Ce:Sm:La Phosphors," *Doklady Acad. Sci. SSSR*, Vol. 58, 779–783, 1947.

633. N. ADIROVICH, "Luminescence of Crystal Phosphors at Constant Excitation in the Region of the Fundamental Absorption Band," *Doklady Acad. Sci. SSSR*, Vol. 55, 25–27, 1947.

634. E. I. ADIROVICH, "Initial Stages of the Phosphorescence of Crystal Phosphors," *Doklady Acad. Sci. SSSR*, Vol. 55, 73–76, 1947; "Theoretical Model of a Crystal Phosphor and the Phenomenon of Stimulation while Cold," *ibid.*, Vol. 58, 1927–1931, 1947; *ibid.*, Vol. 60, 361–364, 1948.

635. S. A. FRIDMAN *et al.*, "The Luminescence of Zinc–Sulphide Phosphors," *Doklady Acad. Sci. SSSR*, Vol. 55, 451–454, 563–566, 1946; Vol. 58, 1344–1347, 1947; *Compt. rend. acad. sci. URSS*, Vol. 59, No. 1, 53–56, 1948.

636. L. F. BESSONOV, "Phosphorescence of Rock Salt," *J. Exp. Theoret. Phys. (USSR)*, Vol. 17, No. 11, 1011–1017, Nov. 1947.

637. M. L. KATS, "Luminescence of Alkali-Halide Crystals, Especially When Excited by X-Rays at Low and High Temperatures," *Doklady Acad. Sci. SSSR*, Vol. 58, 1637–1641, 1935–1939, 1947.

638. B. A. PYATNITSKI, "Decay of the Phosphorescence of Benzene and Several Aromatic Acids at Liquid-Air Temperature," *Doklady Acad. Sci. SSSR*, Vol. 55, 771–773, 1947.

639. A. S. DAVYDOV, "Calculation of the Lower Excited Levels of the Naphthalene Molecule," *J. Exp. Theoret. Phys. (USSR)*, Vol. 17, No. 11, 1106–1110, Nov. 1947.

640. B. YA SVESHNIKOV, "Influence of the Concentration of the Activator on the Phosphorescence of Organic Phosphors," *Doklady Acad. Sci. SSSR*, Vol. 58, 49–52, 1947; *ibid.*, Vol. 60, 571–574, 791–794, 1948.

641. N. P. PENKIN, "Measurement of the Relative Vibrator Strength of the Vibrators in the Multiplets of Chromium," *J. Exp. Theoret. Phys. (USSR)*, Vol. 17, No. 11, Nov. 1947.

642. F. I. VERGUNAS and F. F. GAVRILOV, "Temperature Quenching of the Photoluminescence of Zinc Oxide," *Doklady Acad. Sci. SSSR*, Vol. 55, 31–33, 1947.

643. N. A. TOLSTOI and P. P. FEOFILOV, "Study of the Decay of Luminescence by Means of a Cathode-Ray Oscillograph," *Doklady Acad. Sci. SSSR*, Vol. 58, 389–392, 1947.

644. V. L. LEVSHIN and G. D. SHERMET'EV, "The Period of Establishment of the Stationary Distribution in Excited Molecules of Uranyl Salts," *J. Exp. Theoret. Phys. (USSR)*, Vol. 17, 204–226; *J. Phys. (USSR)*, Vol. 11, No. 2, 1947.

645. V. V. ANTONOV-ROMANOVSKII, "Measurement of the Absolute Efficiency of Phosphorescence Flashes Induced by 'Red' Light," *J. Exp. Theoret. Phys. (USSR)*, Vol. 17, 708–710, 1947.

646. YA. I. LARIONOV, "Photoluminescence Spectra of Solutions of Rare-Earth Salts," *Vestnik Leningrad Univ.*, No. 7, 18–28, 1947.

647. S. A. FRIDMAN and N. O. CHECHIK, "Photoelectric Photometry of Phosphors of Short Duration," *J. Tech. Phys. (USSR)*, No. 1, 35–38, 1948.

648. S. A. FRIDMAN and A. A. CHEREPNEV, "New Types of Zinc–Sulphide Phosphors," *Doklady Akad. Nauk SSSR*, Vol. 59, 53–56, 1948.

649. B. YA. SVESHNIKOV and P. P. DIKUN, "Absorption of Light by the Metastable States in Organic Phosphors and Photochemical Reactions with Quadratic

Dependence on Light Intensity," *Doklady Akad. Nauk SSSR*, Vol. 59, 37–40, 1948.

650. N. A. TOLSTOI and P. P. FEOFILOV, "Decay of Luminescence of Glasses and Uranium Salts," *Doklady Akad. Nauk SSSR*, Vol. 59, No. 2, 235–238, 1948.

651. G. I. BORN, N. RIEHL, and K. G. ZIMMER, "Coefficient of Useful Action on the Excitation of Luminescence of Zinc Sulphide with Beta-Rays," *Doklady Akad. Nauk SSSR*, Vol. 59, No. 7, 1269–1272, 1948.

652. F. I. VERGUNAS and F. F. GAVRILOV, "Dependence of the Relative Quantum Yield of Luminescence of Zinc Oxide on the Wavelength of the Exciting Radiation," *Doklady Akad. Nauk SSSR*, Vol. 59, No. 7, 1273–1275, 1948.

653. D. A. SHKLOVER, "Quantum Yield of Luminescence in Some Silicates, Tungstates, and Borates," *J. Tech. Phys. (USSR)*, Vol. 17, No. 11, 1239–1252, 1947.

654. E. I. ADIROVICH, "Decay Law for the Phosphorescence of Crystals," *J. Exp. Theoret. Phys. (USSR)*, Vol. 18, No. 1, 58–73, 1948.

655. V. L. LEVSHIN, "Nature of the Different Forms of Luminescence of Phosphors with Deep Traps," *J. Exp. Theoret. Phys. (USSR)*, Vol. 18, 82–95, 149–163, 1948; *Izv. Akad. Nauk, SSSR*, Vol. 12, No. 3, 217–238, 1948.

656. V. L. LEVSHIN, "The Influence of the Distribution of Electrons Among the Trapping Levels on the Sequence of Different Processes of Luminescence in cub.-(Ca:Sr)S:Ce:Sm:La Phosphor and in the Number of Repeated Trappings of Electrons," *J. Exp. Theoret. Phys. (USSR)*, Vol. 18, No. 2, 149–163, 1948.

657. M. L. KATS, "The Nature of Centers of Luminescence in Photochemically Colored Alkali-Halide Crystals," *J. Exp. Theoret. Phys. (USSR)*, Vol. 18, 124–139, 164–173, 501–509, 944–950, 1948.

658. F. I. VERGUNAS and F. F. GAVRILOV, "The Luminescence Spectrum of Zinc Oxide," *J. Exp. Theoret. Phys. (USSR)*, Vol. 18, 224–227, 873–877, 1948.

659. R. H. MORGAN, "Characteristics of X-Ray Films and Screens," *Radiology*, Vol. 49, 90–94, July 1947.

660. C. M. SLACK *et al.*, "High-Speed X-Ray Motion Pictures," *Bull. Am. Phys. Soc.*, Vol. 23, No. 3, 35, Apr. 29, 1948.

661. A. D. POGORELYI, "Thermal Dissociation of ZnS and CdS," *J. Phys. Chem. USSR*, Vol. 22, 731–745, 1948.

662. H. W. LEVERENZ, *Cascade Phosphor Screens*, U. S. Pat. 2,452,522, 10-26-48.

663. H. W. LEVERENZ, *Cascade Phosphor Screen*, U. S. Pat. 2,452,523, 10-26-48.

664. G. R. FONDA, *Cascade Screen of ZnS-type Phosphors*, U. S. Pat. 2,435,436, 2-3-48.

665. H. W. LEVERENZ, *Long-persistent Silicate and Germanate Phosphors Containing Tin or Lead*, U. S. Pat. 2,457,054, 12-21-48.

666. H. N. WILSON and J. G. M. BREMNER, "Disproportionation in Inorganic Compounds," *Quart. Revs. (London)*, Vol. 2, 1–24, 1948.

667. J. R. HAYNES and W. SHOCKLEY, *The Trapping of Electrons in Silver Chloride*, Phys. Soc. (London), Report on Bristol Conference on *Strength of Solids*, 151–157, 1948.

668. K. H. HELLWEGE, "Über die Spektren der kristallinen Salze seltener Erden," *Naturwissenschaften*, Vol. 34, No. 8, 225–232, 1947.

669. K. H. HELLWEGE, "Bestimmung von Übergangswahrscheinlichkeiten für strahlende und strahlungslosen Prozesse in Kristalle," *Naturwissenschaften*, Vol. 34, No. 7, 212, 1947.

670. D. T. VIER, M. L. SCHULTZ, and J. BIGELEISEN, "Fluorescence of Uranium Trioxide Hydrates," *J. Optical Soc. Am.*, Vol. 38, No. 9, 811–814, Sept. 1948.

671. O. V. FIALKOVSKAYA, "New Absorption Band in Thallium-activated Alkali-Halide Phosphors," *Doklady Akad. Nauk SSSR*, Vol. 60, 49–52, 1948; "Emission Spectra of Mixed Alkali-halide:Tl Phosphors," *ibid.*, Vol. 60, 575–578, 1948.

672. E. M. BRUMBERG and F. M. PEKERMAN, "New Method of Investigation of Absorption Spectra of Crystal Phosphors," *Doklady Akad. Nauk SSSR*, Vol. 61, 43–46, 1948.

673. Z. BODO, "On the Inadequacy of the Measurement of Emission Characteristics of Fluorescent Substances without Spectral Resolution," *J. Optical Soc. Am.*, Vol. 38, No. 9, 815, Sept. 1948.

674. J. S. ALLEN, "Particle Detection with Multiplier Tubes," *Nucleonics*, Vol. 3, No. 1, 34–39, 1948.

675. G. A. MORTON and J. A. MITCHELL, "Performance of 931-A Type Multiplier in a Scintillation Counter," *RCA Rev.*, Vol. 9, No. 4, 632–642, Dec. 1948; *Nucleonics*, Vol. 4, No. 2, 25–29, Feb. 1949.

676. R. R. NEWTON, "Space Charge Effects in Bombardment Conductivity through Diamond," *Phys. Rev.*, Vol. 75, No. 2, 234–246, Jan. 15, 1949; A. G. CHYNOWETH, "Removal of Space Charge in Diamond Crystal Counters," *Phys. Rev.*, Vol. 76, No. 2, 310, July 15, 1949.

677. S. ZERFOSS, L. R. JOHNSON, and O. IMBER, "Single Crystal Growth of Scheelite," *Phys. Rev.*, Vol. 75, No. 2, 320, Jan. 15, 1949 (Note: Large single crystals of synthetic monocl.-$CdWO_4$:[W], tetr.-$CaWO_4$:[W], and rbhdl.-Al_2O_3:Cr are available from the Linde Air Products Co., New York).

678. H. W. LEVERENZ, "Luminescent Solids (Phosphors)," *Science*, Vol. 109, No. 2826, 183–195, Feb. 25, 1949.

679. A. L. SMITH, "Some New Complex Silicate Phosphors Containing Calcium, Magnesium, and Beryllium," *J. Electrochem. Soc.*, Vol. 96, 1949.

680. J. F. MULLANEY, F. REINES, and H. G. WEISS, "Excitation Time of Silver-Activated Zinc Sulphide on Electron Bombardment," *Phys. Rev.*, Vol. 74, No. 4, 491–492, Aug. 15, 1948.

681. M. L. KATS and A. S. ANDRYANOV, "Thermoluminescence of Colored Single Crystals in the Visible Region," *Doklady Akad. Nauk SSSR*, Vol. 61, 817–820, 1948.

682. Z. M. SVERDLOV and A. N. SEVCHENKO, "Long Photoluminescence of Uranyl Compounds at -185°C," *Doklady Akad. Nauk SSSR*, Vol. 61, 821–823, 1948.

683. K. V. SHALIMOVA, "Intensity Distribution in the Photoluminescence Spectrum of the Sublimate Phosphor KI:Tl," *Doklady Akad. Nauk SSSR*, Vol. 61, 1031–1033, 1948; *J. Exp. Theoret. Phys. (USSR)*, Vol. 18, 1044–1048, 1948.

684. A. F. WELLS, *Alkaline-Earth-Sulphide Phosphors Containing Zn or Cd Compounds*, Brit. Pat. 574,321, 1-1-46.

685. C. G. SHULL and E. O. WOLLAN, "X-Ray, Electron, and Neutron Diffraction," *Science*, Vol. 108, 69–75, July 23, 1948.

686. S. I. PEKAR *et al.*, "New Conception of the Electronic Conduction by Polarons in Ionic Crystals," *J. Exp. Theoret. Phys. (USSR)*, Vol. 18, 105–109, 419–423, 481–486, 818–824, 1948.

687. R. E. SHRADER, "Temperature Dependent Structure in Emission Spectra of Two Crystalline Forms of the Zinc-Silicate:Mn Phosphors," *Bull. Am. Optical Soc.*, 4, Mar. 10, 1949.

688. R. H. Bube, "A Correlation Between Cathodoluminescence Efficiency and Decay," *Bull. Am. Optical Soc.*, 4, Mar. 10, 1949.

689. F. J. Studer and A. Rosenbaum, "The Phosphorescence Decay of Halophosphates and Other Doubly Activated Phosphors," *Bull. Am. Optical Soc.*, 2, Mar. 10, 1949.

690. H. W. Leverenz, *High-Temperature High-Pressure Synthesis of Phosphors*, U. S. Pat. 2,462,517, 2-22-49.

691. W. A. Weyl, J. H. Schulman, R. J. Ginther, and L. W. Evans, "On the Fluorescence of Atomic Silver in Glasses and Crystals," *J. Electrochem. Soc.*, Vol. 95, No. 2, 70–79, Feb. 1949.

692. F. Gutman, "The Electret," *Revs. Modern Phys.*, Vol. 20, No. 3, 457–472, July, 1948; A. Gemant, "Electrets," *Physics Today*, Vol. 2, No. 3, 8–13, Mar. 1949.

693. A. A. Cherepnev et al., "Zinc-Sulphide Phosphors Containing Cobalt or Tin," *Doklady Akad. Nauk SSSR*, Vol. 62, 325–328, 1948; Vol. 62, 767–769, 1948.

694. C. Peyrou, "Luminescence of Zinc-Cadmium-Sulphides," *Ann. Phys., Paris* (*Sér.* 12), Vol. 3, 459–503, Aug. 1948.

695. H. C. Froelich, "Double- and Triple-Activated Magnesium-Pyrophosphate Phosphors," *J. Electrochem. Soc.*, Vol. 95, No. 5, 254–266, May 1949.

696. R. H. Peckham, "The Measurement of the Luminescence of Fluoroscopic X-Ray Screens," *Bull. Am. Optical Soc.*, 2, Mar. 10, 1949.

697. H. P. Steier, J. Kelar, C. T. Lattimer, and R. D. Faulkner, "Development of a Large Metal Kinescope for Television," *RCA Rev.*, Vol. 10, No. 1, 43–58, Mar. 1949.

698. A. H. McKeag and P. W. Ranby, "New Zinc Sulphide Phosphors Activated by Phosphorus," *J. Electrochem. Soc.*, Vol. 96, 85–89, Aug. 1949.

699. E. R. Jervis and A. M. Skellett, "Dark-trace Cathode-ray Tube with Fast Erasure," *IRE Electron Devices Conference, Princeton*, June 20, 1949.

700. C. A. Hutchison, "Paramagnetic Resonance Absorption in Crystals Colored by Irradiation," *Phys. Rev.*, Vol. 75, No. 11, 1769–1770, June 1, 1949.

701. S. F. Kaisel, "Radio-frequency Spectrum Analysis by Non-scanning Techniques," *Dissertation, Stanford Univ.*, May 1949.

702. S. Merkader, H. Haberlandt, and F. Hernegger, "Determination of Europium, Samarium, and Uranium by Fluorescence-emission Analysis," *Akad. Wiss. Wien*, Vol. 149, 2A, No. 7/8, 349–365, 1940; *ibid.*, Vol. 155, 2A, No. 7/8, 359–370, 1947.

703. M. N. Alentsev and V. V. Antonov-Romanovskii, "Measurement of the Absolute Efficiency of Powdered Phosphors," *Doklady Akad. Nauk SSSR*, Vol. 64, 478–486, 1949.

704. H. Brinkman and C. C. Vlam, "The Representation of the Emission Bands of Luminescent Solids by Gaussian Distribution Functions," *Physica*, Vol. 14, No. 10, 650–668, Feb. 1949.

705. V. A. Yastrebov, "Temperature-stability of Luminescence Bands," *J. Exp. Theoret. Phys.* (*USSR*), Vol. 17, 140–154, 1947; *J. Phys.* (*USSR*), Vol. 11, No. 3, 1947.

706. V. M. Kudryatseva and G. I. Sinyapkina, "Thermal Radiation of Zinc Oxide," *Doklady Akad. Nauk SSSR*, Vol. 59, No. 8, 1411–1414, 1948.

707. W. Schikore and E. G. Müller, "Adsorptive Purification in Phosphor Chemistry," *Z. anorg. Chem.*, Vol. 255, 327–330, 1948.

708. M. L. Kats and N. V. Zhukova, "Influence of the Conditions of Excitation on the Temperature Emission of ZnS:Cu Phosphors," *Doklady Akad. Nauk SSSR*, Vol. 63, 247–250, 1948.

709. C. G. A. Hill and H. A. Klasens, "The Influence of Temperature on the Efficiencies of ZnS:Cu:Co Phosphors," *J. Electrochem. Soc.*, Vol. 96, 1949.

710. W. H. Byler and G. P. Kirkpatrick, "On the Decay of Phosphorescence and the Mechanism of Luminescence of Zinc Sulphide Phosphors," *J. Electrochem. Soc.*, Vol. 95, No. 4, 194–204, Apr. 1949.

711. E. S. Krylova, "Slow Growth of the Luminescence of the ZnS:Cu:Co Phosphor," *Doklady Akad. Nauk SSSR*, Vol. 64, 495–498, 1949.

712. N. T. Melamed, "An Investigation of Zinc-sulphide Infrared-quenching Phosphors," *J. Electrochem. Soc.*, Vol. 96, 1949.

713. E. F. Daly, "The Spectral Response of a (Zn:Cd)S:Cu Phosphor to Infrared Radiation," *Proc. Roy. Soc., A*, Vol. 196, No. 1047, 554–562, Apr. 22, 1949.

714. G. F. J. Garlick, A. F. Wells, and M. H. F. Wilkins, "Zinc-sulphide Phosphor Constitution and Its Effect on Electron Traps," *J. Chem. Phys.*, Vol. 17, No. 4, 399–404, Apr. 1949.

715. G. F. J. Garlick and D. E. Mason, "Electron Traps and Infrared Stimulation of Phosphors," *J. Electrochem. Soc.*, Vol. 96, 90–113, Aug. 1949.

716. J. Prener, R. W. Mason, and R. Ward, "Radioactive Tracer Study of Activator Distribution in Infrared Phosphor Systems," *J. Am. Chem. Soc.*, Vol. 71, No. 5, 1803–1805, May 1949.

717. E. Banks, H. I. Yakel, and R. Ward, "Luminescence and Conduction in Solid Solutions of Rare-earth Sulphides in Alkaline-earth Sulphides," *J. Electrochem. Soc.*, Vol. 96, 1949.

718. N. F. Miller, $Cub.-(Ca:Sr)S:Pb(0.0013):Bi(0.023)$, U. S. Pat. 2,458,286, 1-4-49.

719. W. Schikore and G. Redlich, "Luminescence of Spinels," *Z. anorg. Chem.*, Vol. 257, 96–102, 1948.

720. J. T. Anderson and R. S. Wells, "Two New Red-emitting Phosphors," *J. Electrochem. Soc.*, Vol. 95, No. 6, 299–303, June 1949.

721. S. Jones, "Alumina–Lithia–Iron Oxide Phosphor," *J. Electrochem. Soc.*, Vol. 95, No. 6, 295–298, June 1949.

722. E. J. Wood, $ZnS:Cu:Mn:Tb$ Phosphor, U. S. Pat. 2,470,451, 5-17-49.

723. H. W. Leverenz, $(Zn:Cd)F_2:Mn$ Phosphor, U. S. Pat. 2,470,627, 5-17-49.

724. H. W. Leverenz, *Radar Indicator Using Infrared-stimulable Storage*, U. S. Pats. 2,468,452 and 2,468,714, 4-26-49.

725. F. E. Williams, $Cd_2P_2O_7 \cdot 0.02Mn$ Phosphor, U. S. Pat. 2,463,449, 3-1-49.

726. J. Ewles and G. C. Farnell, "The Luminescence of Wetted Solids," *Proc. Phys. Soc., A*, Vol. 62, No. 352A, 216–224, Apr. 1, 1949.

727. I. M. Dikman, "Theory of the Photon and Secondary Electron Emission of Effective Semiconducting Emitters," *J. Tech. Phys., USSR*, Vol. 18, No. 11, 1426–1442, 1948.

728. N. S. Khlebnikov and A. E. Melamid, "Energy-level Structure of the Sb-Cs Cathode," *Doklady Akad. Nauk SSSR*, Vol. 63, 649–651, 1948.

729. W. Żuk, "Phosphorescence of cub.-KCl:Tl," *Ann. Univ. Mariae Curie Sklodowska Lublin-Polonia*, Sect. AA, Vol. 1, 63–89, 1946.

730. M. D. Galanin, "Duration of the Initial Luminescence Process of Phos-

phors," *Doklady Akad. Nauk SSSR*, Vol. 60, No. 5, 783–784, 1948.

731. W. T. DYALL, "A Study of the Persistence Characteristics of Various Cathode-ray-tube Phosphors," *Mass. Inst. Tech. Res. Lab. of Electronics*, TR 56, Jan. 16, 1948.

732. A. LANDÉ, "The Physical Significance of the Reciprocal Lattice of Crystals," *Am. Scientist*, Vol. 37, No. 3, 414–416, July 1949.

733. N. BOHR, "The Penetration of Atomic Particles through Matter," *K. Danske Vidensk. Selsk. Mat.-fys. Medd.*, Vol. 18, No. 8, 144 pp., 1948.

734. R. RITSCHL, "Experiments on the Excitation of ZnS:Cu by Radiation in the Emission Band," *Ann. Physik*, [6], Vol. 1, No. 4/5, 207–218, 1947.

735. W. DE GROOT, "The Effect of Temperature on Light Emission by Zinc Sulphide Phosphors," *Nederland Tijdschr. Natuurk.*, Vol. 14, 229–238, 1948.

736. H. W. LEVERENZ, *High-Efficiency Blue-Emitting (Zn:Cd) (S:Se):Ag Phosphors*, U. S. Pat. 2,477,070, 7-26-49.

737. A. BRIL, "On the Saturation of Fluorescence with Cathode-ray Excitation," *Physica*, Vol. 15, No. 3/4, 361–379, May 1949.

738. W. HOOGENSTRAATEN and F. A. KRÖGER, "The Intensity Dependence of the Efficiency of Fluorescence of Willemite Phosphors," *Physica*, Vol. 15, No. 5/6, 541–556, July 1949.

739. F. A. KRÖGER and W. HOOGENSTRAATEN, "Temperature Quenching and Decay of Fluorescence in Zinc and Zinc–Beryllium Silicates Activated with Manganese," *Physica*, Vol. 15, No. 5/6, 557–568, July 1949.

740. V. V. ANTONOV-ROMANOVSKII et al., "Symposium on Luminescent Solids," *Izvest. Akad. Nauk SSSR*, Ser. Fiz., Vol. 13, 75–207, 1949.

741. G. P. BALIN, "Luminescence of Phosphors with a Layer Structure of the Crystal Lattice," *Doklady Akad. Nauk SSSR*, Vol. 66, 33–36, 1949.

742. A. A. CHEREPNEV and T. S. DOBROLYUBSKAYA, "Formation of the Emission Centers in ZnS:Cu Luminophors," *Doklady Akad. Nauk SSSR*, Vol. 6, No. 4, 621–623, June 1949.

743. N. RIEHL and H. ORTMANN, "New Observations on the Luminescence of Zinc Sulphide," *Doklady Akad. Nauk SSSR*, Vol. 66, No. 4, 613–616, June 1949.

744. K. V. SHALIMOVA, "Dependence of the Absorption, Excitation, and Photoluminescence Spectra of KI:Tl Sublimate Phosphors on the Concentration of the Activator," *Doklady Akad. Nauk SSSR*, Vol. 66, No. 4, 625–628, June 1949.

745. V. A. SOKOLOV, "The Thermal Nature of the Emission of Light in the Oxidation of Zinc Oxide and the Absence of Candoluminescence in Zinc Oxide," *Doklady Akad. Nauk SSSR*, Vol. 66, 45–48, 1949.

746. V. V. ZELINSKII and T. V. TIMOFEEVA, "Phosphors with Ultraviolet Emission," *Doklady Akad. Nauk SSSR*, Vol. 66, No. 2, 187–189, May 1949.

747. G. R. FONDA, "The Enigma of Multiple Band Emission," *J. Electrochem. Soc.*, Vol. 96, No. 2, 42c–44c, Aug. 1949.

748. R. H. BUBE et al., "Symposium on Luminescence," *J. Optical Soc. Am.*, Vol. 39, No. 8, 641–717, Aug. 1949.

749. R. HOFSTADTER, "Scintillation Counters," *Research Revs. (Office of Naval Research)*, 4–8, Sept. 1949.

750. J. FRANCK and R. LIVINGSTON, "Remarks on Intra- and Inter-Molecular Migration of Excitation Energy," *Rev. Modern Phys.*, Vol. 21, No. 3, 505–509, July 1949.

General Recent References

751. H. W. LEVERENZ, "Luminescence," article in *Encyclopaedia Britannica*, 1966.
752. D. G. THOMAS, "Electroluminescence," *Physics Today*, Vol. 21, No. 2, 42–48, February 1968.
753. J. BEESLEY *et al.*, "High Resolution and High Quality Phosphor Screens," pp. 551–581 of *Advances in Electronics and Electron Physics*, Vol. 22A, *Photo-Electronic Image Devices*, Academic Press, London, 1966.
754. P. GOLDBERG, Editor, *Luminescence of Inorganic Solids*, Academic Press, New York, 1966.
755. J. L. OUWELTJES, "Luminescence and Phosphors," pp. 161–257 of *Modern Materials, Advances in Development and Applications*, B. W. Gonser, Editor, Academic Press, New York, Vol. 5, 1965.
756. S. LARACH, Editor, *Photoelectronic Materials and Devices*, Van Nostrand, Princeton, 1965.
757. F. A. KRÖGER, *Chemistry of Imperfect Crystals*, North-Holland, Amsterdam, 1964.
758. H. F. IVEY, *Electroluminescence and Related Effects*, Academic Press, New York, 1963.
759. H. K. HENISCH, *Electroluminescence*, Pergamon Press, New York, 1962.
760. H. KALLMANN and G. M. SPRUCH, Editors, *Luminescence of Organic and Inorganic Materials*, John Wiley, New York, 1962.
761. D. CURIE, *Luminescence Cristalline*, Dunod, Paris, 1960. (English translation by G. F. J. GARLICK, *Luminescence in Crystals*, Methuen, London, 1963.)
762. A. SCHLEEDE, "Leuchtstoffe, anorganische und organische," *Ullman Encyklopädie der technischen Chemie*, Urban & Schwarzenberg, München, 1960.
763. G. F. J. GARLICK, "Luminescence," Chapter in *Handbuch der Physik*, Vol. 26, Springer, Berlin, 1958.
764. A. YARIV, *Quantum Electronics*, John Wiley, New York, 1967.
765. H. M. CROSSWHITE and H. W. MOOS, Editors, *Optical Properties of Ions in Crystals*, John Wiley, New York, 1967.
766. W. V. SMITH and P. P. SOROKIN, *The Laser*, McGraw-Hill, New York, 1966.
767. B. A. LENGYEL, *Introduction to Laser Physics*, John Wiley, New York, 1966.
768. A. K. LEVINE, *Lasers*, Dekker, New York, 1966.
769. K. S. PENNINGTON, "Advances in Holography," *Scientific American*, Vol. 218, No. 2, 40–48, February 1968.
770. H. W. LEVERENZ (Committee Chairman), *et al.*, *Characterization of Materials*, Prepared by the Committee on Characterization of Materials, Materials Advisory Board, National Academy of Sciences/National Research Council, Washington (available as MAB-229-M from the Clearinghouse for Federal Scientific and Technical Information, Springfield, Virginia, 22151), 1967.

FORMULA INDEX

The materials are arranged according to (1) host-crystal composition, (2) activator, and (3) crystal structure, in the given order. Phosphor formulas are given in the usual form: structure-host crystal: activator(s). Cognate phosphors, belonging to the same host-crystal family, are grouped together, despite occasional departures from alphabetical order.

T = table

Inorganic Materials

Organic Luminophors

Acetone, CH_3COCH_3, 61
Acriflavine, $C_{14}H_{14}N_3Cl$, 61
Al salt of 8-hydroxyquinoline, 409, 410
monocl.-Anthracene, $C_6H_4:(CH)_2:C_6H_4$, 61, 250, 252

Benzene, C_6H_6, 61, 252

Ca salt of 8-hydroxyquinoline, 409, 410
Chinolin, 248
rhomb.-Chrysene, $C_{18}H_{12}$, 381
Crystal violet, $[(CH_3)_2NC_6H_4]_3COH$, 252

Eosin, $C_{20}H_8O_5Br_4$, 248, 252

Fluorescein, $C_{20}H_{12}O_5$, 248, 251, 252

H_3BO_3:fluorescein, 391
H_3BO_3:terephthalic acid, 253
H_3BO_3:toluic acid, 253, 254

Mg salt of 8-hydroxyquinoline, 409, 410

monocl.-Naphthalene, $C_{10}H_8$, 61, 98, 250, 252, 381, 425–427, 467, 468, 485
2-Naphthol 3-6 disulphonic acid, $C_{10}H_7O_4NaS_2$, 410

monocl.-Phenanthrene, $C_{14}H_{10}$, 61, 250
Pseudoisocyanin, 248

R salt, 410

monocl.-Stilbene, $C_6H_5CH:CHC_6H_5$, 61, 250, 381, 426, 465

Trypaflavine, 61
See also Acriflavine

Zn salt of 8-hydroxyquinoline, 409, 410

Ab, E. A., 187, 191, 194; *158*
Adams, E. Q., *547*
Addink, N. W. H., *52*
Adirovich, E. I., *634, 654*
Adirovich, N., *633*
Agate, J. N., *139*
Ahearn, A. J., *56*
Aldington, J. N., *543*
Alentsev, M. N., *703*
Allen, A. J., *564*
Allen, A. O., *464*
Allen, J. S., *674*
Allison, S. K., *91*
Alper, T., *466*
Altar, W., *394*
Aminoff, G., *274b*
Anderson, B. W., *54*
Anderson, J. T., *720*
Anderson, R. B., *32*
Andryanov, A. S., *681*
Antonov-Romanovskii, V. V., *166, 179, 239, 373, 629, 645, 703, 740*
Archibald, E. H., *85*
Armistead, W. H., *507*
Armor, A. J., *136, 141*
Auergesellschaft, *210a*

Bachman, C. H., *100b*
Badger, A. E., *408*
Baez, A. V., *397*
Baker, R. F., *322*
Baker, W., *532*
Baker, W. R., *292*
Bakker, J., *199*
Bales, P. D., *510*
Balin, G. P., *741*
Bandow, F., *269*
Banks, E., *717*
Bardeen, J., 115; *287*

Barnes, B. T., *386*
Barnett, C. E., *409; 185, 471*
Bartlett, J. G., *520*
Bartlett, N. R., *603*
Batchelor, J. C., *606*
Bayliss, N. S., *123b*
Becker, A., *240, 241*
Becker, F., *241*
Beese, N. C., *328, 384, 458*
Beggs, E. W., *90*
Bell, P. R., *292*
Berg, O. E., *500*
Berg, W. F., *509*
Bergmann, L., 129; *121*
Berton, A., *280*
Bessonov, L. F., *636*
Bethe, H. A., 158; *3, 8, 51, 301*
Betz, H., *316*
Beutel, E., *425*
Bhagavantam, S., *325*
Bigeleisen, J., *670*
Bijvoet, J. M., *47*
Birus, K., *222, 306, 489*
Blake, G. G., *378*
Blau, M., *502, 613*
Bloembergen, N., *288*
Bodo, Z., *673*
Bohr, N., *733*
Borchert, W., *274a*
Boring, E. G., *551*
Born, G. I., *651*
Born, H. J., *615*
Born, M., 93; *51*
Bose, H. N., 250; *292, 331, 563*
Botden, Th. P. J., *156a, 229a, 390*
Bouma, P. J., *551*
Bouroff, P., *413*
Bowen, E. J., *304, 340, 535*
Bowie, R. M., *585*

Italicized numbers are *not* page numbers; they denote items in the list of references, pages 491–537.

Bowtells, J. N., *182*
Bozorth, R. M., *20*
Bradfield, G., *520*
Bragg, L., *40*
Brandenberger, E., *43*
Brauer, P., 139; *178, 308*
Bremner, J. G. M., *666*
Brentano, J. C. M., *291*
Breunig, E., *483*
Briggs, H. B., *118*
Bril, A., *737*
Brillouin, L., *300*
Brinkman, H., *704*
British CIOS, *208*
British Standards Institute, *472*
Brockman, F. G., *530*
Broomé, B., *274b*
Broser, I., *292*
Brown, B. W., *56*
Brüchersteinkuhl, K., *358*
Bruining, H., *130*
Brumberg, E. M., *672*
Bruninghaus, L., *116a*
Bube, R. H., 177, 263, 266–268; *688, 748*
Buerger, M. J., *50*
Bünger, W., *162*
Bunn, C. W., *47*
Bünnagel, R., *345*
Bunting, E. N., 226
Burdett, R. A., *98*
Burstein, E., *228*
Burton, J. A., *56*
Burton, M., *468*
Burton, W. K., *51*
Butaeva, F. A., *542*
Butement, F. D. S., *70a*
Butkow, K., *431*
Butler, K. H., 227; *247, 389*
Buttolph, L. J., *517*
Byler, W. H., *351, 370, 710*

Calvin, M., *367*
Campbell, J. H., *557*
Campbell, N., *83*
Campbell, P. N., *213b*
Carborundum Co., *88*
Chadwick, J., *499*

Chandrasekharan, V., *55*
Chechik, N. O., *647*
Cherbyntsev, S. V., *496a*
Cherepnev, A. A., *169, 648, 693, 742*
Cherry, W. H., *503*
Chomse, H., *77*
Chynoweth, A. G., *676*
Clapp, R. H., *248*
Clark, C. H. D., *17*
Clark, G. L., *92*
Classen-Nekludova, M. V., *277*
Coeterier, F., *610*
Cohn, B. E., *410, 411*
Cohn, G., *60*
Cole, S. S., *159*
Collins, G. B., 250; *292*
Coltman, J. W., *394, 504, 505, 612a*
Committee on Colorimetry, *552*
Compton, A. H., *91*
Condon, E. U., *9, 26*
Coolidge, W. D., 321
Corning Glass Works, *507, 625*
Corwin, C. H., *620*
Coster, D., *309*
Cotton, H., *519*
Coulson, C. A., *77a*
Coustal, R., *160*
Cox, J. L., *332*
Cunningham, J. A. K., *140*
Curie, M., *80, 232, 333*
Curran, S. C., *292*
Curtiss, L. F., *56*

Dalton, R. H., *507*
Daly, E. F., *713*
Dana, E. S., 485
Danckwortt, P. W., *76*
Daniels, F., 1
Darrow, K. K., *15*
Davey, W. P., *42*
Davidson, N., *278*
Davis, D. H., *291*
Davis, R. H., *292*
Davis, W. L., *114*
Davydov, A. S., *639*
Dawson, L. H., *617*
de Bettencourt, J. T., *609*
DeBoer, J. H., *569, 583, 586*

Italicized numbers are *not* page numbers; they denote items in the list of references, pages 491–537.

Italicized numbers are *not* page numbers; they denote items in the list of references, pages 491–537.

Italicized numbers are *not* page numbers; they denote items in the list of references, pages 491–537.

Italicized numbers are *not* page numbers; they denote items in the list of references, pages 491–537.

Italicized numbers are *not* page numbers; they denote items in the list of references, pages 491–537.

Italicized numbers are *not* page numbers; they denote items in the list of references, pages 491–537.

Italicized numbers are *not* page numbers; they denote items in the list of references, pages 491–537.

Italicized numbers are *not* page numbers; they denote items in the list of references, pages 491–537.

Italicized numbers are *not* page numbers; they denote items in the list of references, pages 491–537.

Italicized numbers are *not* page numbers; they denote items in the list of references, pages 491–537.

SUBJECT INDEX

Page numbers in bold-face type indicate definitions.

Page numbers in bold-face type indicate definitions.

Page numbers in bold-face type indicate definitions.

Page numbers in bold-face type indicate definitions.

Page numbers in bold-face type indicate definitions.

Page numbers in bold-face type indicate definitions.

Page numbers in bold-face type indicate definitions.

Page numbers in bold-face type indicate definitions.

Page numbers in bold-face type indicate definitions.

Page numbers in bold-face type indicate definitions.

Page numbers in bold-face type indicate definitions.

Page numbers in bold-face type indicate definitions.

Page numbers in bold-face type indicate definitions.

Page numbers in bold-face type indicate definitions.

SOME DOVER SCIENCE BOOKS

SOME DOVER SCIENCE BOOKS

WHAT IS SCIENCE?,
Norman Campbell
This excellent introduction explains scientific method, role of mathematics, types of scientific laws. Contents: 2 aspects of science, science & nature, laws of science, discovery of laws, explanation of laws, measurement & numerical laws, applications of science. 192pp. 5⅜ x 8.   Paperbound $1.25

FADS AND FALLACIES IN THE NAME OF SCIENCE,
Martin Gardner
Examines various cults, quack systems, frauds, delusions which at various times have masqueraded as science. Accounts of hollow-earth fanatics like Symmes; Velikovsky and wandering planets; Hoerbiger; Bellamy and the theory of multiple moons; Charles Fort; dowsing, pseudoscientific methods for finding water, ores, oil. Sections on naturopathy, iridiagnosis, zone therapy, food fads, etc. Analytical accounts of Wilhelm Reich and orgone sex energy; L. Ron Hubbard and Dianetics; A. Korzybski and General Semantics; many others. Brought up to date to include Bridey Murphy, others. Not just a collection of anecdotes, but a fair, reasoned appraisal of eccentric theory. Formerly titled *In the Name of Science*. Preface. Index. x + 384pp. 5⅜ x 8.
Paperbound $1.85

PHYSICS, THE PIONEER SCIENCE,
L. W. Taylor
First thorough text to place all important physical phenomena in cultural-historical framework; remains best work of its kind. Exposition of physical laws, theories developed chronologically, with great historical, illustrative experiments diagrammed, described, worked out mathematically. Excellent physics text for self-study as well as class work. Vol. 1: Heat, Sound: motion, acceleration, gravitation, conservation of energy, heat engines, rotation, heat, mechanical energy, etc. 211 illus. 407pp. 5⅜ x 8. Vol. 2: Light, Electricity: images, lenses, prisms, magnetism, Ohm's law, dynamos, telegraph, quantum theory, decline of mechanical view of nature, etc. Bibliography. 13 table appendix. Index. 551 illus. 2 color plates. 508pp. 5⅜ x 8.
Vol. 1 Paperbound $2.25, Vol. 2 Paperbound $2.25,
The set $4.50

THE EVOLUTION OF SCIENTIFIC THOUGHT FROM NEWTON TO EINSTEIN,
A. d'Abro
Einstein's special and general theories of relativity, with their historical implications, are analyzed in non-technical terms. Excellent accounts of the contributions of Newton, Riemann, Weyl, Planck, Eddington, Maxwell, Lorentz and others are treated in terms of space and time, equations of electromagnetics, finiteness of the universe, methodology of science. 21 diagrams. 482pp. 5⅜ x 8.
Paperbound $2.50

CHANCE, LUCK AND STATISTICS: THE SCIENCE OF CHANCE,
Horace C. Levinson
Theory of probability and science of statistics in simple, non-technical language. Part I deals with theory of probability, covering odd superstitions in regard to "luck," the meaning of betting odds, the law of mathematical expectation, gambling, and applications in poker, roulette, lotteries, dice, bridge, and other games of chance. Part II discusses the misuse of statistics, the concept of statistical probabilities, normal and skew frequency distributions, and statistics applied to various fields—birth rates, stock speculation, insurance rates, advertising, etc. "Presented in an easy humorous style which I consider the best kind of expository writing," Prof. A. C. Cohen, Industry Quality Control. Enlarged revised edition. Formerly titled *The Science of Chance*. Preface and two new appendices by the author. Index. xiv + 365pp. 5⅜ x 8.   Paperbound $2.00

BASIC ELECTRONICS,
prepared by the U.S. Navy Training Publications Center
A thorough and comprehensive manual on the fundamentals of electronics. Written clearly, it is equally useful for self-study or course work for those with a knowledge of the principles of basic electricity. Partial contents: Operating Principles of the Electron Tube; Introduction to Transistors; Power Supplies for Electronic Equipment; Tuned Circuits; Electron-Tube Amplifiers; Audio Power Amplifiers; Oscillators; Transmitters; Transmission Lines; Antennas and Propagation; Introduction to Computers; and related topics. Appendix. Index. Hundreds of illustrations and diagrams. vi + 471pp. 6½ x 9¼.
Paperbound $2.75

BASIC THEORY AND APPLICATION OF TRANSISTORS,
prepared by the U.S. Department of the Army
An introductory manual prepared for an army training program. One of the finest available surveys of theory and application of transistor design and operation. Minimal knowledge of physics and theory of electron tubes required. Suitable for textbook use, course supplement, or home study. Chapters: Introduction; fundamental theory of transistors; transistor amplifier fundamentals; parameters, equivalent circuits, and characteristic curves; bias stabilization; transistor analysis and comparison using characteristic curves and charts; audio amplifiers; tuned amplifiers; wide-band amplifiers; oscillators; pulse and switching circuits; modulation, mixing, and demodulation; and additional semiconductor devices. Unabridged, corrected edition. 240 schematic drawings, photographs, wiring diagrams, etc. 2 Appendices. Glossary. Index. 263pp. 6½ x 9¼.   Paperbound $1.25

GUIDE TO THE LITERATURE OF MATHEMATICS AND PHYSICS,
N. G. Parke III
Over 5000 entries included under approximately 120 major subject headings of selected most important books, monographs, periodicals, articles in English, plus important works in German, French, Italian, Spanish, Russian (many recently available works). Covers every branch of physics, math, related engineering. Includes author, title, edition, publisher, place, date, number of volumes, number of pages. A 40-page introduction on the basic problems of research and study provides useful information on the organization and use of libraries, the psychology of learning, etc. This reference work will save you hours of time. 2nd revised edition. Indices of authors, subjects, 464pp. 5⅜ x 8.
Paperbound $2.75

THE RISE OF THE NEW PHYSICS (formerly THE DECLINE OF MECHANISM), *A. d'Abro*
This authoritative and comprehensive 2-volume exposition is unique in scientific publishing. Written for intelligent readers not familiar with higher mathematics, it is the only thorough explanation in non-technical language of modern mathematical-physical theory. Combining both history and exposition, it ranges from classical Newtonian concepts up through the electronic theories of Dirac and Heisenberg, the statistical mechanics of Fermi, and Einstein's relativity theories. "A must for anyone doing serious study in the physical sciences," *J. of Franklin Inst.* 97 illustrations. 991pp. 2 volumes.

Vol. 1 Paperbound $2.25, Vol. 2 Paperbound $2.25,
The set $4.50

THE STRANGE STORY OF THE QUANTUM, AN ACCOUNT FOR THE GENERAL READER OF THE GROWTH OF IDEAS UNDERLYING OUR PRESENT ATOMIC KNOWLEDGE, *B. Hoffmann*
Presents lucidly and expertly, with barest amount of mathematics, the problems and theories which led to modern quantum physics. Dr. Hoffmann begins with the closing years of the 19th century, when certain trifling discrepancies were noticed, and with illuminating analogies and examples takes you through the brilliant concepts of Planck, Einstein, Pauli, de Broglie, Bohr, Schroedinger, Heisenberg, Dirac, Sommerfeld, Feynman, etc. This edition includes a new, long postscript carrying the story through 1958. "Of the books attempting an account of the history and contents of our modern atomic physics which have come to my attention, this is the best," H. Margenau, Yale University, in *American Journal of Physics.* 32 tables and line illustrations. Index. 275pp. 5⅜ x 8.

Paperbound $1.75

GREAT IDEAS AND THEORIES OF MODERN COSMOLOGY, *Jagjit Singh*
The theories of Jeans, Eddington, Milne, Kant, Bondi, Gold, Newton, Einstein, Gamow, Hoyle, Dirac, Kuiper, Hubble, Weizsäcker and many others on such cosmological questions as the origin of the universe, space and time, planet formation, "continuous creation," the birth, life, and death of the stars, the origin of the galaxies, etc. By the author of the popular *Great Ideas of Modern Mathematics.* A gifted popularizer of science, he makes the most difficult abstractions crystal-clear even to the most non-mathematical reader. Index. xii + 276pp. 5⅜ x 8½. Paperbound $2.00

GREAT IDEAS OF MODERN MATHEMATICS: THEIR NATURE AND USE, *Jagjit Singh*
Reader with only high school math will understand main mathematical ideas of modern physics, astronomy, genetics, psychology, evolution, etc., better than many who use them as tools, but comprehend little of their basic structure. Author uses his wide knowledge of non-mathematical fields in brilliant exposition of differential equations, matrices, group theory, logic, statistics, problems of mathematical foundations, imaginary numbers, vectors, etc. Original publications, 2 appendices. 2 indexes. 65 illustr. 322pp. 5⅜ x 8. Paperbound $2.00

THE MATHEMATICS OF GREAT AMATEURS, *Julian L. Coolidge*
Great discoveries made by poets, theologians, philosophers, artists and other non-mathematicians: Omar Khayyam, Leonardo da Vinci, Albrecht Dürer, John Napier, Pascal, Diderot, Bolzano, etc. Surprising accounts of what can result from a non-professional preoccupation with the oldest of sciences. 56 figures. viii + 211pp. 5⅜ x 8½. Paperbound $1.50

COLLEGE ALGEBRA, *H. B. Fine*
Standard college text that gives a systematic and deductive structure to algebra; comprehensive, connected, with emphasis on theory. Discusses the commutative, associative, and distributive laws of number in unusual detail, and goes on with undetermined coefficients, quadratic equations, progressions, logarithms, permutations, probability, power series, and much more. Still most valuable elementary-intermediate text on the science and structure of algebra. Index. 1560 problems, all with answers. x + 631pp. 5⅜ x 8.   Paperbound $2.75

HIGHER MATHEMATICS FOR STUDENTS OF CHEMISTRY AND PHYSICS, *J. W. Mellor*
Not abstract, but practical, building its problems out of familiar laboratory material, this covers differential calculus, coordinate, analytical geometry, functions, integral calculus, infinite series, numerical equations, differential equations, Fourier's theorem, probability, theory of errors, calculus of variations, determinants. "If the reader is not familiar with this book, it will repay him to examine it," *Chem. & Engineering News*. 800 problems. 189 figures. Bibliography. xxi + 641pp. 5⅜ x 8.   Paperbound $2.50

TRIGONOMETRY REFRESHER FOR TECHNICAL MEN, *A. A. Klaf*
A modern question and answer text on plane and spherical trigonometry. Part I covers plane trigonometry: angles, quadrants, trigonometrical functions, graphical representation, interpolation, equations, logarithms, solution of triangles, slide rules, etc. Part II discusses applications to navigation, surveying, elasticity, architecture, and engineering. Small angles, periodic functions, vectors, polar coordinates, De Moivre's theorem, fully covered. Part III is devoted to spherical trigonometry and the solution of spherical triangles, with applications to terrestrial and astronomical problems. Special time-savers for numerical calculation. 913 questions answered for you! 1738 problems; answers to odd numbers. 494 figures. 14 pages of functions, formulae. Index. x + 629pp. 5⅜ x 8.
Paperbound $2.00

CALCULUS REFRESHER FOR TECHNICAL MEN, *A. A. Klaf*
Not an ordinary textbook but a unique refresher for engineers, technicians, and students. An examination of the most important aspects of differential and integral calculus by means of 756 key questions. Part I covers simple differential calculus: constants, variables, functions, increments, derivatives, logarithms, curvature, etc. Part II treats fundamental concepts of integration: inspection, substitution, transformation, reduction, areas and volumes, mean value, successive and partial integration, double and triple integration. Stresses practical aspects! A 50 page section gives applications to civil and nautical engineering, electricity, stress and strain, elasticity, industrial engineering, and similar fields. 756 questions answered. 556 problems; solutions to odd numbers. 36 pages of constants, formulae. Index. v + 431pp. 5⅜ x 8.   Paperbound $2.00

INTRODUCTION TO THE THEORY OF GROUPS OF FINITE ORDER, *R. Carmichael*
Examines fundamental theorems and their application. Beginning with sets, systems, permutations, etc., it progresses in easy stages through important types of groups: Abelian, prime power, permutation, etc. Except 1 chapter where matrices are desirable, no higher math needed. 783 exercises, problems. Index. xvi + 447pp. 5⅜ x 8.   Paperbound $3.00

FIVE VOLUME "THEORY OF FUNCTIONS" SET BY KONRAD KNOPP

This five-volume set, prepared by Konrad Knopp, provides a complete and readily followed account of theory of functions. Proofs are given concisely, yet without sacrifice of completeness or rigor. These volumes are used as texts by such universities as M.I.T., University of Chicago, N. Y. City College, and many others. "Excellent introduction . . . remarkably readable, concise, clear, rigorous," *Journal of the American Statistical Association.*

ELEMENTS OF THE THEORY OF FUNCTIONS,
Konrad Knopp
This book provides the student with background for further volumes in this set, or texts on a similar level. Partial contents: foundations, system of complex numbers and the Gaussian plane of numbers, Riemann sphere of numbers, mapping by linear functions, normal forms, the logarithm, the cyclometric functions and binomial series. "Not only for the young student, but also for the student who knows all about what is in it," *Mathematical Journal*. Bibliography. Index. 140pp. 5⅜ x 8.   Paperbound $1.50

THEORY OF FUNCTIONS, PART I,
Konrad Knopp
With volume II, this book provides coverage of basic concepts and theorems. Partial contents: numbers and points, functions of a complex variable, integral of a continuous function, Cauchy's integral theorem, Cauchy's integral formulae, series with variable terms, expansion of analytic functions in power series, analytic continuation and complete definition of analytic functions, entire transcendental functions, Laurent expansion, types of singularities. Bibliography. Index. vii + 146pp. 5⅜ x 8.   Paperbound $1.35

THEORY OF FUNCTIONS, PART II,
Konrad Knopp
Application and further development of general theory, special topics. Single valued functions. Entire, Weierstrass, Meromorphic functions. Riemann surfaces. Algebraic functions. Analytical configuration, Riemann surface. Bibliography. Index. x + 150pp. 5⅜ x 8.   Paperbound $1.35

PROBLEM BOOK IN THE THEORY OF FUNCTIONS, VOLUME 1.
Konrad Knopp
Problems in elementary theory, for use with Knopp's *Theory of Functions,* or any other text, arranged according to increasing difficulty. Fundamental concepts, sequences of numbers and infinite series, complex variable, integral theorems, development in series, conformal mapping. 182 problems. Answers. viii + 126pp. 5⅜ x 8.   Paperbound $1.35

PROBLEM BOOK IN THE THEORY OF FUNCTIONS, VOLUME 2,
Konrad Knopp
Advanced theory of functions, to be used either with Knopp's *Theory of Functions,* or any other comparable text. Singularities, entire & meromorphic functions, periodic, analytic, continuation, multiple-valued functions, Riemann surfaces, conformal mapping. Includes a section of additional elementary problems. "The difficult task of selecting from the immense material of the modern theory of functions the problems just within the reach of the beginner is here masterfully accomplished," *Am. Math. Soc.* Answers. 138pp. 5⅜ x 8.
                                    Paperbound $1.50

NUMERICAL SOLUTIONS OF DIFFERENTIAL EQUATIONS,
H. Levy & E. A. Baggott
Comprehensive collection of methods for solving ordinary differential equations of first and higher order. All must pass 2 requirements: easy to grasp and practical, more rapid than school methods. Partial contents: graphical integration of differential equations, graphical methods for detailed solution. Numerical solution. Simultaneous equations and equations of 2nd and higher orders. "Should be in the hands of all in research in applied mathematics, teaching," *Nature*. 21 figures. viii + 238pp. 5⅜ x 8. Paperbound $1.85

ELEMENTARY STATISTICS, WITH APPLICATIONS IN MEDICINE AND THE BIOLOGICAL SCIENCES, *F. E. Croxton*
A sound introduction to statistics for anyone in the physical sciences, assuming no prior acquaintance and requiring only a modest knowledge of math. All basic formulas carefully explained and illustrated; all necessary reference tables included. From basic terms and concepts, the study proceeds to frequency distribution, linear, non-linear, and multiple correlation, skewness, kurtosis, etc. A large section deals with reliability and significance of statistical methods. Containing concrete examples from medicine and biology, this book will prove unusually helpful to workers in those fields who increasingly must evaluate, check, and interpret statistics. Formerly titled "Elementary Statistics with Applications in Medicine." 101 charts. 57 tables. 14 appendices. Index. vi + 376pp. 5⅜ x 8. Paperbound $2.00

INTRODUCTION TO SYMBOLIC LOGIC,
S. Langer
No special knowledge of math required — probably the clearest book ever written on symbolic logic, suitable for the layman, general scientist, and philosopher. You start with simple symbols and advance to a knowledge of the Boole-Schroeder and Russell-Whitehead systems. Forms, logical structure, classes, the calculus of propositions, logic of the syllogism, etc. are all covered. "One of the clearest and simplest introductions," *Mathematics Gazette*. Second enlarged, revised edition. 368pp. 5⅜ x 8. Paperbound $2.00

A SHORT ACCOUNT OF THE HISTORY OF MATHEMATICS,
W. W. R. Ball
Most readable non-technical history of mathematics treats lives, discoveries of every important figure from Egyptian, Phoenician, mathematicians to late 19th century. Discusses schools of Ionia, Pythagoras, Athens, Cyzicus, Alexandria, Byzantium, systems of numeration; primitive arithmetic; Middle Ages, Renaissance, including Arabs, Bacon, Regiomontanus, Tartaglia, Cardan, Stevinus, Galileo, Kepler; modern mathematics of Descartes, Pascal, Wallis, Huygens, Newton, Leibnitz, d'Alembert, Euler, Lambert, Laplace, Legendre, Gauss, Hermite, Weierstrass, scores more. Index. 25 figures. 546pp. 5⅜ x 8. Paperbound $2.25

INTRODUCTION TO NONLINEAR DIFFERENTIAL AND INTEGRAL EQUATIONS,
Harold T. Davis
Aspects of the problem of nonlinear equations, transformations that lead to equations solvable by classical means, results in special cases, and useful generalizations. Thorough, but easily followed by mathematically sophisticated reader who knows little about non-linear equations. 137 problems for student to solve. xv + 566pp. 5⅜ x 8½. Paperbound $2.00

AN INTRODUCTION TO THE GEOMETRY OF N DIMENSIONS,
D. H. Y. Sommerville
An introduction presupposing no prior knowledge of the field, the only book
in English devoted exclusively to higher dimensional geometry. Discusses
fundamental ideas of incidence, parallelism, perpendicularity, angles between
linear space; enumerative geometry; analytical geometry from projective and
metric points of view; polytopes; elementary ideas in analysis situs; content of
hyper-spacial figures. Bibliography. Index. 60 diagrams. 196pp. 5⅜ x 8.
                                                        Paperbound $1.50

ELEMENTARY CONCEPTS OF TOPOLOGY, *P. Alexandroff*
First English translation of the famous brief introduction to topology for the
beginner or for the mathematician not undertaking extensive study. This unusually useful intuitive approach deals primarily with the concepts of complex,
cycle, and homology, and is wholly consistent with current investigations.
Ranges from basic concepts of set-theoretic topology to the concept of Betti
groups. "Glowing example of harmony between intuition and thought," David
Hilbert. Translated by A. E. Farley. Introduction by D. Hilbert. Index. 25
figures. 73pp. 5⅜ x 8.   Paperbound $1.00

ELEMENTS OF NON-EUCLIDEAN GEOMETRY,
D. M. Y. Sommerville
Unique in proceeding step-by-step, in the manner of traditional geometry.
Enables the student with only a good knowledge of high school algebra and
geometry to grasp elementary hyperbolic, elliptic, analytic non-Euclidean geom-
etries; space curvature and its philosophical implications; theory of radical
axes; homothetic centres and systems of circles; parataxy and parallelism;
absolute measure; Gauss' proof of the defect area theorem; geodesic representation; much more, all with exceptional clarity. 126 problems at chapter endings
provide progressive practice and familiarity. 133 figures. Index. xvi + 274pp.
5⅜ x 8.   Paperbound $2.00

INTRODUCTION TO THE THEORY OF NUMBERS, *L. E. Dickson*
Thorough, comprehensive approach with adequate coverage of classical literature, an introductory volume beginners can follow. Chapters on divisibility,
congruences, quadratic residues & reciprocity. Diophantine equations, etc. Full
treatment of binary quadratic forms without usual restriction to integral coefficients. Covers infinitude of primes, least residues. Fermat's theorem. Euler's
phi function, Legendre's symbol, Gauss's lemma, automorphs, reduced forms,
recent theorems of Thue & Siegel, many more. Much material not readily
available elsewhere. 239 problems. Index. I figure. viii + 183pp. 5⅜ x 8.
                                                        Paperbound $1.75

MATHEMATICAL TABLES AND FORMULAS,
compiled by Robert D. Carmichael and Edwin R. Smith
Valuable collection for students, etc. Contains all tables necessary in college
algebra and trigonometry, such as five-place common logarithms, logarithmic
sines and tangents of small angles, logarithmic trigonometric functions, natural
trigonometric functions, four-place antilogarithms, tables for changing from
sexagesimal to circular and from circular to sexagesimal measure of angles, etc.
Also many tables and formulas not ordinarily accessible, including powers,
roots, and reciprocals, exponential and hyperbolic functions, ten-place logarithms of prime numbers, and formulas and theorems from analytical and
elementary geometry and from calculus. Explanatory introduction. viii +
269pp. 5⅜ x 8½.   Paperbound $1.25

A SOURCE BOOK IN MATHEMATICS,
D. E. Smith
Great discoveries in math, from Renaissance to end of 19th century, in English translation. Read announcements by Dedekind, Gauss, Delamain, Pascal, Fermat, Newton, Abel, Lobachevsky, Bolyai, Riemann, De Moivre, Legendre, Laplace, others of discoveries about imaginary numbers, number congruence, slide rule, equations, symbolism, cubic algebraic equations, non-Euclidean forms of geometry, calculus, function theory, quaternions, etc. Succinct selections from 125 different treatises, articles, most unavailable elsewhere in English. Each article preceded by biographical introduction. Vol. I: Fields of Number, Algebra. Index. 32 illus. 338pp. 5⅜ x 8. Vol. II: Fields of Geometry, Probability, Calculus, Functions, Quaternions. 83 illus. 432pp. 5⅜ x 8.
Vol. 1 Paperbound $2.00, Vol. 2 Paperbound $2.00,
The set $4.00

FOUNDATIONS OF PHYSICS,
R. B. Lindsay & H. Margenau
Excellent bridge between semi-popular works & technical treatises. A discussion of methods of physical description, construction of theory; valuable for physicist with elementary calculus who is interested in ideas that give meaning to data, tools of modern physics. Contents include symbolism; mathematical equations; space & time foundations of mechanics; probability; physics & continua; electron theory; special & general relativity; quantum mechanics; causality. "Thorough and yet not overdetailed. Unreservedly recommended," *Nature* (London). Unabridged, corrected edition. List of recommended readings. 35 illustrations. xi + 537pp. 5⅜ x 8.　　　　　　　　　　　　　　Paperbound $3.00

FUNDAMENTAL FORMULAS OF PHYSICS,
ed. by D. H. Menzel
High useful, full, inexpensive reference and study text, ranging from simple to highly sophisticated operations. Mathematics integrated into text—each chapter stands as short textbook of field represented. Vol. 1: Statistics, Physical Constants, Special Theory of Relativity, Hydrodynamics, Aerodynamics, Boundary Value Problems in Math, Physics, Viscosity, Electromagnetic Theory, etc. Vol. 2: Sound, Acoustics, Geometrical Optics, Electron Optics, High-Energy Phenomena, Magnetism, Biophysics, much more. Index. Total of 800pp. 5⅜ x 8.
Vol. 1 Paperbound $2.25, Vol. 2 Paperbound $2.25,
The set $4.50

THEORETICAL PHYSICS,
A. S. Kompaneyets
One of the very few thorough studies of the subject in this price range. Provides advanced students with a comprehensive theoretical background. Especially strong on recent experimentation and developments in quantum theory. Contents: Mechanics (Generalized Coordinates, Lagrange's Equation, Collision of Particles, etc.), Electrodynamics (Vector Analysis, Maxwell's equations, Transmission of Signals, Theory of Relativity, etc.), Quantum Mechanics (the Inadequacy of Classical Mechanics, the Wave Equation, Motion in a Central Field, Quantum Theory of Radiation, Quantum Theories of Dispersion and Scattering, etc.), and Statistical Physics (Equilibrium Distribution of Molecules in an Ideal Gas, Boltzmann Statistics, Bose and Fermi Distribution. Thermodynamic Quantities, etc.). Revised to 1961. Translated by George Yankovsky, authorized by Kompaneyets. 137 exercises. 56 figures. 529pp. 5⅜ x 8½.
Paperbound $2.50

MATHEMATICAL PHYSICS, *D. H. Menzel*
Thorough one-volume treatment of the mathematical techniques vital for classical mechanics, electromagnetic theory, quantum theory, and relativity. Written by the Harvard Professor of Astrophysics for junior, senior, and graduate courses, it gives clear explanations of all those aspects of function theory, vectors, matrices, dyadics, tensors, partial differential equations, etc., necessary for the understanding of the various physical theories. Electron theory, relativity, and other topics seldom presented appear here in considerable detail. Scores of definition, conversion factors, dimensional constants, etc. "More detailed than normal for an advanced text . . . excellent set of sections on Dyadics, Matrices, and Tensors," *Journal of the Franklin Institute.* Index. 193 problems, with answers. x + 412pp. 5⅜ x 8. Paperbound $2.50

THE THEORY OF SOUND, *Lord Rayleigh*
Most vibrating systems likely to be encountered in practice can be tackled successfully by the methods set forth by the great Nobel laureate, Lord Rayleigh. Complete coverage of experimental, mathematical aspects of sound theory. Partial contents: Harmonic motions, vibrating systems in general, lateral vibrations of bars, curved plates or shells, applications of Laplace's functions to acoustical problems, fluid friction, plane vortex-sheet, vibrations of solid bodies, etc. This is the first inexpensive edition of this great reference and study work. Bibliography, Historical introduction by R. B. Lindsay. Total of 1040pp. 97 figures. 5⅜ x 8. Vol. 1 Paperbound $2.50, Vol. 2 Paperbound $2.50,
The set $5.00

HYDRODYNAMICS, *Horace Lamb*
Internationally famous complete coverage of standard reference work on dynamics of liquids & gases. Fundamental theorems, equations, methods, solutions, background, for classical hydrodynamics. Chapters include Equations of Motion, Integration of Equations in Special Gases, Irrotational Motion, Motion of Liquid in 2 Dimensions, Motion of Solids through Liquid-Dynamical Theory, Vortex Motion, Tidal Waves, Surface Waves, Waves of Expansion, Viscosity, Rotating Masses of Liquids. Excellently planned, arranged; clear, lucid presentation. 6th enlarged, revised edition. Index. Over 900 footnotes, mostly bibliographical. 119 figures. xv + 738pp. 6⅛ x 9¼. Paperbound $4.00

DYNAMICAL THEORY OF GASES, *James Jeans*
Divided into mathematical and physical chapters for the convenience of those not expert in mathematics, this volume discusses the mathematical theory of gas in a steady state, thermodynamics, Boltzmann and Maxwell, kinetic theory, quantum theory, exponentials, etc. 4th enlarged edition, with new material on quantum theory, quantum dynamics, etc. Indexes. 28 figures. 444pp. 6⅛ x 9¼. Paperbound $2.75

THERMODYNAMICS, *Enrico Fermi*
Unabridged reproduction of 1937 edition. Elementary in treatment; remarkable for clarity, organization. Requires no knowledge of advanced math beyond calculus, only familiarity with fundamentals of thermometry, calorimetry. Partial Contents: Thermodynamic systems; First & Second laws of thermodynamics; Entropy; Thermodynamic potentials: phase rule, reversible electric cell; Gaseous reactions: van't Hoff reaction box, principle of LeChatelier; Thermodynamics of dilute solutions: osmotic & vapor pressures, boiling & freezing points; Entropy constant. Index. 25 problems. 24 illustrations. x + 160pp. 5⅜ x 8. Paperbound $1.75

CELESTIAL OBJECTS FOR COMMON TELESCOPES,
Rev. T. W. Webb
Classic handbook for the use and pleasure of the amateur astronomer. Of inestimable aid in locating and identifying thousands of celestial objects. Vol I, The Solar System: discussions of the principle and operation of the telescope, procedures of observations and telescope-photography, spectroscopy, etc., precise location information of sun, moon, planets, meteors. Vol. II, The Stars: alphabetical listing of constellations, information on double stars, clusters, stars with unusual spectra, variables, and nebulae, etc. Nearly 4,000 objects noted. Edited and extensively revised by Margaret W. Mayall, director of the American Assn. of Variable Star Observers. New Index by Mrs. Mayall giving the location of all objects mentioned in the text for Epoch 2000. New Precession Table added. New appendices on the planetary satellites, constellation names and abbreviations, and solar system data. Total of 46 illustrations. Total of xxxix + 606pp. 5⅜ x 8.   Vol. 1 Paperbound $2.25, Vol. 2 Paperbound $2.25
The set $4.50

PLANETARY THEORY,
E. W. Brown and C. A. Shook
Provides a clear presentation of basic methods for calculating planetary orbits for today's astronomer. Begins with a careful exposition of specialized mathematical topics essential for handling perturbation theory and then goes on to indicate how most of the previous methods reduce ultimately to two general calculation methods: obtaining expressions either for the coordinates of planetary positions or for the elements which determine the perturbed paths. An example of each is given and worked in detail. Corrected edition. Preface. Appendix. Index. xii + 302pp. 5⅜ x 8½.   Paperbound $2.25

STAR NAMES AND THEIR MEANINGS,
Richard Hinckley Allen
An unusual book documenting the various attributions of names to the individual stars over the centuries. Here is a treasure-house of information on a topic not normally delved into even by professional astronomers; provides a fascinating background to the stars in folk-lore, literary references, ancient writings, star catalogs and maps over the centuries. Constellation-by-constellation analysis covers hundreds of stars and other asterisms, including the Pleiades, Hyades, Andromedan Nebula, etc. Introduction. Indices. List of authors and authorities. xx + 563pp. 5⅜ x 8½.   Paperbound $2.50

A SHORT HISTORY OF ASTRONOMY, *A. Berry*
Popular standard work for over 50 years, this thorough and accurate volume covers the science from primitive times to the end of the 19th century. After the Greeks and the Middle Ages, individual chapters analyze Copernicus, Brahe, Galileo, Kepler, and Newton, and the mixed reception of their discoveries. Post-Newtonian achievements are then discussed in unusual detail: Halley, Bradley, Lagrange, Laplace, Herschel, Bessel, etc. 2 Indexes. 104 illustrations, 9 portraits. xxxi + 440pp. 5⅜ x 8.   Paperbound $2.75

SOME THEORY OF SAMPLING, *W. E. Deming*
The purpose of this book is to make sampling techniques understandable to and useable by social scientists, industrial managers, and natural scientists who are finding statistics increasingly part of their work. Over 200 exercises, plus dozens of actual applications. 61 tables. 90 figs. xix + 602pp. 5⅜ x 8½.
Paperbound $3.50

PRINCIPLES OF STRATIGRAPHY,
A. W. Grabau
Classic of 20th century geology, unmatched in scope and comprehensiveness. Nearly 600 pages cover the structure and origins of every kind of sedimentary, hydrogenic, oceanic, pyroclastic, atmoclastic, hydroclastic, marine hydroclastic, and bioclastic rock; metamorphism; erosion; etc. Includes also the constitution of the atmosphere; morphology of oceans, rivers, glaciers; volcanic activities; faults and earthquakes; and fundamental principles of paleontology (nearly 200 pages). New introduction by Prof. M. Kay, Columbia U. 1277 bibliographical entries. 264 diagrams. Tables, maps, etc. Two volume set. Total of xxxii + 1185pp. 5⅜ x 8.   Vol. 1 Paperbound $2.50, Vol. 2 Paperbound $2.50,
The set $5.00

SNOW CRYSTALS, W. A. Bentley and W. J. Humphreys
Over 200 pages of Bentley's famous microphotographs of snow flakes—the product of painstaking, methodical work at his Jericho, Vermont studio. The pictures, which also include plates of frost, glaze and dew on vegetation, spider webs, windowpanes; sleet; graupel or soft hail, were chosen both for their scientific interest and their aesthetic qualities. The wonder of nature's diversity is exhibited in the intricate, beautiful patterns of the snow flakes. Introductory text by W. J. Humphreys. Selected bibliography. 2,453 illustrations. 224pp. 8 x 10¼.   Paperbound $3.25

THE BIRTH AND DEVELOPMENT OF THE GEOLOGICAL SCIENCES,
F. D. Adams
Most thorough history of the earth sciences ever written. Geological thought from earliest times to the end of the 19th century, covering over 300 early thinkers & systems: fossils & their explanation, vulcanists vs. neptunists, figured stones & paleontology, generation of stones, dozens of similar topics. 91 illustrations, including medieval, renaissance woodcuts, etc. Index. 632 footnotes, mostly bibliographical. 511pp. 5⅜ x 8.   Paperbound $2.75

ORGANIC CHEMISTRY, F. C. Whitmore
The entire subject of organic chemistry for the practicing chemist and the advanced student. Storehouse of facts, theories, processes found elsewhere only in specialized journals. Covers aliphatic compounds (500 pages on the properties and synthetic preparation of hydrocarbons, halides, proteins, ketones, etc.), alicyclic compounds, aromatic compounds, heterocyclic compounds, organophosphorus and organometallic compounds. Methods of synthetic preparation analyzed critically throughout. Includes much of biochemical interest. "The scope of this volume is astonishing," Industrial and Engineering Chemistry. 12,000-reference index. 2387-item bibliography. Total of x + 1005pp. 5⅜ x 8.   Two volume set, paperbound $4.50

THE PHASE RULE AND ITS APPLICATION,
Alexander Findlay
Covering chemical phenomena of 1, 2, 3, 4, and multiple component systems, this "standard work on the subject" (Nature, London), has been completely revised and brought up to date by A. N. Campbell and N. O. Smith. Brand new material has been added on such matters as binary, tertiary liquid equilibria, solid solutions in ternary systems, quinary systems of salts and water. Completely revised to triangular coordinates in ternary systems, clarified graphic representation, solid models, etc. 9th revised edition. Author, subject indexes. 236 figures. 505 footnotes, mostly bibliographic. xii + 494pp. 5⅜ x 8.
Paperbound $2.75

A Course in Mathematical Analysis,
Edouard Goursat
Trans. by E. R. Hedrick, O. Dunkel, H. G. Bergmann. Classic study of fundamental material thoroughly treated. Extremely lucid exposition of wide range of subject matter for student with one year of calculus. Vol. 1: Derivatives and differentials, definite integrals, expansions in series, applications to geometry. 52 figures, 556pp. Paperbound $2.50. Vol. 2, Part 1: Functions of a complex variable, conformal representations, doubly periodic functions, natural boundaries, etc. 38 figures, 269pp. Paperbound $1.85. Vol. 2, Part 2: Differential equations, Cauchy-Lipschitz method, nonlinear differential equations, simultaneous equations, etc. 308pp. Paperbound $1.85. Vol. 3, Part 1: Variation of solutions, partial differential equations of the second order. 15 figures, 339pp. Paperbound $3.00. Vol. 3, Part 2: Integral equations, calculus of variations. 13 figures, 389pp.   Paperbound $3.00

Planets, Stars and Galaxies,
A. E. Fanning
Descriptive astronomy for beginners: the solar system; neighboring galaxies; seasons; quasars; fly-by results from Mars, Venus, Moon; radio astronomy; etc. all simply explained. Revised up to 1966 by author and Prof. D. H. Menzel, former Director, Harvard College Observatory. 29 photos, 16 figures. 189pp. 5⅜ x 8½.   Paperbound $1.50

Great Ideas in Information Theory, Language and Cybernetics,
Jagjit Singh
Winner of Unesco's Kalinga Prize covers language, metalanguages, analog and digital computers, neural systems, work of McCulloch, Pitts, von Neumann, Turing, other important topics. No advanced mathematics needed, yet a full discussion without compromise or distortion. 118 figures. ix + 338pp. 5⅜ x 8½.
                                                Paperbound $2.00

Geometric Exercises in Paper Folding,
T. Sundara Row
Regular polygons, circles and other curves can be folded or pricked on paper, then used to demonstrate geometric propositions, work out proofs, set up well-known problems. 89 illustrations, photographs of actually folded sheets. xii + 148pp. 5⅜ x 8½.   Paperbound $1.00

Visual Illusions, Their Causes, Characteristics and Applications,
M. Luckiesh
The visual process, the structure of the eye, geometric, perspective illusions, influence of angles, illusions of depth and distance, color illusions, lighting effects, illusions in nature, special uses in painting, decoration, architecture, magic, camouflage. New introduction by W. H. Ittleson covers modern developments in this area. 100 illustrations. xxi + 252pp. 5⅜ x 8.
                                                Paperbound $1.50

Atoms and Molecules Simply Explained,
B. C. Saunders and R. E. D. Clark
Introduction to chemical phenomena and their applications: cohesion, particles, crystals, tailoring big molecules, chemist as architect, with applications in radioactivity, color photography, synthetics, biochemistry, polymers, and many other important areas. Non technical. 95 figures. x + 299pp. 5⅜ x 8½.
                                                Paperbound $1.50

The Principles of Electrochemistry,
D. A. MacInnes

Basic equations for almost every subfield of electrochemistry from first principles, referring at all times to the soundest and most recent theories and results; unusually useful as text or as reference. Covers coulometers and Faraday's Law, electrolytic conductance, the Debye-Hueckel method for the theoretical calculation of activity coefficients, concentration cells, standard electrode potentials, thermodynamic ionization constants, pH, potentiometric titrations, irreversible phenomena. Planck's equation, and much more. 2 indices. Appendix. 585-item bibliography. 137 figures. 94 tables. ii + 478pp. 5⅝ x 8⅜.
Paperbound $2.75

Mathematics of Modern Engineering,
E. G. Keller and R. E. Doherty

Written for the Advanced Course in Engineering of the General Electric Corporation, deals with the engineering use of determinants, tensors, the Heaviside operational calculus, dyadics, the calculus of variations, etc. Presents underlying principles fully, but emphasis is on the perennial engineering attack of set-up and solve. Indexes. Over 185 figures and tables. Hundreds of exercises, problems, and worked-out examples. References. Two volume set. Total of xxxiii + 623pp. 5⅜ x 8. Two volume set, paperbound $3.70

Aerodynamic Theory: A General Review of Progress,
William F. Durand, editor-in-chief

A monumental joint effort by the world's leading authorities prepared under a grant of the Guggenheim Fund for the Promotion of Aeronautics. Never equalled for breadth, depth, reliability. Contains discussions of special mathematical topics not usually taught in the engineering or technical courses. Also: an extended two-part treatise on Fluid Mechanics, discussions of aerodynamics of perfect fluids, analyses of experiments with wind tunnels, applied airfoil theory, the nonlifting system of the airplane, the air propeller, hydrodynamics of boats and floats, the aerodynamics of cooling, etc. Contributing experts include Munk, Giacomelli, Prandtl, Toussaint, Von Karman, Klemperer, among others. Unabridged republication. 6 volumes. Total of 1,012 figures, 12 plates, 2,186pp. Bibliographies. Notes. Indices. 5⅜ x 8½.
Six volume set, paperbound $13.50

Fundamentals of Hydro- and Aeromechanics,
L. Prandtl and O. G. Tietjens

The well-known standard work based upon Prandtl's lectures at Goettingen. Wherever possible hydrodynamics theory is referred to practical considerations in hydraulics, with the view of unifying theory and experience. Presentation is extremely clear and though primarily physical, mathematical proofs are rigorous and use vector analysis to a considerable extent. An Engineering Society Monograph, 1934. 186 figures. Index. xvi + 270pp. 5⅜ x 8.
Paperbound $2.00

Applied Hydro- and Aeromechanics,
L. Prandtl and O. G. Tietjens

Presents for the most part methods which will be valuable to engineers. Covers flow in pipes, boundary layers, airfoil theory, entry conditions, turbulent flow in pipes, and the boundary layer, determining drag from measurements of pressure and velocity, etc. Unabridged, unaltered. An Engineering Society Monograph. 1934. Index. 226 figures, 28 photographic plates illustrating flow patterns. xvi + 311pp. 5⅜ x 8.
Paperbound $2.00

APPLIED OPTICS AND OPTICAL DESIGN,
A. E. Conrady
With publication of vol. 2, standard work for designers in optics is now complete for first time. Only work of its kind in English; only detailed work for practical designer and self-taught. Requires, for bulk of work, no math above trig. Step-by-step exposition, from fundamental concepts of geometrical, physical optics, to systematic study, design, of almost all types of optical systems. Vol. 1: all ordinary ray-tracing methods; primary aberrations; necessary higher aberration for design of telescopes, low-power microscopes, photographic equipment. Vol. 2: (Completed from author's notes by R. Kingslake, Dir. Optical Design, Eastman Kodak.) Special attention to high-power microscope, anastigmatic photographic objectives. "An indispensable work," *J., Optical Soc. of Amer.* Index. Bibliography. 193 diagrams. 852pp. 6⅛ x 9¼.
Two volume set, paperbound $7.00

MECHANICS OF THE GYROSCOPE, THE DYNAMICS OF ROTATION,
R. F. Deimel, Professor of Mechanical Engineering at Stevens Institute of Technology
Elementary general treatment of dynamics of rotation, with special application of gyroscopic phenomena. No knowledge of vectors needed. Velocity of a moving curve, acceleration to a point, general equations of motion, gyroscopic horizon, free gyro, motion of discs, the damped gyro, 103 similar topics. Exercises. 75 figures. 208pp. 5⅜ x 8. Paperbound $1.75

STRENGTH OF MATERIALS,
J. P. Den Hartog
Full, clear treatment of elementary material (tension, torsion, bending, compound stresses, deflection of beams, etc.), plus much advanced material on engineering methods of great practical value: full treatment of the Mohr circle, lucid elementary discussions of the theory of the center of shear and the "Myosotis" method of calculating beam deflections, reinforced concrete, plastic deformations, photoelasticity, etc. In all sections, both general principles and concrete applications are given. Index. 186 figures (160 others in problem section). 350 problems, all with answers. List of formulas. viii + 323pp. 5⅜ x 8.
Paperbound $2.00

HYDRAULIC TRANSIENTS,
G. R. Rich
The best text in hydraulics ever printed in English . . . by former Chief Design Engineer for T.V.A. Provides a transition from the basic differential equations of hydraulic transient theory to the arithmetic integration computation required by practicing engineers. Sections cover Water Hammer, Turbine Speed Regulation, Stability of Governing, Water-Hammer Pressures in Pump Discharge Lines, The Differential and Restricted Orifice Surge Tanks, The Normalized Surge Tank Charts of Calame and Gaden, Navigation Locks, Surges in Power Canals—Tidal Harmonics, etc. Revised and enlarged. Author's prefaces. Index. xiv + 409pp. 5⅜ x 8½. Paperbound $2.50

Prices subject to change without notice.

Available at your book dealer or write for free catalogue to Dept. Adsci, Dover Publications, Inc., 180 Varick St., N.Y., N.Y. 10014. Dover publishes more than 150 books each year on science, elementary and advanced mathematics, biology, music, art, literary history, social sciences and other areas.